±1100kV UHV DC Power Transmission Technology

Zehong Liu

±1100kV UHV DC Power Transmission Technology

中国电力出版社
CHINA ELECTRIC POWER PRESS

Springer

Zehong Liu
State Grid Corporation of China
Beijing, China

ISBN 978-981-96-3711-9 ISBN 978-981-96-3712-6 (eBook)
https://doi.org/10.1007/978-981-96-3712-6

Jointly published with China Electric Power Press
The print edition is not for sale in China (Mainland). Customers from China (Mainland) please order the print book from: China Electric Power Press.
Original Chinese edition ISBN: 978-7-5198-6655-6

The original submitted manuscript has been translated into English. The translation was done using artificial intelligence. A subsequent revision was performed by the author(s) to further refine the work and to ensure that the translation is appropriate concerning content and scientific correctness. It may, however, read stylistically different from a conventional translation.

This Springer imprint is published by the registered company Springer Nature Singapore Pte Ltd.
The registered company address is: 152 Beach Road, #21-01/04 Gateway East, Singapore 189721, Singapore

Preface

The construction of the new power system with clean energy as the main power source is a major deployment made by China with a view to improving ecological environment, ensuring national energy security, and achieving sustainable development, which is of great significance to the transformation and development of China's energy and power. Accelerating the construction of a strong backbone power grid to ensure the consumption of clean energy is a key task in building the new power system and achieving China's "dual-carbon" goal. In this context, UHV DC transmission projects, as the main transmission channels for clean energy, are particularly important.

UHV DC refers to ± 800 kV and above voltage level direct current transmission and related technologies, which have the characteristics of large capacity, long distance, low loss, and high efficiency. It is a key technology to promote the large-scale development and transmission of clean energy and to achieve wide-range complementary and mutual benefit, which can realize inter-regional and cross-border energy interconnection. Since 2004, China has been focusing on the research, development, and demonstration of UHV DC technology and has successively built dozens of ± 800 kV UHV DC projects, making it possible for clean power from the western and northern parts of China to be delivered to the power load centers in the east and south, which has effectively promoted the large-scale development and efficient consumption of clean energy.

With the continuous westward and northward shift of China's energy development focus, the power transmission distance needs to extend to more than 3000 km, exceeding the economic transmission distance of ± 800 kV power transmission technology. In order to fully exploit the technical and economic advantages of UHV DC in ultra-large capacity and ultra-long distance transmission and achieve a wider range of energy allocation, the State Grid Corporation of China has for the first time in the world raised the power transmission voltage level to ± 1100 kV, increased the transmission capacity to 12 GW with a transmission distance of 3324 km, directly connected the sending end with the 750 kV Alternating Current (AC) power grid, and layered connected the receiving end with the 500 kV and 1000 kV AC grids, achieving a comprehensive increase in "DC voltage, AC voltage, and transmission capacity". Compared to the ± 800 kV technology, the ± 1100 kV technology has achieved the increase of rated voltage by 37.5% and transmission capacity by 50%, respectively. The Changji-Guquan ± 1100 kV UHV DC project built on the basis of the technology is currently one of the most advanced power transmission projects in the world, with the highest voltage level, largest transmission capacity, and longest transmission distance. In the context of the large-scale development of clean energy and the accelerated interconnection of power grids around the world, ± 1100 kV UHV DC transmission technology has a broad application prospect.

This monograph starts from the development overview of UHV DC transmission technology, with the characteristics, key technologies, equipment development difficulties, and engineering applications of ± 1100 kV technology as the main line, combined with the author's in-depth thinking on the basic theory, key technology, and engineering construction of DC transmission, comprehensively summarizes the research and development (R&D) process and engineering practice of ± 1100 kV UHV DC transmission technology.

This book consists of six chapters, including an overview, key technologies, equipment in converter station, general layout of converter station, key construction technologies, and commissioning. It covers multiple aspects of technical R&D and engineering applications such as basic theory, scheme research, engineering design, equipment development, engineering construction, and engineering commissioning. It can be used as references for researchers, equipment manufacturers, and project construction and maintenance personnel engaged in UHV DC and also as references for the development and innovation of UHV DC transmission technology in the future.

This monograph brings together the collective wisdom and rich experience of research, design, equipment, and commission technicians in the field of DC transmission. During the writing process, the author also received support and assistance from colleagues such as Wang Shaowu, Huang Yong, Guo Xianshan, Zhang Jin, Fu Ying, Yu Jun, Ding Yongfu, Zhu Quanle, Yang Wankai, Li Tianjiao, Wang Liyan, Jiang Mingze, Lu Yajun, Zhao Zheng, Li Peng, Zhang Jian, Wu Chuanqi, Bai Guangya, Yang Hengjie, Liang Hongsheng, Xing Kezheng, You Shaohua, Zhou Chunyu, Zeng Weiwen, Chen Qi, Ni Yu, Chen Chen, Liu Zhe, Xue Haoyan. It is anticipated that practitioners in the field around the world will get help from this monograph and contribute together to the development of DC power transmission technology.

Due to time constraints, there are inevitably inaccuracies in the book, and the author sincerely invites readers to criticize and correct.

Beijing, China
April 2023

Zehong Liu

Contents

Chapter 1
Introduction

The distribution of energy resources and economic activities in China is highly unbalanced. Wind, solar, hydro, and coal resources are mainly distributed in the north, northwest, and southwest, while energy demand is distributed in the east and central regions. Accelerating the development of UHV DC transmission technology is a necessary path to promote the upgrade of the power grid, boost the low-carbon transition of energy, and contribute to realizing China's carbon peak and neutrality goals. UHV DC can transmit clean energy from the western region to the load centers in the eastern and central regions, achieving large-scale and long-distance optimization of energy resources, and playing an important role in the grid integration and consumption of clean energy.

With the acceleration of the development of energy bases in western China, the transmission distance need to extend to 3000 km and above. The economy of $\pm$ 800 kV transmission technology has been seriously challenged. Therefore it is necessary to accelerate the research of longer-distance, larger-capacity transmission technology to promote a wider range of energy resource optimization. The $\pm$ 1100 kV UHV DC transmission technology has the advantages of longer economic transmission distance, larger transmission capacity, and lower transmission loss. It can achieve large-scale optimization of energy resources from 3000 to 5000 km, and will play an important role in promoting energy transformation, helping to achieve "dual carbon" goals, and the construction of the Global Energy Interconnection.

Z. Liu, *±1100kV UHV DC Power Transmission Technology*,
https://doi.org/10.1007/978-981-96-3712-6_1

1 Overview of the Development of UHV DC Power Transmission

From 1990 to 2007, China successively completed a number of ± 500 kV HVDC power transmission projects such as Gezhouba-Shanghai, Three Gorges-Changzhou, Three Gorges-Guangdong, and Three Gorges-Shanghai. In 2011, China completed the Ningdong-Shandong ± 660 kV HVDC power transmission project.

On the basis of HVDC power transmission technology, the State Grid Corporation of China (SGCC) started to carry out comprehensive research on ± 800 kV UHV DC transmission technology in 2004. The SGCC Simulation Center, the R&D Center for System Design of UHV DC Transmission Engineering, and UHV DC test bases have been established. Through systematic and in-depth research, it has fully grasped the planning, design, construction, operation, and maintenance technology of UHV DC, laying a solid foundation for the construction of the UHV DC projects.

In 2010, the successful completion and operation of the Xiangjiaba-Shanghai ± 800 kV UHV DC transmission demonstration project (referred to as the XS project) marked that China's DC transmission technology had reached the world's leading level. With a rated capacity of 6400 MW and a DC line length of 1891 km, this project was the most advanced DC transmission project at that time. Subsequently, SGCC realized one breakthrough after another in the field of UHV DC transmission, and successively built a number of ± 800 kV UHV DC transmission projects such as Jinping-Suzhou (JS), Haminan-Zhengzhou (HZ), Xiluodu-Zhexi (XZ), Ningdong-Zhejiang (NZ), Jiuquan-Hunan (JH), Ximeng-Taizhou (XT), Zhalute-Qingzhou (ZQ), Shanghaimiao-Linyi (SL). The rated transmission capacity has been increased from 6400 MW to 7200, 8000 and 10000 MW, and the voltage of the integrated AC system has been increased from 500 to 750 kV and 1000 kV. The XT, ZQ, and SL projects have achieved layered connected to the 500 and 1000 kV AC systems at the receiving converter station. As of October 2022, SGCC had cumulatively completed and put into operation fourteen ± 800 kV UHV DC transmission projects. In view of the significant advantages of the UHV DC technology in the development and utilization of energy resources and its outstanding contribution to China's social and economic development, it was recognized as the 2017 National Scientific and Technology Progress Special Award.

With the development of China's energy bases continuously moving westward and northward, the distance from the new energy bases in Xinjiang and Xizang to the eastern and central parts of the country exceeds 3000 km. As the transmission distance increases, the existing ± 800 kV UHV DC technology has a loss of more than 10%, which cannot meet the demand for efficient and economical power transmission. It is necessary to develop DC transmission technology of higher voltage levels. At the beginning of 2008, SGCC initiated research related to ± 1000 kV level UHV DC technology. In December 2010, based on summarizing relevant scientific research results and comprehensively analyzing and comparing the technical and economic aspects of the three voltage levels of ± 1000, ± 1100, and ± 1200 kV, it was determined that ± 1100 kV is the appropriate DC voltage level above ±

800 kV. SGCC decided to conduct further research on key technologies, equipment, and engineering application relied on the demonstration project of power transmission of Zhundong energy bases in Xinjiang. Since 2011, SGCC had carried out a large amount of R&D activities on key issues of ± 1100 kV technology, such as overvoltage suppression, addressing pollution, icing external insulation, air gap, electromagnetic environment, full voltage real-type equipment testing in the aspect of key technologies; engineering preliminary design and valve hall concept design in terms of system design.[1] For key equipment, the converter transformer model prototype, converter valve, wall bushing prototype, and most of the DC yard equipment were successfully developed.

In December 2015, the National Energy Administration of China officially approved the construction of the Changji-Guquan ± 1100 kV UHV DC transmission project (referred as CG project). The project took 4 years to build and was officially put into operation in September 2019. The project starts from the Changji converter station in Xinjiang and ends at the Guquan converter station in Anhui, passing through provinces of Gansu, Ningxia, Shaanxi, and Henan, with a transmission distance of up to 3324 km, setting a series of world records such as the highest voltage level, largest transmission capacity, and longest transmission distance. The project has greatly improved the capacity of outbound delivery of Xinjiang's power and the utilization rate of new energy, further satisfied the power demand of the load center in the eastern and central parts, which promotes the construction of a clean, low-carbon, safe and efficient energy system in China.

2 Technical Characteristics of ± 1100 kV UHV DC Power Transmission

With development through years, China's HVDC power transmission system has increased the voltage level from ± 500 to ± 800 kV, and further increased to ± 1100 kV, the transmission power has increased from 3000 to 8000 MW, 10,000 MW, and then to 12000 MW. The ± 1100 kV UHV DC technology has achieved both increase in DC voltage and transmission capacity, increasing the economic transmission distance from 2000 km of ± 800 kV to 3000–5000 km, and the transmission loss per kilometer from about 3% of ± 800 kV to about 1.5%.

CG project has achieved many innovative breakthroughs. This project is the world's first ± 1100 kV UHV DC transmission project, with a total length of about 3324 km, a transmission power of up to 12000 MW, and an annual transmission electricity of up to 80 TWh to East China. The starting converter station Changji is connected to the Northwest 750 kV AC grid, and the ending converter station Guquan is layered connected to the East China 500 and 1000 kV AC grids. This project is an important milestone in the continuous innovation of SGCC in the field of UHV

[1] Including single line diagram, system studies, control and protection strategies, requirement specification for all main circuit equipment, etc.

transmission, refreshing the new height of grid technology in the world, opening a new era of UHV DC transmission technology, and has a significant demonstration effect on the development of the Global Energy Interconnection.

Meanwhile, higher voltage levels and superior transmission capacity of ± 1100 kV UHV DC system impose higher requirements on the system research, single line diagram (SLD), insulation coordination, control protection strategy, equipment development, converter station layout, construction technology, and testing and commissioning skills. The construction and implementation of ± 1100 kV UHV DC transmission project is facing great challenges and a series of major innovations must be carried out.

I. Features of Technical Scheme

Due to the significant increase in the DC voltage, transmission capacity, and the voltage of the AC system connected, many technologies in the ± 800 kV UHV DC transmission projects are no longer applicable. It is necessary to carry out a brand new design and formulate a technically feasible and economically reasonable overall technical scheme.

(1) In terms of the connected AC system, due to the huge power capacity of the ± 1100 kV UHV DC project, the interaction between the AC and DC systems has a significant impact on the security and stability characteristics of the system. To ensure the reliability of the DC transmission system, it is necessary to adopt various means such as offline simulation, hybrid digital-analog power system real-time simulation, etc., and focus on the stability characteristics and control strategies of the connected power system from multiple perspectives. Finally, the layered connected scheme of the Guquan converter station to the 500 and 1000 kV AC grids is proposed. In order to reduce the difficulty of equipment manufacturing and improve technical and economic efficiency, it is proposed to connect the AC side of the low voltage converter (referred as LV converter) that withstands lower DC voltage to the higher AC voltage 1000 kV grid, and the AC side of the high voltage converter (referred as HV converter) that withstands higher DC voltage to the lower AC voltage 500 kV grid.
(2) In terms of converter connection, due to the significant increase in voltage and capacity of the ± 1100 kV system compared to the ± 800 kV system, the wiring of the converter will directly affect the reliability and economy of the entire system. By comparing the technical and economic aspects of the double 12-pulse series connection scheme (550 + 550 kV), three 12-pulse series connection scheme I (400 + 400 + 300 kV), and another three 12-pulse series connection scheme II (367 + 367 + 367 kV), it is decided to adopt the double 12-pulse series connection scheme (550 + 550 kV). On this basis, considering the impact of factors such as converter valve short-circuit current, transportation conditions, commutation requirements, reactive power consumption, insulation level, etc., the short-circuit impedance of the sending end converter transformer is set to 20%, and the that of the receiving end converter transformer is set to 22%.

(3) In terms of insulation coordination, since the overvoltage level of ± 1100 kV system is greatly increased compared with that of ± 800 kV system, and the economic loss of power outage brought by insulation breakdown accidents and the impact of disturbance on the system are more serious, it is necessary to propose more effective overvoltage suppression measures and insulation coordination schemes. Based on actual engineering conditions such as transmission distance, transmission capacity, and converter structure and parameters, various overvoltage situations that may occur in the system are simulated, and the overvoltage amplitude and waveform generated on the electrical equipment of the converter station are calculated. Then the arrester configuration scheme is proposed, which adds protection for the HV converter transformer valve-side Y-connected arrester AH and the LV converter transformer valve-side Y-connected arrester AL. By these means, the switching impulse insulation level of the smoothing reactor on the valve side of the DC pole line is controlled at 2100 kV. At the same time, by optimizing the type and parameters of the DB arrester, and reducing the voltage ratio of the arrester, the switching impulse insulation level of the of the DC pole line is reduced to be consistent with the valve side of the smoothing reactor.
(4) In terms of control and protection strategies, the increase in voltage level, transmission capacity, and transmission distance pose great challenges to the dynamic performance, strategy design, parameter selection, and margin control of the control and protection system. Due to the increased voltage level, the rate of change of fault current is large, and the problem of pole-to-pole coupling becomes very prominent. When one pole has a ground fault, the current and voltage of the other pole surge. In 3–5 ms after fault occurred on the pole, the other non-fault pole will also experience a commutation failure. Conventional commutation failure prediction strategies cannot effectively work in such a short time. To meet the rapidity requirements of commutation failure prediction, a new type of prediction function based on DC voltage sudden change is researched and applied, solving the problem of commutation failure caused by pole-to-pole coupling. Due to the extension of the transmission line, the delay in inter-station communication increases, which adds difficulty to the coordinated control of the sending and receiving converter controllers. To address this issue, by optimizing the coordination logic of the control and protection of sending and receiving ends, the large disturbances generated during transient processes such as manual and protective valve group withdrawal are eliminated.

II. Features of Equipment R&D

All DC equipment for the ± 1100 kV UHV DC transmission project needs to be re-developed, facing great challenges in design, production, testing, and quality control.

(1) In terms of design, the equipment needs to have high withstand voltage, large current, and large size and weight, and it is a great challenge to comprehensively solve the constraints of electromagnetic field distribution, insulation design, current heat dissipation and mechanical stress. By checking key parameters,

refining design requirements, strengthening test verification, and optimizing design margin, the difficult problem of equipment design is solved.

Ensure that the converter transformer iron-core has sufficient window height (HV side of the sending end and receiving end $\geq$ 3.6 m, LV side of the receiving end $\geq$ 3.3 m). Control the current density and magnetic flux density in the converter transformer (current density $\leq$ 3.2 A/mm^2, the magnetic flux density $\leq$ 1.75 T). The insulation electric field design of the converter transformer is strictly verified, with a safety margin of more than 1.2 times, and not less than that of previous $\pm$ 800 kV converter transformers. The test assessment requirements for converter transformer components are improved, the valve-side bushing impulse test voltage is set at 1.1 times the winding insulation, and the withstand time of external DC voltage increase from 120 to 180 min. The converter transformer adopts enhanced magnetic shielding design to reduce losses, the winding adopts refined oil channels, and the temperature rise is strictly controlled. The valve-side bushing of the HV converter transformer exits from the top of the box, reducing the difficulty of the outgoing line and body design, and ensuring the clearance to the ground. The structural strength of the converter transformer oil tank is increased (the bottom is 40 mm, the wall is 12 mm, and the oil tank of valve-side bushing elevated seat area is $\geq$ 40 mm). The elevated seat of the bushing is equipped with a reinforced support frame, and the reinforcement spacing is reduced, thus improving the mechanical margin. The design of the DC wall bushing equalizing electrode and clearance value is carried out in coordination, improving the outdoor arcing distance (12.3 m). The internal insulation clearance is about 44% more than the 800 kV wall bushing. A series of measures are proposed to prevent SF_6 from liquefying at low temperatures: controlling SF_6 pressure ($\leq$ 0.57 MPa), ensuring the minimum temperature of the indoor field ($\geq$ 10 °C), controlling the installation angle of the bushing ($\geq$ 5°), and adding the bushing electric heating device. The maximum electric field strength of the 1100 kV DC equipment in the indoor field is controlled to be $\leq$ 14 kV/cm, and the sparkover voltage of the key equipment $U_{50\%}$ is $\geq$ 2600 kV.

(2) A large number of newly developed components and new production processes are used, leading to major challenges in ensuring that the selection of raw materials, processing capabilities, production processes and environment meeting the quality requirements of equipment. The development and testing of key equipment requires new tooling and testing facilities, and there are major challenges in ensuring that the technical solutions and transformation progress meet the requirements. To adapt to the requirements of 1100 kV DC equipment, the technical requirements, transformation plans and implementation plans for tooling facilities, testing equipment, and large-scale transportation of each equipment factory have been determined.

Due to the increase in size and weight of the 1100 kV converter transformer and the thickening of insulation materials, the corresponding coil winding, core processing, body assembly, gas phase drying, lifting and transportation, and processing lathe all need to be upgraded. Among them, the vertical coil winding machine has a diameter of 4.2 m and a winding weight of 20 t, the size of kerosene vapor phase drying tank is 15 m × 6 m × 7 m, while the minimum power is 600 kW. The coil pressure bed has a diameter of 4.5 m with the pressure of 200–450 kN. The air cushion car has a maximum load of 850 t, and the assembly workshop crane is 450 t. For electric devices, the series resonance device has a rated voltage of 1800 kV and a rated capacity of 25,200 kVA, the DC voltage generator has a rated voltage of 2250 kV and a rated current of 35 mA, and the impulse voltage generator has a rated voltage of 6000 kV and an energy requirement of 900 kJ.

In the production of a certain 1100 kV valve-side bushing, issues such as inadequate control of the epoxy curing process, unbalanced heat reaction, faster internal reaction and slower external reaction, and insufficient epoxy curing leading to thermal stress breakdown and multiple instances of capacitor core cracking were addressed. This was achieved through new processes such as reducing the aluminum tube wall thickness from 8 to 6 mm, increasing the transition layer from 2 to 4 mm, embedding temperature sensors to establish process control speed and standards, and using heat pipe technology internally to further release excess heat. These measures controlled the temperature rise speed of the epoxy curing process of the bushing and solved the problem of bushing core cracking after a significant increase in casting volume.

For the HV converter transformer bushings at the Changji converter station, where the diameter and length of the capacitor core have significantly increased and the test voltage has increased dramatically, the use of conventional vacuum oil injection process can easily leave tiny bubbles inside, causing partial discharge during testing. By adopting a new high vacuum oil injection process and control indicators, it is possible to achieve no visible bubbles in the oil at 40 Pa and 60 °C, and the gas content is less than 0.1%, which is an order of magnitude higher than the current standard. Products using this process have passed batch testing in the factory.

(3) Ensuring the successful development and timely delivery of ± 1100 kV UHV DC equipment presents a significant challenge. It is necessary to adopt strict process requirements, improve the test and evaluation, enhance the quality control of components and raw materials, and compile a risk prevention and control manual for key equipment.

SGCC has compiled the enterprise standard Q/GDW 11666—2017, entitled "Guidelines for Supervision and Manufacturing of Main Electrical Equipment for Converter Stations of ± 1100 kV UHV DC Project". It specifies the requirements for the supervision and manufacture of the main electrical equipment of ± 1100 kV UHV DC converter stations during the manufacturing process, and puts forward guiding opinions on the contents of the equipment supervision and manufacture (including raw materials and components, production and assembly, routine test, packaging and shipment, and other processes), as well as the important risk points and management measures.

Considering the characteristics of ± 1100 kV UHV DC equipment, combined with the quality problems of previous UHV DC equipment, 140 technical risks of key DC equipment have been identified, and pre-control measures for the risks have been formulated, clarifying the requirements for equipment in terms of raw material selection, key process control, routine testing, and type testing.

Specialized research on topics such as converter transformer electromagnetic wire, insulation materials, silicon steel sheets, oil tank manufacturing, iron core assembly and grounding system, body drying, bushing and riser assembly, vacuum and dispose of transformer oil, routine testing, etc., has been conducted. 61 quality improvement measures have been proposed and their implementation has been tracked.

Special comparisons of various equipment have also been carried out, including the oil tank, electromagnetic wire, insulation oil test of the converter transformer, and the noise of the converter valve thyristor and AC filter capacitor, etc. Differences in components, equipment in terms of raw material manufacturing, key process indicators, key performance parameters among different manufacturers have been compared, and targeted improvement measures have been proposed.

III. Features of General Layout

The Changji converter station is arranged in accordance with "750 kV AC filter bank—750 kV AC switchyard (east)—valve hall and converter transformer courtyard (center)—DC switchyard (west)—AC filter bank (east)". The 750 kV AC lines go out to the east, and the ± 1100 kV UHV DC lines go out to the west. The 750 kV AC distribution device adopts an external GIS, arranged on the east side of the station area, and exits eastward. The areas of converter transformer and valve hall adopt "face-to-face" layout at the HV and LV sides, with the main control building and auxiliary control building closely arranged in the middle of the station area. The ± 1100 kV pole line and DC filters adopt an internal arrangement, with the "L" shaped indoor yard closely attached to the HV valve hall. The neutral line is arranged outdoors between the two indoor DC yards, and the pole line and grounding pole line exit westward. Two banks of AC filters are centrally arranged on the southeast sides of the station area, two banks of AC filters are centrally arranged on the north side of the station area, another two banks of AC filters are centrally arranged on the southeast side of the station area, and are connected through the GIS pipeline. The AC protection adopts a down arrangement. The administration building, comprehensive water pump room, garage, etc. are arranged on the southwest side of the station area, and the entrance road enters the station from the southwest side of the station area. The whole station is square and compact, with a small footprint, clear zoning within the converter station, and a reasonable layout. The area within the wall is about 26.51 hm^2.

The Guquan converter station is arranged according to the process flow of "DC yard—valve hall and converter transformer—AC yard" from west to east. The DC yard is arranged on the west side of the station. ± 1100 kV pole line and DC filters are arranged indoors. Other DC yard equipment items are arranged outdoors, and the ± 1100 kV UHV DC line goes out to the west. The AC grid side is layered connected with the 1000 and 500 kV AC grids. The 1000 kV AC yard is arranged in the north-central part of the station area, and the 1000 kV AC lines exit northward. The 500 kV AC yard is arranged on the south side of the station area, and the 500 kV AC lines exit southward. The 1000 kV AC filter yard and 500 kV AC filter yard are arranged on the east side of the station area. Two sets of condensers are located in the southeast corner of the station area. The valve hall and converter transformer area are arranged between the AC yard and the DC yard, adopting a single-line form layout, 2 auxiliary control buildings are arranged at the angle between the two pole valve halls and the indoor DC yard. The main control building is arranged in the open area of the 1000 kV AC yard, which is convenient for operation and maintenance. The entrance road is connected from the south side of the station area, leading directly to the converter transformer courtyard, which is conducive to the transportation and maintenance of the converter transformer. The general layout of the entire station is compact and reasonable, with clear functional zoning, and smooth connections between each distribution device and between them. The area within the wall is about 27.43 hm^2.

IV. Features of Construction Technology

CG project has achieved comprehensive innovative breakthroughs in the construction technology of ± 1100 kV UHV transmission projects, marking that China has mastered a full set of intellectual property rights of UHV project construction technology, and has set a precedent for the construction of subsequent power transmission projects. As the first project of ± 1100 kV voltage level of UHV DC power transmission project, it has no precedent to learn from at the beginning. Besides, the following special conditions have brought challenges for engineering implementation.

(1) Harsh natural environment. The sending end converter station is located in the Gobi area of Xinjiang, with large sandstorms in spring and autumn and extreme temperatures in summer and winter. The lowest temperature could reach below -30 °C and last for a long time, causing large fluctuations in the surface temperature of the steel structure, which leads to difficulty in steel structure construction due to material deformation. The rainy season at the receiving end converter station is long, with rain falling 40% of a year. Combined with relatively stringent requirements for the installation environment of equipment, the available construction time is short. The harsh natural environment poses higher technical requirements for the installation of converter transformers, GIS, and converter valve, as well as for the concrete construction and maintenance.
(2) The indoor DC yard contains a large number of equipment items with considerable size. The operating space is limited. The DC wall bushings and pole bus smoothing reactors are arranged in the indoor DC yard. The single hundred-ton equipment and 500 t cranes are transported from outdoor to indoor, and hoisting operations are carried out indoor, requiring sufficient turning radius and installation distance. The smoothing reactor weighs 106 t, with an installation height of 21 m, docking accuracy of ± 1 mm, and angular error of 0.1°. The DC wall bushing weighs 18 t with a length of 28 m, requiring millimeter-level installation errors. The construction accuracy requirement is like "threading a needle through a hundred meters".
(3) The installation of HV converter transformer is facing several difficulties. A single HV converter transformer weighs 909 t, and the positioning deviation

needs to be less than 5 mm. The requirement of fire resistance for valve-side plugging should be 4 h. At the same time, due to the huge workload of oil service, heavy bushing and valve-side outlet devices, and different equipment structure, the construction difficulty and accuracy requirements are higher than those of previous ± 800 kV UHV DC transmission projects.

(4) Main current loop heating and point discharge. When the DC current is increased to 5455 A, the AC current is increased accordingly. The possibility of heating of the AC and DC main current loops, equipment shell, and wall conductor surroundings increases. The increase of DC voltage level to 1100 kV brings a series of problems. Firstly, the uneven shape of equipment and fittings is more prone to point discharge. Secondly, higher requirements are put forward for air distance, especially the combination gap. Thirdly, construction errors will lead to insufficient air distance and cause sparkover, therefore the construction process standards must be improved.

(5) Uncertainty in the arrival time of the main equipment. As the uncertainty of the research and development, production, and routine tests of main equipment exists, the project schedule may be affected. If the equipment research and development and production go smoothly, they maybe arrive at the site in advance, and the overall schedule is adjusted forward. However, if the equipment production and transportation are delayed, it is necessary to arrange arrival collectively and cross-installation on site in order to ensure the target of project progress. It would place higher demands on process scheduling, resource input and quality assurance measures of construction.

(6) The steel structure of the indoor DC yard is difficult in construction. It has a large span, high height, and some roof frames are asymmetrical, combined with the valve hall layout. It is the largest span steel structure building of a power substation to date. The truss beam has a maximum span of 80 m and a maximum hoisting height of 47.5 m, which is the first in UHV projects. The technical requirements for the hoisting process are higher than the world advanced level.

(7) Less design and construction experience. Compared with the conventional ± 800 kV UHV DC transmission project, CG project is technically more difficult, has obvious design differences, and requires a higher level of construction technology. There are no directly applicable standards, nor mature technology and experience to refer to. It is likely to encounter design problems and pre-construction planning inadequacies due to insufficient design consideration and lack of engineering practice experience, which directly leads to increased construction difficulty.

(8) There are many participants, and the construction safety risk is high. Some HV equipment in the converter station is installed for the first time, including converter transformers, 1100 kV DC yard equipment, and condensers are all newly developed equipment. They are large in volume and mass, high in value, and the installation safety risk increases accordingly. During the construction of converter station projects, the peak number of participating personnel reached nearly 2000. More than ten construction units and manufacturers were working at the same time. There are a lot of height and cross operations on site, which

increase the construction safety risk. In order to ensure the effective implementation of safety and quality assurance measures, the level of construction technology is required to be high.

(9) The goal of promoting and applying new technologies is clear. On the basis of steadily achieving the established construction goals of "zero settlement exceeding the standard, zero circuit heating, zero sparkover of fittings, zero wiring errors, zero software false alarms, one time to pass the withstand voltage test and one time be successfully charged", in order to adapt to the scientific and technological innovation in power construction, the project have clarified the relevant requirements for the research and application of the new engineering construction technologies, and implemented the responsibilities and strengthened the relevant management measures. In this way, the wide application of new technologies, new processes, new procedures, new equipment, and new materials in CG project can be promoted, further improving the technical level of construction.

V. Features of Commissioning

The on-site commission of the $\pm$ 1100 kV UHV DC transmission project is divided into five stages: control & protection combination test (referred to as factory acceptance test, FAT), equipment test, subsystem test, station system test, and the whole system test. The characteristics and functions of commissioning are as follows.

(I) Verification of Project Design and Equipment Performance

1. FAT

The FAT of the control & protection system refers to the simulation of the power system in a digital or physical way, connecting the main equipment of the real DC control protection system of the project through interface equipment such as power amplifiers, forming a closed-loop test system. The FAT is the basis for the normal operation of the converter station control & protection according to the system design. As the actual operating conditions of the DC control & protection system can be fully shown, the interface characteristics of each component can be comprehensively checked, as well as the overall function and performance of the DC control and protection system. Before the DC control & protection system arrives at the site, internal function modules and interfaces can be tested, and problems can be quickly solved using the favorable conditions of reproducible faults of the FAT and better working environment, greatly reducing the time for on-site commissioning and shortening the construction period of the project. In general, the combination test of the converter station includes control tests, protection tests, interface tests, special tests, verification of problems existing in past projects, etc. The combination test needs to verify the above contents one by one.

CG project has the engineering characteristics of layered receiving end, which needs higher requirements to the design of the DC control & protection system. In the process of the FAT, relevant norms and procedures were strictly implemented, as well as the supervision and inspection of the test results. The problems existing in the control & protection related logic were dealt with in time, and the results of the system design were verified, ensuring the safety and reliability during the system commissioning period and in normal operation.

2. Equipment test

The purpose of equipment commissioning is to check whether the equipment meets the design requirements or has been damaged in transit, and whether the installation quality on site meets the requirements. It is the final check to judge whether the equipment can be put into operation safely, and is the key part before putting the DC transmission project into operation. At the same time, judgement is made to check whether the equipment can be safely charged, loaded or started, and whether the performance and operation of the equipment comply with the requirements of the contract and the technical specification. Equipment commissioning lays a good foundation for the commissioning of sub-systems of the converter station.

The equipment test has compiled the equipment debugging test plan according to national and industry standards. The detailed test operation guide has been formulated. Through careful arrangement, all equipment test processes and results meet the requirements of relevant procedures.

3. Subsystem test

The subsystem test is the charging or start-up test of all independent subsystems of the converter station. It ensures that after a single device is connected to the system, it can operate normally in combination with several other related devices or components as a subsystem. It is the basis for the normal operation of the entire converter station system, as well as to check whether the functions and performances satisfy the requirements of the contract and the technical specification of the project.

The commissioning of the DC converter station subsystem can be divided into eight parts, namely, the converter valve, the converter transformer, the AC yard, the AC filter yard, the DC yard, the station power subsystem, the auxiliary and other subsystems, and the control & protection equipment. The subsystem test is carried out separately for these eight parts.

In the process of subsystem tests, the relevant norms and regulations is compiled. Results of equipment tests are well supervised and inspected. All kinds of equipment defects are dealt with in a timely manner, which ensure that the subsystem tests were completed on schedule.

4. Station system test

Station system test is a validation of the converter station equipment and its performance, which check equipment and sequence control of converter station, DC line insulation performance, and equipment protection. It also prepares for end-to-end system commissioning.

Station system test includes the following six items. ① AC bus and AC filter charging test, while checking the primary and secondary equipment wiring and protection; ② Sequence operation test; ③ Last line disconnection test; ④ Converter transformer and valve group charging test; ⑤ Anti-interference test; ⑥ DC line open circuit test (no-load voltage test).

The process of station system test includes the following five parts. ① Prepare the station commissioning plan; ② Prepare the dispatch plan for station commissioning accordingly; ③ Report the station commissioning plan and dispatching plan to the Acceptance Start-up Committee for approval; ④ Station commissioning is jointly responsible by the station test responsible entity and the contractor, with the operating entity carrying out the operation, and the cooperation of the construction and supervision entities; ⑤ The system commissioning entity is responsible for supervising and checking the station systems, including whether the test items and test reports and data are complete, and whether the test results of each station test meet the requirements of the contract and the technical specifications, and gives a supervision and checking report. At the same time, it can also undertake some other parts of the test.

According to the station system commissioning plan, the tests of the two converter stations are completed, and the summary report and technical supervision report of station system commissioning are formulated, which create good technical conditions for system commissioning.

5. System test

System test is the last process before the DC transmission project is put into operation. Through the system test of the project, all equipment and its functions of the DC transmission project are comprehensively assessed, whether the performance indicators of the system meet the indicators specified in the contract and technical specifications are verified. The test ensures the safety and reliability of the project during operation, helps the operation and maintenance personnel to familiarize themselves with the characteristics of the interconnected power grid, and makes a comprehensive and correct evaluation of the performance of the project.

System test includes four parts. ① System calculation and analysis; ② Formulation of system commissioning plan, dispatch plan and test plan; ③ System on-site commissioning; ④ Summary of system commissioning.

Through the system commissioning, a more comprehensive assessment of primary and secondary system equipment was carried out. The primary equipment of a single converter withstood the assessment of the rated load and 1.05 times overload test, the primary equipment of a single pole withstood the assessment of the rated load test, and the assessment of various operations and AC/DC system short circuit tests were also carried out. At the same time, it fully exposed the defects of the secondary system equipment. For the discovered defects, the equipment manufacturers and the construction entities took the time to solve the problem, ensuring the smooth progress of system commissioning and laying the foundation for the safe and stable operation of the project after it is put into operation.

(II) Quality Inspection of Engineering Design and Construction

The commissioning of a power transmission and project is a strict inspection of the construction quality, which is related to whether the project can be put into operation on schedule and the safe operation after commissioning. Taking the DC transmission project as an example, when the equipment of the converter station arrives at the site, the manufacturers and the construction and installation entities firstly inspect the equipment, and then install the equipment. The scope of equipment installation includes: ① Converter valve hall main equipment; ② Valve cooling equipment; ③ Converter transformers (including 1 backup); ④ Smoothing reactors (including 1 backup); ⑤ Station control, DC pole control, protection system; ⑥ Station power system (including primary and secondary equipment); ⑦ Power cables, control cables and related auxiliary facilities; ⑧ Outdoor lighting for the whole station, integrated pump room; ⑨Communication and integrated automation system; ⑩ Image surveillance system for the whole station, large screen and call system; ⑪ Water spray system for converter transformers and smoothing reactors; ⑫ Fire alarm system for the whole station; ⑬ Structural supports for the whole station; ⑭ Lightning protection and grounding; ⑮ DC passive filter bank; ⑯ DC yard distribution device; ⑰ 500/750/1000 kV GIS AC distribution device, AC outgoing line; ⑱ 500/750/1000 kV AC filter bank. The construction entity and the equipment manufacturer install the equipment one by one according to the engineering design drawings, and then connect these equipment together to form a converter station system, which is a large and tedious task.

Every stage of the commissioning of a DC transmission project is a test of the engineering design and construction quality. In the commissioning, the problems found are analyzed and studied. For problems belonging to the design part, the designer is required to correct them on time. For wiring errors or loose joints, the construction entity is required to correct them immediately. For some difficult problems, it is necessary to carry out thematic analysis and research before solving them, so as to ensure that the construction period of the project meets the requirements, but also to ensure the construction quality.

(III) Laying the Foundation for Operation

The commissioning of a power transmission project is to ensure the safe and reliable operation, and to lay the technical foundation and provide technical assurance for operation.

During the commissioning, the procedures and regulations are strictly followed. In order to ensure the safety of the tests, the "two permissions and three regulations"[2] and "double issuance" regulations are strictly implemented during the process. All experiments are done by the test entity by filling in the work ticket. The personnel of the test or construction entity act as the person in charge of the work and the issuer of the work ticket, with the participation of the commissioning entity, and with the permission of the operator to issue it finally.

[2] A system used in China for the safe production and operation of substations and power plants, including work permission, operation permission, shift handover regulation, inspection regulation, and equipment periodic test rotation regulation.

For the problems found during the commissioning process, the commissioning entity or owner's representative presides over the analysis of the problem and study the solution, with the participation of the operator, constructor and manufacturer personnel.

Whether it is the "two permissions and three regulations" or the analysis and solution of problems, it is the training and exercise of operation personnel during the commissioning stage, which cultivates technical forces for the future operation of the project.

The operation entity and its personnel need to take advantage of the valuable time of equipment installation and commissioning of the converter station, strengthen personnel professional training, improve the entity's technical level, actively compile various procedures and regulations, and prepare for the smooth takeover of the converter station.

During the commissioning, the operation personnel should closely track the equipment defects, discover the problems in time, and submit them to the commissioning entity and the owner. The commissioning entity and the owner need to arrange timely elimination of defects, and strive to ensure the normal commissioning of the equipment, aiming at zero-defect in the reception of the equipment.

Chapter 2
Key Technologies of ± 1100 kV UHV DC Power Transmission

For the ± 1100 kV UHV DC power transmission project, currently there is no mature design experience and complete theoretical system support in the world. Focusing on the construction and implementation needs, SGCC has organized a series of new technology research, and has achieved a number of key technological breakthroughs in DC systems studies, DC SLD and main circuit design, DC overvoltage insulation coordination, air gaps and DC external insulation, DC control protection and electromagnetic environment, providing strong technical support for the implementation of the ± 1100 kV UHV DC transmission project. This chapter provides a detailed introduction to the CG project.

1 Research on Sending and Receiving Ends of DC System

The starting position of CG project is in the eastern part of Changji, Xinjiang, and the ending position is in the southern area of Anhui Province. The line is approximately 3324 km long, with a rated power of 12 GW. It was put into operation at the bipolar low voltage mode in 2018 and fully in operation in 2019. The grid-side voltage of the sending end Changji converter station is 750 kV, and three outgoing lines are connected to the Wucaiwan 750 kV substation through cables. The power supply of CG project was set to be 6.6 GW during the "13th Five-Year Plan" period (2016–2020). The Wucaiwanbei I Power Plant (4 units) and Wucaiwanbei II Power Plant (4 units) are both connected to the Changji converter station through two 750 kV lines. The receiving end Guquan converter station is connected to the system at 1000/500 kV levels: the newly built Guquan-Wuhu 1000 kV double-circuit line and Guquan-E'xi 500 kV double-circuit line. The Fanchang and Jingting 500 kV substation are connected to the converter station via double-circuit lines. The schematic diagram of the near-region power grid structure of the CG project converter stations is shown in Fig. 1.

Z. Liu, *±1100kV UHV DC Power Transmission Technology*,
https://doi.org/10.1007/978-981-96-3712-6_2

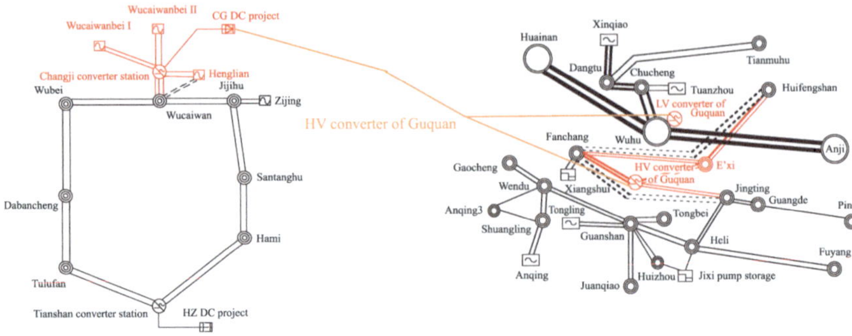

Fig. 1 Schematic diagram of the near-region power grid structure of the CG project converter stations

The following will introduce the research on the mutual influence of AC and DC systems, and the safety and stability control measures of the sending and receiving end power systems after the commissioning of the CG project.

I. Research on the Mutual Influence of AC and DC Systems

(I) Research Purpose

The operation of the CG project will show significant impacts on the safety and stability characteristics of both sending and receiving end systems. The Changji converter station at the sending end is located in Xinjiang, and is connected to UHV DC projects such as NZ, HZ, JH, as well as the large-scale wind power bases in Hami through the Northwest 750 kV system. The integration of DC project has affected the power flow and transmission capacity of interconnected channels between Xinjiang and the Northwest main grids. Besides, there is a correlation with the operation of NZ, HZ, JH DC projects and Hami wind power bases. Therefore, it is necessary to conduct a comprehensive and in-depth research on the mutual influence of AC and DC Systems, to provide guidelines for the construction, commissioning and operation of the project.

(II) Research Content

1. Interaction impact on sending end system

   (1) When the CG project has a blocking fault, the power flow transfer will cause some generation units of Xinjiang and the Xinjiang-Northwest interconnection channels (referred to as the interconnection channels) to lose stability in the power angle with the Northwest power grid (as shown in Fig. 2), which may lead to the disconnection of the interconnection channels. The power flow transfer after the blocking fault will also cause low voltage on the 330 kV Chaidamu bus, causing the CL DC power to fluctuate and affect safe operation. After the blocking fault, a large amount of surplus power will cause the system frequency to rise sharply, which may trigger high-frequency generator tripping action.

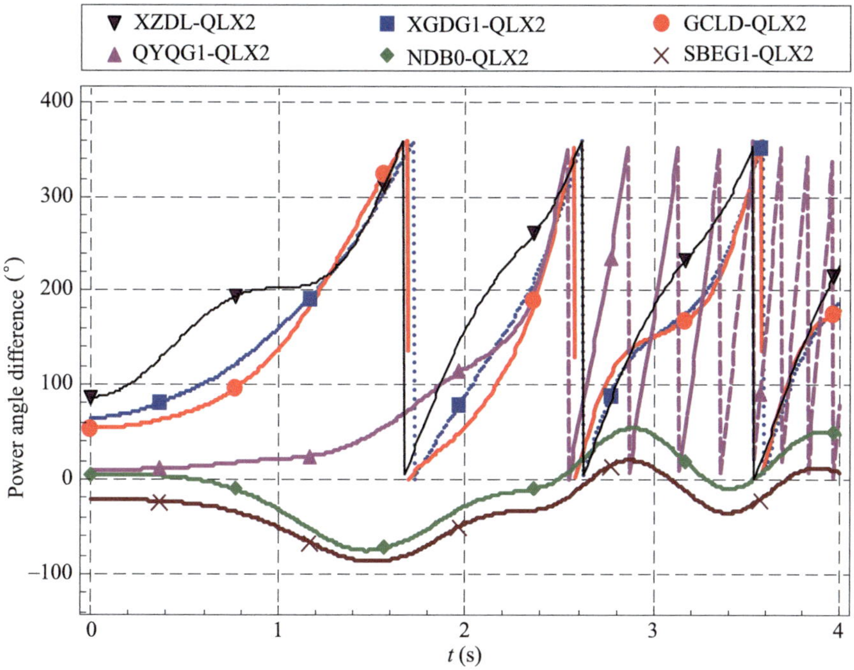

Fig. 2 Power angle curve of Northwest units under bipolar blocking of the CG project

(2) If the CG project experiences more than twice commutation failures or once and more bipolar restarts, the power impulse will cause some units in Xinjiang and the interconnection channel to lose stability in the power angle of the northwest main grids, which may lead to the disconnection of the interconnection channels. The power shock will also cause low voltage on the 330 kV Chaidamu bus (as shown in Fig. 3), leading to fluctuations in the CL DC, which will affect safe operation.

(3) When an $N - 1$ three-phase short circuit fault occurs on the 750 kV line near the Changji converter station, the system power angle is stable, the bus voltage is within the normal range, and the line is not overloaded. When an $N - 2$ three-phase short circuit fault occurs on the Wubei-Wucaiwan 750 kV lines, a large amount of power flow is transferred through the 750 kV Tianshandong ring grids, causing the bus voltage of some loads in the Tulufan area to drop significantly to below 0.8 p.u. When an $N - 2$ three-phase short circuit fault occurs on the Changji converter station-Wucaiwan 750 kV line, the remaining line is overloaded.

2. Interaction impact on receiving end system

(1) Under the heavy load mode of the East China power grid, CG project can maintain stable operation after a bipolar blocking. Under the small load

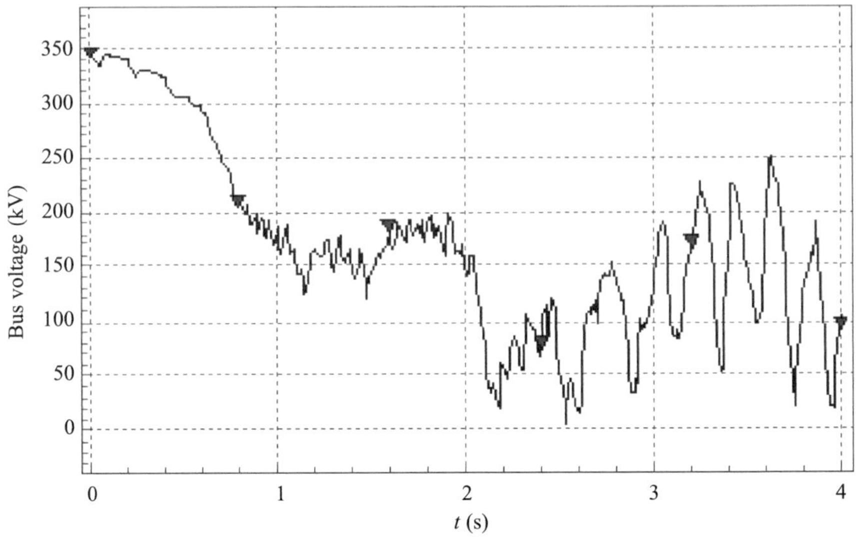

Fig. 3 Voltage curve of Chaidamu 330 kV bus under four consecutive commutation failures of the CG project

mode of the East China power grid, the frequency drops below 49 Hz after a bipolar blocking in the CG project (as shown in Fig. 4), causing the low-frequency load shedding device of the East China power grid to operate.

(2) If the CG project experiences twice consecutive commutation failures or once successful bipolar restart, the system can maintain stable operation. If the CG project experiences once failed bipolar restart, the frequency of the East China power system under the small load mode drops below 49 Hz, causing the low-frequency load shedding device to operate.

(3) After an $N-1$ three-phase short circuit fault occurs on the 1000 and 500 kV lines near the Guquan converter station, the system power angle is stable, the bus voltage is within the normal range, and the line is not overloaded. After a $N-2$ three-phase short circuit fault occurs on the Guquan converter station-Wuhu 1000 kV lines, the converter station is disconnected from the East China main grids, and the East China power system is stable. When a three-phase short circuit occurs on the Wuhu-Anji 1000 kV line, the Wuhu's main transformer is overloaded. When a three-phase short circuit fault occurs on the Huainan-Xuyi 1000 kV line, the units of Huaihu loses power angle stability.

3. Simultaneous faults of multiple DC projects

If CG Project and NZ Project experienced twice consecutive commutation failures at the same time, the system can maintain stable operation. If the CG project and NZ project first experienced twice consecutive commutation failures, and then three

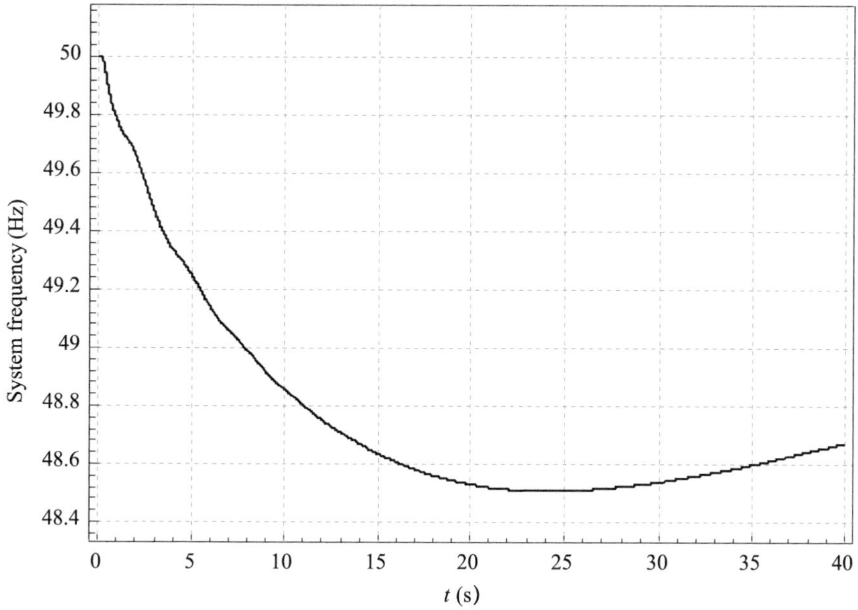

Fig. 4 Frequency curve of East China power system under bipolar blocking of the CG project

consecutive commutation failures occurred at intervals of 1.5–2 s, the power fluctuation would impact on the interconnected channel, leading to power angle instability of the Xinjiang unit relative to the Northwest main grid, as shown in Fig. 5.

The short-circuit fault in the near-area of the sending end triggers different levels of power transient drop in the JH, HZ and CG DC projects at the same time. Since the DC power does not drop to 0 MW simultaneously, the system can maintain stable operation, and the system frequency reaches a maximum of 50.8 Hz.

(III) Research Results and Applications

The research results reveal the influence of the CG project's blocking, commutation failure, restart and other faults or disturbances on the safety and stability characteristics of the AC system at the sending and receiving ends, as well as the characteristics and influence of the DC commutation failure and power drop caused by the $N-1$ or $N-2$ three-phase short circuit faults of the AC system at the sending and receiving ends of the CG Project.

The research results have laid a solid foundation for subsequent specialized research on safety and stability control measures and operation strategies at various stages, ensuring the safe and stable operation of the sending and receiving systems after the CG project is put into operation.

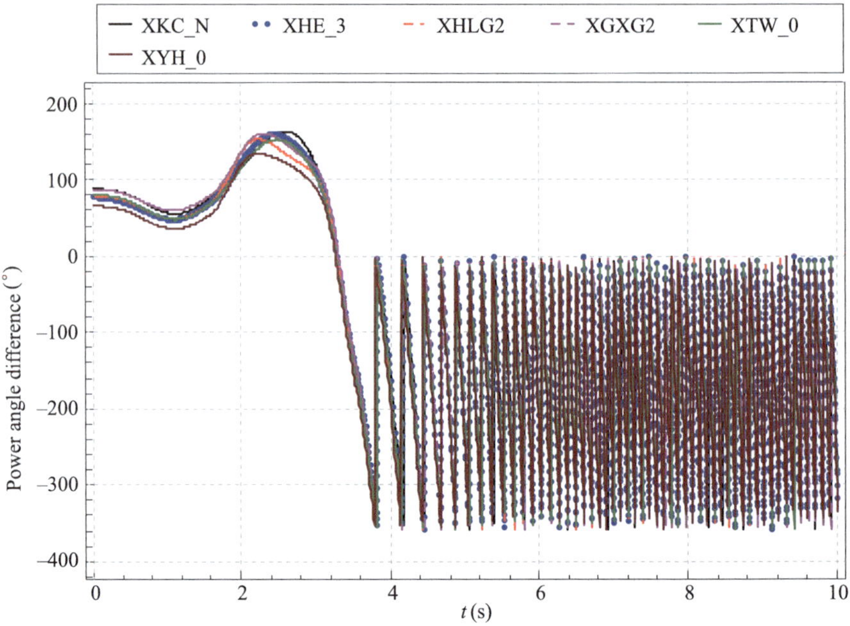

Fig. 5 Power angle curve of Xinjiang units to Northwest main grids under CG project and NZ project if experienced consecutive commutation failures

II. Research on Safety and Stability Control Measures for Sending and Receiving End Systems

(I) Research Purpose

The sending end of the CG project is integrated to the Northwest power grid of China, with a relatively small system scale and a weak grid structure. DC faults can bring a large power impulse to the Northwest power grid. The receiving end is connected to the East China grid, nearby the UHV AC channels within the grid. Due to the large DC transmission power, a large amount of power flow needs to be transferred in the near-region of the converter station through the UHV AC and the 500 kV channels, which may lead to a large impact on the system if the near-region AC/DC system is seriously faulted.

(II) Research Content

1. Stability characteristics and control measures of the sending end system of CG project

   (1) Under the mode of 8-generator as power supply, the short-circuit current of the Changji converter station is slightly less than the minimum short-circuit current of 33.5 kA in the system design document. Under the mode of 0-generator as power supply, the short-circuit current of the Changji converter

station does not meet the minimum short-circuit current requirement of 33.5 kA. When the Changji converter station switches one bank of filters, the steady-state voltage change is about 5 kV, and the steady-state voltage change rate is about 0.6%.

(2) After the $N - 2$ three-phase short circuit fault occurs on the Changji converter station—Wucaiwan 750 kV lines, the remaining line is overloaded. If the same fault occurs on the Wubei-Wucaiwan 750 kV lines, the voltage of some load buses in the region of Tulufan drops significantly to below 0.8 p.u.

(3) If no safety and stability control measures are taken after a single pole or bipolar blocking failure occurs in the CG project, the power flow transfer will cause some units in Xinjiang and the interconnection channels to lose stability in the power angle to the Northwest main grid, which may lead to the disconnection of the interconnection channels. The power flow transfer will also cause the voltage of the Chaidamu 330 kV bus to drop below 315 kV, causing fluctuations and unsafe operation in the CL DC project. Besides, the large amount of surplus power will cause the system frequency to rise significantly, possibly triggering high-frequency generator-tripping action.

(4) If the CG project experiences three or more consecutive commutation failures or once and more bipolar restarts, the power impulse of the commutation failure process will cause some units in Xinjiang and the interconnection channels to lose stability in the power angle to the northwest main grid, which may lead to the disconnection of the interconnection channel. The power impulse will also cause low voltage in the Chaidamu 330 kV bus, causing power fluctuations in the CL DC project.

(5) For different modes of CG project sending end system, such as the Northwest main grid sending to Xinjiang or vice versa, near-region AC line maintenance, etc., DC faults such as DC blocking, commutation failure, restart, power drop, and severe AC faults in the near area such as an $N - 2$ three-phase short circuit fault, it is necessary to take safety and stability control measures such as cutting off supporting power supply or units of Xinjiang main grid or Haixi PV power, and quickly reducing CG project DC power. The specific safety and stability control measures are shown in Table 1.

2. Stability characteristics and control measures of the receiving end system of CG Project

(1) When the CG project was put into operation, the UHV AC northern half ring of East China power grid had not yet been formed, and Wuhu substation only had one main transformer. Due to the overload constraint of the main transformer of Wuhu after an $N - 2$ three-phase short circuit fault occurs in the Wuhu-Anji lines, it is necessary to take measures to cut off the sending end generation units of Huaihu. Since the maximum number of tripping generators was 4, the transmission capacity of the CG project was limited to 6 GW. After the northern half ring grid was completed in 2019, the power

Table 1 Safety and stability control measures of sending end system of CG project

Operation modes	Faults	Measures
Power supply: 8 generators, Power transmission direction: Xinjiang to Northwest grid, CG project power: 12 GW	Bipolar blocking of CG project	(1) Tripping 1000 MW of Haixi PV, 3960 MW of supporting power supply and 6600 MW of Xinjiang units (2) Increasing the DC power of HZ Project by 400 MW; tripping 1000 MW of Haixi PV, 3960 MW of supporting power supply, and 6200 MW of Xinjiang units
	Three consecutive commutation failures of CG project	(1) If the DC power of CG project does not exceed 7000 MW, blocking immediately when the 3rd commutation failure occurs; tripping the supporting power supply of 3960 and 1600 MW of Xinjiang units (2) If the DC power of CG project does not exceed 8000 MW, blocking immediately when the 3rd commutation failure occurs; tripping the supporting power supply of 3960 and 4200 MW of Xinjiang units
	Two single pole restart failures of CG project	Tripping the supporting power supply of 3960 and 300 MW of Xinjiang units
	One single pole of CG project restart failure, while the other pole blocking	(1) Tripping 1000 MW of Haixi PV, 3960 MW of supporting power supply, and 7800 MW of Xinjiang units (2) Increasing the DC power of HZ project by 400 MW, tripping 1000 MW of Haixi PV, 3960 MW of supporting power supply, and 7300 MW of Xinjiang units
	Sudden power drop of CG project exceeds 1.6 GW	Adopting safety and stability control measures for generator tripping

(continued)

Table 1 (continued)

Operation modes	Faults	Measures
	N− 2 three-phase short circuit fault occurs on Changji-Wucaiwan lines	Rapid decrease 1100 MW DC power of CG project
	N− 2 three-phase short circuit fault occurs on Wubei-Wucaiwan lines	Tripping 2000 MW of Xinjiang units

transmission channel has been increased, but Wuhu still only has one main transformer. Subject to the above constraints, the transmission capacity of the CG project can only be increased to 9 GW. Considering the expansion of the second main transformer in Wuhu and the upgrade project of the near-region 500 kV Chucheng-Dangtu double lines, E'xi-Huifeng double lines, Dangtu-Tianmuhu double lines, Guquan-Jingting double lines, Guangde-Pingyao single line, etc., the transmission capacity of the CG project can reach 12 GW.

(2) In the 2018 operation mode, the short-circuit current of the 500 kV bus at Guquan converter station is about 36.8 kA, and that of the 1000 kV bus is about 22.8 kA. In the 2019 operation mode, the short-circuit current of the 500 kV bus at Guquan converter station is about 36.7 kA, and that of the 1000 kV bus is about 23.3 kA, all of which meet the design requirements. The steady-state voltage change rate is less than 1% when one bank of filters is switched at the LV or HV side of Guquan converter station.

(3) In the operation mode from 2018 to 2019, the peak load of East China power grid was 155 GW. When the CG project experienced a bipolar blocking fault, the frequency dropped to below 49 Hz, which would trigger the low-frequency load shedding device. It is necessary to implement control measures, such as increasing power of other DC projects by 2500 MW and tripping pumped storage load by 3500 MW. The amount is below the maximum value of the existing frequency control system in the East China power grid (increasing DC by 2500–3000 MW, tripping pumped storage load by 9800 MW).

(4) In response to an $N-2$ three-phase short-circuit fault occurring in 1000/500 kV lines in the near-region under different maintenance modes of primary and secondary AC outgoing lines of Guquan converter station, it is necessary to take safety control measures such as quickly reducing the power of the CG project, blocking HV/LV converters, and tripping the Huaihu sending end units. The specific safety and stability control measures are shown in Table 2.

(III) Research Results and Applications

The research results reveal the safety and stability characteristics, the transmission capacity of important channels and related constraints of the CG project's integrated

Table 2 Safety and stability control measures of receiving end system of CG project

Operation modes	Faults	Measures
All	Bipolar blocking of CG project	Increasing the DC power of other projects by 260 MW; tripping pump storage load by 3500 MW
N − 1 maintenance mode of other 1000 kV lines except Guquan-Wuhu	N− 2 three-phase short circuit on Guquan-Wuhu 1000 kV lines	Blocking double converters of the low voltage end of the CG project
N − 1 maintenance mode of Chucheng-Dangtu 500 kV lines	N− 2 three-phase short circuit on Wuhu-Anji 1000 kV lines	Tripping 4 units of Yuanzhuang/Pingxu, pre-control power of CG project not exceeding 2300 MW
N − 1 maintenance mode of Huainan-Wuhu 1000 kV lines	N− 2 three-phase short circuit on Huainan-Xuyi 1000 kV lines	Tripping 3 units of Yuanzhuang/Pingxu
N − 1 maintenance mode of E'xi-Huifeng 500 kV lines		Pre-control power of CG project not exceeding 3600 MW, rapid decrease 1600 MW DC power
N − 1 maintenance mode of Guquan-Fanchang 500 kV lines	N− 2 three-phase short circuit on E'xi-Guquan converter station 500 kV lines	Pre-control power of CG project not exceeding 3000 MW, rapid decrease 1600 MW DC power
N − 1 maintenance mode of Guquan-Jingting 500 kV lines	N− 2 three-phase short circuit on Huifeng-E'xi 500 kV lines	Pre-control power of CG project not exceeding 4000 MW, rapid decrease 1600 MW DC power
N − 1 maintenance mode of E'xi-Huifeng 500 kV lines	N− 2 three-phase short circuit on Dangtu-Tianmu 500 kV lines	Pre-control power of Dangtu-Tianmu double lines not exceeding 3400 MW, and E'xi-Huifeng single line not exceeding 1100 MW
N − 1 maintenance mode of E'xi-Huifeng 500 kV lines	N− 2 three-phase short circuit on Fuyang-Heli 500 kV lines	Pre-control power of Fuyang-Heli double lines not exceeding 1800 MW, and E'xi-Huifeng single line not exceeding 1400 MW
N − 1 maintenance mode of Jingting-Guangde 500 kV lines		Pre-control power of Fuyang-Heli double lines not exceeding 1800 MW, and Guangde-Pingyao single line not exceeding 1400 MW

(continued)

Table 2 (continued)

Operation modes	Faults	Measures
N − 1 maintenance mode of Fangchang-Feixi 500 kV lines	N− 2 three-phase short circuit on Zhaoguan-Dangtu 500 kV lines	Pre-control power of Dangtu-Zhaoguan double lines not exceeding 2200 MW, and Fanchang-Feixi single line not exceeding 1200 MW
N − 1 maintenance mode of E'xi-Huifeng 500 kV lines		Pre-control power of Zhaoguan-Dangtu double lines not exceeding 1300 MW, and E'xi-Huifeng single line not exceeding 1300 MW
Other maintenance modes	N− 2 three-phase short circuit on Jingting-Heli 500 kV lines	Pre-control power of CG project not exceeding 5400 MW, rapid decrease 1600 MW DC power

power systems under different operation modes. It also proposes safety and stability control measures for the sending and receiving power systems under faults such as blocking, commutation failure, restart, and severe faults in the near-region AC system.

The research results have been practically applied to the preliminary design of the safety and stability control system of the CG project's sending and receiving ends. Based on this, dispatch centers at all levels have formulated detailed safety and stability control strategies, effectively enhancing the ability to withstand severe faults.

2 DC Single Line Diagram and Main Circuit Design

In the ± 1100 kV UHV DC transmission system, the SLD of converter station has a significant impact on the main DC equipment research, project implementation difficulty, reliability level, project economy, AC power loss, etc. It is also the basis and prerequisite for project system design. The key issue of the scheme is to determine the technical scheme of the converter station connection lines and the specific configuration of the AC and DC equipment based on this structure. It needs to be determined after research and demonstration from the aspects of technical scheme feasibility and project reliability and economy. After the SLD is determined, detailed main circuit parameter calculations can be carried out according to the equipment configuration and connection form. Then, the SLD type can be optimized based on the result of the main circuit parameter calculations.

I. Research on the SLD

(I) Recommendation of Converter Scheme

The type selection of the converter connection type is a key issue that needs to be resolved in determining the DC SLD. The current capacity of the 6in thyristor in China, which has undergone technical upgrades, is as high as 6250 A. The current of the ± 1100 kV UHV DC transmission project is within the allowable range of the 6in thyristor, so there is no need to consider the converter parallel scheme, and the multiple 12-pulse converter series connection scheme can meet the demand. According to the current technical level, ± 1100 kV UHV DC transmission project is technically feasible with either two 12-pulse scheme (HV + LV) or three 12-pulse scheme (HV + MV + LV) per pole. Based on the research results and engineering practice experience of the ± 800 kV UHV DC transmission, three different SLD types are preliminarily recommended. Their topological structures are shown in Fig. 6, and their AC system access schemes are shown in Table 3.

Among them, Scheme 1 is a typical double 12-pulse converter series scheme, which is adopted in all previous UHV DC transmission projects in China. Except for the difference in voltage of each single 12-pulse, the operation mode, converter station layout and so on are exactly the same as the ± 800 kV UHV DC transmission project. Scheme 2 is a three 12-pulse equal division scheme. Compared with the ± 800 kV transmission project, the voltage of each 12-pulse is slightly lower than 400 kV, but the voltage of the three 12-pulses are exactly the same (the voltage to ground is different). The operation mode of this scheme is more flexible, any one or two 12-pulse can be combined to operate, but the number of converter transformers is 1/3 more, and the layout and area are completely different from the conventional ± 800 kV UHV DC project. Scheme 3 also adopts a three 12-pulse scheme, and the lower voltage end is completely consistent with the conventional ± 800 kV UHV DC project, and the higher voltage end is only 300 kV. The advantage of this scheme is that the lower voltage end can be fully drawn from the equipment and operation experience of the conventional ± 800 kV UHV DC project. As the higher voltage end is lower than 400 kV, the equipment capacity and insulation level between terminals are relatively reduced, thus greatly reducing the manufacturing difficulty. However, as the voltages of the three 12-pulse converters are different, the spare parts are not interchangeable, and the parameters between equipment terminals are also different, compared with that of Scheme 2, the equipment manufacturing and operation mode of Scheme 3 are more complex.

(II) Converter Scheme Selection

1. Main circuit parameters of each scheme

The ideal no-load DC voltage under the SLD of the double 12-pulse converter in Scheme 1 is shown in Table 4, that of Scheme 2 is shown in Table 5, and that of Scheme 3 in both sending and receiving ends are shown in Tables 6 and 7, respectively.

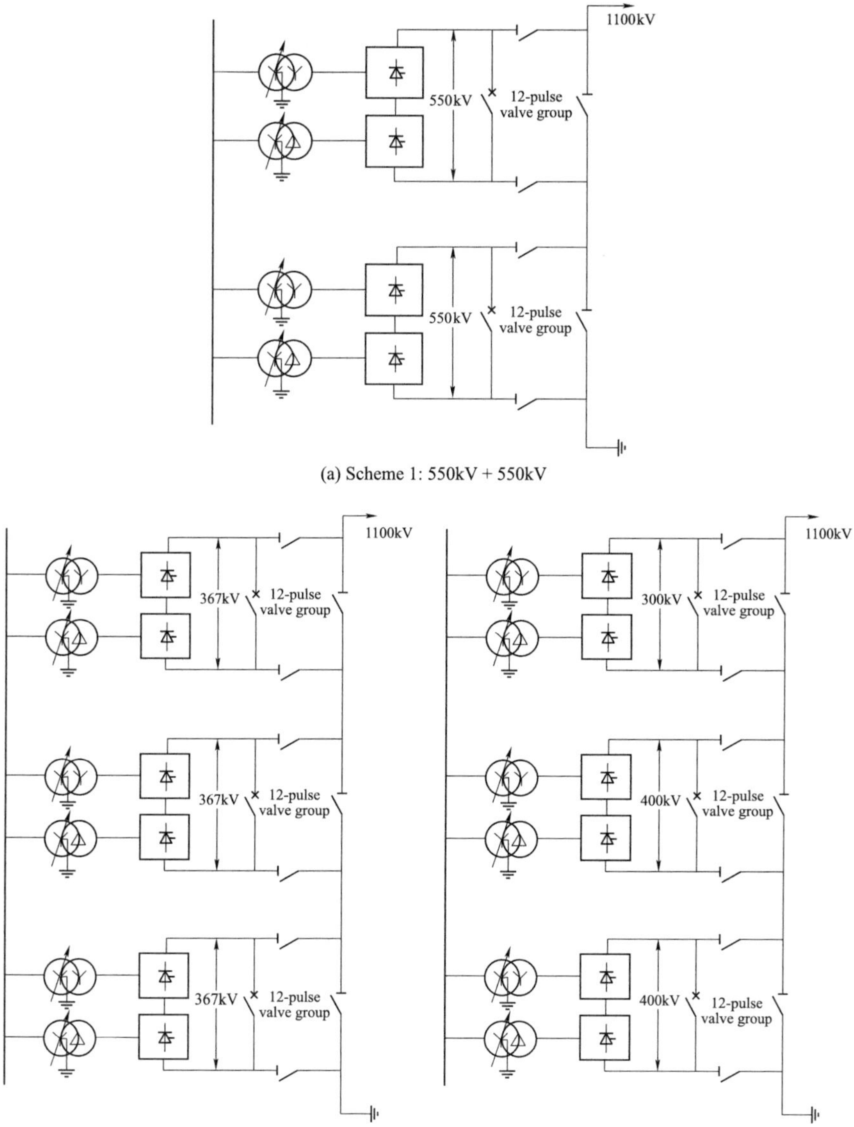

(a) Scheme 1: 550kV + 550kV

(b) Scheme 2: 367kV + 367kV + 367kV

(c) Scheme 3: 400kV + 400kV + 300kV

Fig. 6 Topology of 12-pulse converter series connection scheme

2. Converter transformer parameters of each scheme

The converter transformer is one of the most important equipment in the DC transmission project, and its cost accounts for more than 1/5 of the total project. Its technical parameters, design, manufacturing, transportation, and insulation level are the most important factors affecting the construction of the project. Therefore, analyzing the

Table 3 Scheme of converter integration with AC systems

No.	Combination of 12-pulse (kV)	Converter	Voltage of receiving end AC system (kV)
1	550 + 550	12-pulse HV converter	500
		12-pulse LV converter	1000
2	367 + 367 + 367	12-pulse HV converter	500
		12-pulse MV converter	500
		12-pulse LV converter	1000
3	400 + 400 + 300	12-pulse HV converter	500
		12-pulse MV converter	500
		12-pulse LV converter	1000

Table 4 Ideal no-load DC voltage for double 12-pulse converter series mode of Scheme 1 (kV)

Ideal no-load DC voltage	Changji station connected with 750 kV grid	HV side of Guquan station connected to 500 kV grid	LV side of Guquan station connected to 1000 kV grid
U_{di0N}	319.04	308.33	308.33
U_{di0min}	287.97	288.53	288.53
U_{di0max}	321.13	318.89	318.89
$U_{\mathrm{di0maxOLTC}}$	321.13	312.45	312.45
U_{di0G}	321.13	313.11	315.89
U_{di0L}	325.25	318.89	318.89
$U_{\mathrm{di0absmax}}$	328.50	322.08	322.08

Note U_{di0N} is the rated no-load DC voltage; U_{di0min} is the minimum no-load DC voltage. U_{di0max} is the maximum no-load DC voltage. $U_{\mathrm{di0maxOLTC}}$ is the maximum no-load DC voltage used for tap calculation. U_{di0G} is the lower voltage limit for normal adjustment of the tap to increase the no-load dc voltage U_{di0}. U_{di0L} is the upper voltage limit for normal adjustment of the tap

Table 5 Ideal no-load DC voltage for three 12-pulse converter series mode of Scheme 2 (kV)

Ideal no-load DC voltage	Changji station connected with 750 kV grid	HV side of Guquan station connected to 500 kV grid	LV side of Guquan station connected to 1000 kV grid
U_{di0N}	213.0	201.1	201.1
U_{di0min}	192.8	183.3	183.3
U_{di0max}	213.9	204.7	204.7
$U_{\mathrm{di0maxOLTC}}$	213.9	204.7	204.7
U_{di0G}	213.9	200.9	202.7
U_{di0L}	216.6	204.7	204.7
$U_{\mathrm{di0absmax}}$	219.0	207.0	207.0

Table 6 Ideal no-load DC voltage for three 12-pulse converter series mode of Scheme 3 (sending end) (kV)

Ideal no-load DC voltage	Changji station connected with 750 kV grid	HV side of Guquan station connected to 500 kV grid	LV side of Guquan station connected to 1000 kV grid
U_{di0N}	174.1	231.9	231.9
U_{di0min}	157.6	210.0	210.0
U_{di0max}	174.8	232.9	232.9
$U_{\text{di0maxOLTC}}$	174.8	232.9	232.9
U_{di0G}	174.8	232.9	232.9
U_{di0L}	177.0	235.9	235.9
$U_{\text{di0absmax}}$	179.0	239.0	239.0

Table 7 Ideal no-load DC voltage for three 12-pulse converter series mode of Scheme 3 (receiving end) (kV)

Ideal no-load DC voltage	Changji station connected with 750 kV grid	HV side of Guquan station connected to 500 kV grid	LV side of Guquan station connected to 1000 kV grid
U_{di0N}	163.8	218.3	218.3
U_{di0min}	148.5	198.1	198.1
U_{di0max}	166.8	222.4	222.4
$U_{\text{di0maxOLTC}}$	166.8	222.4	222.4
U_{di0G}	163.7	220.3	220.3
U_{di0L}	166.8	222.4	222.4
$U_{\text{di0absmax}}$	169.0	225.0	225.0

technical parameters and insulation level of the converter transformer corresponding to each SLD type, as well as the resulting design and manufacturing cost, transportation cost, spare parts, safety and reliability, etc., plays an important role in the selection of the SLD. According to the results of the main circuit calculation, the key technical parameters of the converter transformer corresponding to the three recommended SLDs can be obtained, as shown in Tables 8, 9, 10, and 11.

For the sending end converter transformer, the AC side is connected to the 750 kV grid. While for the receiving end, to reduce manufacturing difficulty, both the HV and medium voltage (MV) converter transformers are connected to the 500 kV grid, and the LV converter transformer is connected to the 1000 kV grid. Therefore, the focus is on the valve-side parameters and capacity. From Table 8, 9, 10, and 11, it can be seen that for the double 12-pulse series scheme, since there are only two 12-pulse converters, the valve-side voltage and capacity of each converter transformer are significantly higher than those of the three 12-pulse series scheme. Among them, the valve-side voltage of the sending end is up to 83% higher, the valve-side voltage of the receiving end converter connected to the 500 kV grid is up to 88% higher, and

Table 8 Parameters of converter transformer in double 12-pulse series mode (2 × 550 kV)

Item	Sending end connected to 750 kV grid HV and LV converter transformers			Receiving end connected to 500 kV grid HV converter transformer			Receiving end connected to 1000 kV grid LV converter transformer		
	Grid-side winding	Valve-side winding		Grid-side winding	Valve-side winding		Grid-side winding	Valve-side winding	
Type of converter transformer winding	Y/Y0	Y	D	Y/Y0	Y	D	Y/Y0	Y	D
Rated phase voltage (tap 0) (kV, rms)	447.5	136.4	236.2	294.4	131.8	228.3	606.2	131.8	228.3
Maximum steady state phase voltage (kV, rms)	461.9	140.6	243.6	303.1	138.1	239.2	617.8	138.1	239.2
Nominal Capacity (MVA)	607.5	607.5	607.5	587.1	587.1	587.1	587.1	587.1	587.1
Current at Tap 0 without cooling equipment (A, rms)	1357.6	4453.6	2571.3	1993.8	4453.6	2571.3	968.4	4453.6	2571.3
Tap position	+25/−5			+25/−5			+20/−10		

Table 9 Parameters of converter transformer in three 12-pulse series mode (3 × 367 kV)

Item	Sending end connected to 750 kV grid HV, MV and LV converter transformer			Receiving end connected to 500 kV grid HV and MV converter transformer			Receiving end connected to 1000 kV grid LV converter transformer		
	Grid-side winding	Valve-side winding		Grid-side winding	Valve-side winding		Grid-side winding	Valve-side winding	
Type of converter transformer winding	Y/Y0	Y	D	Y/Y0	Y	D	Y/Y0	Y	D
Rated phase voltage (tap 0) (kV, rms)	447.5	91.06	157.72	294.4	85.97	148.91	606.2	85.97	148.91
Maximum steady state phase voltage (kV, rms)	461.9	93.57	162.17	303.1	88.47	153.28	617.8	88.47	153.28
Nominal capacity (MVA)	405.55	405.55	405.55	382.89	382.89	382.89	382.89	382.89	382.89
Current at Tap 0 without cooling equipment (A, rms)	910	4453.6	2571.3	1300	4453.6	2571.3	630	4453.6	2571.3
Tap position	+25/−5			+25/−5			+20/−10		

Table 10 Parameters of sending end converter transformer in three 12-pulse series mode (400 + 400 + 300 kV)

Item	Sending end connected to 750 kV grid HV converter transformer			Sending end connected to 750 kV grid MV converter transformer			Sending end connected to 750 kV grid LV converter transformer		
	Grid-side winding	Valve-side winding		Grid-side winding	Valve-side winding		Grid-side winding	Valve-side winding	
Type of converter transformer winding	Y/Y0	Y	D	Y/Y0	Y	D	Y/Y0	Y	D
Rated phase voltage (tap 0) (kV, rms)	447.5	74.43	128.92	447.5	99.14	171.72	447.5	99.14	171.72
Maximum steady state phase voltage (kV, rms)	461.9	76.53	132.55	461.9	102.17	176.97	461.9	102.17	176.97
Nominal Capacity (MVA)	331.48	331.48	331.48	441.53	441.53	441.53	441.53	441.53	441.53
Current at Tap 0 without cooling equipment (A, rms)	740	4453.6	2571.3	990	4453.6	2571.3	990	4453.6	2571.3
Tap position	+25/−5			+25/−5			+25/−5		

Table 11 Parameters of receiving end converter transformer in three 12-pulse series mode (400 + 400 + 300 kV)

Item	Receiving end connected to 500 kV grid HV converter transformer			Receiving end connected to 500 kV grid MV converter transformer			Receiving end connected to 1000 kV grid LV converter transformer		
	Grid-side winding	Valve-side winding		Grid-side winding	Valve-side winding		Grid-side winding	Valve-side winding	
Type of converter transformer winding	Y/Y0	Y	D	Y/Y0	Y	D	Y/Y0	Y	D
Rated phase voltage (tap 0) (kV, rms)	294.4	70	121.29	294.4	93.33	161.65	606.2	93.33	161.65
Maximum steady state phase voltage (kV, rms)	303.1	72.25	125.14	303.1	96.2	166.61	617.8	96.2	166.61
Nominal Capacity (MVA)	311.87	311.87	311.87	415.64	415.64	415.64	415.64	415.64	415.64
Current at Tap 0 withot cooling equipment (A, rms)	1060	4453.6	2571.3	1412	4453.6	2571.3	690	4453.6	2571.3
Tap position	+25/−5			+25/−5			+19/−11		

the valve-side voltage of the receiving end connected to the 1000 kV grid is up to 53% higher. The capacity of the sending end converter transformer is 83% higher, and the capacity of the receiving end is 88% higher. It can also be seen that the double 12-pulse main connection is much larger than the three 12-pulse connection in terms of manufacturing and transportation difficulties, but its technical and economic performance is better than that of the three 12-pulse.

From Tables 12, 13, 14, 15, and 16, it can be seen that the insulation level of the valve-side winding of the HV converter transformer of the double 12-pulse scheme is significantly lower than that of the three 12-pulse scheme. Especially when the switching voltage is higher than 2000 kV, a small increase in voltage will lead to a significant increase in insulation design difficulty, volume, and cost. Therefore, from the perspective of insulation level, the double 12-pulse series scheme is much better than the three 12-pulse series scheme.

3. Recommended scheme

Based on the above analysis, considering the impact on system reliability, DC control protection system complexity, AC system power loss, equipment manufacturing cost, converter station construction cost, DC system operation and other factors, it is recommended that the ± 1100 kV UHV DC transmission project adopt the main connection scheme of double 12-pulse (2 × 550 kV) series, the finalized determined SLD schematic is shown in Fig. 7.

II. Main Circuit Design and Key Equipment Parameters

(I) Main Circuit Parameters

Based on the recommended SLD, taking the CG project as an example, the rated voltage of the converter bus at the Changji station is 775 kV, the rated voltage of the converter bus of LV converter connected to the 1000 kV AC system at the Guquan station is 1050 kV, and that of the HV converter connected to the 500 kV AC system is 510 kV. The nominal resistance of the DC line is 9.41 Ω. The nominal resistance of the grounding electrode line and grounding electrode of the Changji converter station is 1.70 Ω, and that of the Guquan converter station is 1.07 Ω. The main operating characteristic and parameters of the system are as follows (the reactive power balance of the converter station is not considered).

(1) The power direction is from Changji to Guquan, with bipolar full voltage operation. The converter bus voltage at the Changji station is 775 kV, and the converter bus voltage at the Guquan station is 510/1050 kV. The line resistance is 9.41 Ω. The operating parameters of the rectifier and inverter with power factor ranging from 0.1 to 1.05 p.u. are shown in Table 17.
(2) The power direction is from Changji to Guquan, with full voltage operation of monopolar ground return. The converter bus voltage at the Changji station is 775 kV, and the converter bus voltage at the Guquan station is 510/1050 kV. As the total resistance of the DC line and ground return is 12.18 Ω, the operating parameters are shown in Table 18.

Table 12 Insulation level of converter transformer in the mode of double 12-pulse series connection (2 × 550 kV) (kV)

Type of impulse		HV converter transformer at sending end			LV converter transformer at sending end			HV converter transformer at receiving end			LV converter transformer at receiving end		
		Grid-side winding	Valve-side winding		Grid-side winding	Valve-side winding		Grid-side winding	Valve-side winding		Grid-side winding	Valve-side winding	
		Y/Y0	Y	D	Y/Y0	Y	D	Y/Y0	Y	D	Y/Y0	Y	D
Lightning impulse (LI)	Terminal 1	1950	2300	1980	1950	1350	1240	1550	2300	1930	2250	1350	1235
	Terminal 2	185	–	–	185	–	–	185	–	–	185	–	–
Switching impulse (SI)	Terminal 1	1550	–	–	1550	–	–	1175	–	–	1800	–	–
	Terminal 2	–	–	–	–	–	–	–	–	–	–	–	–
	Terminal 1 + 2	–	2100	1840	–	1250	1175	–	2100	1785	–	1250	1165

Table 13 Insulation level of sending end converter transformer in the mode of three 12-pulse series connection (3 × 367 kV) (kV)

Type of impulse		HV converter transformer			MV converter transformer			LV converter transformer		
		Grid-side winding	Valve-side winding		Grid-side winding	Valve-side winding		Grid-side winding	Valve-side winding	
		Y/Y0	Y	D	Y/Y0	Y	D	Y/Y0	Y	D
Lightning impulse (LI)	Terminal 1	1950	2450	2300	1950	1800	1550	1950	1250	1150
	Terminal 2	185	–	–	185	–	–	185	–	–
Switching impulse (SI)	Terminal 1	1550	–	–	1550	–	–	1550	–	–
	Terminal 2	–	–	–	–	–	–	–	–	–
	Terminal 1 + 2	–	2200	2025	–	1475	1250	–	1100	975

Table 14 Insulation level of receiving end converter transformer in the mode of three 12-pulse series connection (3 × 367 kV) (kV)

Type of impulse		HV converter transformer			MV converter transformer			LV converter transformer		
		Grid-side winding	Valve-side winding		Grid-side winding	Valve-side winding		Grid-side winding	Valve-side winding	
		Y/Y0	Y	D	Y/Y0	Y	D	Y/Y0	Y	D
Lightning impulse (LI)	Terminal 1	1550	2450	2300	1550	1800	1550	2250	1250	1150
	Terminal 2	185	–	–	185	–	–	185	–	–
Switching impulse (SI)	Terminal 1	1175	–	–	1175	–	–	1800	–	–
	Terminal 2	–	–	–	–	–	–	–	–	–
	Terminal 1 + 2	–	2200	2025	–	1475	1250	–	1100	975

Table 15 Insulation level of sending end converter transformer in the mode of three 12-pulse series connection (400 + 400 + 300 kV) (kV)

Type of impulse		HV converter transformer			MV converter transformer			LV converter transformer		
		Grid-side winding	Valve-side winding		Grid-side winding	Valve-side winding		Grid-side winding	Valve-side winding	
		Y/Y0	Y	D	Y/Y0	Y	D	Y/Y0	Y	D
Lightning impulse (LI)	Terminal 1	1950	2450	2350	1950	1900	1625	1950	1300	1175
	Terminal 2	185	–	–	185	–	–	185	–	–
Switching impulse (SI)	Terminal 1	1550	–	–	1550	–	–	1550	–	–
	Terminal 2	–	–	–	–	–	–	–	–	–
	Terminal 1 + 2	–	2200	2050	–	1675	1360	–	1175	1050

Table 16 Insulation level of receiving end converter transformer in the mode of three 12-pulse series connection (400 + 400 + 300 kV) (kV)

Type of impulse		HV converter transformer			MV converter transformer			LV converter transformer		
		Grid-side winding	Valve-side winding		Grid-side winding	Valve-side winding		Grid-side winding	Valve-side winding	
		Y/Y0	Y	D	Y/Y0	Y	D	Y/Y0	Y	D
Lightning impulse (LI)	Terminal 1	1550	2450	2350	1550	1900	1625	2250	1300	1175
	Terminal 2	185	–	–	185	–	–	185	–	–
Switching impulse (SI)	Terminal 1	1175	–	–	1175	–	–	1800	–	–
	Terminal 2	–	–	–	–	–	–	–	–	–
	Terminal 1 + 2	–	2200	2050	–	1675	1360	–	1175	1050

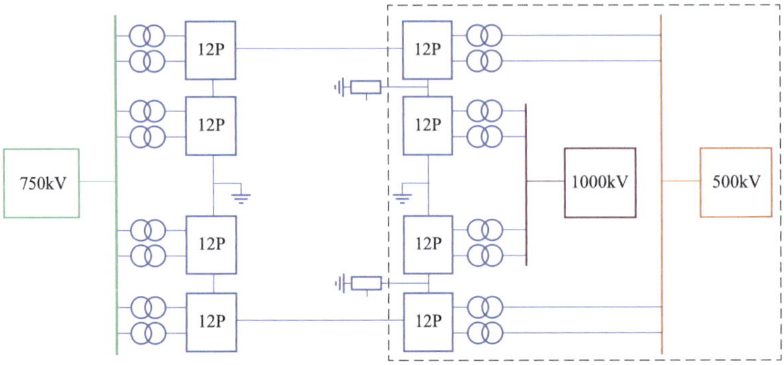

Fig. 7 Recommended SLD

(3) The power direction is from Changji to Guquan, with full voltage operation of monopolar metallic return. The converter bus voltage at the Changji station is 775 kV, and the converter bus voltage at the Guquan station is 510/1050 kV. As the total resistance of the DC line and metallic return is 18.82 Ω, the operating parameters are shown in Table 19.

(4) The power direction is from Changji to Guquan, with bipolar reduced voltage operation. The converter bus voltage at the Changji station is 775 kV, and the converter bus voltage at the Guquan station is 510/1050 kV. The line resistance is 9.41 Ω. The operating parameters of the rectifier and inverter when the voltage is reduced by 80% to 880 kV are shown in Table 20.

(5) The power direction is from Guquan to Changji, with bipolar full voltage operation. The converter bus voltage at the Changji station is 775 kV, and the converter bus voltage at the Guquan station is 510/1050 kV. As the DC line resistance is 9.41 Ω, the operating parameters are shown in Table 21.

(II) Design of AC and DC filters

1. AC filter design

   The AC filter and reactive power compensation device in the converter station are generally considered together, with the specific configuration as follows.

   (1) Changji converter station. 5 banks of BP11/13 filters, each with a capacity of 305 Mvar; 4 banks of HP24/36 filters, each with a capacity of 305 Mvar; 3 banks of HP3 filters, each with a capacity of 305 Mvar; 8 banks of SC parallel capacitors, each with a capacity of 380 Mvar.

   (2) 500 kV side of Guquan converter station. 8 banks of HP12/24 filters, each with a capacity of 285 Mvar; 1 bank of HP3 filters, with a capacity of 285 Mvar; 5 banks of SC parallel capacitors, each with a capacity of 285 Mvar.

   (3) 1000 kV side of Guquan converter station. 10 banks of HP12/24 filters, each with a capacity of 340 Mvar; 2 banks of HP3 filters, each with a capacity of 340 Mvar. These two banks are adopted with the same grouping scheme.

Table 17 Parameters of bipolar full voltage operation in forward direction

P_{dR} (MW)	I_{dc} (kA)	U_{dR} (kV)	U_{di0R} (kV)	U_{di0IH} (kV)	U_{di0IL} (kV)	α (°)	γ_H (°)	γ_L (°)	*TCR*	*TCIH*	*TCIL*
1200	0.55	1100	288.41	289.36	289.36	15.0	17.0	17.0	12.3	5.2	10.1
6000	2.73	1100	302.02	297.79	297.79	15.0	17.0	17.0	6.6	2.8	5.4
12 000	5.45	1100	319.03	308.33	308.33	15.0	17.0	17.0	0.0	0.0	0.0
12 600	5.77	1092.5	319.04	307.58	307.58	15.0	17.0	17.0	0.0	0.2	0.4

Note P_{dR} is DC power; I_{dc} is DC current; U_{dR} is DC voltage; U_{di0R} is no-load DC voltage at the valve side of the rectifier; U_{di0IH} is no-load DC voltage at the HV valve side of the inverter; U_{dioIL} is no-load DC voltage at the LV valve side of the inverter; α is firing angle; γ_H is extinction angle of HV valve of the inverter; γ_L is extinction angle of LV valve of the inverter; *TCR* is tap position of the rectifier; *TCIH* is tap position of HV valve of the inverter; *TCIL* is tap position of LV valve of the inverter

Table 18 Parameters of full voltage operation of monopolar ground return in forward direction

P_{dR} (MW)	I_{dc} (kA)	U_{dR} (kV)	U_{di0R} (kV)	U_{di0IH} (kV)	U_{di0IL} (kV)	α (°)	γ_H (°)	γ_L (°)	*TCR*	*TCIH*	*TCIL*
600.0	0.55	1100	288.41	288.96	288.96	15.0	17.0	17.0	12.3	5.4	10.3
3000.0	2.73	1100	302.02	295.82	295.82	15.0	17.0	17.0	6.6	3.4	6.5
6000.0	5.45	1100	319.03	304.38	304.38	15.0	17.0	17.0	0.0	1.0	2.0
6300.0	5.77	1092.5	319.04	303.41	303.41	15.0	17.0	17.0	0.0	1.3	2.5

Table 19 Parameters of full voltage operation of monopolar metallic return in forward direction

P_{dR} (MW)	I_{dc} (kA)	U_{dR} (kV)	U_{di0R} (kV)	U_{di0IH} (kV)	U_{di0IL} (kV)	α (°)	γ_H (°)	γ_L (°)	*TCR*	*TCIH*	*TCIL*
600.0	0.55	1100	288.41	288.02	288.02	15.0	17.0	17.0	12.3	5.6	10.9
3000.0	2.73	1100	302.02	291.08	291.08	15.0	17.0	17.0	6.6	4.7	9.1
6000.0	5.45	1100	319.03	294.91	294.91	15.0	17.0	17.0	0.0	3.6	7.0
6300.0	5.77	1092.5	319.04	293.40	293.40	15.0	17.0	17.0	0.0	4.1	7.8

2. DC filter design

In order to dampen the impact of DC side harmonics, DC transmission projects that use overhead lines are generally equipped with DC filters. In the past, when designing DC filters, it was necessary to consider the impact of factors such as restrictions of DC equivalent interference current I_{eq} and prevention of DC line resonance. Despite the mature technology and significant effects, DC filters are costly and have a large footprint, and the height of their high-voltage capacitor towers is often the controlling factor for the height of the entire DC yard. With the increase in voltage level and transmission capacity of the ± 1100 kV UHV DC transmission project, this contradiction is more prominent, and it is urgent to carry out simplified design of DC filters.

As communication lines in China have basically been fibre-optic, and are almost unaffected by the harmonic interference current of the DC transmission line, the DC filter can be designed without regarding the effect of the equivalent interference current, and only other functions of the DC filter, i.e., limiting the harmonic voltage at the outlet and preventing the resonance of the DC line, should be considered. The main idea of simplification is to relax the restriction on the effective value of the equivalent interference current I_{eq}, cancel the high-frequency branch of the DC filter, and reduce the overall cost of the equipment. According to the research conclusions of insulation coordination, as long as the constant continuous operating voltage of the DC pole line *CCOV* (the sum of the DC voltage and harmonic voltage) does not exceed 1170 kV, the insulation level of the pole line operation can be maintained at 2100 kV. When checking the filtering effect of the simplified DC filter, this is used as the evaluation basis.

Table 20 Parameters of bipolar reduced voltage operation in forward direction

P_{dR} (MW)	I_{dc} (kA)	U_{dR} (kV)	U_{di0R} (kV)	U_{di0IH} (kV)	U_{di0IL} (kV)	α (°)	γ_H (°)	γ_L (°)	*TCR*	*TCIH*	*TCIL*
1200	0.68	880	262.58	234.92	272.86	31.3	18.9	35.5	25.0	25.0	20.0
6000	3.41	880	262.58	242.91	272.86	23.5	17.0	31.6	25.0	21.5	20.0
9600	5.45	880	262.58	250.82	272.86	15.4	17.0	28.5	25.0	18.3	20.0

Table 21 Parameters of bipolar full voltage operation in reverse direction

P_{dR} (MW)	I_{dc} (kA)	U_{dR} (kV)	U_{di0RH} (kV)	U_{di0RL} (kV)	U_{di0I} (kV)	α_H (°)	α_L (°)	γ (°)	*TCRH*	*TCRL*	*TCI*
1200	0.571	1050.6	276.01	276.01	276.37	15.0	15.0	17.0	9.5	18.3	18.0
6000	2.856	1050.6	291.14	291.14	284.51	15.0	15.0	17.0	4.9	9.4	14.1
9600	4.569	1050.6	302.49	302.49	290.61	15.0	15.0	17.0	1.7	3.2	11.4

Only the 2nd and 12th branch are configured in the simplified DC filter, and a dual-tuned filter is used for implementation. The comparison between the simplified and conventional DC filter design scheme is shown in Table 22.

The simplified DC filter scheme reduces the main capacitance from 1.45 to 0.6 μF, greatly reducing the manufacturing difficulty of the DC filter, and the cost is reduced to 50% of the conventional scheme, saving a total investment of more than RMB 100 million.

3. Key equipment parameters

The key technical parameters of the ± 1100 kV converter transformer are shown in Table 23, those of the ± 1100 kV converter valve are shown in Table 24, and those of the smoothing reactor are shown in Table 25.

III. Application of Results

The research results of the SLD and main circuit design of ± 1100 kV UHV DC transmission have been successfully applied to the CG project. The project adopts the main connection of double 12-pulse converter in series for each pole, with a total of 28 converter transformers (including 4 spare transformers).

Table 22 Comparison between simplified and conventional DC filter design scheme

Component	Conventional scheme		Simplified scheme
	Filter bank type		Filter type
	12/24	2/30	2/12
Total number of filter banks	1	1	1
Harmonic frequency (Hz)	600/1200	100/1500	100/600
C_1 (μF)	0.35	1.45	0.6
L_1 (mH)	89.35	28.028	408.4
C_2 (μF)	0.810	0.564 614	0.267
L_2 (mH)	48.86	1243	2 719.7
R_1 (Ω)	10 000	6000	4130
R_2(Ω)	–	–	9560

Table 23 Key technical parameters of the ± 1100 kV converter transformer

Converter transformer winding	Changji converter station			Guquan converter station		
	Line-side winding	Valve-side winding		Line-side winding	Valve-side winding	
Y/D		Y	D		Y	D
Rated phase voltage (tap 0) (kV, rms)	447.5	136.4	236.2	294.4	131.8	228.3
Maximum steady-state phase voltage (kV, rms)	461.9	140.6	243.6	303.1	138.1	239.2
Rated capacity of each one (MVA)	607.5	607.5	607.5	587.1	587.1	587.1
Rated current without cooled equipment (A, rms)	1357.6	4453.6	2571.3	1993.8	4453.6	2571.3
Tap position	+25/−5			+25/−5		
Tap adjustment step (%)	0.86			1.25		
Impedance at rated tap position (tap 0) (%)	20			22		
Maximum error in relative inductive voltage drop of the converter transformer (%)	± 1.0			± 1.1		

3 DC Overvoltage and Insulation Coordination

The increase in voltage level of UHV DC transmission projects would lead to a significant increase in the insulation level of the system and equipment. A reasonable insulation coordination scheme can ensure the safety of the project while considering its economy. This section proposes a reasonable scheme for the ± 1100 kV UHV DC transmission project through the study of arrester parameters, configuration scheme, and overvoltage simulation calculations, effectively reducing the requirements of insulation level of the equipment.

I. Characteristics of ± 1100 kV Insulation Coordination

The significant increase in the insulation level of equipment poses a great challenge to the design and construction of the project. If the insulation coordination method of the ± 800 kV transmission project is linearly extrapolated to that of ± 1100 kV

Table 24 Key technical parameters of the ± 1100 kV converter valve

No	Item		Changji converter station	Guquan converter station (HV/LV)
1	*Rated current*			
1.1	Rated DC current (I_{dN}) (A)		5455	5455
1.2	Minimum continuous operation DC current (A)		550	550
1.3	Maximum continuous operation DC current at rated power (A)		5523	5523
2	*Rated voltage*			
2.1	Rated DC voltage, pole-to-neutral (U_{dRN}) (kV)		1100	1100
2.2	Maximum continuous DC voltage (kV)		1122	1122
2.3	No-load DC voltage	Rated no-load DC voltage (U_{dioN}) (kV)	319.04	308.33/308.33
		Maximum no-load DC voltage ($U_{dioabsmax}$) (kV)	329	323/323
		Minimum no-load DC voltage (U_{diomin}) (kV)	287.97	288.53/288.53
2.4	Transient overvoltage load shedding factor (p.u.)		1.4 blocking 1.3 < 0.06 s 1.2 > 0.06 s	1.4 blocking 1.3 < 0.06 s 1.2 > 0.06 s
3	*Control angle*			
3.1	Firing angle during rectification operation α	Rated value (α_N) (°)	15	
		Minimum value at rated power (°)	12.5–0.5	
		Maximum value at rated power (°)	17.5 + 0.5	
		Minimum value (°)	5	
3.2	Extinction angle during contravariant operation γ	Rated value (γ_N) (°)		17
		Minimum value at rated power (°)		17 − 1
		Maximum value at rated power (°)		17 + 1
		Minimum value (°)		9
4	*Inductive voltage drop d_x*	Normal state (%)	10	11
		Minimum value (−10%) (%)	9	9.9
		Maximum value (+ 5%) (%)	10.5	11.55
5	*Transient current in valve*	Short circuit level of valve (MVA) at the following system short circuit capacities	Maximum value	Maximum value
			60,016	60,016/120032
5.1	Peak short-circuit current ($\alpha_{min} = 5°$, $f = 49.8$ Hz)	Single short-circuit current peak with subsequent blocking (kA)	50.18	42.92/45.28

(continued)

Table 24 (continued)

No	Item		Changji converter station	Guquan converter station (HV/LV)
		0.06s peak short circuit current without subsequent blocking (kA)	53.07	45.81/48.16
5.2	Recovery time with subsequent blocking	(ms)	< 2	< 1.8
5.3	Off-state voltage with subsequent blocking (excluding test factor of 1.05)	(kV, rms)	268	255/246
6	*Insulation withstand level*			
6.1	Cross-valve	*SIWL* (kV, peak)	627	614
		LIWL (kV, peak)	606	605
		FWWL (kV, peak)		
6.2	Upper 12-pulse bridge DC bus to ground	*SIWL* (kV, peak)	2100	2100
		LIWL (kV, peak)	2550	2550
6.3	HV terminal of upper 12-pulse bridge valve connected to converter transformer secondary Y winding to ground	*SIWL* (kV, peak)	2100	2100
		LIWL (kV, peak)	2550	2550
6.4	Upper 12-pulse bridge valve midpoint bus to ground	*SIWL* (kV, peak)	1639	1639
		LIWL (kV, peak)	1845	1843
6.5	Double 12-pulse bridge midpoint DC bus to ground	*SIWL* (kV, peak)	1180	1140
		LIWL (kV, peak)	1313	1267
6.6	Lower 12-pulse bridge valve midpoint bus to ground	*SIWL* (kV, peak)	761	761
		LIWL (kV, peak)	842	842
6.7	Neutral bus to ground	*SIWL* (kV, peak)	503	503
		LIWL (kV, peak)	571	571

project, the insulation level of the HV converter transformer will reach 2200 kV. Considering the impact of simplified DC filters, the insulation level of the line-side pole equipment of the smoothing reactor will even reach 2268 kV. The research and development capabilities of equipment in the short term are difficult to meet such high insulation level requirements, and the economic cost is also unbearable.

On the basis of overvoltage and insulation coordination studies in previous UHV DC transmission projects, a new coordination scheme of surge arrester and insulation is proposed for the CG project, which significantly reduces the insulation level requirements of key equipment, and the new scheme is verified to be reasonable through simulation calculations.

Table 25 Key technical parameters of the ± 1100 kV smoothing reactor

Type		Dry-type air-core
Rated inductance (mH)		75
Rated current	DC current at rated power under ideal conditions I_{dN} (A)	5455
	Maximum continuous DC current I_{mcc} (A)	5523
	2h overload current I (A)	5839
	Transient fault current (kA, peak)	40
Rated voltage	DC voltage to ground at rated power under ideal conditions U_{dN} (kV)	1100
	Maximum continuous DC voltage to ground U_{dmax} (kV)	1122, simultaneous overlay harmonics 75 kV (rms)
Insulation level and test voltage (total level in series connection)	*SIWL* between terminals (kV, peak)	2100
	SIWL between terminal and ground (kV, peak)	2100
	LIWL between terminals, full wave (kV, peak)	2600
	LIWL between terminal and ground, full wave (kV, peak)	2580
	LIWL between terminals, chopped wave (kV, peak)	2860
	DC withstand level to ground (kV, 120 min)	1755

II. Research Conditions

(I) DC System Conditions

The CG project has a bipolar rated transmission capacity of 12 GW, with a rated operating current of 5454 A. The sending end converter station is connected to the 750 kV AC system, and the receiving end converter station is layered connected into the 500 and 1000 kV AC systems. The impedance of the sending end converter transformer is 20%, and that of the receiving end converter transformer is 22%. Two 75 mH dry-type smoothing reactors are arranged on the pole bus and neutral bus respectively. The rated resistance of the DC line is 9.41 Ω. Under the metallic return mode, the receiving end station is the grounding station. The maximum ideal no-load DC voltages of the sending and receiving ends are 329 and 323 kV respectively.

(II) AC System Conditions

The Changji converter station is connected to the AC system with a steady-state maximum operating voltage of 800 kV, a steady-state minimum operating voltage of 750 kV, and a rated frequency of (50 ± 0.2) Hz. The HV side of the Guquan converter station is connected to the AC system with a steady-state maximum operating voltage of 525 kV, a steady-state minimum operating voltage of 490 kV. And the LV side is connected to the AC system with a steady-state maximum operating voltage of 1070 kV, a steady-state minimum operating voltage of 1000 kV, and a rated frequency of (50 ± 0.1) Hz. The parameters of the AC system of the sending and receiving end converter stations are shown in Table 26.

(III) DC Filter Parameters

To reduce the project cost, a simplified DC filter scheme is adopted for CG project. The maximum DC voltage peak on the line side of the smoothing reactor can reach 1170 kV. The parameters of the DC filter components are shown in Table 27.

(IV) AC Filter Parameters

The AC filters of Changji converter station are divided into 4 banks and 20 sub-banks. Among them, there are 5 banks of BP11/13 AC filters, each with a capacity of 305 Mvar; 4 banks of HP24/36 AC filters, each with a capacity of 305 Mvar; 3 banks of

Table 26 Parameters of the AC system of the sending and receiving end converter stations

Parameters	Changji converter station	Guquan converter station	
		HV	LV
Maximum short-circuit current (kA)	63	63	63
X/*R*	7	7	7
Minimum short-circuit current (kA)	33.5	36.2	22.8

Table 27 Parameters of the DC filter components

Components	Filter bank type
	2/12
C_1(μF)	0.6000
L_1(mH)	408.4
C_2(μF)	0.2670
L_2(mH)	2720
R_1(Ω)	4130
R_2(Ω)	9560
Quality factor (inductance)	100
tanδ of capacitor (50 Hz)	0.0002

HP3 AC filters, each with a capacity of 305 Mvar; and 8 banks of SC, each with a capacity of 380 Mvar.

The AC filters of Guquan converter station 500 kV side are divided into 3 banks and 14 sub-banks. Among them, there are 8 banks of HP12/24 AC filters, each with a capacity of 285 Mvar; 1 bank of HP3 AC filters, each with a capacity of 285 Mvar; and 5 banks of SC, each with a capacity of 285 Mvar.

The AC filters of Guquan converter station 1000 kV side are divided into 2 banks and 12 sub-banks. Among them, there are 10 banks of HP12/24 AC filters, each with a capacity of 340 Mvar; 2 banks of HP3 AC filters, each with a capacity of 340 Mvar.

The detailed parameters of the AC filters are shown in Tables 28, 29 and 30.

III. Configuration Scheme of Arrester Protection

(I) Impact of the Simplified DC Filter

With the continuous increase of the DC voltage level, the cost of the DC filter has greatly increased. On the basis of ensuring project design reliable, in order to improve the economy of project construction, a DC filter configuration scheme of 2/12 with main capacitance 0.6 μF has been proposed through the comparison of schemes such as retaining the complete DC filter, completely canceling the DC filter, and configuring different types of simplified DC filters.

If the complete DC filter configuration is adopted, the voltage at the DC system pole line exit can be controlled at 1122 kV. If the simplified DC filter configuration

Table 28 Parameters of the AC filters of Changji converter station

Components	Filter bank type			
	BP11/BP13	HP24/36	HP3	SC
C_1 (μF)	0.8015/0.8034	1.614	1.616	2.014
L_1 (mH)	104.5/74.62	7.568	783.5	1.000
C_2 (μF)	–	9.391	12.93	–
L_2 (mH)	–	1.199	–	–
R_1/R_2	8000/8000	300.0	1313	–

Table 29 Parameters of the AC filters of Guquan converter station 500 kV side

Components	Filter bank type		
	HP12/24	HP3	SC
C_1 (μF)	3.469	3.488	3.488
L_1 (mH)	11.46	363.1	0.9000
C_2 (μF)	6.116	27.90	–
L_2 (mH)	5.228	–	–
R_1/R_2	400.0	912.6	–

Table 30 Parameters of the AC filters of Guquan converter station 1000 kV side

Components	Filter bank type	
	HP12/24	HP3
C_1 (μF)	0.976 2	0.981 6
L_1(mH)	38.04	1290
C_2 (μF)	1.763	7.853 1
L_2 (mH)	19.89	–
R_1/R_2	1500	2162

is adopted, it is known from the steady-state simulation results that the peak voltage at the pole line exit can reach 1170 kV. Since the pole line and neutral line are evenly arranged in the UHV DC transmission project, the midpoint voltage of the double 12-pulse converter is no longer a pure DC voltage, but a DC voltage containing ripples with a peak value of 585 kV.

The increase in DC voltage of pole line and double 12-pulse converter will affect the design of the HV valve hall and arrester of pole line side.

(II) New Configuration Scheme of Arrester

The principle of insulation coordination design of the UHV DC converter station is:

(1) The overvoltage on the AC side should be limited as much as possible by the arrester installed on the AC side, and the overvoltage on the DC side should be limited by the arrester or combination of arresters on the DC side.
(2) The key components of the converter equipment (such as valves) should be directly protected by the arrester closely connected to the component. The ± 1100 kV UHV DC converter station adopts two 12-pulse converter units per pole, retaining most of the arresters of the ± 800 kV UHV DC transmission project, as shown in Fig. 8.

As can be seen from Fig. 8, compared with the arrester configuration scheme of ± 800 kV transmission, scheme of ± 1100 kV transmission retains the MH, CBL2, ML, and DB arresters. Furthermore, due to the adoption of the indoor DC yard in the CG project, there is no issue of lightning wave intrusion near the valve hall pole line bushing, the CBH arrester is cancelled. For the first time, AH and AL arresters are added, which are placed at the HV and LV Yy converter transformer valve side, respectively. By adopting direct protection, the valve-side operation insulation levels of the HV and LV Yy converter transformer can be reduced by 6% and 17%, respectively.

In addition, the DB arrester adopts a multi-column parallel scheme. Compared with the linear extrapolation scheme of the ± 800 kV pole line DB arrester, the new scheme reduces the operation insulation level of the pole line equipment on the smoothing reactor line side by 8%, which is consistent with the valve side of the smoothing reactor (2100 kV), basically offsetting the impact of simplified DC filter on the insulation level.

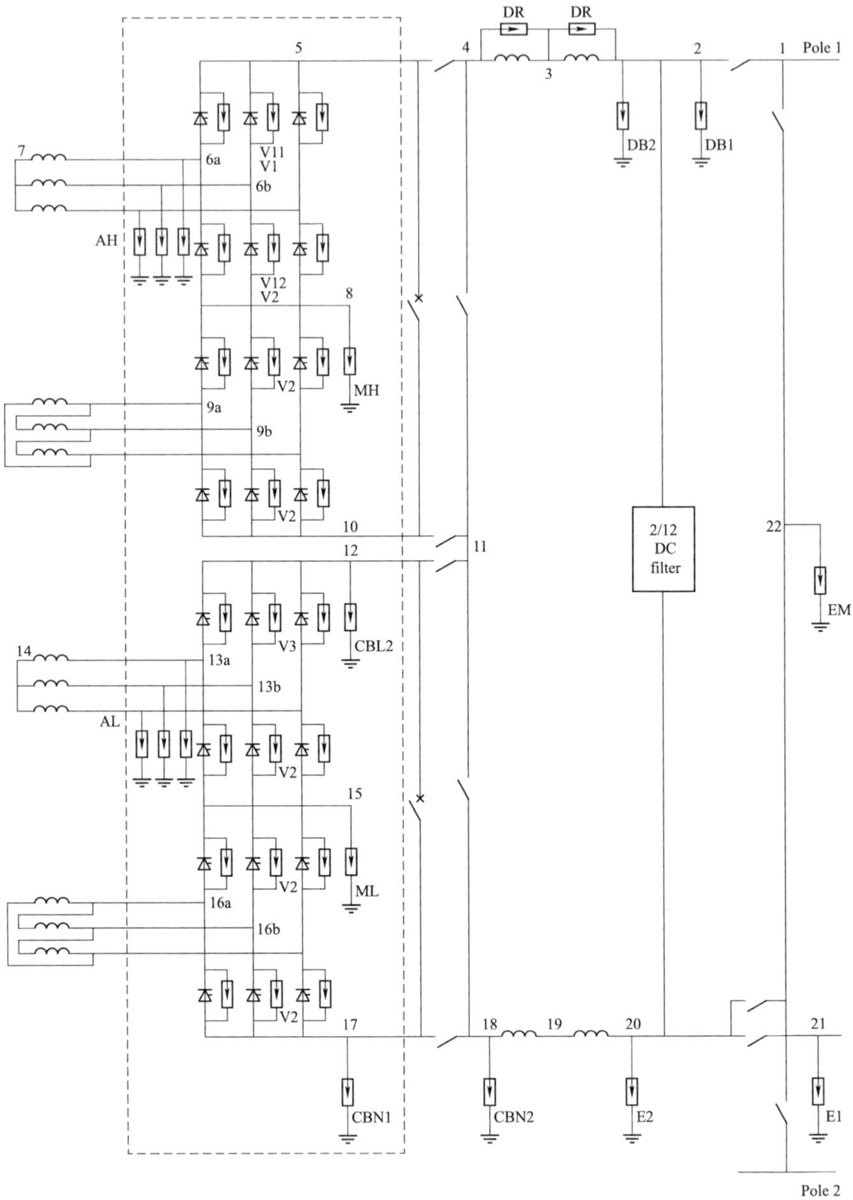

Fig. 8 Configuration of the arresters of converter station

IV. Selection of Chargeability of AH, AL Arrester

Through PSCAD simulation, the voltages of the AH and AL arresters in steady state are obtained as shown in Figs. 9 and 10. As can be seen from Fig. 8, there are three groups of AH (AL) arresters (one for each phase) on the valve side of the converter transformer, connected to the top of the valve which are conducting cyclically. Each arrester actually only bears a higher voltage for half a cycle, and the actual voltage is very low for the other half cycle due to the valve group connected to the valve top being in a non-conductive state. At this time, the leakage current of the arrester is almost zero. A relatively high chargeability can be chosen as the AH (AL) arresters do not have issues of heating and aging. According to the conclusions of a large number of aging tests conducted in cooperation with Chinese arrester manufacturers, it is safe to take a value of chargeability of 0.9 for the AH (AL) arrester, while that for other pure DC arresters in the valve hall is generally 0.82. To ensure project reliability, the value of the AH (AL) arrester in the practical engineering is considered to be within 0.85.

V. Selection of Main Arresters

(I) Selection of Rated (Reference) Voltage for Main Arresters

1. Valve arrester

The selection of the arrester's rated value is determined according to recognized experience. The peak continuous operating voltage of the valve arrester *CCOV* is calculated by formula (1)

$$CCOV = U_{\text{dioabsmax}} \times \frac{\pi}{3} \tag{1}$$

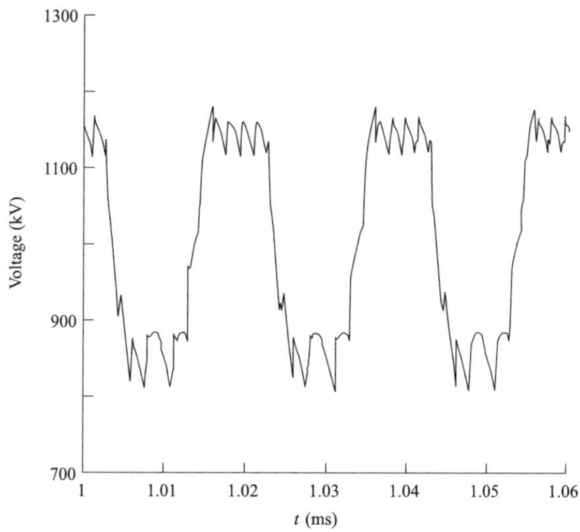

Fig. 9 Steady state voltage of AH arrester

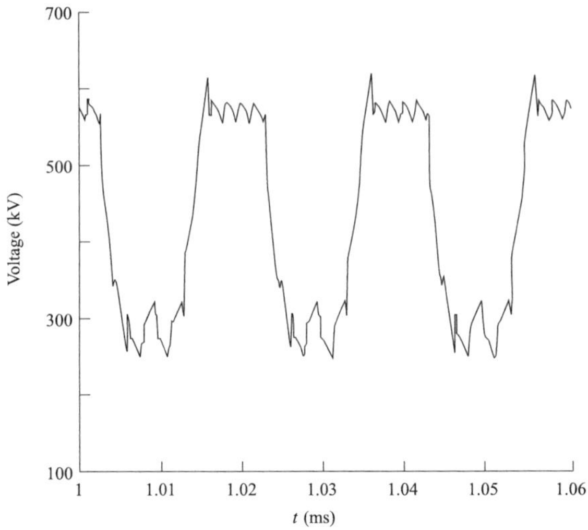

Fig. 10 Steady state voltage of AL arrester

Considering commutation overshoot of 16% to 19%, the peak value of the continuous operating voltage (*PCOV*) of the V1 valve arrester at the Changji converter station is 399.3 kV, and the value of *PCOV* of the V2 and V3 arresters is 409.3 kV. The value of chargeability of the valve is set to 1.0.

2. AH arrester

The AH arrester is used for direct protection of valve-side connected equipment of the HV Yy converter transformer. It is composed of a DC voltage component and an AC component, and its peak is almost equivalent to the peak voltage of the smoothing reactor valve side. When $(\alpha + \mu)$ is very small, theoretically, the value of *CCOV* can be calculated by formula (2).

$$CCOV = 2 \times U_{\text{diomax}} \times \frac{\pi}{3} \times \cos 15^{\circ} + U_{\text{offset}} \tag{2}$$

where U_{offset} is the midpoint voltage of the double 12-pulse converter in kV.

U_{offset} is set as 585 kV, so the vlaue of *CCOV* of the AH arrester is 1225 kV. Through simulation studies, if the commutation overshoot at AH is 6%, the value of *PCOV* is 1298 kV.

In order to limit the switching overvoltage level of the HV Yy converter transformer valve side, the value of chargeability of the AH arrester is set to 0.85.

3. AL arrester

The AL arrester is used for direct protection of valve-side connected equipment of the LV Yy converter transformer. It is also composed of a DC voltage component and an AC component. When the low voltage is operating in single 12-pulse, its peak is equivalent to the peak voltage borne by the CBL arrester. When $(\alpha + \mu)$ is very small, theoretically, the value of *CCOV* can be calculated by formula (3).

$$CCOV = 2 \times U_{\text{diomax}} \times \frac{\pi}{3} \times \cos 15^{\circ} \quad (3)$$

Therefore, the operating voltage *CCOV* of the AL arrester is 641 kV. Through simulations, if the commutation overshoot at AL arrester is 12%, the value of *PCOV* is 718 kV.

The switching impulse level of the LV Yy converter transformer valve side is not a key factor limiting equipment development. In order to ensure the arrester has higher reliability, the value of chargeability of the AL arrester is set to 0.83.

4. DB arrester

In ± 800 kV UHV DC transmission projects, according to previous operation experience, the insulation level of pole line equipment is generally about twice the rated voltage of the project, and the value of chargeability of DB pole line arrester is generally 0.85.

On one hand, due to the simplification of the DC filter design, the operating voltage *CCOV* of the DB arrester in the ± 1100 kV UHV DC project is 1170 kV, deviating from the rated DC voltage 1100 kV by 6%. On the other hand, due to the limitations of equipment development level, the insulation level of the equipment should not exceed 2100 kV. The parameters of the DB arrester maintain 1 kA coordinating current unchanged, and a larger valve disc and multi-column parallel method are adopted to reduce the voltage ratio to meet the insulation level requirements of this project.

5. Other arresters

The operating voltage and chargeability of other arresters are the same as the selection principles of those in the ± 800 kV UHV DC converter station.

(II) Parameters of Arresters

According to theoretical calculations and simulation analysis, the final determined operating voltage and reference voltage U_{ref} are shown in Table 31.

VI. Insulation Coordination of DC Equipment

(I) Insulation Margin

According to relevant international standards and the insulation level requirements of equipment for UHV DC transmission project, the minimum insulation margin of the equipment shall not be less than the values in Table 32.

Table 31 Parameters of the arresters in converter station

Arrester	*PCOV* (kV)	*CCOV* (kV)	U_{ref} (kV)	Energy (MJ)
V11/V12	399.3	344.5	282.4 (rms)	21
V2/V3	409.3	344.5	289.5 (rms)	9
ML	–	400	487.8	5
MH	970	906	1105	22
CBL2	718	641	782	15
AH	1298	1225	1441	39
AL	718	641	772	22
DB1	–	1170	1376	32
DB2	–	1170	1376	32
CBN1	212	165	333	4.4
CBN2	212	165	304	60
E	–	90	304	4
EL	–	20	202	8
EM	–	90	278	75

Table 32 Minimum insulation margin of the equipment (%)

Item	Oil insulation (valve-side)	Air insulation	Single valve
Steep wave	25	25	15
Lightning	20	20	10
Switching	15	15	10

(II) Arrester Protective Level and Coordinating Current

The protective level including lightning impulse protective level (*LIPL)* and switching impulse protective level (*SIPL*), and coordinating current including lightning protection level current (I_{LIPL}) and switching protection level current (I_{SIPL}) of the selected converter station arrester are shown in Table 33.

(III) Equipment Insulation Level

According to the equipment protected by each arrester, considering the above insulation margin, the protection level and insulation level of the equipment, including lightning impulse withstand level (*LIWL)* and switching impulse withstand level (*SIWL*) are shown in Table 34.

As shown in Table 34, compared with the traditional MH + V and ML + V scheme, the switching impulse insulation level of the HV Yy converter transformer valve side is reduced from 2214 to 2093 kV, and the lightning impulse insulation level is reduced from 2485 to 2420 kV. When it comes to the LV Yy converter transformer valve side, the switching impulse insulation level is reduced from 1415 to 1182 kV, and the lightning impulse insulation level is from 1504 to 1297 kV. The

Table 33 Arrester protection level and coordinating current

Arrester	*LIPL* (kV)	I_{LIPL} (kA)	*SIPL* (kV)	I_{SIPL} (kA)
V11	520.7	2	534.1	4
V12	534.1	2	552.2	4
			499.4	0.2
V2/V3	550.6	2	569.3	4
ML	701.7	2	661.6	0.5
MH	1536	2	1425	0.2
CBL2	1094	2	1026	0.2
AH	2018	2	1820	0.5
AL	1081	2	1028	0.5
DB1	2150	20	1826	1
DB2	2150	20	1826	1
CBN1	476	2	–	–
CBN2	426	2	448	14
E	478	3	–	–
EL	311	10	303	8
EM	431	20	398	7

Note The coordinating current of a converter station is mainly affected by the lightning line above the converter station areas. In specific projects, it is implemented by designer according to the system design requirements. Inside the converter station, it is generally taken as about 2 kA. In the converter station pole line, grounding pole line, metallic return exit, it is generally taken as 10 kA, or 20 kA in conservative design cases

scheme significantly reduces the manufacturing difficulty and cost of the converter transformer.

At the same time, according to the configuration method of the ± 800 kV UHV DC pole line DB arrester, the insulation level of the smoothing reactor line side pole line will reach 2268 kV. The study reduces the voltage ratio of the arrester through reasonable configuration of DB arrester parameters, and the switching impulse level of pole line equipment on the smoothing reactor line side is controlled at 2100 kV, consistent with the valve side.

VII. Simulation of DC Overvoltage

Based on the AC system conditions, the actual AC/DC equipment situation, and the system topology structure, the detailed PSCAD model of the ± 1100 kV UHV DC transmission project is built. Under different DC operating modes (bipolar symmetric operation, monopolar ground operation, monopolar metallic operation; double converter full voltage operation, single converter half voltage operation; minimum power operation, maximum power operation, etc.), considering pole line grounding faults (including opposite pole), smoothing reactor valve side grounding faults, ground flashover of HV/LV Yy converter transformer valve side bushing,

Table 34 Protection level and insulation level of the equipment in the sending end converter station (kV)

Location	Combination of arresters	*LIPL*	*LIWL*	*SIPL*	*SIWL*
Both sides of valve bridge	max (V11/V12/V2/V3)	551	606	569	627
DC line (smoothing reactor side)	max (DB1, DB2)	2150	2580	1826	2100
High voltage 12-pulse Bridge	max (V11, V12) + V2	1085	1302	1122	1290
Upper converter transformer Yy valve side, relative to ground	AH	2018	2420	1820	2093
Upper converter transformer Yy valve side, relative to ground	MH + V2	2070	2485	1925	2214
Upper converter transformer Yy valve side, neutral point	A' + MH	–	–	1790	2059
Upper 12-pulse bridge midpoint bus	MH	1536	1845	1425	1639
Upper converter transformer Yd valve side, relative to ground	V2 + CBL2	1645	1974	1596	1835
Midpoint between upper and lower 12-pulse bridges	CBL2	1094	1313	1026	1180
Lower converter transformer Yy valve side, relative to ground	AL	1081	1297	1028	1182
Lower converter transformer Yy valve side, relative to ground	ML + V2	1253	1504	1231	1415
Lower converter transformer Yy valve side, neutral point	A' + ML	–	–	1027	1181
Lower 12-pulse bridge midpoint bus	ML	701.7	842	661.6	761
Lower converter transformer Yd valve side, relative to ground	max (V2 + CBN1, V2 + CBN2)	1027	1233	1018	1171
Ground pole bus	EL	311	373	303	348
Metallic return bus	EM	431	517	398	458

AC system single/three-phase faults, commutation failure, etc., a comprehensive simulation of DC system overvoltage was carried out. The simulation results of key operating conditions are given below, as shown in Tables 35, 36, and 37, and the corresponding waveform diagrams are shown in Figs. 11, 12, 13, and 14.

According to the overvoltage simulation results in Tables 35, 36, and 37, it can be seen that the insulation coordination scheme and design parameters of arresters proposed in this study meet the system operation requirements.

VIII. R&D of DC Line Arrester

The world's first ± 1100 kV UHV DC line arrester was developed in the CG project. The development involves research and data collection on lightning protection design and lightning sparkover characteristics of DC lines, determination of arrester operating characteristics and overall electrical parameters, assessment of mechanical performance of super-high and super-long line arresters, research on installation methods, and some other aspects, which are extremely challenging.

According to the focus of the research on UHV DC line arresters, combined with the R&D and grid operation experience of line arrester of 1000 kV and below AC and ± 500 kV DC, the main difficulties are as follows.

(1) Determination of electrical performance parameters of arrester. For UHV DC transmission lines, due to the longer transmission line insulator string, which is greater than the minimum air gap distance of the tower head, the shortest distance of lightning impulse flashover is no longer the insulator string but the tower window, and it varies with different tower types. Therefore, the lightning impulse insulation level of different tower types of UHV DC should be first studied and determined, so as to preliminarily identify the performance parameters of the line arrester. Then a large number of experiments and simulation calculations should be carried out to determine the electrical performance parameters of the

Table 35 Protection level and coordinating current of AH arrester

Item	Fault description	Property of arrester	U_{max} (kV)	I (kA)	Energy (MJ)
AH.1	Bipolar operation, three-phase short circuit in AC grid	max	1413.7	0.000	0.066
Design parameters of arrester			1820	0.500	39

Table 36 Protection level and coordinating current of AL arrester

Item	Fault description	Property of arrester	U_{max} (kV)	I (kA)	Energy (MJ)
AL.1	Bipolar operation, three-phase short circuit in AC grid	max	877.5	0.000	0.024
Design parameters of arrester			1028	0.500	22

Table 37 Protection level and coordinating current of AL arrester

Item	Fault description	Property of arrester	U_{max} (kV)	I (kA)	Energy (MJ)
DB.1	Bipolar operation, three-phase short circuit in AC grid	max	1299.7	0.002	0.211
DB.2	DC line fault in opposite pole		1452.1	0.260	1.11
Design parameters of arrester			1826	1	32

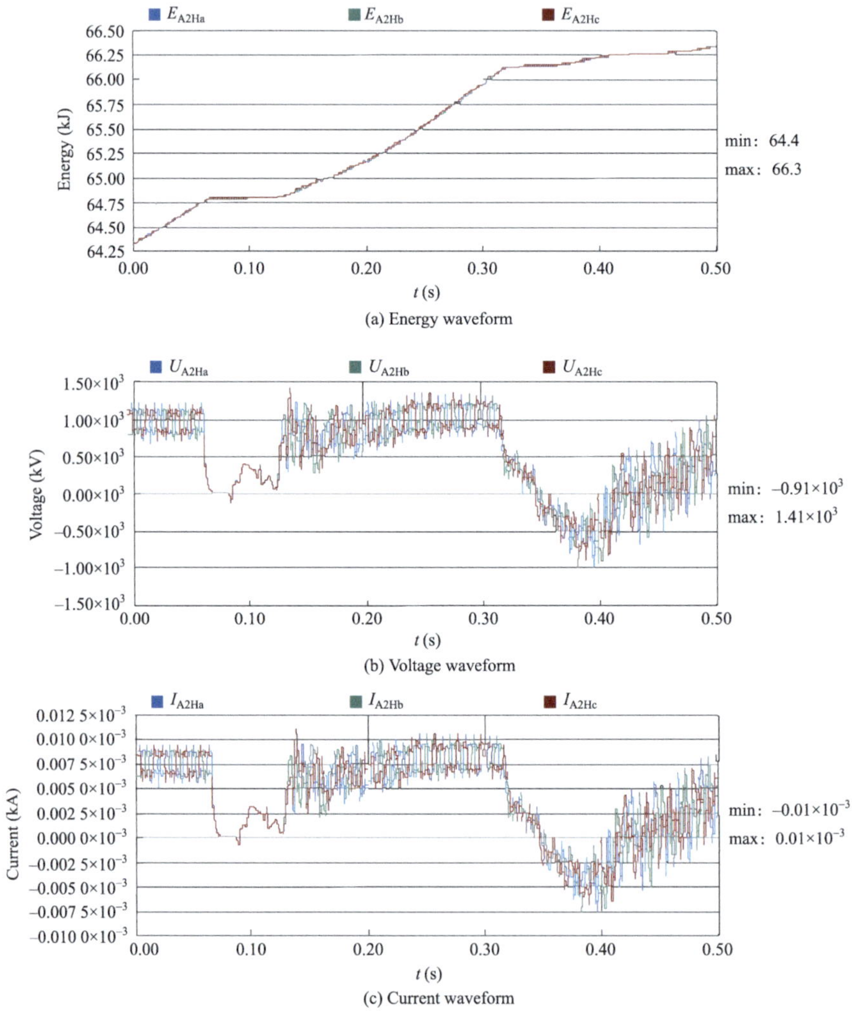

Fig. 11 Energy, voltage, and current of AH arrester under three-phase short-circuit in AC grid in bipolar operation

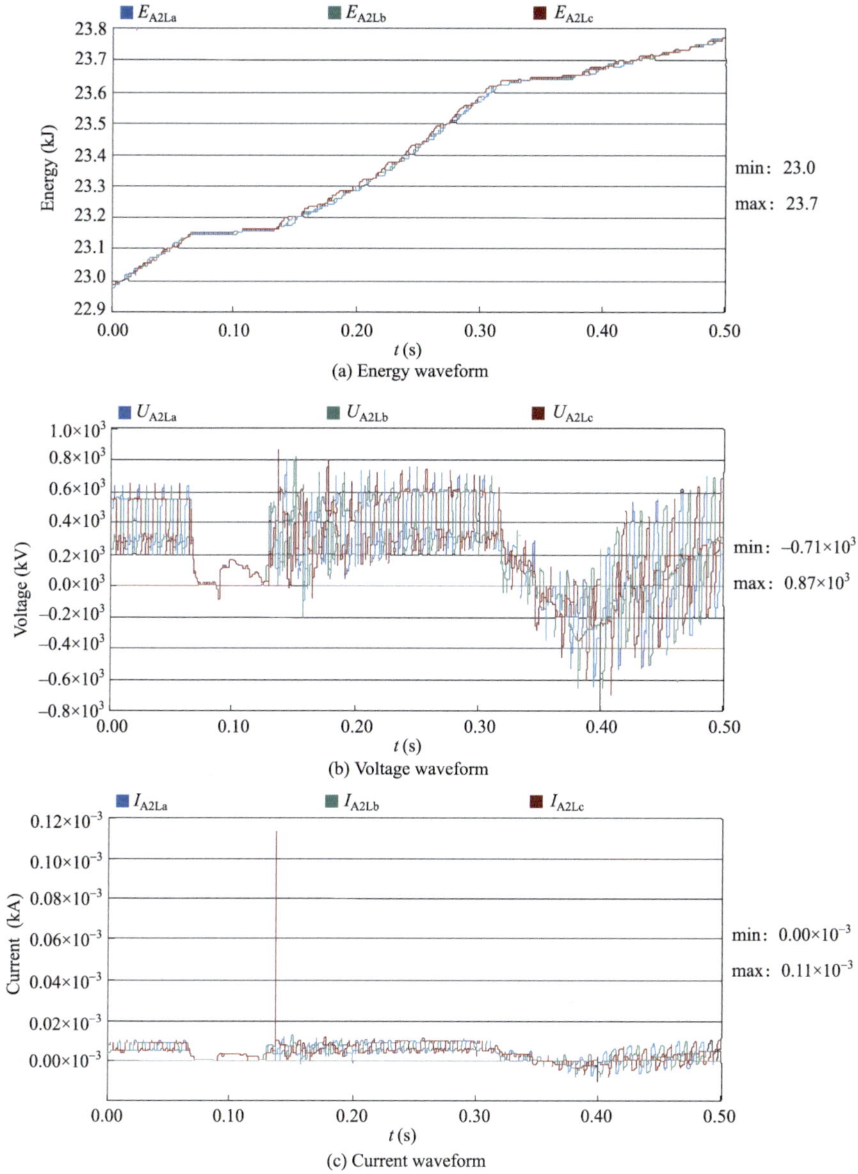

Fig. 12 Energy, voltage, and current of AL arrester under three-phase short-circuit in AC grid in bipolar operation

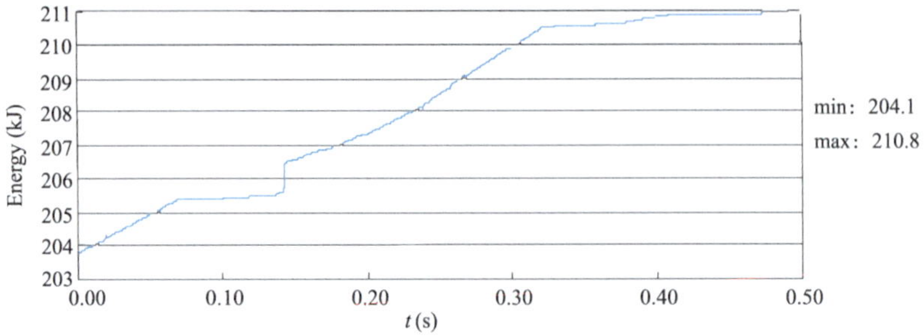

(a) Energy waveform

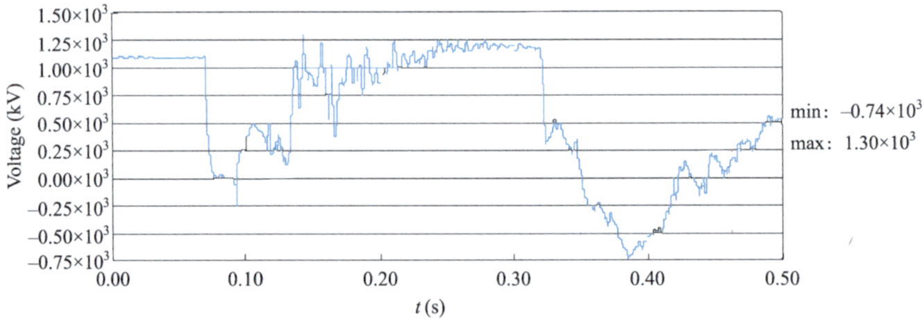

(b) Voltage waveform

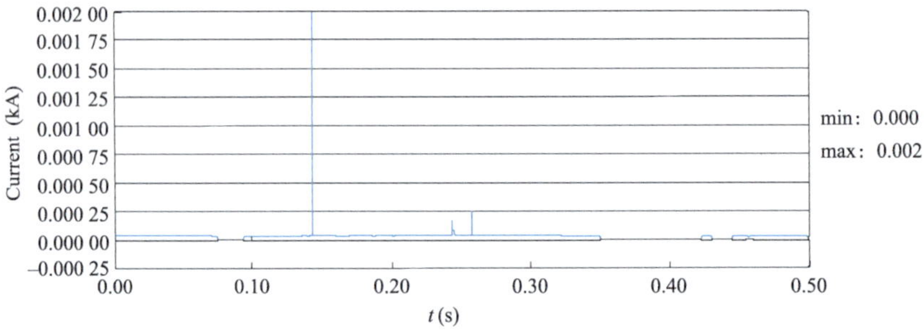

(c) Current waveform

Fig. 13 Energy, voltage, and current of DB arrester under three-phase short-circuit in AC grid in bipolar operation

arrester, including the reference voltage, sparkover characteristics, protection level, follow current breaking and energy absorption of the arrester.

(2) Structural design of arrester. Due to the high voltage level and large volume of UHV DC transmission lines, the weight is much higher than that of 1000 kV and below AC and ± 500 kV DC. Therefore, higher requirements are put forward for the mechanical performance of UHV DC line arrester. This requires the adequate design of the mechanical structure of the arrester, if necessary, using new design technology of the current polymer-housed DC arresters, and verified through calculations and tests.

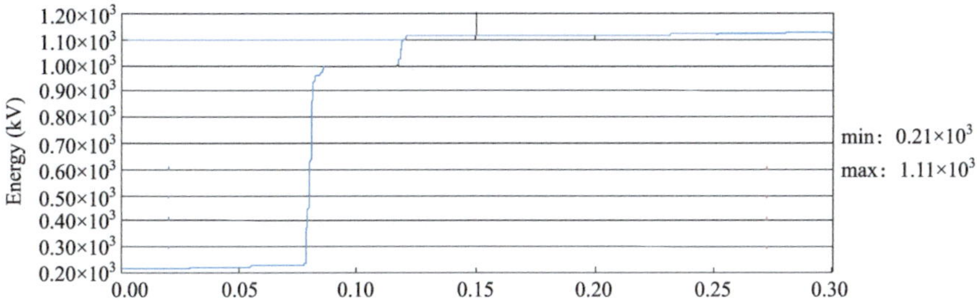

(a) Energy waveform

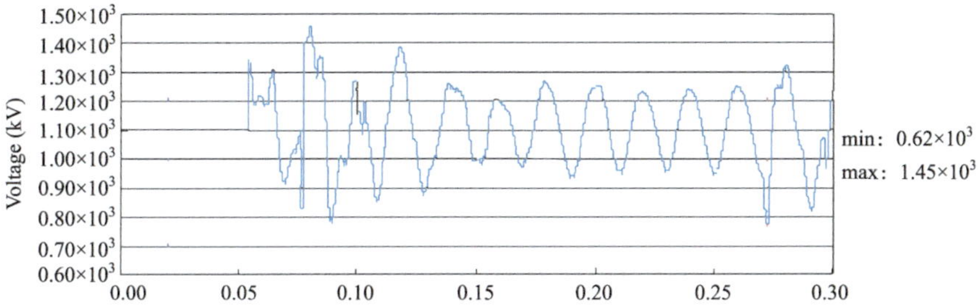

(b) Voltage waveform

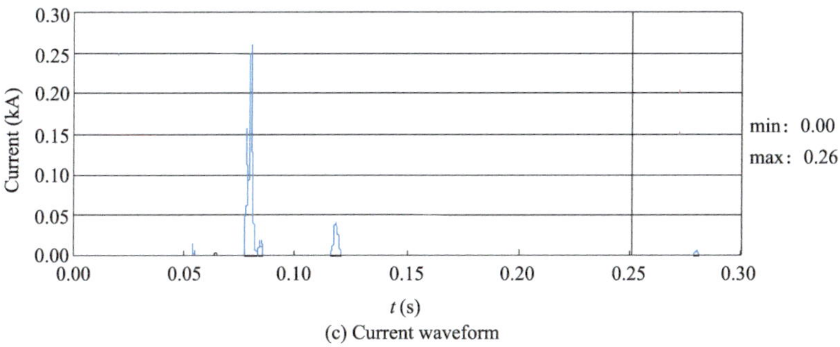

(c) Current waveform

Fig. 14 Energy, voltage, and current of DB arrester under DC line fault in opposite pole

(3) Installation method of arrester. The weight of the line arrester is more than 600 kg, the body length is more than 7 m, and the mechanical performance of the arrester itself should be considered during installation, as well as the impact on the safety and stability of the transmission line tower. In the meantime, the reasonable coordination of the length and the installation space should be considered. All of the above need to be focused on.

(I) Research on Performance Parameters of ± 1100 kV UHV DC Line Arrester

1. Reference voltage of arrester body

Calculations show that the operating voltage of the ± 1100 kV DC transmission system does not exceed 1122 kV. Therefore, when the DC reference voltage of the arrester body is not less than 1122 kV, and the current through the arrester body after a lightning impulse does not exceed its DC reference current (1 mA or 2 mA). Therefore, the internal current of the arrester has little effect on the breaking of the follow current arc.

At the same time, the short-term chargeability of resistors allowed under DC voltage is more than 90%, and some manufacturers' resistors' long-term chargeability allowed under DC voltage can reach 90%. Therefore, it is determined that the DC reference voltage of the arrester body should not be less than 1122 kV / 90% = 1247 kV.

Subsequent calculations show that a DC reference voltage of 1247 kV can meet the requirements for the protection level.

2. Selection of 50% lightning impulse sparkover voltage of line arrester

Line arresters are mainly used to protect transmission line insulators or air gaps from flashover under lightning impulses which might cause line tripping. The current relevant standards require that line arresters reliably operate under lightning overvoltage, do not operate under operating overvoltage as far as possible, and reliably withstand DC overvoltage.

The lightning impulse sparkover voltage of the ± 1100 kV UHV DC transmission line arrester should be less than the sparkover voltage of the transmission line tower gap or insulator, and the arrester can reliably operate during a lightning strike.

The minimum air gap distance of the ± 1100 kV UHV DC line tower is designed to be 9 m. According to the 50% sparkover characteristic of lightning, it is about 580 kV/m (peak value). Its lightning impulse sparkover voltage is about 5220 kV (peak value).

The coefficient of variation of lightning impulse is 0.03, so the 50% sparkover voltage of the line arrester under lightning impulse should not exceed 82% of that of the minimum air gap. Furthermore, from the experience of 500 kV and below AV line arresters, the 50% sparkover voltage of the arrester can be selected as 75% of the tower head gap or insulator sparkover voltage, i.e., the sprakover voltage of the ± 1100 kV UHV DC line arrester should not exceed 3915 kV (peak value). Conservatively, the value should not exceed 3900 kV (peak value).

3. Selection of the switching impulse withstand voltage of line arrester

The maximum relative ground switching overvoltage along the ± 1100 kV DC line is generally not more than 1.7 p.u. (1907 kV), so the line arrester should be able to reliably withstand that.

It is suggested that the switching withstand voltage of the ± 1100 kV DC line arrester should not be less than 2000 kV.

4. Selection of the DC withstand voltage for the line arrester

The selection of the withstand voltage for AC line arresters in the power frequency has a wealth of practical and theoretical data. For example, the overvoltage protection and insulation coordination design specifications for AC electrical devices states that measures should be taken to reduce the power frequency overvoltage. Generally, the main method is to install parallel reactors on the line. The level of power frequency overvoltage of the system should generally not exceed 1.3 p.u. on the substation side of the line circuit breaker, and 1.4 p.u. on the line side. Therefore, the power frequency overvoltage that the AC line arrester should be able to withstand should not be less than 1.4 p.u.

The principle of selecting the DC withstand voltage for the ± 1100 kV DC line arrester is the same as that for the AC surge arrester. It should be able to reliably withstand the possible maximum DC voltage (1122 kV).

5. Study on the protection level of line arrester

The determination of the residual voltage under the nominal discharge current of the line arrester should consider the influence of factors such as insulation coordination of the entire arrester with the parallel insulator or air gap, and lightning arc voltage drop during discharge.

Under the nominal lightning impulse current (30 kA), the sum of the protection level of arrester and the arc voltage drop should be less than 70% of the lightning impulse spark voltage. According to the minimum gap distance of 9 m and the impulse sparkover voltage is of 580 kV/m, preliminary calculations show that the residual voltage of the arrester (including the arc voltage drop) under lightning impulse should not exceed $9 \times 580 \times 70\% = 3654$ kV (peak value).

The gap distance of the ± 1100 kV DC line arrester should be in the range of 2–3 m, but currently there is no measured or accurate calculation data for the arc voltage of such a gap distance. In some literature, it is set as 10 Ω in overvoltage insulation coordination study, so the arc voltage drop under 30 kA is about 300 kV (peak value). Therefore, the lightning impulse residual voltage should not exceed 3354 kV (peak value) to meet the requirements of insulation coordination.

At the same time, considering the current manufacturing level of resistors, the residual voltage of resistors that may be applied to ± 1100 kV DC line arresters under 30 kA lightning impulse current is generally not greater than 1.8. Corresponding to the arrester body with the DC reference voltage around 1247 kV, its lightning protection level generally does not exceed 2245 kV (peak). Since the production capacity of the manufacturer and the dispersion of the resistors, the residual voltage of the 30 kA lightning impulse is set to 2500 kV (peak).

The residual voltage of 30 kA steep wave impulse is 1.1 times that of the lightning impulse, i.e. 2750 kV (peak).

6. Research on the external insulation parameters of arrester body

For the external insulation withstand capability of the arrester body, the relatively harsh conditions are gap discharge. At the moment of discharge, the arrester body has to withstand the lightning impulse voltage, that is, the residual voltage of the body under the discharge current. Before the flow current interruption, the arrester body has to withstand the system's continuous operating voltage.

When the arrester series gap is broken down, without considering the arc voltage drop, the lightning impulse voltage that the external insulation of the arrester body can withstand will not exceed the residual voltage of the arrester body. The short-term DC voltage it can withstand will not exceed the system's highest operating voltage, that is, 1122 kV (rms).

Considering both reliability and economy, the book proposes that the housing of arrester body should be able to withstand a lightning impulse voltage not less than 1.40 times the lightning protection level (corresponding to the nominal current). The DC voltage (wet) withstand capability is considered with a safety margin of 1.5 times, which can be taken as not less than 1122 kV × 1.5, that is, 1683 kV.

7. Selection of the energy absorption requirements of line arrester body

According to the design principle, the line arrester does not act under switching overvoltage. Considering extreme situations, such as arrester gap short circuit and enduring switching overvoltage during this period. The energy absorbed by the arrester under in this situation can be obtained through calculation, and its endurance performance can be assessed through the 2 ms energy withstand capability of the resistor.

It can be obtained through calculation that the 2 ms square wave impulse current corresponding to the ± 1100 kV line arrester absorbing 2.59 MJ energy is about 799 A, that is, the single 2 ms square wave impulse current that the arrester can withstand should not be less than 799 A. According to the relevant standard, the arrester's capability to absorb energy should not be less than 799 A/2 = 400 A (for consecutive 3 times).

Considering the high safety and reliability requirements of the ± 1100 kV DC transmission line, combined with the experience of AC line arresters, it is determined that the energy absorption capability of the ± 1100 kV DC line arrester to be 2000 A.

8. Key technical parameters of line arrester

Combining the above analysis, the book has proposed the key technical parameters of the ± 1100 kV DC line arrester, shown in Table 38.

(II) Research on the Installation Method of UHV DC Line Arrester

The difficulty in installation is a major factor affecting the widespread promotion of line arrester. The body length of the ± 1100 kV UHV DC line arrester is more than 7 m, the mass is more than 600 kg, and the air gap length is more than 2 m. These parameters are much larger than the corresponding parameters of the ± 500 kV DC and 1000 kV and below AC line arresters. Therefore, the installation is more difficult, requires consideration of more factors, and is a topic worthy of in-depth study.

1. Research on the installation distance between insulator and line arrester

Table 38 Key technical parameters of ± 1100 kV DC line arrester

No.	Item			Parameter
1	Arrester body	Nominal voltage of system (kV, rms)		± 1100
2		Rated voltage of arrester (kV, rms)		1247
3		Nominal discharge current (kA, peak)		30
4		Reference DC voltage (kV)		≥ 1247
5		Leakage current test at 0.75 times DC reference voltage (μA)		≤ 100
6		Protection level	Residual voltage of 30 kA lightning impulse (kV, peak)	2500
			Residual voltage of 30 kA steep wave impulse (kV, peak)	2750
7		4/10 μs high current withstand capability (kA, peak)		100
8		2 ms square wave impulse current withstand capability (A, peak)		2000
9		Dieletric test	1 min wet withstand capability of the whole DC (kV, RMS)	1683
			Dry withstand capability of positive lightning impulse (kV, peak)	1.4 times of residual voltage
10		Creepage distance check (mm)		22,440
11		Pressure release test	Value of high current test (kA, peak)	50
			Value of low current test (A, peak)	800
12		Internal partial discharge test (pC)		10
13	Whole	Lightning impulse sparkover voltage (positive) (kV, peak)		3900
14		Switching impulse sparkover voltage (positive) (kV, peak)		2000
15		DC withstand	Body normal, withstand 1 min (kV, rms)	1683
			Body short circuit, withstand 1 min (kV, rms)	1122

As mentioned above, the main function of line arrcster is to reduce the lightning tripping rate of the transmission line by avoiding ground flashover of insulators or air gaps, thereby ensuring the stability of the system and improving the quality of power transmission. Line arresters are generally installed near the insulators on the transmission tower, in parallel with the insulators.

Table 39 Distance between insulator and line arrester for various voltage levels

Voltage (kV)	110	220	1000
Typical distance (m)	0.8	1.4	4.0

The distance between line arrester and insulator is a topic worth studying. If the distance is too close, the installed line arrester will affect the electric field near the insulator, accelerating the aging of the insulator shed material. Meanwhile, the insulator will also affect the electric field distribution of the line arrester, reducing the effectiveness of the type test and the undermining safety and reliability. In severe cases, sparkover may occur between the insulator and the line arrester under overvoltage conditions. The line arrester not only cannot play a role in reducing the lightning tripping rate, but also become a potential fault spot of line tripping. If the distance is too far, it will increase the requirements for the mechanical strength of the electric power fitting, thus increasing the installation difficulty as well as the quality standard of the fitting.

Table 39 provides the typical values of the installation distance between insulator and line arrester for the voltage levels of 500 kV and below in China.

Actual tests have verified that when the installation distance is 5.3 m, the discharge path is between the upper and lower gaps of the line arrester, and there will be no discharge through the gap between the arrester and insulator.

Based on calculations and actual test results, it is recommended that the installation distance between the ± 1100 kV UHV DC line arrester and insulator should not be less than 5.5 m.

2. Installation method of the line arrester

After analyzing and comparing the advantages and disadvantages of the current two different gap structures of line arrester, that of the UHV DC line arrester is finally determined to be a pure air gap.

The typical installation method of the 500 kV and below line arrester with pure air gap has many years of experience and is currently recognized as a reliable, reasonable, and simple installation method. However, compared with the line arresters of the 500 kV and below, the ± 1100 kV UHV DC line arrester has its uniqueness in the installation.

The line arrester of the 500 kV and below is installed in parallel, used to protect the line insulators from lightning damage. The sum of the pure air gap line arrester body and gap length is generally not much different from the parallel insulator, which provides some convenience for the installation.

However, for UHV DC transmission lines, due to the long insulator string (generally more than 9 m), which is greater than the minimum air gap distance of the tower head, the shortest distance of lightning flashover is no longer the insulator string but the tower window, and it varies with different tower types. The design value of minimum air gap distance for UHV double-circuit tower is 6.7–7 m, which is much lower than the length of the line insulator.

The total length of the UHV line arrester is less than the length of the parallel insulator. Therefore, in the design and installation of armour clamp, not only the impact on the surrounding electric field, the impact on the mechanical strength of the tower, and the mechanical strength of the arrester itself should be considered, but also special transition fitting should be made to ensure that the gap distance of the line arrester is within the specified range.

The role of the transition fitting is mainly to ensure that the gap distance can be adjusted to within the specified range. Since the designed arrester body length is about 8.1 m, the gap distance is about 2.7 m, and the insulator length is more than 12 m, the transition fitting length should be more than 1 m.

There are two possible types of transition fitting. In type A, the transition fitting is located between the arrester body and the high voltage electrode of the discharge gap. In type B, it is located between the tower cross arm and the arrester body. The mechanical tension borne by the fitting in type A is small, and the requirements for its mechanical strength are not high, which is conducive to reducing the overall weight and installation difficulty of the arrester. Its disadvantage is that when the difference in the length of the transition fitting is too large, it will affect the discharge performance of the arrester and the surrounding electric field. In type B, the transition fitting has to bear the tension caused by the self-weight of the arrester body and the wind pressure in the horizontal direction, and its mechanical strength requirements are higher, which may increase the overall weight of the arrester. Its advantage is the change in the length of the fitting will not affect the electric field distribution and discharge performance of the arrester, ensuring the consistency and effectiveness of the actual installation on site and the installation during type testing.

Considering the advantages and disadvantages of the two installation types comprehensively, type B is finally chosen.

IX. Conclusion

(1) Based on the configuration scheme of the arrester in the ± 800 kV UHV DC converter station, the book proposes a new insulation coordination scheme. On one hand, the scheme retains conventional arresters such as MH, CBL2, ML, and DB. On the other hand, since the ± 1100 kV UHV DC transmission project adopts an indoor DC yard, there is no problem of lightning wave intrusion near the valve hall pole line bushing, the CBH arrester is cancelled. To control the insulation level of the valve side of the HV and LV Yy converter transformers, AH and AL arresters are used.

(2) Compared with the conventional arrester configuration scheme, the new insulation coordination scheme reduces the switching impulse level of the valve side of the HV Yy converter transformer from 2214 to 2093 kV; reduces that of the LV Yy converter transformer from 1415 to 1182 kV, with reductions of 6 and 17% respectively. The lightning impulse level is also greatly reduced, which greatly facilitates the development of equipment and reduces the cost of development and manufacturing.

(3) After adopting the simplified DC filter scheme, the pole line voltage at the outlet of the smoothing reactor can reach up to 1170 kV. Extrapolating

from the insulation level of the previous ± 800 kV pole line equipment to ± 1100 kV equipment, its insulation level can reach 2268 kV. Through reasonable configuration of the DB arrester, the DC pole line switching impulse level of the line-side of the smoothing reactor can be controlled to 2100 kV, reduced by more than 8%.

(4) Through the simulation calculation of DC system overvoltage, the new insulation coordination scheme has been verified to be reasonable.

4 Air Gap and DC External Insulation

A series of problems arise when the DC voltage of UHV transmission project is raised to ± 1100 kV. In terms of air gap and DC external insulation, compared to other voltage levels, the size of ± 1100 kV DC equipment is getting larger and the difference in the shape of electrodes for the terminal fittings is also increasing. The switching impulse withstand level of ± 1100 kV DC equipment is 31.25% higher than that of ± 800 kV DC equipment, reaching 2100 kV, and the switching impulse discharge voltage enters the deep saturation area of long gap discharge. To ensure the reliable design of the air gap and DC external insulation, in this section, the tests of discharge characteristic for the DC yard and valve hall of the converter station are firstly conducted, and the minimum gap distance required for the DC yard and valve hall is proposed. Then we conduct pollution prediction and external insulation configuration research for the converter station, and propose the pollution level and external insulation configuration suggestions for the sending and receiving converter stations. To ensure the safety of inspectors, the ± 1100 kV UHV DC transmission project adopts a double-layer shielding channel for the first time, and design suggestions for the double-layer shielding channel are proposed. Based on the above results, considering multiple factors, the selection design of the ± 1100 kV DC yard and valve hall is proposed. Finally, the lightning protection design of the building roof of the indoor DC yard and valve hall is analyzed.

I. Typical Gap Discharge Characteristics of DC Yard and Valve Hall

With the increase of the voltage level of the DC project, the size of the equipment is getting larger and the difference in the shape of the electrodes is also increasing. The switching impulse withstand voltage level required by ± 1100 kV equipment is 31.25% higher than that of ± 800 kV. It is preliminarily estimated that the external insulation size of ± 1100 kV DC equipment will reach more than 14 m.

The ± 1100 kV converter station may be equipped with indoor DC yard. In this case, the air gap distance required by the switching impulse discharge voltage will directly determine the safety clearance of the DC equipment from the surrounding grounded body, thereby determining the size of the indoor yard, which will greatly affect the design economy of the indoor DC yard.

(I) Analysis of Typical Gaps in DC Yard

The DC part of the converter station usually includes two parts: the valve hall and the DC yard. The former is indoor, and the latter is mostly outdoor. In areas with particularly heavy pollution, there are also indoor ones for DC yard.

Whether the DC yard is designed as indoor or outdoor, the main equipment of the DC yard includes smoothing reactors, filters, disconnectors, arresters, optical current transformers, DC voltage dividers, etc., as well as the HVDC pole busbars.

The above equipment is generally equipped with corona ring fittings for shielding and grading. In laboratory, typical corona ring-ground gaps are usually used for air gap discharge characteristic tests.

(II) Analysis of Typical Gaps in Valve Hall

The air gap structure in the valve hall is very complex. When designing the valve hall, dozens of discharge paths usually need to be considered. The space in the UHV valve halls is large, making it difficult to conduct true type tests on all flashover paths. Therefore, it is necessary to select a typical discharge path with a high overvoltage level, a possibility of flashover, and a direct impact on the volume of the valve hall for study.

According to the preliminary layout plan of the ± 1100 kV valve hall, the layout of the ± 1100 kV valve hall is basically the same as that of previous UHV/EHV projects, with one set of 12-pulse converter valve groups arranged in each valve hall.

The typical gaps in the valve hall of the converter station mainly include the grading sphere—ground and grading sphere—wall/ground gap.

(III) Results of Typical Gap Discharge Test

The arrangement of the samples during the test is shown in Fig. 15. The discharge characteristic curves of the tubular busbar ends installed with a φ2.0 m grading sphere to the ground and to the wall/ground, obtained through the switching impulse discharge test are shown in Figs. 16 and 17.

We conducted the switching impulse discharge characteristic test of a corona ring with a diameter of 2 m and a tube diameter of 300 mm vertically to the ground. The obtained relationship of 50% switching impulse discharge voltage and gap distance is shown in Fig. 18.

Fig. 15 Photos of discharge test of the grading sphere (tubular busbar) to ground

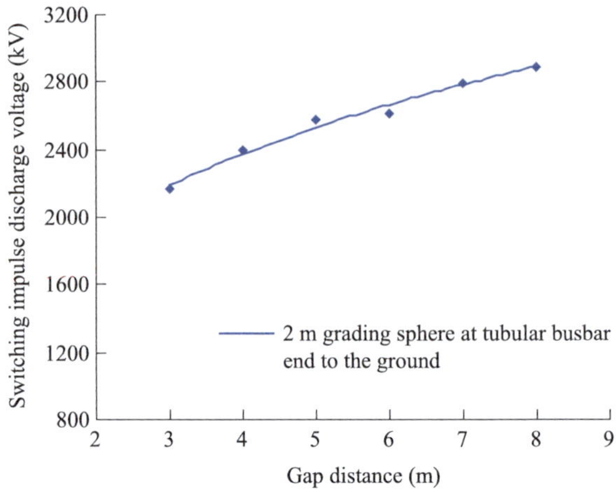

Fig. 16 The switching impulse discharge curve of the tubular busbar with a ϕ 2 m grading sphere at each end to the ground

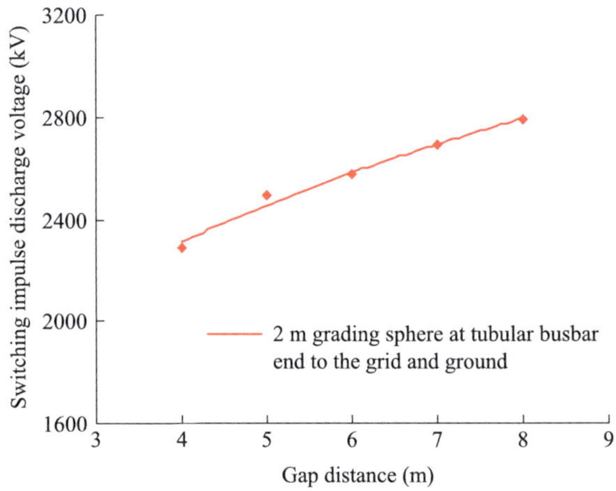

Fig. 17 The switching impulse discharge curve of the tubular busbar with a ϕ 2 m grading sphere at each end to the grid and wall

A grading sphere with a diameter of ϕ 2 m is installed at each end of the tubular busbar for a discharge test to ground. The test results show that when the distance between the sphere and the ground is 5 m (the distance between the tubular busbar and the ground is 5.82 m) or less, the discharge path is mainly the tubular busbar to ground. While when the distance between the sphere and the ground is 6–8 m (the distance between the tubular busbar and the ground is 6.82–8.82 m), the discharge path is mainly the grading sphere to ground. However, when the grading sphere at one end is discharged to the net and to the ground, the discharge path is mainly the sphere to the net and to the ground.

(IV) Selection of Typical Gaps in DC Yard

According to the equipment layout of the indoor DC yard and considering the parallelism of multiple gaps in the DC yard and the proximity effect of surrounding

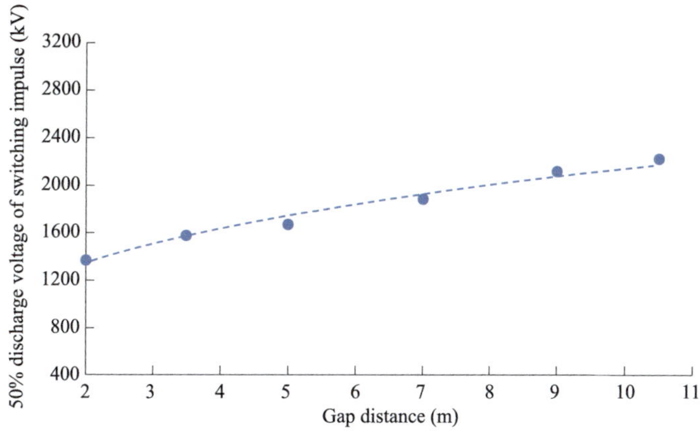

Fig. 18 The switching impulse discharge characteristic of a grading ring with a diameter of 2 m and a tube diameter of 300 mm vertically to the ground

grounded items, due to the impact on the surface of the fittings during installations, connections, or other processes, the relevant design margin is considered as 20%. The minimum distance of typical gaps in different positions of the ± 1100 kV indoor DC yard is shown in Table 40.

(V) Selection of Terminal Fittings of Typical Equipment in DC Yard

The insulation withstand voltage of air gap shows a clear saturation trend with the increase of gap size. The traditional method of increasing the withstand voltage by simply increasing the air gap distance is not effective, resulting in large size of indoor DC yard equipment and difficulty in design and manufacturing. Through electric field simulation and switching impulse flashover characteristic test of full-scale equipment, after optimizing the structure type and size of the terminal fittings in the design, most of the equipment has a design margin of 10% or more for its switching impulse flashover voltage.

Table 40 Minimum distance of typical gaps in different positions of the ± 1100 kV indoor DC yard

Typical gap position	*Switching impulse discharge voltage* $U_{50\%}$ (kV)	Type of test gap	Minimum gap distance (m)	
			Test value	Recommend value
Corona ring-ground	2386	2 m diameter, 300 mm pipe diameter corona ring—ground	10.5	12.6

Note The distance between the ring or sphere and the wall or ground in the table refers to the part of the ring/sphere closest to the wall/ground

Fig. 19 Comparison of ± 1100 kV disconnector terminal fittings before and after optimization

For the terminal fittings of the ± 1100 kV disconnector, by optimizing the traditional double-layer corona ring electrode shape into a fully-enclosed mushroom-shaped shielding structure (as shown in Fig. 19), the insulation withstand voltage increased from 2395 to 2505 kV with the terminal fitting height reduced by 0.3 m and the fracture distance reduced by 0.5 m.

The terminal fittings of the ± 1100 kV smoothing reactor are generally adopted with a multi-layer corona ring. By increasing the diameter of the ring, lowering the installation position, and adding a smaller corona ring at the flange of the HV end of the post insulator, the insulation withstand voltage can reach 2650 kV.

The terminal fittings of the ± 1100 kV AH arrester are optimized from a structure of three-layer ring + ball to cylindrical ring. The terminal fittings of the RI capacitor are optimized from double-layer ring to ball + double-layer ring. The terminal fittings of the DC filter are optimized from a groined shape to a shield. All of these have significantly improved the insulation withstand level, thereby reducing the difficulty of equipment manufacturing and promoting the localized production of DC equipment.

(VI) Selection of Typical Gaps in the Valve Hall

Since the extreme temperature and humidity inside the valve hall will be considered in the design, the temperature and humidity of the valve hall in this study are also considered under the severe conditions. The maximum dry temperature t_d is taken as 60 °C, the minimum relative humidity r_h is taken as 5%, and the air pressure p is taken as 101.3 kPa.

Considering the test results after temperature correction, the minimum distance of typical gaps in different positions in the valve hall is shown in Table 41.

It should be noted that the recommended clearances in Table 41 are based on test data from well-preserved grading spheres or rings in the laboratory. In order to achieve high repeatability of the test results, the corresponding test samples were carefully arranged and any defects in the test samples were eliminated as much as possible. During the actual installation of the equipment in the valve hall, it is not

Table 41 Minimum distance of typical gaps in different positions of the ± 1100 kV valve hall

Typical gap position	Switching impulse discharge voltage $U_{50\%}$(kV)	Type of test gap	Minimum gap distance (m)	
			Test value	Recommend value
Top of the valve	2386	2 m sphere of tubular busbar to ground	7.0	9.8
		2 m sphere of tubular busbar to disconnector	4.5	6.3
		2 m sphere of tubular busbar to disconnector (9 m from the wall)	5.5	7.7
Y-connected winding terminal on the valve side of the high voltage transformer	2386	2 m sphere of tubular busbar to wall	6.2	8.7
		2 m sphere of tubular busbar to wall/ground	6.7	9.4
		2 m sphere of pillar insulator to wall	6.8	9.5
		2 m sphere of pillar insulator to wall/ground	7.1	9.9

possible to ensure that the surface of the grading sphere or ring is in an optimal state due to the interconnection of the equipment. Besides, due to the influence of on-site construction techniques, defects or damage may exist on the surface or connection parts of large-size grading spheres, which will reduce the discharge voltage of the sphere-ground gap to a certain extent. Therefore, when making relevant clearance recommendations, appropriate margins should be considered based on the test results.

II. Converter Station Pollution Forecast and External Insulation Configuration

(I) Converter Station Pollution Forecast

The design of external insulation for UHV DC transmission mainly depends on the pollution performance of the insulator under operating voltage. Therefore, determining the pollution level of the area where the converter station is located is the premise of external insulation design. The operating experience of the areas where the two converter stations of the CG project are located is lacking, and there is no pollution zone distribution map or representative pollution monitoring data for design reference. In addition, the Changji converter station is close to the heavy pollution area of the energy and chemical base, and the external insulation capability is facing a severe test.

1. Site environment

Local power transmission and distribution equipment configuration and operation, environmental emissions, meteorology, industrial planning, etc. are collected, and

(a) The site of Changji converter station (b) The site of the Guquan converter station

Fig. 20 Site environment of converter stations

on-site samples are taken for analysis and testing of saline-alkali soil. The site environment of the station is shown in Fig. 20.

2. Calculation of pollution level at the converter station

Using the model of atmospheric pollutant diffusion and insulator surface pollution, the pollution level of AC equipment at the local converter station is calculated.

3. Calculation of DC/AC deposition ratio and pollution level of station equipment

Based on the research results of the DC/AC contamination deposition ratio in China, combined with the distribution of atmospheric pollutant particle size and local meteorological conditions in the station area, the DC/AC deposition ratio is calculated. On this basis, the pollution level of the equipment in the DC yard of the converter station is determined.

4. Selection of DC yard and configuration of external insulation

Based on the characteristics of pollution flashover obtained from the UHV DC test, the required creepage distances for the post and bushing is calculated, and on this basis, the following recommendations are made for the selection of the DC yard and insulation configuration.

(1) The proposed Changji converter station is located in Changji, Xinjiang. Due to the influence of surrounding industrial pollution sources and saline-alkali land, the annual equivalent salt density of the station's XP-type insulators can be 0.09 mg/cm^2. The proposed Guquan converter station is located on the border between Xuancheng and Wuhu, Anhui. Due to the influence of surrounding industrial pollution sources, the annual equivalent salt density of the station's XP-type insulators can be 0.06 mg/cm^2.
(2) Based on the results of the particle size analysis of the polluted particles and the prediction of the AC/DC deposition ratio, as well as the analysis and calculation of the effective salt density coefficient, the annual equivalent salt density of the

outdoor DC yard post insulator at Changji converter station is 0.08 mg/cm^2, and that of Guquan converter station is 0.07 mg/cm^2.

(3) If the proposed Changji converter station adopts an outdoor DC yard, composite insulation can be used in the DC yard. The creepage distance design value of the large and small umbrella type composite post insulator and vertical composite bushing in the DC yard can be taken as 47–53 mm/kV. If the proposed Guquan converter station adopts an outdoor DC yard, composite insulation can also be used. The creepage distance design value of the large and small umbrella type composite post insulator and vertical composite bushing in the DC yard can be taken as 46.5–52.5 mm/kV.
(4) If the proposed converter station adopts a well-sealed indoor DC yard, the design value of the creepage distance should not be less than 25 mm/kV. If the indoor DC yard designed as an incompletely sealed environment, the relative humidity should be controlled within 70%. Therefore the creepage distance of the DC yard post insulator and bushing type equipment is recommended to be not less than 36 mm/kV.
(5) To prevent uneven flashover in rain, the external insulation of the valve hall bushing uses composite hollow insulators. Since most of the bushings are installed horizontally, their external insulation surface area is less polluted than vertical ones, the creepage distance of the DC yard wall bushing should not be less than 45 mm/kV.
(6) It is recommended to use high-temperature silicone rubber for the composite insulator of the DC divider and the valve hall bushing. Its hydrophobicity needs to meet the requirements of Chinese industrial standard DL/T 376—2019 "General technical requirements of insulation materials for umbrella skirt and sheath of polymeric insulators". The size of the grading shield ring and the depth of the cover should be reasonable.
(7) The specific design values for the creepage distance of neutral line DC post insulators and bushings are as follows: 56 mm/kV for Changji converter station, and 52 mm/kV for Guquan converter station. The results of the artificial pollution test of the composite post insulator are shown in Fig. 21.

(II) External Insulation Configuration of the Converter Station

The external insulation pollution issue of DC equipment is the key to the design of UHV DC transmission project. The issue of DC equipment is more prominent than that of AC equipment. It is more prone to accumulate dirt under DC voltage than AC, and the DC flashover voltage under the same pollution conditions is lower than the AC flashover voltage, so the design of external insulation of DC transmission lines is an important aspect to ensure the safe and stable operation of the line.

(1) Under different salt densities (0.03, 0.05, 0.08, 0.10 mg/cm^2), a single DC porcelain post insulator was subjected to an artificial pollution test, and it was found that the 50% flashover voltage of the post insulator decreased as the salt density increased. Based on past experience, an exponential function with

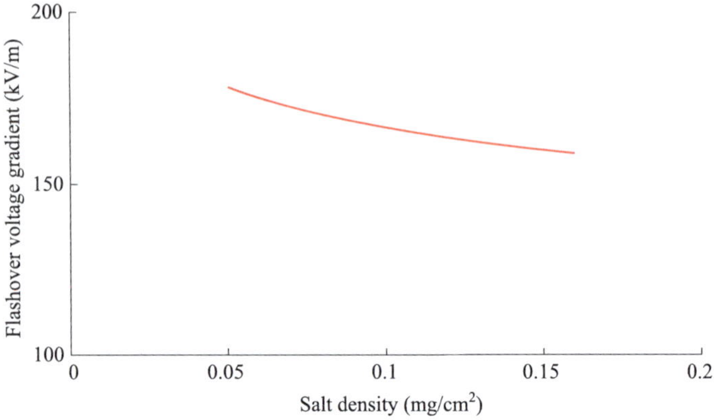

Fig. 21 Results of the artificial pollution test of the composite post insulator

negative base was used for fitting, and the fitting results were good, as shown in Fig. 22.

The fitting curve in Fig. 22 corresponds to the following formula:

$$U_{50\%} = 25.15S^{-0.267} \tag{4}$$

where

$U_{50\%}$ is 50% flashover voltage, kV;
S is salt density, mg/cm^2.

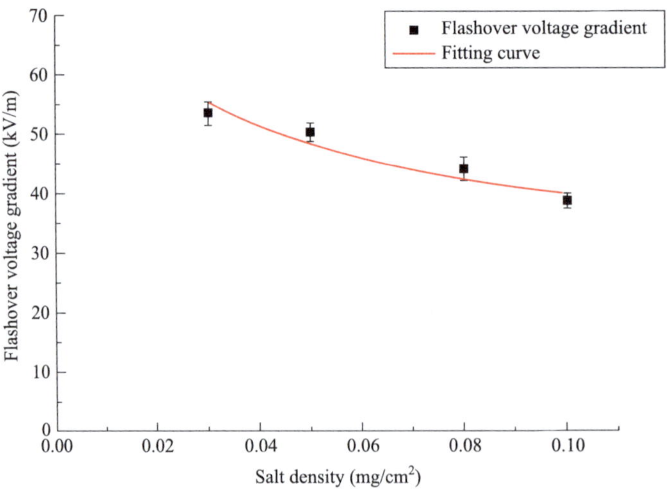

Fig. 22 Relation between salt density and pollution flashover voltage gradient of DC porcelain post insulator

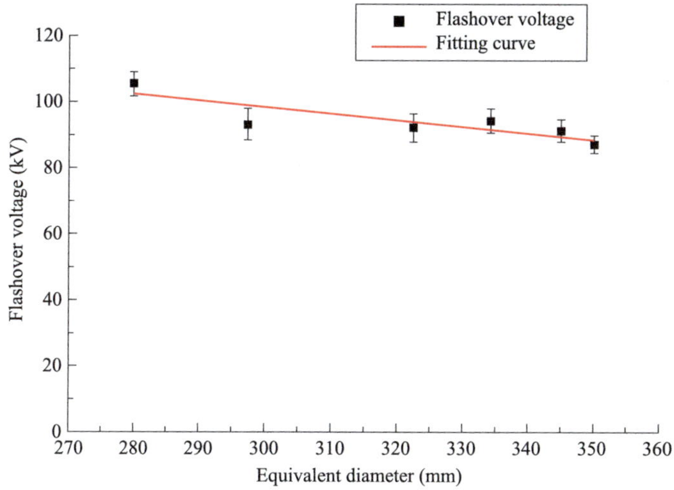

Fig. 23 Relation between the equivalent diameter of a single section post insulator and the flashover voltage

(2) Artificial pollution tests were conducted on post insulators of different diameters, and it was found that the 50% flashover voltage of the post insulator decreased with the increase of the equivalent diameter, roughly showing a linear relation, as shown in Fig. 23.

The linear fitting formula in Fig. 23 is

$$U_{50\%} = 160.28 - 0.20D \tag{5}$$

where D is the equivalent diameter of the post, mm.

According to the linear fitting formula of the equivalent diameter, the test results of the flashover voltage under different structural heights were corrected, and it was found that the 50% flashover voltage of the post insulator also showed a linear relation with its structural height. The corrected fitting curve is shown in Fig. 24.

The linear fit of Fig. 24 is quite good. As the structure height increases, its 50% pollution flashover voltage increases proportionally.

(3) Under different salt densities (0.03, 0.05, 0.08, 0.10 mg/cm^2), artificial pollution tests were conducted on single DC porcelain insulator coated with PRTV under weak hydrophobic conditions. It was found that the 50% pollution flashover voltage of the insulator also decreases as the salt density increases. The fitting curve and comparison with the test results of the insulator without PRTV coating are shown in Fig. 25.

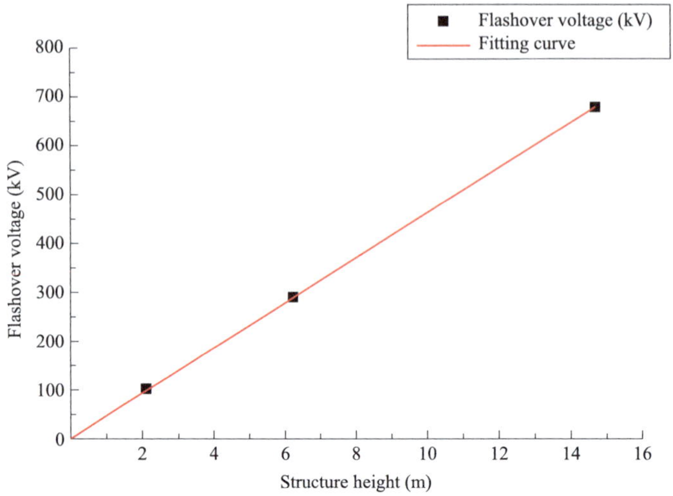

Fig. 24 Relation between the structure height of the post insulator and the flashover voltage

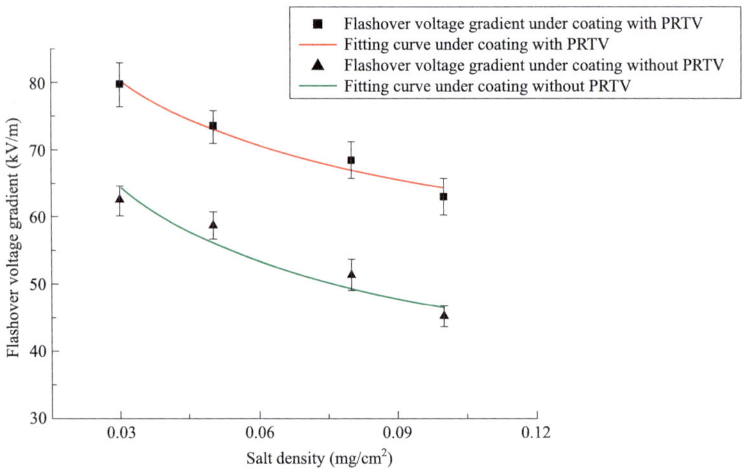

Fig. 25 Comparison of the flashover voltage gradient between coating with PRTV and without PRTV under weak hydrophobic conditions

After coating with PRTV, the flashover voltage of insulator under weak hydrophobic conditions changes with the salt density in the same way as before coating, showing a negative power exponential relationship, and the flashover voltage is significantly higher than before coating.

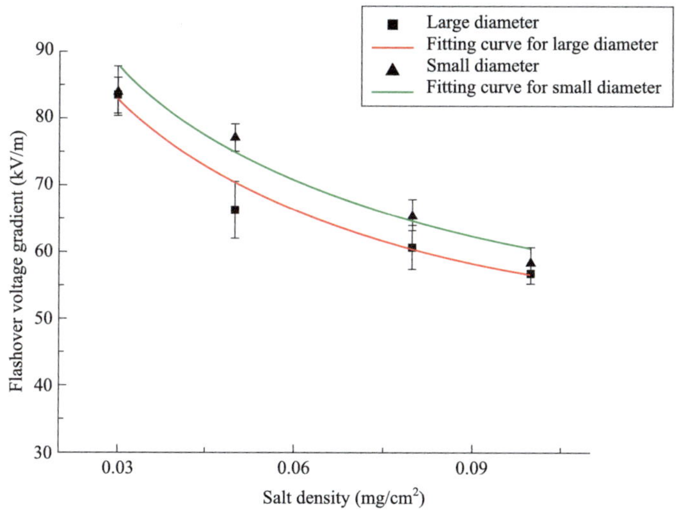

Fig. 26 Withstand voltage for insulators with two diameters and their fitting curves

(4) Under different salt densities (0.03, 0.05, 0.08, 0.10 mg/cm^2), artificial pollution tests were conducted on DC yard composite insulators of different hydrophobicities and diameters. The comparison of pollution flashover voltage for the two diameters is shown in Fig. 26.

The test results show that as the equivalent diameter increases, the pollution flashover voltage shows a linear decreasing trend, further proving the conclusion that the larger the diameter, the lower the pollution flashover voltage.

For composite insulators of the same diameter under both hydrophilic and weak hydrophobic condition, the comparison of 50% withstand voltage curves are shown in Fig. 27. The test results show that the flashover voltage of the hydrophobic post insulator is higher than that of the weak hydrophilic post insulator, with an increase of 9.95–14.79%.

(5) The scheme for the creepage distance of DC equipment at Changji and Guquan converter stations is proposed. The design value of the composite external insulation creepage distance of the umbrella-type post insulator in the DC yard at Guquan converter station can be taken as 46.5–52.5 mm/kV, and the creepage distance of the horizontal bushing can be taken as 45 mm/kV. The design value of the composite external insulation creepage distance of the umbrella-type post insulator and bushing in the DC yard at Changji converter station can be taken as 47–53 mm/kV, and the creepage distance of the horizontal bushing can be taken as 45 mm/kV to meet operational requirements.

III. Research on Double-Layer Shielding Corridor Design

To protect personnel safety, the valve hall of the ± 800 kV UHV DC project adopts a single-layer shielding corridor. With the increase in voltage level, the switching

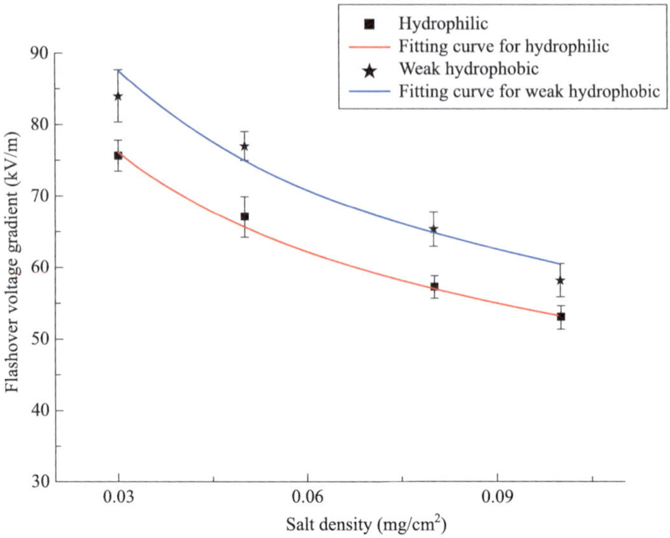

Fig. 27 Withstand voltages for insulators with hydrophilic and weak hydrophobic and their fitting curves

overvoltage also increases. The switching impulse withstand level required by the ± 1100 kV project is as high as 2100 kV. The air gap structure in the converter valve hall is very complex. To protect the safety of the maintenance and operation staffs, a new double-layer shielding structure needs to be proposed. As the design of the ± 800 kV project is of little reference significance, and it is necessary to study and determine the double-layer shielding corridor's grounding, layer spacing, grid size, etc.

(1) Thirteen simulation models with different parameters were established to obtain the maximum potential difference in each model. The results of the typical model (Model 1) is illustrated in Fig. 28, and the results of conductor potential under different models are shown in Table 42.

(2) We investigated the instantaneous safe voltage standard for the human body under short-term overvoltage, calculated the limit values applicable to switching overvoltage, all shown in Table 43. The safety of the transient potential difference of different calculation models was analyzed.

(3) The influence of model parameters on transient potential differences is analyzed, and design recommendations are given as follows.

1. Whether the inner and outer shielding corridors are directly connected and the spacing between them has little effect on the calculation results of the potential difference of the inner shielded channel conductors. From the perspective of insulation, the gap distance between the inner and outer cages needs to withstand

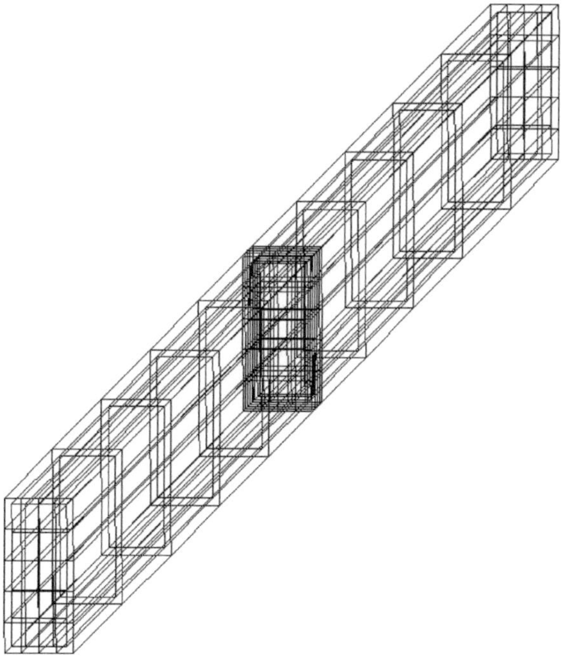

Fig. 28 Typical model for simulation

a potential rise caused by a switching impulse voltage of no more than 100 kV. The gap distance can be 15 cm, and the two layers of shielding can be supported by short post insulators, only supporting the lower plane, while the other planes are suspended or supported by insulators.

2. The grounding type of the inner and outer shielding corridors has a certain impact on the conductor potential difference of the inner part. It can be adopted with single-point grounding on both sides, multi-point grounding, etc., according to the project design requirements, avoiding single-point grounding on the same side as much as possible.
3. The size of the shielding cage has little effect on the calculation results of the potential difference, and can be designed according to that of the ± 800 kV project.
4. For the conductor potential difference of the inner shielding corridor, if it is in accordance with the IEEE standard and the effect of soil resistivity ρ is not considered, the conductor potential difference under the considered calculation model is less than the voltage limit when the duration of fault current is less than 0.003 s.

IV. DC yard selection and valve hall design

1. Analysis of factors affecting the ground distance of DC pole line equipment

Table 42 Calculation results of conductor potential under different models

Model	Maximum potential difference (V)
Model 1—Single-point grounding on both sides of the inner and outer cage	1747
Model 2—Single-point grounding on the same side of the inner and outer cage	2531
Model 3–Both sides of the inner and outer cage are grounded	1866
Model 4—Both ends of the inner and outer cage and the middle are grounded	1707
Model 5—Direct connection between inner and outer cage	1747
Model 6—Direct connection of inner and outer cage at intervals of 10 m	1783
Model 7—No dense conductors within ± 1 m of the outer cage	1710
Model 8—Current amplitude of 15 kA	2619
Model 9—Cage length of 60 m	1845
Model 9—Cage length of 40 m	1239
Model 11—Cage length of 130 m with inner cage of ± 1 m dense grid	662
Model 12—Cage length of 130 m with inner cage of ± 1 m sparse grid	665
Model 13—Cage length of 130 m with inner cage of ± 15 m dense grid	905

Table 43 Standard of instantaneous safety voltage for the human body under short-term over-volatge

Standard	Voltage limit (V)	0.003 s voltage limit (V)
DL/T 5340—2015 ≪Technical code for protective designing of DC overhead transmission lines to occur danger and interference effects to telecommunication lines≫	U_{RSIWV}	3177
GB 50065—2011 ≪Code for design of AC electrical installations earthing≫	σ	3177
IEEE 80—2000 ≪IEEE Guide for Safety in AC Substation Grounding≫	σ	2867
DL/T 5340—2015 ≪Technical code for protective designing of DC overhead transmission lines to occur danger and interference effects to telecommunication lines≫	$t < 0.05$, 3000	$t < 0.05$, 3000

Note 1. t is the clearing time for a short-circuit fault on the transmission line, s.
2. ρ is soil resistivity, Ω m.

A series of problems arise when the DC voltage of UHV transmission project is raised to ± 1100 kV. The selection of the grounding distance of the DC yard pole equipment will directly affect the design difficulty and even the feasibility of ± 1100 kV outdoor DC yard equipment. Therefore, it is necessary to study the selection of the grounding distance of the ± 1100 kV DC yard pole equipment, and then determine the reasonable type of DC yard by combining technical and economic analysis.

When selecting the grounding distance of the DC pole line equipment, the main considerations are: ① ensuring a safe air clearance to the ground under switching impulse condition; ② ensuring the creepage distance requirements of the insulators; ③ ensuring that the maximum total electric field at ground level generated by the DC yard conductors and equipment is within the control index range.

2. Air clearance requirement

The air clearance to the ground of the DC yard pole line equipment under switching impulse condition is mainly related to the amplitude and waveform of the switching impulse voltage and the gap factor.

(1) Insulation level of DC yard pole line equipment. According to the research conclusions of insulation coordination, the switching impulse withstand level (*SIWL*) of DC pole line equipment will reach 2100 kV, which is 31% higher than the 1600 kV insulation level of ± 800 kV DC pole line. When the switching withstand voltage reaches 2100 kV, the air gap discharge curve enters the deep saturation region, which means that the air gap requirements will increase nonlinearly compared to those of the ± 800 kV project, posing great challenges to project design and construction.
(2) Calculation of 50% switching impulse discharge voltage of DC yard pole line equipment. In the ± 1100 kV project, according to the insulation coordination conclusion, 15% insulation margin is taken, and the switching impulse withstand voltage is 2100 kV. The 50% discharge voltage can be obtained by formula (6)

$$U_{50\%} = \frac{U_{\mathrm{RSIWV}}}{1 - n\sigma} \tag{6}$$

where

$U_{50\%}$ is 50% discharge voltage, kV;
U_{RSIWV} is withstand voltage, kV;
σ is the relative standard deviation, with 0.03 for lightning impulse and 0.06 for switching impulse.

In the formula, n determines the discharge probability under withstand voltage condition. For ordinary gaps, it is selected according to 2σ, i.e. n = 2. At this time, the discharge probability under withstand voltage condition is about 2%. The obtained 50% discharge voltage $U_{50\%}$ is 2387 kV.

When determining the air clearance, the selection of the switching impulse voltage waveform is important. The change in the front time of the operating impulse voltage waveform will affect the insulation strength of the clearance. For a certain clearance, there is a critical front time. Figure 29 shows the characteristic curve of the rod-plane switching impulse discharge voltage. It can be seen that the lowest point of the characteristic curve corresponds to the critical front time, which increases with the growth of the clearance distance. For a given gap arrangement, the critical wave determines the minimum discharge voltage.

IEC 60071-2 "Insulation coordination Part 2: Application guidelines" recommends the calculation formula for the critical wave rod-plane gap below 25 m as follows.

$$U_{50\%} = 1080\ln(0.46d + 1) \tag{7}$$

where d is the gap length, m.

The gap coefficient K under the switching impulse condition is

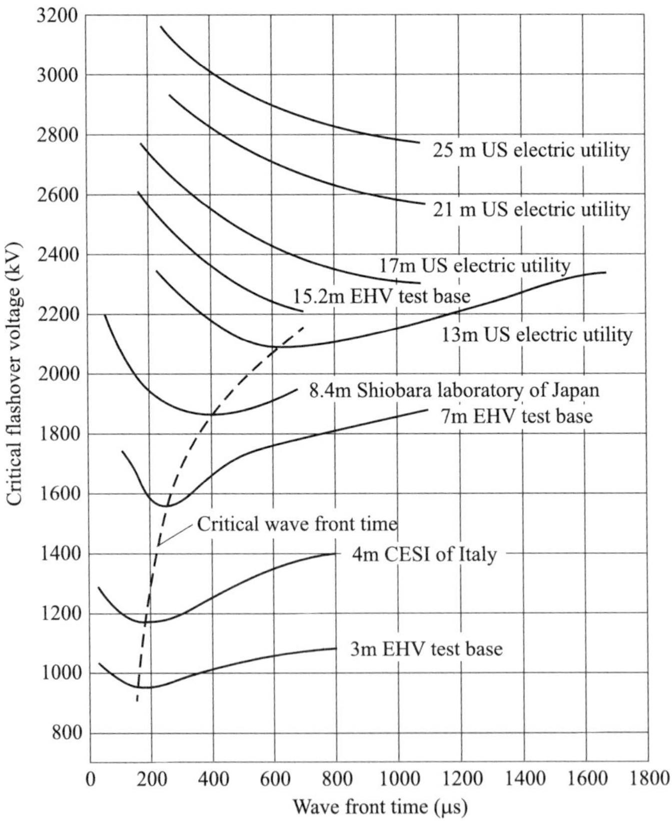

Fig. 29 Critical discharge voltage of rod-plane gap

$$K = U_{50\%} / U_{50\%\ \mathrm{RP}} \tag{8}$$

where $U_{50\%\ \mathrm{RP}}$ is the 50% discharge voltage of the rod-plate gap, kV.

(3) Selection of ground air clearance for outdoor DC yard equipment. Considering the actual layout conditions, the following factors will affect the discharge voltage of the outdoor DC yard gap.

1. The effect of rain on the discharge voltage. Since rainwater is similar to a conductor, when the outdoor insulator is rained on, the shape of the rainwater dripping at the end of the insulator is equivalent to a high-voltage electrode, and its equivalent diameter is much smaller than the actual fittings. Therefore, it can be concluded that rain will destroy the electrode shape under dry conditions, affect the electric field distribution under dry conditions, and greatly reduce the gap switching impulse discharge voltage.
2. The effect of multi-column insulators parallel connection on the discharge voltage. When multi-column insulators are connected in parallel, flashover of each insulator is considered as an independent event. If one column discharges, it is considered that the entire gap discharges. If the number of insulators connected in parallel is *m*, the single column discharge probability is $P(U)$, 50% flashover voltage is $U_{50\%}$, the probability of *m* insulators in parallel gap discharging is Pm (U), and the 50% flashover voltage is $U_{50\%m}$. Then $Pm\,(U) = 1 - [1 - P\,(U)]m$. After m insulators are connected in parallel, we obtain

$$U_{50\%\ m} = U_{50\%} - 4U_{50\%}\,\sigma\left(1 - \frac{1}{\sqrt[5]{m}}\right) \tag{9}$$

where

σ is the relative standard deviation of an individual gap;
m is the number of parallel insulators;

The change in discharge voltage when multiple insulators are paralleled is shown in Fig. 30.

3. The influence of nearby equipment support structures and other ground structures on the discharge voltage. When there are other supports and structures around the equipment, they will affect the distribution of the electric field and reduce the discharge voltage. The closer and denser the supports and structures around the equipment are, the greater the impact on the discharge voltage.

Considering the effects of rain, parallel multi-column insulators, and the influence of nearby equipment support structures and other ground structures on the gap discharge voltage, the discharge voltage of outdoor DC yard line equipment will decrease by

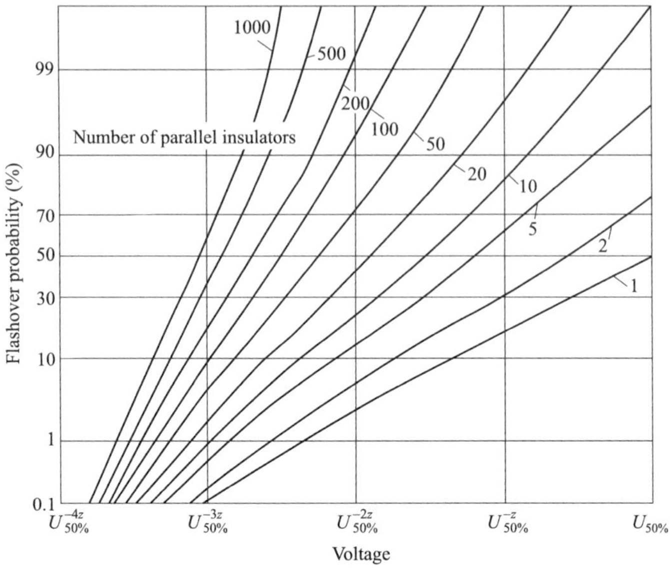

$U_{50\%}$—50% flashover voltage of a single gap;
z —Habitual deviation of a single gap

Fig. 30 Changes in discharge voltage when multiple insulators are parallel connected

about 4%. According to the recommended formula for calculating the rod-plane gap below 25 m of critical wave, the ground distance of ± 1100 kV outdoor DC yard line equipment required for switching impulse air clearance is about 20.6 m.

(4) Selection of ground air clearance for indoor DC yard pole line equipment. The distribution of the electric field has a significant impact on gap discharge. When analyzing the insulation strength or breakdown voltage of different gap structures, the impact of insulation distance needs to be considered. Also, the impact of the unevenness of the electric field needs to be given more consideration. Generally, under the same gap distance, the more uneven the electric field, the lower the breakdown voltage; the more uniform the electric field, the higher the breakdown voltage. When determining the safe clearance required for switching impulse of UHV project, optimizing the electrode shape by setting grading spheres or rings to improve the electric field distribution, increasing the breakdown voltage and thereby reducing the safe air gap is an effective method. For outdoor DC yard, the effect of these measures is small due to the impact of rain. However, for indoor DC yard, which are not affected by rain and wind, the minimum air gap required for the same switching impulse voltage can be reduced by optimization of the electrode shape.

Based on actual layout conditions, considering the effects of parallel multi-column insulators and the influence of nearby equipment support structures and other ground structures on the gap discharge voltage, the discharge voltage of indoor DC yard pole line equipment will increase by about 1%. According to the recommended formula for calculating the rod-plane gap below 25 m of critical wave, after optimizing the electrode shape, the gap factor is taken as 1.2, and the ground distance of ± 1100 kV indoor DC yard pole line equipment required for switching impulse air clearance is about 13 m, which is basically consistent with the test results. At the same time, for the air clearance of the existing parallel gap (such as when the electrode is near multiple walls), in order to reduce the mutual influence of multiple walls on the electric field distribution, the other gaps except the main gap are considered as 1.3 m × 13 m.

3. Creepage distance requirement

When an external DC yard is adopted, according to the conclusions of pollution prediction and external insulation configuration research, the creepage distance requirement for large-diameter vertical composite insulators is 53 mm/kV. The highest DC operating voltage is considered to be 1122 kV, and the creepage distance required for the pole line equipment insulator is 53 mm/kV × 1122 kV = 59,466 (mm). The ratio of the creepage distance of the composite insulator to the dry arc distance is considered to be 4, resulting in an insulator dry arc distance of about 14.9 m. Considering the total height of the multi-section post insulator metal flange is about 2 m, the total height of the insulator is about 16.9 m.

When an internal DC yard is adopted, due to the better environmental conditions, the creepage distance requirement will be significantly reduced, considered to be 25 mm/kV. The highest DC operating voltage is considered to be 1122 kV, and the creepage distance required for the pole line equipment insulator is 25 mm/kV × 1122 kV = 28,050 (mm). The ratio of the creepage distance of the composite insulator to the dry arc distance is considered to be 4, resulting in an insulator dry arc distance of about 7 m. Considering the total height of the multi-section post insulator metal flange is about 2 m, the total height of the insulator is about 9 m.

4. Requirement of total electric field at ground level

The ± 1100 kV indoor/outdoor DC yards both adopt the patrol corridor scheme of the Faraday cage. As the Faraday cage has a good electromagnetic shielding effect, the maximum total electric field at ground level is no longer the decisive factor controlling the ground height of the pole line equipment.

5. Selection of the ground distance of the pole line equipment

Based on the above analysis, it can be concluded that for both the ± 1100 kV outdoor DC yard and the indoor DC yard, the ground distance of the pole line equipment is determined by the air clearance required by the switching impulse, with the outdoor DC yard being 20.6 m and the indoor DC yard being 13 m.

6. Technical and economic comparison of indoor/outdoor DC yards

Based on the above conclusions, the technical and economic comparison of indoor and outdoor DC yards is shown in Tables 44 and 45, respectively.

Table 44 Technical comparison of indoor and outdoor DC yards

No	Content	Indoor	Outdoor
1	Layout	± 1100 kV equipment, DC filter high-voltage capacitor tower and some of the ± 550 kV bypass equipment are arranged indoor, and the other DC yard equipment is arranged outdoor	All DC yard equipment is located outdoor
2	Development difficulty of DC equipment	The post insulators only need to be partially modified for the existing ± 800 kV outdoor insulators in accordance with the 13 m dry arc distance requirement	The post insulators need to meet the requirements of a dry arc distance of 17–18 m, which is about 50% longer than the existing ± 800 kV insulators. All equipment with increased dry arc distance needs to be fully demonstrated for reliability. The impact of heavy windstorms on equipment at the sending end also needs to be addressed
3	External insulation creepage distance requirement for DC equipment	The requirement for external insulation is relatively low, and DC equipment is easier to manufacture	The external insulation requirement is relatively high, and the DC equipment is difficult to manufacture
4	Seismic performance	The equipment is small in height and has relatively high mechanical performance	The equipment is large in height and has relatively poor mechanical performance
5	Operating conditions for DC equipment	The operating environment is relatively good, which can reduce the maintenance and cleaning frequency of DC yard equipment and reduce the damage rate of high-voltage capacitor components	The degree of equipment contamination is greatly affected by external factors. The operating environment is relatively poor, and extreme working conditions such as strong winds, low temperatures, and sandstorms need to be considered
6	Occupied area of DC yard	Slightly larger than the outdoor field, increased by about 5%	About 5% less than indoor field
7	Differences in electric equipment	Additional two ± 1100 kV, two ± 550 kV and four neutral line bushings to the entire station The number of ± 1100 kV insulators is small, and the cost of individual equipment is reduced	The number of insulators increases, and ± 1100 kV outdoor insulators require double-column parallel connection

(continued)

Table 44 (continued)

No	Content	Indoor	Outdoor
8	Additional civil construction	The converter station at the sending end requires the addition of two L-shaped single-story factory buildings with a length, width, and height of 118 m × 119 m × 43 m, as well as corresponding fire protection and HVAC equipment. The receiving station requires the addition of two rectangular single-story factory buildings with a length, width, and height of 100 m × 96 m × 35 m, as well as corresponding fire protection and HVAC equipment	None
9	Station power load	Due to the adoption of the indoor DC yard, the station's power load increases	No additional

Table 45 Economic comparison of indoor and outdoor DC yards (in 10,000 RMB)

No.	Content	Layout of DC yard	
		Indoor	Outdoor
1	Civil construction cost	+ 10,660	0
2	HVAC cost	+ 1372	0
3	Fire protection cost	+ 248	0
4	Lightning cost	+ 772	0
	Civil engineering price difference, subtotal	+ 13,052	0
5	Price difference for main equipment		
5.1	High-voltage capacitor tower of DC filter	– 3510	0
5.2	1100 kV bushings	+ 3000	0
5.3	1100 kV disconnetors	– 1500	0
5.4	1100 kV post illustrators	– 1014	0
5.5	550 kV bushings	+ 800	0
5.6	Neutral line bushing	+ 900	0
5.7	Smoothing reactor	– 2310	0
5.8	Difference of other equipment insulation	– 1333	0
	Equipment price difference, subtotal	– 4867	0
6	Total	+ 8185	0

From Table 45, it can be seen that the initial investment of the indoor DC yard construction will increase by about 81.85 million yuan, but the probability of failure and maintenance time will be greatly reduced. Calculated at an electricity fee of 0.2 yuan/kWh, a power outage loss of 34 h can make up for the initial investment. Therefore, the economy and reliability of the indoor DC yard are better than the outdoor field. It is recommended that the indoor DC yard scheme is adopted for ± 1100 kV DC parts, and the outdoor DC yard scheme is for ± 550 kV and neutral line part.

V. Lightning Protection Design for Building Roofs

The height of the ± 1100 kV DC valve hall and indoor DC yard is higher than that of the ± 800 kV DC. After the height is increased, the station's lightning protection net cannot effectively protect it, and the lightning protection design of the valve hall and indoor DC yard must be carried out separately.

(1) The valve hall and indoor DC yard of Changji converter station can be designed according to the third category of lightning protection buildings. The LV valve hall of Guquan converter station can be designed according to the third category of lightning protection buildings, while the HV valve hall and indoor DC yard can be designed according to the second category of lightning protection buildings. However, considering the importance of UHV DC projects, the valve halls and indoor DC yards of both Changji and Guquan converter stations are proposed to be strictly designed according to the second category of lightning protection buildings. The specific calculation process is as follows.

The size of the HV valve hall of Pole 1 is 125.5 m × 47.5 m = 5961.25 m^2; the size of the LV valve hall of Pole 1 is 83.5 m × 27.5 m = 2296.25 m^2; the size of the LV valve hall of Pole 2 is 83.5 m × 27.5 m = 2296.25 m^2; the size of the HV valve hall of Pole 2 is 125.5 m × 47.5 m = 5961.25 m^2. The size of the indoor DC yard of Pole 1 is 109.1 m × 63.4 m + 42 m × 19.4 m = 7731.74 m^2; the size of the indoor DC yard of Pole 2 is 109.1 m × 63.4 m + 42 m × 19.4 m = 7731.74 m^2.

The cloud-to-ground lightning density of the location of Changji converter station is < 0.78 times/(km^2 year). The expected number of lightnings for the LV valve hall, HV valve hall, and indoor DC yard of Pole 1/Pole 2 at Changji converter station is 0.003, 0.008, and 0.01 times/year respectively. According to the standard, they all belong to the third category of lightning protection buildings. The size of the HV valve hall of Pole 1 is 125 m × 46 m = 5750 m^2, and the size of the LV valve hall is 99.75 m × 27.5 m ≈ 2743.13 m^2; the size of the LV valve hall of Pole 2 is 99.75 m × 27.5 m ≈ 2743.13 m^2, and the size of the HV valve hall is 125 m × 46 m = 5750 m^2. The size of the indoor DC yard of Pole 1 is 125 m × 62.5 m = 7812.5 m^2, and the size of the indoor DC yard of Pole 2 is 125 m × 62.5 m = 7812.5 m^2.

The cloud-to-ground lightning density of the location of Guquan converter station is between [5.0, 7.98) times/(km^2·year). Through calculation, the expected number of lightnings for the LV valve hall, HV valve hall, and indoor DC yard of Pole 1/Pole 2 at Guquan converter station is [0.024, 0.037) times/year, [0.049, 0.078) times/year,

Table 46 Rolling sphere radius and grid size for lightning protection building verification (m)

Building		Rolling sphere radius h_r
Changji converter station	Valve hall	41
	Indoor DC yard	40
Guquan converter station	Valve hall	39
	Indoor DC yard	38
Maximum value		41(< 45)

and [0.066, 0.105) times/year respectively. According to the standard, the LV valve hall belongs to the third category of lightning protection buildings, while the HV valve hall and indoor DC yard belong to the second category of lightning protection buildings.

(2) The valve halls and indoor DC yards of both Changji and Guquan converter stations are pure steel structures, with a roof thickness of 0.8 mm. According to GB 50057—2010 "Design Code for Structures against Lightning", their roofs can be used as air-termination system, and there is no need to separately set up lightning rods or lightning wire columns.
(3) Given the importance of the ± 1100 kV project, the valve hall of the converter station and the indoor DC yard adopted a reinforced lightning protection design with lightning lines and belts laid on the roof. The rolling sphere method and the piecewise linear method were used to check the protection range of the lightning conductors and lightning belts. The maximum rolling sphere radius of the valve hall and indoor DC yard of Changji converter station and Guquan converter station is 41 m, which is less than the standard requirement of 45 m for the second category of lightning protection buildings, as shown in Table 46. The check results using the piecewise linear method show that both the valve hall and the indoor DC yard are within the protection range of the lightning conductors and lightning belts.
(4) By arranging a combination of lightning conductors and lightning belts on the top of the valve hall, the maximum equivalent impulse resistance calculated is about 2.78 Ω, as shown in Table 47, which meets the GB 50057—2010 standard requirement that the impulse resistance along each down lead of the second category of lightning protection buildings should not exceed 10 Ω.

5 Control and Protection

In the ± 1100 kV UHV DC transmission system, the control and protection scheme, as the core part, is of great significance to the system safe and stable operation. A reasonable control and protection scheme should be adapted to the SLD and control

Table 47 Impulse grounding resistance of valve hall and indoor DC yard in Changji and Guquan converter station (Ω)

Building		Maximum impulse grounding resistance
Changji converter station	Valve hall	2.03
	Indoor DC yard	2.51
Guquan converter station	Valve hall	2.48
	Indoor DC yard	2.78
Maximum value		2.78

requirements of the DC transmission system, and should be able to effectively disconnect the faulty components under various fault conditions to ensure the continuous operation of the intact system as much as possible.

I. Overview

The UHV DC system is layered connected to 500 and 1000 kV AC grids at the receiving end at different voltage levels, which facilitates the optimization of the grid structure, the balanced distribution of power flow, the reasonable distribution and consumption of higher DC power, and the improvement of the utilization efficiency of the 1000 kV AC grid. It can also improve the power balance under fault conditions, enhancing the safety and stability level of the receiving end grid.

Figure 31 shows the schematic diagram of the main circuit of the converter station connected to the AC grid at different voltage levels. The converter station at the sending end is a converter station without layered connection, and the HV and LV converters are connected to the same AC grid, which is the same as the conventional UHV DC system. The receiving converter station is a layered connected converter station, and the HV and LV converters in series are connected to two AC systems, with the following characteristics.

(1) The HV and LV converters of the receiving end are connected to the 500 and 1000 kV AC grids respectively. After the 1000 kV grid is reduced to 500 kV by the step-down transformer, it is connected to the 500 kV grid outside the station through the line. Therefore there is a certain degree of coupling between the 500 and 1000 kV AC grids.
(2) The 500 and 1000 kV AC grids are equipped with two independent AC filter yards. The reactive power of two AC grids is independently controlled.
(3) The HV and LV converters are connected to the 500 and 1000 kV AC grids respectively. There are differences in synchronous voltage, frequency, phase, converter transformer impedance, tap position, and tap adjustment step.
(4) Voltage measurement points have been added between the HV and LV converters of the layered accessed converter station for the balance control of the converter voltages.

Since the two AC systems each have different system parameters and operating characteristics, it is required that the DC control should be able to control and respond

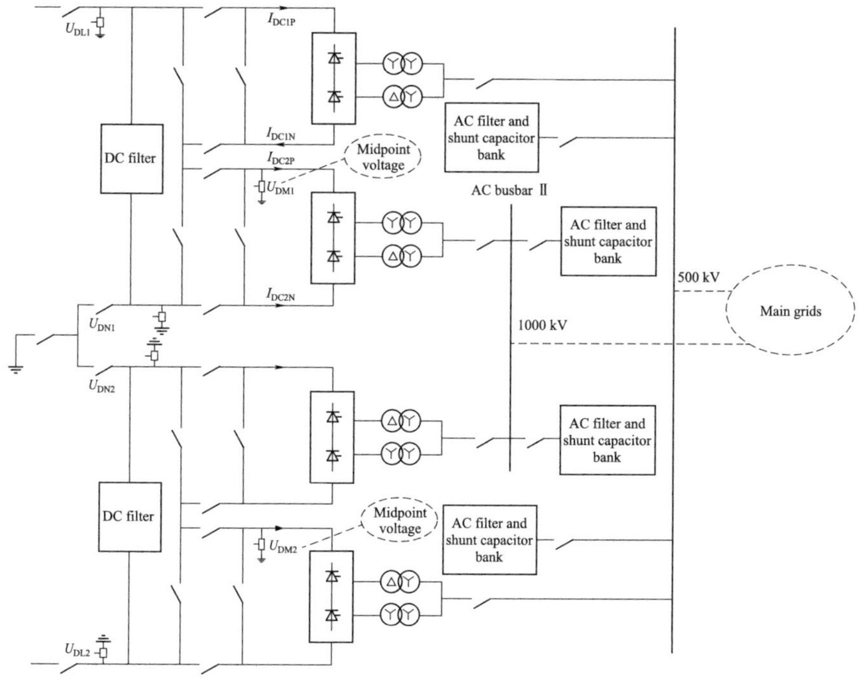

Fig. 31 Schematic diagram of the main circuit of the converter station connected to the AC grids at different voltage levels

to the power, voltage, frequency, and control commands of the two AC systems relatively independently, and coordinate operating according to the needs of the system. The complexity of the DC control system is greatly increased compared to conventional DC, and the layered connection on the receiving end proposes the following requirements for DC control protection.

(1) Balancing the DC voltage of the HV and LV converters. The HV and LV converters are layered connected to 500 and 1000 kV AC systems. The adjustment difference and operating gear of the HV and LV converter transformers are different, and the AC system voltage and phase angle are inconsistent, which may cause a difference in the operating voltage of the HV and LV converters. The imbalance of DC voltage may cause over-stress of the converter, and at the moment when a converter exits, large disturbances will occur due to the large DC voltage deviation of the remaining converters at both ends. Therefore, a control function for balancing the DC voltage of the HV and LV converters should be configured to ensure that the DC voltage of the two converters remains balanced whether the layered accessed converter station operates as a rectifier or an inverter.
(2) Reactive power control of the two AC systems. The HV and LV converters of the converter station are layered connected to 500 and 1000 kV AC systems.

It is required that the DC control should be able to independently control and respond to the reactive power and AC voltage of the two AC systems, and achieve coordination according to the coupling degree of the two AC systems.

(3) Power transfer and distribution. The HV and LV converters of the converter station are layered connected to 500 and 1000 kV AC systems. The function of power transfer and distribution is more complex than that of systems connected to one grid. Due to the series connection of the HV and LV converters, increasing or decreasing the power of the converter connected to one AC system will affect the power transmission of the other AC system. In addition, after a converter is disconnected, the transfer of lost power is not only limited by the pole power control mode, but also needs to consider the power transmission of the two AC systems on the receiving end. Furthermore, layered access calls for higher requirements on additional control.

(4) DC protection. In terms of fault characteristic analysis, the mutual influence between the AC and DC systems on the receiving end needs to be considered, such as the impact of a fault in one AC system on the DC system and the other AC system, and the DC protection function and criteria should be adjusted accordingly.

From the analysis of the characteristics of the layered accessed system, the key points of control and protection are summarized as follows.

1. Measurement point configuration scheme of main circuit.
2. Control system layered structure adapted to layered access.
3. Control strategy of the firing angle of the HV and LV converters on the receiving end and the tap of the converter transformer.
4. Configuration of reactive power control function for receiving end layered connected to two AC grids.
5. Balance control of voltage of series converters on the receiving end.
6. Online entry and exit control strategy for converter stations.
7. Power transfer and distribution adapted to layered access.
8. Strategy to resist commutation failure at the receiving end.
9. DC protection configuration adapted to layered access.
10. Measures to deal with the fault of the DC voltage divider between the HV and LV converters on the receiving end.

The overall structure and function of the control and protection system need to be redesigned based on the foundation of conventional UHV DC projects, so as to adapt to the new main circuit structure and control requirements.

II. Key Strategy of Control System

(I) Basic Control Strategy

Under normal operating conditions, the rectifier side of the UHV DC system maintains a constant DC current by quickly adjusting the firing angle via the current controller, and the inverter side controls the voltage or the extinction angle. The tap changer control of the rectifier side converter transformer maintains the firing angle within ± 2.5°. When the inverter side adopts voltage control, the tap changer control of the converter transformer maintains the extinction angle within a certain range. When the extinction angle control is adopted, the tap changer control of the converter transformer maintains the DC voltage on the rectifier side within the reference range.

Both the rectifier side and the inverter side are equipped with closed-loop current controllers and voltage controllers, but they are configured with different operating parameters. Through the coordination of the controllers on both sides, the rectifier side controls the current while the inverter side controls the voltage under normal operating conditions. This control method is achieved by subtracting a current margin from the current command on the inverter side (the current margin value is usually chosen as 10% of the rated value), and the effective command of the inverter side is lower than that of the rectifier side. In summary, the converter station with a high current command operates as a rectifier station, and the other station operates as an inverter station.

Figure 32 shows the DC voltage-current static characteristic of the inverter side using predictive control of extinction angle (modified fixed extinction angle control).

In the case of abnormal AC voltage, the inverter side may gain current control rights. At this time, after the rectifier side enters the minimum firing angle control, the closed-loop current regulator on the inverter side controls the current. When current

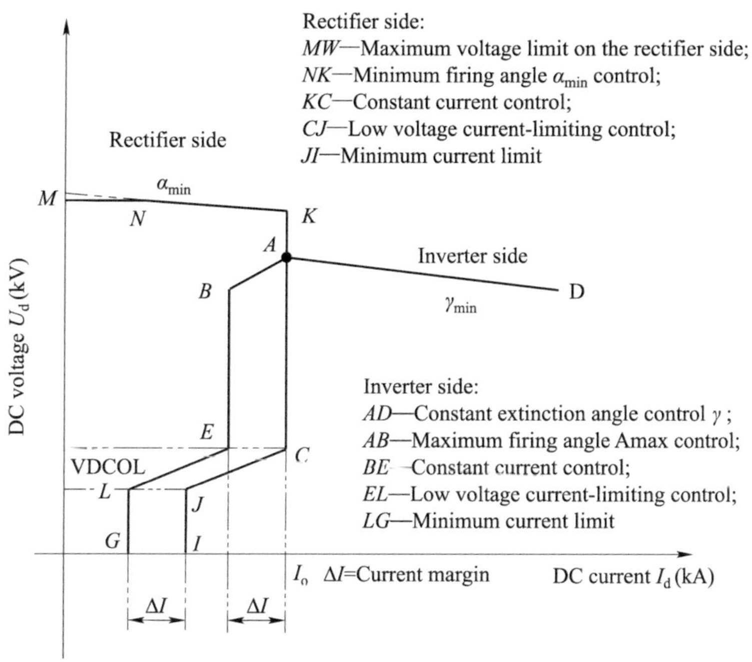

Fig. 32 DC voltage-current static characteristic of the inverter side using predictive control of extinction angle

control is transferred to the inverter side, current margin compensation control can prevent the DC current from dropping due to the current margin.

Both the rectifier and inverter sides are equipped with low voltage current-limiting control links. When the DC voltage drops, the DC current command is limited to help the DC system recover quickly after an AC/DC fault.

The basic control strategy of the receiving end is the same as that of the conventional DC transmission system. However, because the HV and LV converters of the receiving end are connected to different AC grids, the adjustment differences of converter transformers are different, such as the tap changer difference of the 500 kV converter transformer is 1.25%, and the tap changer difference of the 1000 kV converter transformer is 0.65%. Although the DC currents of the HV and LV converters are the same, the DC voltages may be unbalanced due to the different voltages and phase angles of the two AC systems.

The DC system has two power transmission directions, forward and reverse. The DC system with layered connection to the AC grid at the receiving end has completely different connections and operating characteristics during forward and reverse operation, resulting in different DC voltage unbalance control strategies. During forward operation, the inverter side is layered into different AC grids. During reverse operation, the rectifier side is layered into different AC grids. There are the following differences in control and protection strategies for forward and reverse operation modes.

(1) Forward operation mode. The rectifier side controls the current. Since the HV and LV converters of the rectifier side are connected to the same AC grid, the voltages of the two 12-pulse converters in series are naturally balanced. Therefore the voltage balance function does not need to be added. The inverter side controls the voltage or the extinction angle. Since they are connected to two different grids, a voltage balance strategy needs to be adopted to keep the DC voltages of the HV and LV converters consistent.
(2) Reverse operation mode. The rectifier side is adopted with a constant current control. Since it is connected to two power grids separately, a voltage balancing strategy is also needed to maintain the balance of DC voltage of the HV and LV converters. The inverter side is adopted with a constant voltage or constant extinction angle control, since the HV and LV converters are connected to the same AC power grid, the voltage of the two 12-pulse converters in series naturally balances, and no additional voltage balancing control function is needed.

For the safety and reliability of the DC system, the converter stations that are not layered aceessed should adopt mature control strategies from past projects, whether they are operating as rectifiers or inverters. The HV and LV converters can be controlled independently or uniformly. Those of the layered accessed converter stations should be controlled independently, that is, independent closed-loop control of current, voltage, and extinction angle. Besides, voltage balance control should be added between the HV and LV converters. The following is a typical control strategy and software function configuration scheme, in which the inverter side uses predictive extinction angle control.

1. Control strategy for forward operation mode.

    a. Rectifier side. The same control strategy as previous UHV DC transmission projects is adopted, that is, the rectifier side adopts constant current control, and the HV and LV converters control the output of the same firing angle. There is no need to set up voltage balancing control between the two converters. As long as the tap positions corresponding to the two converters are the same, the voltage of the two converters naturally balances.
    b. Inverter side. Each converter adopts constant extinction angle control, and the voltage balance of the converter is achieved through its own tap control. The tap of each inverter controls the voltage at half the rated voltage minus the line voltage drop.

2. Control strategy for reverse operation mode.

    a. Rectifier side. Both HV and LV converters adopt constant current control, and each has its own output firing angle. Since they are connected to different AC systems, a voltage balancing control function is set up between the two converters. The voltage deviation between the converters is summed to the current or firing angle reference value after passing through the control loop to maintain balance of converter voltage.
    b. Inverter side. Each converter adopts constant extinction angle control. Since the HV and LV converters are connected to the same AC power grid, there is no need to set up voltage balancing control between the two converters. The function of the inverter side tap control is the same as that of previous UHV DC transmission projects, with the goal of controlling the DC voltage on the rectifier side to the rated value.

Figure 33 shows the schematic diagram of the basic control strategy. Under the aforementioned control strategy, the software function configuration of the non-layered accessed converter station is the same as that of previous UHV DC projects, i.e., current control, voltage control, extinction angle control, and other regulators are configured in pole control, which outputs the firing angle. The function of tap control DC voltage is also realized in pole control. The software function configuration of the layered accessed converter station is different from previous UHV DC projects, with current control, voltage control, extinction angle control, and other regulators independently configured in the HV and LV converter control. The pole control outputs current commands to the HV and LV converters, which output their own firing angles. Independent tap control functions are configured in the HV and LV converter control. The control strategy is characterized by the following: the rectifier side programs of the forward and reverse operation modes are different, so is the inverter side. The program is quite complex, but the advantage is that the converter station with non-layered accessed uses the same software function configuration and control strategy as previous projects, and has mature operational experience.

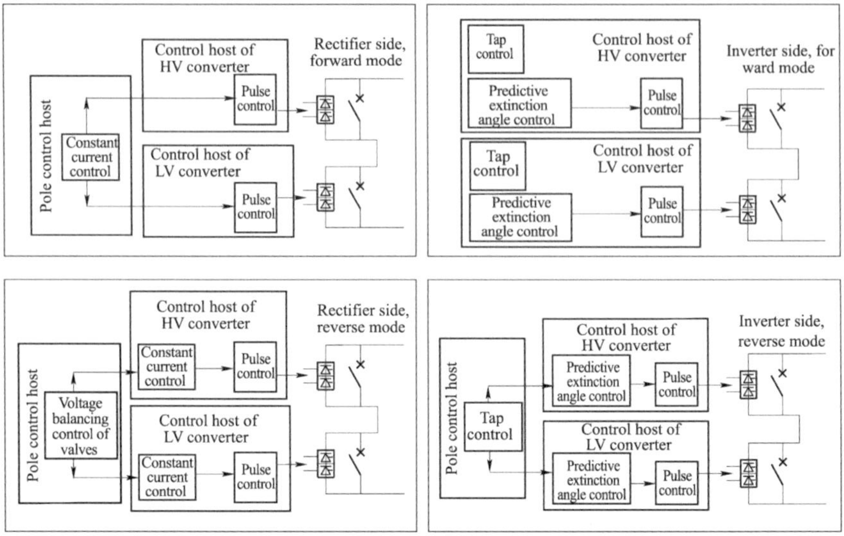

Fig. 33 Schematic diagram of the basic control strategy

(II) DC Voltage Balance Control

From the functional perspective, the UHV DC transmission project control system is divided into the pole-level control system and the converter-level control system. The pole-level control system mainly achieves the control functions of the bipolar and pole level, and the converter-level control system is for the trigger control of the 12-pulse converter and the adjustment of the converter transformer tap. The DC voltage balance control is configured at the pole-level control. DC voltage transformers are installed between the HV and LV converters of each pole in the layered accessed converter stations to measure the DC voltage.

For the layered accessed converter station, the principle of voltage balance control when operating as a rectifier is shown in Fig. 34. The difference between the voltage of the HV converter and the LV converter is proportionally integrated, and the obtained reference value adjustment of current or firing angle is sent to the current control loop of the two converters, thereby achieving the purpose of controlling the voltage balance of the two converters. The voltage balance control function only works when both converters are put into operation. During the process of disconnecting a converter, the voltage balance control function also exits. During the process of putting a converter into operation, the voltage balance control function is also disabled until the converter is put into operation. Under the fault condition of the 500 kV or 1000 kV AC system, the voltage balance control function should be activated.

When the layered accessed converter station operates as an inverter, the voltage balance of the converter is achieved through its own tap control, as shown in Fig. 35.

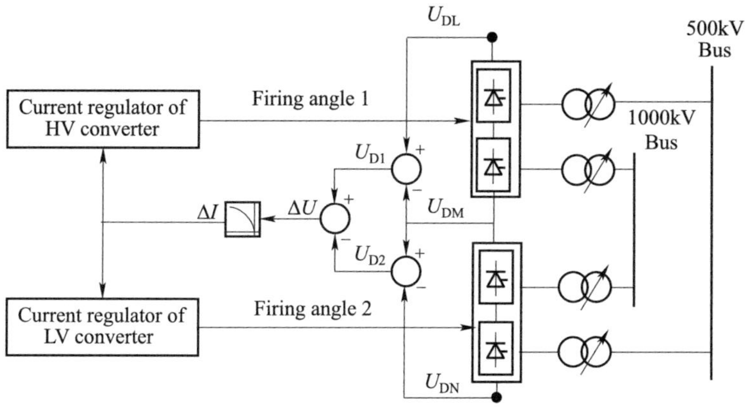

Fig. 34 Schematic diagram of voltage balance control when the layered accessed converter station is operating as the rectifier

The voltage balance of the converter is implemented by setting the voltage reference value of the converter transformer tap. The tap control of each converter controls the 12-pulse converter terminal voltage reference value U_{ref} to be half of the voltage obtained by subtracting the line voltage drop from the rated DC voltage U_{dN}, that is, $U_{\text{ref}} = (U_{\text{dN}} - RI_{\text{d}})/2$. Where R is the DC line resistance, I_{d} is the DC current.

(III) Tap Control of Converter Transformer

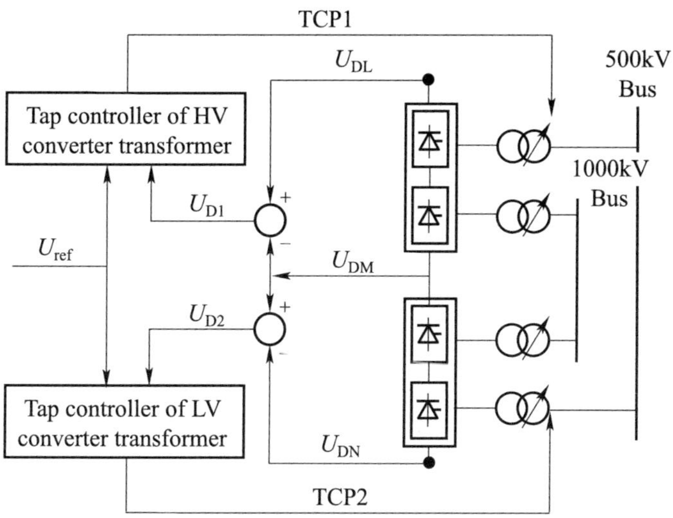

Fig. 35 Schematic diagram of voltage balance control when the layered accessed converter station is operating as the inverter

The tap control of the converter transformer is a slow control in conjunction with the converter control, divided into manual and automatic control mode. The goal of the rectifier side tap control is to maintain the firing angle within a certain range, and the goal of the inverter side tap control is to maintain the DC voltage at the output of smoothing reactor. For the layered accessed DC system, it is to maintain the DC voltage of each converter as half of the pole line voltage on the inverter side. When the converter station is in the state of blocking and open-circuit test, the converter transformer tap control set the no-load DC voltage of the valve side to a set value.

1. Basic control strategy of converter transformer tap

   (1) Angle control. Under normal operating conditions, the rectifier side converter transformer tap control maintains the firing angle within the range of 15° ± 2.5° of the rated angle. The converter transformer tap controller compares the measured firing angle with the reference value to get the angle difference. When the angle difference exceeds the upper limit of the dead zone, a command to reduce the tap position is issued. When the angle difference exceeds the lower limit of the dead zone, a command to raise the tap position is issued. There is a certain delay in implementing the converter transformer tap-raising and tap-reducing commands to avoid the tap position changing during AC and DC voltage disturbances. For layered accessed DC system, the main target of the tap control is still to maintain the firing angle within 12.5°–17.5°.

   (2) DC voltage control. The HV and LV converters are layered connected to two power grids respectively, and the taps of the HV and LV converters are independently controlled to achieve the goal of controlling the DC voltage on the rectifier side. Since the step difference of the 500 kV converter transformer is 1.25% and that of the 1000 kV converter transformer is 0.65%, the voltage change produced on the valve side by adjusting one gear of the 1000 kV converter transformer is about half of that of the 500 kV. Therefore, under normal operation, the positions of the HV and LV converter transformer taps are no longer the same. The dead zones of the tap-change operation vary depending on the step difference of the converter transformer.

When the AC voltage is abnormal, the rectifier side exits constant current control mode and the inverter side enters current control mode, the tap control of HV and LV converter transformers at inverter side adopts a constant firing angle strategy. There will be a difference in the voltages of two converters, and the balance operation of the converter is achieved by adjusting the converter firing angle through the voltage balance function.

(3) No-load voltage control on the valve side of the converter transformer. The no-load voltage control on the valve side of the converter transformer tap is used for the converter station blocking and open-circuit test situations. The no-load voltage controller controls the tap position of the transformer to the following set positions.

1. If the converter transformer loses power, the converter transformer tap moves to the set gear position before charging.
2. If the converter transformer is powered, but not in the open-circuit test state, the converter transformer tap establishes the ideal no-load DC voltage U_{di0} based on the minimum operating current allowed.
3. During the open-circuit test, the required no-load voltage U_{di0} should be used as the reference value.

2. Automatic synchronization function of the tap

The automatic synchronization function of the tap changer for layered accessed UHV DC system has three types: ① Synchronization of the 12-pulse converter transformer tap; ② Synchronization of the taps of HV and LV converter transformers; ③ Synchronization of the taps of the bipolar converter transformers. The automatic synchronization function is only effective in automatic control mode, and the three synchronization functions are as follows.

(1) Synchronization function of the 12-pulse converter transformer taps. When the positions of the taps of the 12-pulse converter transformer are inconsistent, an alarm signal is generated to the converter station monitoring system. At this time, the automatic synchronization function can resynchronize the tap of the converter transformer.

The automatic synchronization function tries to synchronize the positions of the converter transformer taps. If the synchronization function is unsuccessful, it will issue an alarm signal and prohibit automatic control.

(2) Synchronization function of the HV and LV converter transformer taps. The parameters of the HV and LV converters and their transformers for each pole of the non-layered accessed converter station are exactly the same, consistent with previous UHV DC projects. When the two converters of each pole operate simultaneously, the tap control of each converter aims to control the same firing angle command (operating as the rectifier) or the same DC voltage (operating as the inverter), generally ensuring that the tap positions of the converter transformers corresponding to the two converters are consistent, achieving balanced operation of the two converters.

To prevent the special situations where the tap positions of the converter transformers corresponding to the two converters are inconsistent, the non-layered accessed converter station has set up a tap position synchronization function for the converters. When the tap changer positions of the two converters deviate by two or more gears, the positions of the two converters are automatically synchronized.

The HV and LV converters of the layered accessed converter station are connected to the 500 and 1000 kV AC grids respectively. The total number of tap positions and the voltage value of each gear of the converter transformers are different. To ensure voltage balance when the two converters operate simultaneously, it is necessary to adjust the tap to ensure that the DC voltages of the two converters are the same, so the tap positions of the HV and LV converter transformers are not required to be synchronized.

(3) Synchronization function of the bipolar converter transformer taps. When operating in bipolar mode and both poles are operating in bipolar power control mode, the non-layered accessed converter station is the same as previous UHV DC projects, with the configuration of the bipolar converter transformer tap synchronization function.

The converter transformers of the HV and LV converters of the layered accessed converter station have different characteristics. The balanced synchronization function for the four converters cannot be adopted. It is only necessary to synchronize taps according to principles of connecting with the same AC grids and the converter transformers with the same characteristics, that is, to configure the bipolar taps to control the HV converter and LV converter separately and synchronously within the converters of same voltages.

(IV) Reactive Power Control

Reactive power control is used to control the AC filters/reactive power compensators of the entire station. Its main purpose is to calculate the reactive power consumption of the entire station based on the current operating mode and conditions. Through the control the switching of all reactive devices to ensure that the reactive power exchange between the entire station and the AC system is within the allowable range or the AC bus voltage is within the safe operating range. The safety of AC filter equipment and the reduction of harmonic impact on the AC system are also functions that reactive control must achieve.

For the layered accessed DC system at the receiving end converter station, the HV and LV converters of the sending end converter station are connected to the same AC grid. The reactive power control method and function are the same as in previous projects, with unified control of the reactive power of the entire station, with the reactive power exchange or voltage of the AC bus as the control target.

The HV and LV converters of the layered accessed converter station are respectively connected to 500 and 1000 kV AC grids. Each AC grid is equipped with an AC filter yard. The reactive power control functions need to be configured separately. The reactive power controls of the 500 and 1000 kV systems have their own control objects and control logic, with the reactive power exchange or AC voltage of their respective AC buses as the control target. Independent control is carried out according to the reactive power demand of the LV converter and the HV converter, respectively, and the difference between the reactive power consumption of the converter connected to each bus and the compensation of the AC filter is controlled within the dead zone range, specifically as follows.

$$|Q_{\text{acf}-500} - Q_{\text{cov11}} - Q_{\text{cov21}}| \leq Q_{\text{deadband}-500} \tag{10}$$

$$|Q_{\text{acf}-1000} - Q_{\text{cov12}} - Q_{\text{cov22}}| \leq Q_{\text{deadband}-1000} \tag{11}$$

where $Q_{acf-500}$ is reactive power generated by the 500 kV AC filter, MVar; $Q_{acf-1000}$ is reactive power generated by the1000 kV AC filter, MVar; Q_{cov11} is reactive power consumed by high HV converter of Pole 1; Q_{cov21} is reactive power consumed by HV converter of Pole 2; Q_{cov12} is reactive power consumed by low voltage converter of Pole 1; Q_{cov22} is reactive power consumed by low voltage converter of Pole 2; $Q_{deadband-500}$ is dead zone of 500 kV reactive power control; $Q_{deadband-1000}$ is dead zone of 1000 kV reactive power control.

The switching actions of the reactive power units of the two AC systems are asynchronous. With each filter switching, it may cause some disturbance to the voltage of the other AC system. Considering the short electrical distance between the two AC grids, there will be some mutually coupled effects. During DC system is operating, if the AC filter installed in any grid does not meet the absolute minimum filter requirements designed, power reduction will be initiated. Converters connected to different buses on the same pole are connected in series. If a large group of filters is lost on one bus, causing power reduction, it will affect the power of the converter connected to the other bus.

The HV and LV converters of each pole of the layered accessed converter station are connected to 500 and 1000 kV AC grids respectively. For this kind of main circuit, the switching operation of the converter may cause changes in the distribution of DC power between the two AC grids. Especially under high power conditions, it can cause a sudden change in the active power received by the two grids, that is, a sudden increase in the active power received by one grid, and a sudden decrease in the active power received by another grid. The change in active power causes a sudden surplus or deficit of reactive power exchanged between the converter stations and the AC grids. The surplus of reactive power may cause AC overvoltage, and the deficit of reactive power may cause AC undervoltage, as well as excessive harmonics caused by insufficient minimum filters.

In response to the above situation, the reactive power control module can provide such a control strategy. When it detects that a converter is being switched on or off, and the reactive power deficit is large, it can start the fast connecting filter function, shorten time interval of the reactive power control switching-on, and quickly connect the AC filter to compensate for the reactive power deficit. For cases of excess reactive power, the overvoltage fast removal filter function of previous UHV DC transmission projects can be used, and no additional control strategy is required.

(V) Stability Control

The stability control of DC system includes power modulation, frequency control, etc. When the AC system is disturbed, the stability control function restores stable operation as soon as possible by adjusting the transmission power of the DC system.

1. Power modulation

The active power modulation function is an additional control function of the DC pole control system. By adding a modulation value to the DC power command, the actual power transmitted by the DC transmission system can be affected to improve the performance of the entire AC/DC system.

The input signal for power modulation comes from the system's safety and stability control device, or is generated by monitoring the system's AC voltage and frequency.

All power modulation functions have corresponding enable and disable buttons on the operation interface, which can be used by operators to switch on or off as needed. The pole control system uses inter-station communication to send the stability modulation value generated on the inverter to the rectifier, and adds it to the modulation value generated on the rectifier to form the final stable control reference value. When inter-station communication between the two sides fails, the stability control function on the inverter is blocked.

(1) Power Increase. When the AC system on the inverter side loses power generation or the rectifier side occurs a load shedding fault, it may be required to quickly increase the power of the DC system to improve the performance of the AC system. Since the pole control system and the safety and stability control device use digital fiber optic communication, the power increase function receives the power increase signal and value sent by the safety and stability control device, and can achieve continuous modulation.

The power increase function is applied to the power command or current command, and increases the transmitted power by the selected increment. The power increase function can be used in both monopolar operation and bipolar operation. According to the combination of bipolar power control modes, the principles of power increase are as follows.

1) In the case where both poles are in bipolar power control mode, the increased power is distributed according to the voltage ratio of the two poles.
2) In the case where one pole is in pole current control or pole power independent control, and the other pole is in bipolar power control, if the pole in bipolar power control mode can meet the power increase requirement, only the power of that pole is increased; if that pole cannot meet the power increase requirement, the remaining power is supplied by the pole in pole current control or pole power independent control mode.
3) In the case where both poles are either pole current control or pole power independent control, the increased power is distributed according to the voltage ratio of the two poles, so that the unbalanced current does not exceed the level before the power increase.

When both poles operate in emergency pole current control mode, the power increase function can also be used, but only effective on the rectifier side. When the rectifier side receives the power increase signal and the increment, it increases its current command.

(2) Power reduction. When the AC system on the rectifier side loses power generation or the inverter side occurs a load shedding fault, it may be required to automatically reduce the DC transmission power. Whether in monopolar operation or in bipolar operation, the power reduction function can be used. Similar to power increase, the principles of power reduction are as follows.

1) In the case where both poles are in bipolar power control mode, the reduced power is distributed according to the voltage ratio of the two poles.
2) In the case where one pole is in pole current control or pole power independent control, and the other pole is in bipolar power control, if the pole in bipolar power control mode can meet the power reduction requirements, only the power of that pole is reduced. If the bipolar power control pole is cannot meet the power reduction requirements, the remaining power is supplied by the pole in pole current control or pole power independent control.
3) In the case where both poles are either pole current control or pole power independent control, the reduced power is distributed according to the voltage ratio of the two poles, so that the unbalanced current does not exceed the level before the power reduction.

During emergency pole current control, the rectifier side converter station reduces the current reference value at a safe rate.

For the layered accessed DC system, the HV and LV converters are connected to two AC grids of 500 and 1000 kV respectively, and the stability control of the two AC systems is relatively independent. The power modulation outputs of the two independent stability control modules are summed as the total power modulation.

2. Frequency control

During the operation of the DC transmission project, the AC grids at both ends may experience system frequency offset, higher or lower than the rated value of 50 Hz. For frequency offset, the system provides a frequency control function. This function automatically increases or decreases the DC transmission power according to the frequency deviation to maintain system stability. The frequency control function can be put in or out by operators on the operation interface.

The frequency control function is essentially a closed-loop proportional integral controller of difference between the actual system frequency and the rated frequency. When the system frequency exceeds the set frequency dead zone, the frequency limit controller calculates the power adjustment value in real time, controls the DC transmission power in real time, and finally controls the frequency to the dead zone through this closed-loop control.

The controller parameters of the frequency control function will be determined based on the additional control function research, and the coordination with the control parameters of the generator governor and other related controls should be taken into consideration.

The frequency control functions of the rectifier side and the inverter side are logically independent. The operators on both sides can control the switching on and off of the frequency control function on each side through the buttons on the operation interface. When the frequency control function of both stations is put into operation, the power modulation values output by the integral controllers of the frequency difference on both sides are accumulated on the rectifier side to obtain the final frequency control power.

For the layered accessed DC system, the converters of the non-layered accessed end connect the same AC grid, and the corresponding grid frequency is selected for frequency control; the converter of the layered accessed end connect different AC grids, and the grid frequency with larger frequency deviation is selected for frequency control.

(VI) Power Transfer and Rapid Reduction

1. Power transfer

In previous UHV DC transmission projects, the power transfer between converters and between poles is determined by the control mode.

(1) In the single pole independent current control mode, the switching of a converter on the same pole will not cause a change in the current of the pole, and blocking or converter switching of the opposite pole should not cause a change in the current of this pole.
(2) In the single pole independent power control mode, if a converter on the same pole is disconnected, in order to maintain the power of the pole unchanged, the power of the disconnected converter will be transferred to another operating converter on the same pole. If a converter on the same pole is put into operation, the current of the pole will decrease to maintain the power unchanged.
(3) In the bipolar power control mode, the way of power transfer depends on the power control mode of the opposite pole, which is more complex. If the opposite pole is also in bipolar power control, the power transfer of the switching converter will be distributed among all operating converters of the two poles according to the voltage ratio, to maintain the balance of bipolar current. If the opposite pole is also in single pole power control or current control, the pole will supply the total power command minus the power of the opposite pole.

For HV converter and LV converter connected to AC grids of different voltage levels, if the power transfer distribution method mentioned above is followed, it may cause one grid to lose some delivered power, while the transmission power of another grid increases. For example, when the pole is in the single pole power control mode and the HV converter is disconnected, according to the previous principle, the power loss will be transferred to the LV converter, which will cause the power loss of the 500 kV grid to be transferred to the 1000 kV grid. It should be systematically investigated whether the power transfer has an impact on the stability of the two AC grids, and the power transfer distribution control function should be adjusted accordingly.

2. Power rapid reduction

For a layered accessed DC system at the receiving end, when the absolute minimum filter of a certain layer does not meet the requirements causing a power drop, due to the series connection of HV and LV converters of the same pole, it will cause another layer to drop power. In order to reduce the impact on the power of another pole or another layer, certain measures should be taken. In the bipolar half-voltage (single

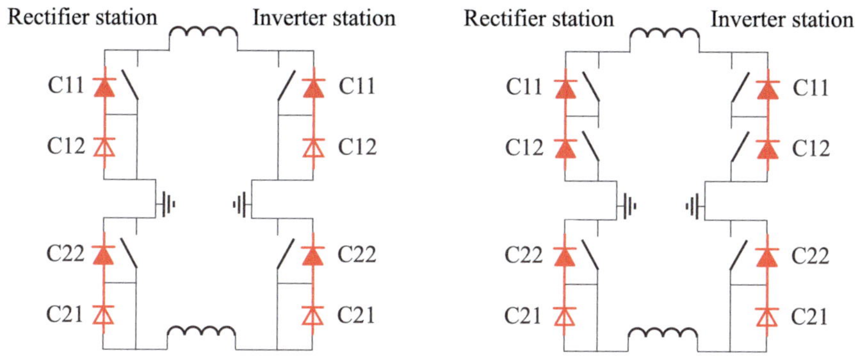

(a) Bipolar half-voltage

(b) Bipolar mixed-voltage

Fig. 36 Schematic diagram of bipolar half-voltage and bipolar mixed-voltage operating modes of the DC system

converter on both poles) and bipolar mixed-voltage (double converters on one pole, single converter on the other) operating modes, when the absolute minimum filter does not meet the requirements for power drop, the grounding pole current balance requirement can be disregarded, and the pole of power reduction can be converted to single pole power control mode. Figure 36 shows the schematic diagram of bipolar half-voltage and bipolar mixed-voltage operating modes of the DC system.

(1) Bipolar half-voltage operation.

   1) When the absolute minimum filter of the 500 kV AC grid does not meet requirement for power drop, convert Pole 1 to single pole power control mode, only reduce the power of Pole 1, and the power of Pole 2 remains unchanged.
   2) When the absolute minimum filter of 1000 kV AC grid does not meet the requirement for power drop, convert Pole 2 to single pole power control mode, only reduce the power of Pole 2, and the power of Pole 1 remains unchanged.

(2) Bipolar mixed-voltage operation.

   1) When the absolute minimum filter of the 500 kV AC grid does not meet requirement for power drop, convert Pole 1 to single pole power control mode, prior reduce Pole 1's power, and the power of Pole 2 remains unchanged. When the absolute minimum filter does not meet the requirement and a converter must be blocked, prior block the Pole 2 HV converter to ensure that the grounding pole current is minimized.

2) When the absolute minimum filter of the 1000 kV AC grid does not meet requirement for power drop, convert Pole 2 to single pole power control mode, only reduce the power of Pole 2 while the power of Pole 1 remains unchanged.

(VII) Converter Online Switching

In UHV DC transmission projects, the online conversion of operation modes can be achieved through the switching sequence of the converters.

The online switching of the converter should not interrupt the normal operation of another converter. At the same time, the disturbance to the DC power transmission should be as small as possible to avoid excessive impact on the entire power grid.

When inter-station communication is normal, the converter switching commands are issued by the master controller, and the control sequence of the two converter stations is coordinated through inter-station communication.

When the converter is put into operation without inter-station communication, the two stations issue converter unlock commands separately. And the operation personnel coordinate the unlocking sequence of the two stations by telephone. The rectifier station unlocks first, and the inverter station unlocks later. When there is no inter-station communication, if a single converter is disconnected due to faults, the other station will automatically disconnect the LV converter of the pole through the converter imbalance protection function.

The basic strategy for the online converter switching in the layered accessed DC system at the receiving end is the same as that in conventional projects. However, since the receiving end is layered accessed to 500 and 1000 kV AC grids, in order to minimize the operation of the 1000 kV AC circuit breaker, under normal circumstances, if the double converters are required to exit a single for any reason, the HV converter will be disconnected by default, but at the same time, the pre-selection function for the converter disconnection should be added. Therefore, the complexity of the valve group switching has increased.

(1) When the non-layered accessed converter station disconnects a converter, the converter station with layered access should be able to choose whether to use the pre-selection function of the converter disconnection.
(2) After the pre-selection function is put into operation, the converter station with layered access defaults to disconnect the HV converter, and can also choose to disconnect the LV converter.
(3) In the case of the deactivation of the pre-selection function, the principle of converter disconnection is consistent regardless of whether the converter stations are connected in layered access manner, i.e. "HV converter for HV converter, LV converter for LV converter."
(4) If the converter connection is unsuccessful, the converter station with layered access disconnects the converter according to the principle of "the last one in, the first one out".
(5) In the past projects, when the converter disconnection was unsuccessful, the converter was blocking according to the principle of "HV converter for HV

converter, LV converter for LV converter". For the converter station with layered access, this principle may cause changes in the power of the two AC grids of 500 and 1000 kV. Therefore, if the converter connection is unsuccessful at the converter station with layered access, the converter is disconnected according to the principle of "the last one in, the first one out".

(VIII) Open Line Test

In order to test the insulation level of the DC pole after a long period of shutdown or maintenance, the DC pole control system has an open line test function, which has the following operating modes.

(1) Open line test of a single converter without line.
(2) Open line test of a single converter with line.
(3) Open line test of two converters in series in single pole without line.
(4) Open line test of two converters in series in single pole with line.

The DC voltage U_{d} produced after the rectification of the 12-pulse bridge peak on the rectifier side can be represented as

$$U_{\mathrm{d}} = \frac{4\pi}{3\sqrt{3}} U_{\mathrm{di0}} \cos(\alpha - 60) \tag{12}$$

Formula 12 indicates that DC voltage starts to rise when α is 150°, and the voltage reaches its maximum when α is 60°. The formula only holds when there is no line (the DC current is zero). If an open line test is conducted with a line, the corona loss and other losses will reduce the DC voltage, and closed-loop control will decrease α to compensate for the voltage drop.

The open line test for pole of conventional UHV DC project is controlled by a unified pole, calculating the firing angle and sending it to the HV and LV converters. Since the parameters of the HV and LV converters are identical, the voltages automatically maintain balanced operation.

The HV and LV converters of the layered accessed converter station are connected to AC grids of different voltage levels. The corresponding converter transformers have different numbers of tap positions and steps, so they cannot be controlled by a unified pole. The HV and LV converters are independently controlled, and their voltage control targets are half of no-load voltage target value for each pole, ensuring balanced operation of the two converters.

(IX) Commutation Failure Prediction

The HV and LV converters of the layered accessed UHV DC system at the receiving end are connected to two AC grids of 500 and 1000 kV, respectively. The two AC grids are connected through an AC interconnecting transformer, creating a certain degree of coupling.

When the two AC grids of the layered access are tightly coupled, an AC fault in one grid will affect the other, distorting the AC voltage of both grids. Using the zero-sequence voltage method and α/β transformation method for commutation failure prediction control, similar with conventional DC transmission projects, the two converters connected to the two AC grids can synchronously detect the AC fault, start the commutation failure prediction control simultaneously, and increase the extinction angle to prevent commutation failure.

When the two AC grids of the layered access have low coupling or no coupling, an AC fault in one grid will not significantly affect the other. Using the zero-sequence voltage method and α/β transformation method for commutation failure prediction control, similar with conventional DC transmission projects, the converter connected to the AC grid where the fault occurred can start the commutation failure prediction control in time, while the converter connected to the other AC grid cannot start the commutation failure prediction control simultaneously.

Since the HV and LV converters are connected in series, when there is a fault in the inverter side, the DC current will increase rapidly, which will cause the commutation overlap angle of the other converter to increase and the extinction angle to decrease. If the converter cannot start the commutation failure prediction control in time, it will increase the probability of commutation failure in that converter.

The ± 1100 kV UHV DC transmission project has a longer distance than the ± 800 kV projects. When operating in bipolar mode, there is a more severe electromagnetic coupling relationship between the two pole lines on the same tower. This coupling relationship does not have a serious impact on the steady-state operation of DC, but in the event of a single pole fault, the DC voltage of the non-fault pole will experience a large disturbance, causing a large fluctuation in the DC current, increasing the probability of commutation failure in the converter of the non-fault pole.

In response to the above situation, it is necessary to make corresponding modifications to the commutation failure prediction control function, and the following two functions can be added to the converter control of the layered accessed converter station.

(1) Simultaneously configure two commutation failure prediction control module. The voltage signals of the two AC grids at the receiving end are simultaneously connected to the HV and LV converters, and the control hosts of the converters are each equipped with two commutation failure prediction control modules, which simultaneously detect faults in the two AC grids. When any module commutation failure prediction control is started, the extinction angle is immediately increased to prevent the occurrence of commutation failure. For the two closely coupled AC grids at the receiving end, the angle increment takes the larger value of the two prediction module calculations.
(2) Adopt a strategy based on the difference between the measured value and the command value of the DC current. As the impact of the DC current caused by a single converter or single pole failure, which causes the commutation failure of another converter in series or another pole operating in the same tower. In the layered accessed DC system, a commutation failure prediction function is added, which uses the difference between the measured DC current and the

set current value as the criterion. After the criterion is activated, the extinction angle is immediately increased, thereby increasing the converter's commutation margin and improving the converter's ability to resist commutation failures.

III. Research on DC Protection Configuration

(I) Protection Zone and Function Configuration of Layered Accessed UHV DC Project

The protection of the layered accessed UHV DC system is divided into DC protection, converter transformer protection, and AC filter protection. Among them, DC protection can be further divided into converter zone protection, pole zone protection, bipolar zone protection, DC line zone protection, and DC filter zone protection. As shown in Figs. 37, the converter transformer protection zone (① and ②), converter protection zone (③, ④, and bypass switch of ⑥), pole protection zone (⑤, ⑦, and converter connection zone of ⑥), bipolar protection zone (⑨ and ⑫), DC line protection zone (⑩ and ⑪), DC filter protection zone (⑧), and AC filter protection zone (⑬ and ⑭).

1. Converter zone protection

Converter zone protection mainly includes valve short-circuit protection, converter commutation failure protection, converter differential protection, converter overcurrent protection, converter transformer valve side neutral shift protection, converter bypass switch protection, converter bypass pair overload protection, converter harmonic protection, and converter DC overvoltage protection. There are also some protective control functions configured in the control host, including converter voltage overstress protection, converter trigger abnormal protection, converter large angle monitoring, converter thyristor junction temperature monitoring, and AC system low voltage detection, etc.

The typical converter zone measurement points and protection configuration are shown in Fig. 38. The protection is configured according to the HV and LV converter configuration, which are identical and distributed in their respective protection host. The measurement points in this protection zone include: I_{DC1P} (DC current of HV converter at high-voltage side), I_{DC1N} (DC current of HV converter at low-voltage side), I_{VYH} (Three-phase AC current on the Y-connected valve side of the HV converter transformer), I_{VDH} (Three-phase AC current on the D-connected valve side of the HV converter transformer), U_{VYH} (Three-phase voltage of AC bushing on the Y-connected valve side of the HV converter transformer) and U_{VDH} (Three-phase AC bushing voltage on the D-connected valve side of the HV converter transformer) as the measurement points for the HV converter zone; I_{DC2P} (DC current of LV converter at high-voltage side), I_{DC2N} (DC current of LV converter at low-voltage side), I_{VYL} (Three-phase AC current on the Y-connected valve side of the LV converter transformer), I_{VDL} (Three-phase AC current on the D-connected valve side of the LV converter transformer), U_{VYL} (Three-phase AC bushing voltage on the Y-connected valve side of the LV converter transformer) and U_{VDL} (Three-phase

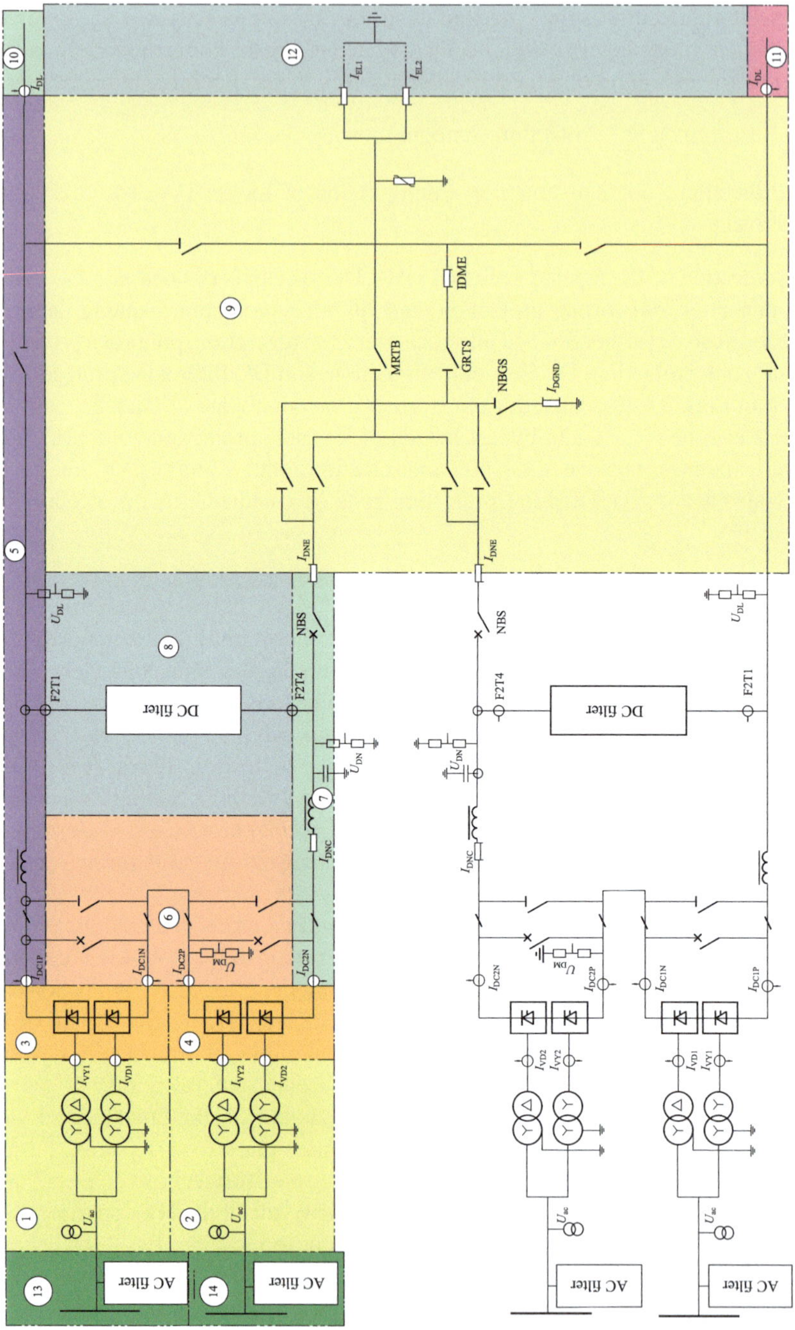

Fig. 37 Protection zones of layered accessed UHV DC transmission project

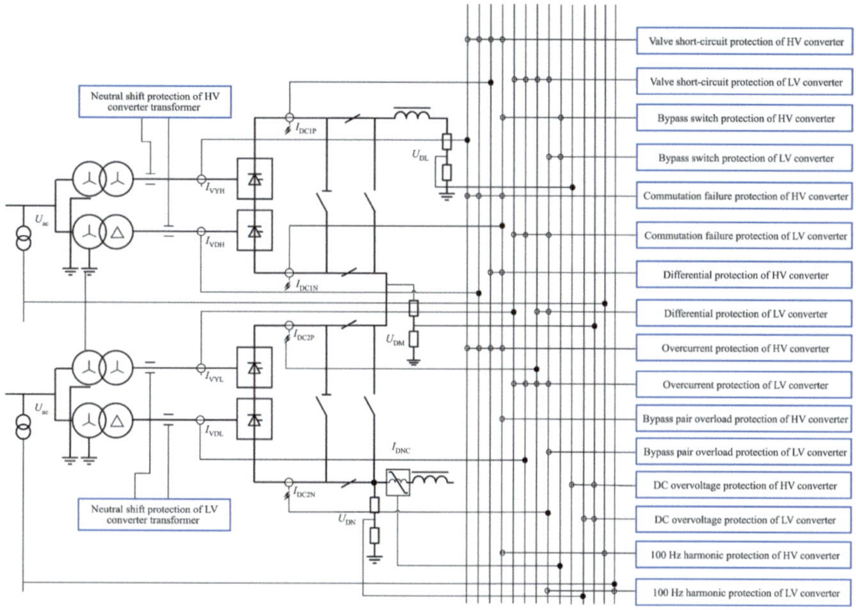

Fig. 38 Protection measuring points and configuration in the converter zone

AC bushing voltage on the D-connected valve side of the LV converter transformer) as the measurement points for the low voltage converter area; and I_{DNC} (Current on the smoothing reactor side in the pole neutral bus).

2. Pole zone protection

Pole zone protection mainly includes pole bus differential protection, pole neutral bus differential protection, converter connection line differential protection, pole differential protection, grounding pole line open-circuit protection, 50/100 Hz harmonic protection, neutral bus switch protection, DC overvoltage protection, DC low voltage protection, open-circuit test protection, surge capacitor overcurrent protection and AC/DC line touch supervision. The typical measurement points and configuration of the pole zone protection are shown in Fig. 39. The measurement points of this protection area include: I_{DC1P} (high-voltage side DC current of HV converter), I_{DC1N} (low-voltage side DC current of HV converter), I_{DC2P} (high-voltage side DC current of low voltage converter), I_{DC2N} (low-voltage side DC current of low voltage converter), I_{DL} (DC line current), U_{DL} (DC line voltage), U_{DN} (neutral bus voltage), I_{DNE} (neutral bus switch current), I_{ZxT1} (DC filter head-end current), I_{ZxT2} (DC filter tail-end current), I_{AN} (neutral bus arrester current), I_{CN} (neutral bus surge capacitor current).

3. Bipolar zone protection

Bipolar zone protection mainly includes bipolar neutral bus differential protection, station grounding overcurrent protection, backup station grounding overcurrent

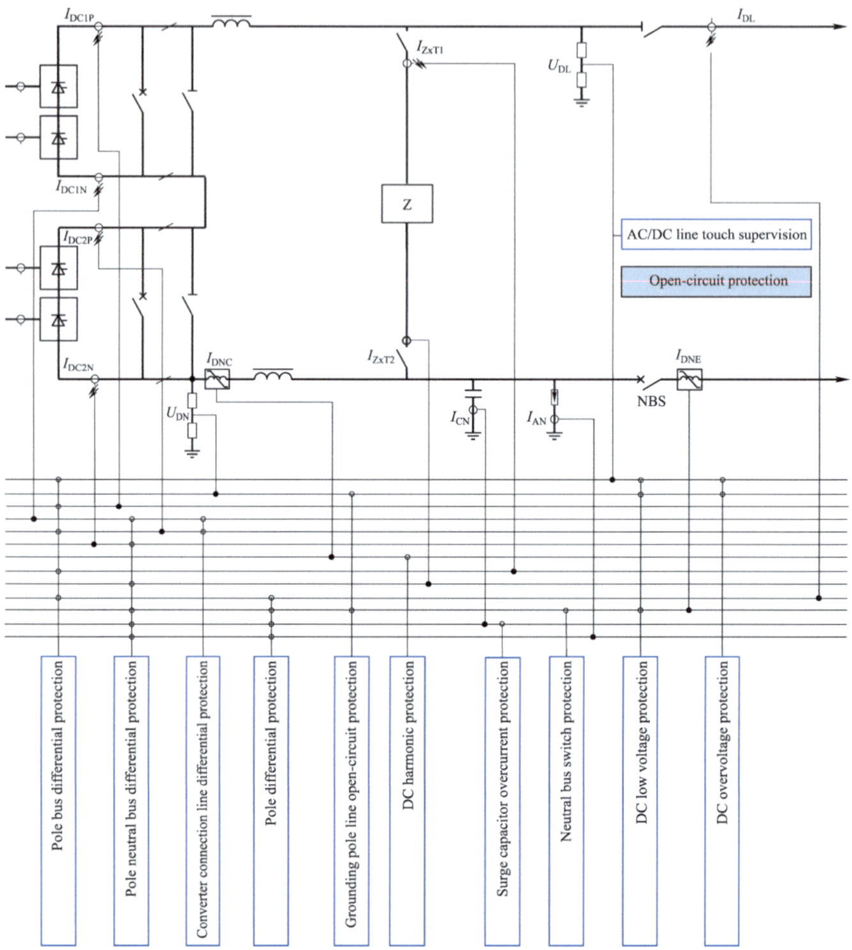

Fig. 39 Protection measuring points and configuration in the pole zone

protection, station earthing switch protection, ground return transfer switch protection, metallic return line transfer breaker protection, metallic return line grounding protection, metallic return line transverse differential protection, metallic return line longitudinal differential protection, grounding electrode line overcurrent protection, grounding electrode line unbalance protection, and grounding electrode line differential protection.

The typical measurement points and configuration of the bipolar zone protection are shown in Figs. 40. The measurement points used in this protection zone include: I_{DL} (DC line current), I_{DNE} (neutral bus switch current), I_{DNE_OP} (opposite pole neutral bus switch current), I_{DME} (metallic return line switch current), I_{DGND} (station grounding switch current), I_{DEL1} (zero flux current of grounding electrode line 1),

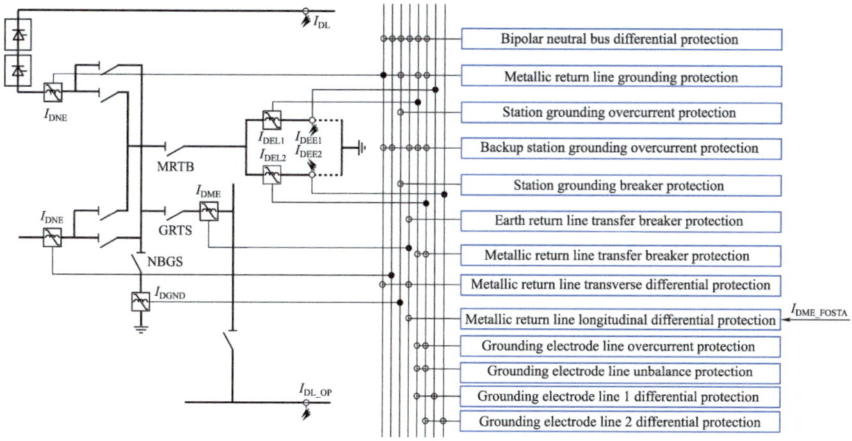

Fig. 40 Protection measuring points and configuration in the bipolar zone

I_{DEL2} (zero flux current of grounding electrode line 2), I_{DEE1} (optical TA current of grounding electrode line 1), I_{DEE2} (optical TA current of grounding electrode line 2).

4. DC line zone protection

DC line zone protection mainly includes traveling wave protection, derivative protection, DC line low voltage protection, and DC line longitudinal differential protection.

The typical protection measurement points and configuration in the DC line zone are shown in Fig. 41. The measurement points used in this protection zone include: I_{DL} (DC line current), U_{DL} (DC line voltage), I_{DL_OP} (opposite pole DC line current), U_{DL_OP} (opposite pole DC line voltage), I_{DL_FOSTA} (opposite station DC line current).

5. DC filter zone protection

DC filter zone protection mainly includes differential protection, high voltage capacitor grounding protection, high voltage capacitor imbalance protection, resistance overload protection, reactance overload protection, and DC filter detuning protection. The typical protection measurement points and configuration in DC filter zone are shown in Fig. 42.

6. Converter transformer zone protection

Converter transformer zone protection can be divided into five major categories: differential protection, overcurrent protection, overvoltage protection, overexcitation protection, and saturation protection. Differential protection is the main protection, and the others are all backup protections. Among them, differential protection can be divided into large differential protection, small differential protection, lead differential protection, and winding differential protection. Overcurrent protection can be divided into switch overcurrent protection, grid-side overcurrent protection,

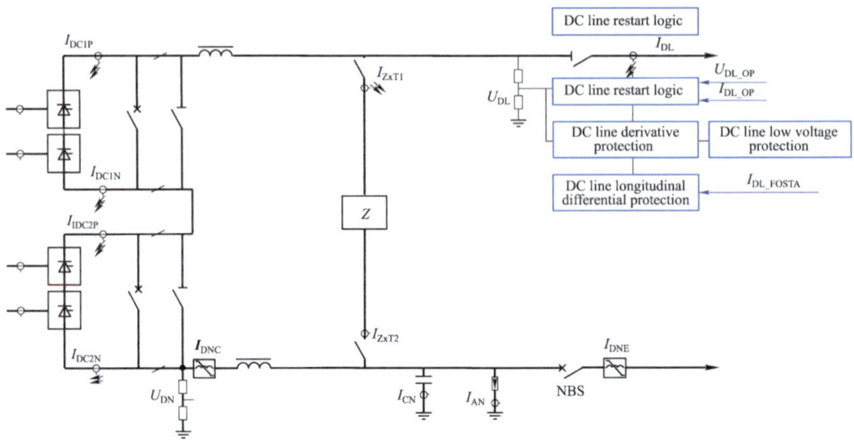

Fig. 41 Protection measuring points and configuration in the DC line zone

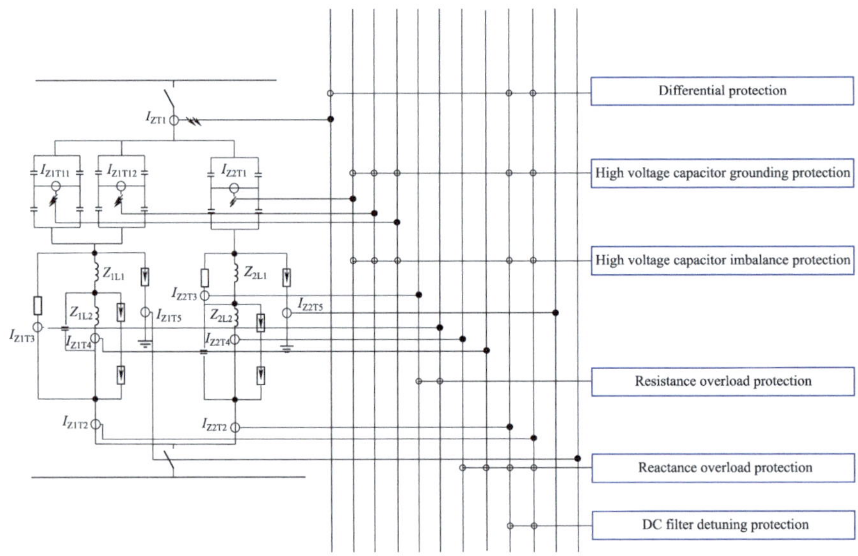

Fig. 42 Protection measuring points and configuration in the DC filter zone

and zero-sequence overcurrent protection. Over-excitation protection can be divided into definite-time over-excitation protection and anti-time over-excitation protection. Small differential protection can also be divided into Y-connection and delta-connection. Winding differential protection can also be divided into YY grid side, YD grid side, YY valve side, and YD valve side. Grid-side overcurrent protection and zero-sequence overcurrent can both be divided into YY and YD.

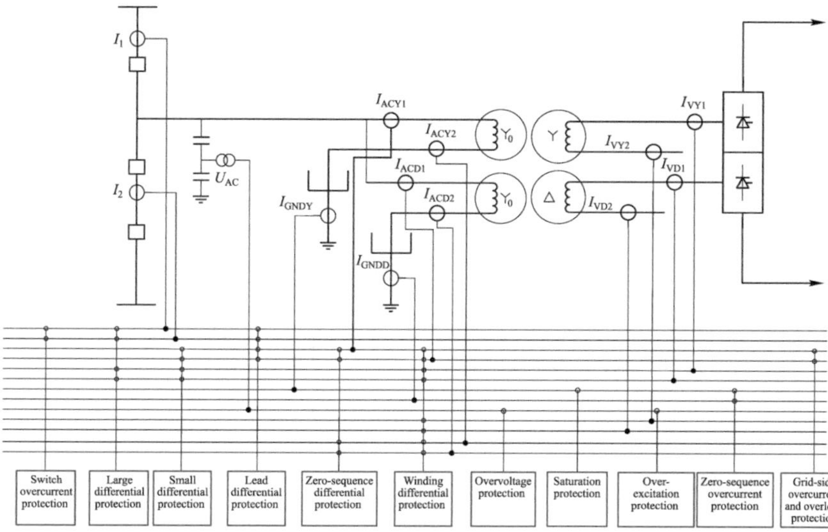

Fig. 43 Protection measuring points and configuration in the converter transformer zone

The principles of large differential protection and small differential protection are similar, both using concepts of differential instantaneous tripping, proportional differentiation, and power frequency variation differentiation. Winding differential protection is divided into phase differential protection and zero-sequence differential protection. Lead differential and YY/YD valve-side winding differential protection do not configure zero-sequence section, because the harmonic content inside the protection zone is relatively large, which is easy to misoperate. Figure 43 shows the measurement points and protection configuration in the converter transformer zone.

7. AC filter zone protection

AC filter zone protection includes AC filter sub-bank protection and AC filter bus protection, where the functions of sub-bank protection include differential protection, overcurrent protection, zero-sequence overcurrent protection, capacitor imbalance protection, resistance overload protection, reactance overload protection, and detuning alarm. The typical protection measurement points and configuration in the AC filter sub-bank are shown in Fig. 44.

AC filter bus protection mainly includes bus differential protection, bus overvoltage protection, bus overcurrent protection, and failure protection. The typical protection measurement points and configuration in the AC filter bus are shown in Fig. 45.

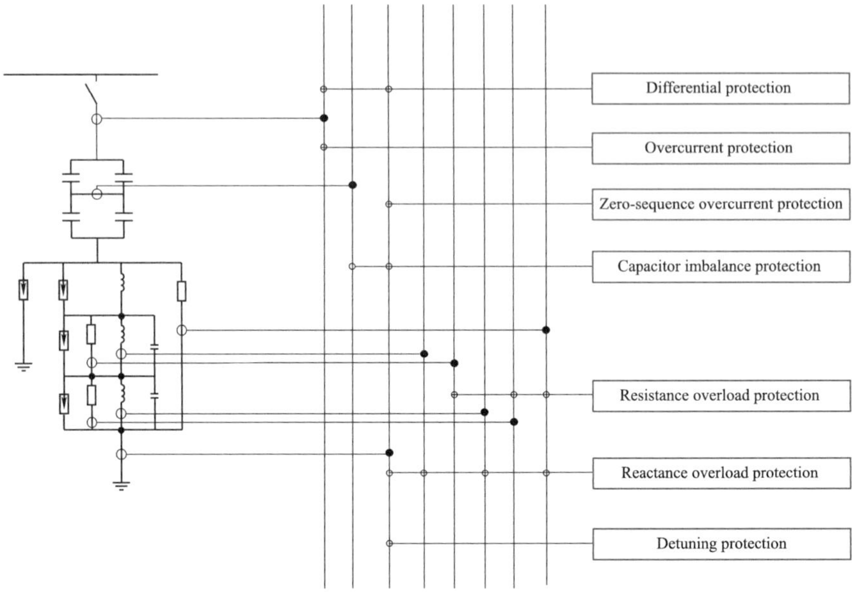

Fig. 44 Protection measuring points and configuration in the AC filter sub-bank

(II) Configuration of Special Protection Functions for Layered Accessed UHV DC Project

The DC protection zoning, measurement point configuration, and protection functions of the layered accessed UHV DC project is basically consistent with conventional UHV DC projects. The difference is that the HV and LV converters of the layered accessed converter station are connected to AC systems of different voltage levels, and a voltage measurement point has been added in the midpoint of HV and LV converters within a pole. Due to the specificity of layered access, adaptive changes has been made in DC protection in two aspects: first, the converter DC overvoltage protection has been added in the converter zone of the layered accessed converter station, replacing the pole overvoltage protection; second, in order to prevent the expansion of the scope of fault action, the DC harmonic protection function of the pole area is assigned to the converter protection level, and only the converter connected to the faulted AC system is cleared when occurring faults. The protection function of the converter station without layered access is consistent with conventional UHV DC projects.

1. Converter DC overvoltage protection

The layered accessed converter station is equipped with a midpoint divides between HV converter and LV converter, which can realize real-time calculation of the DC voltage at both ends of the converter. Therefore, the converter DC overvoltage protection is added at the converter level. The principle is shown in Table 48.

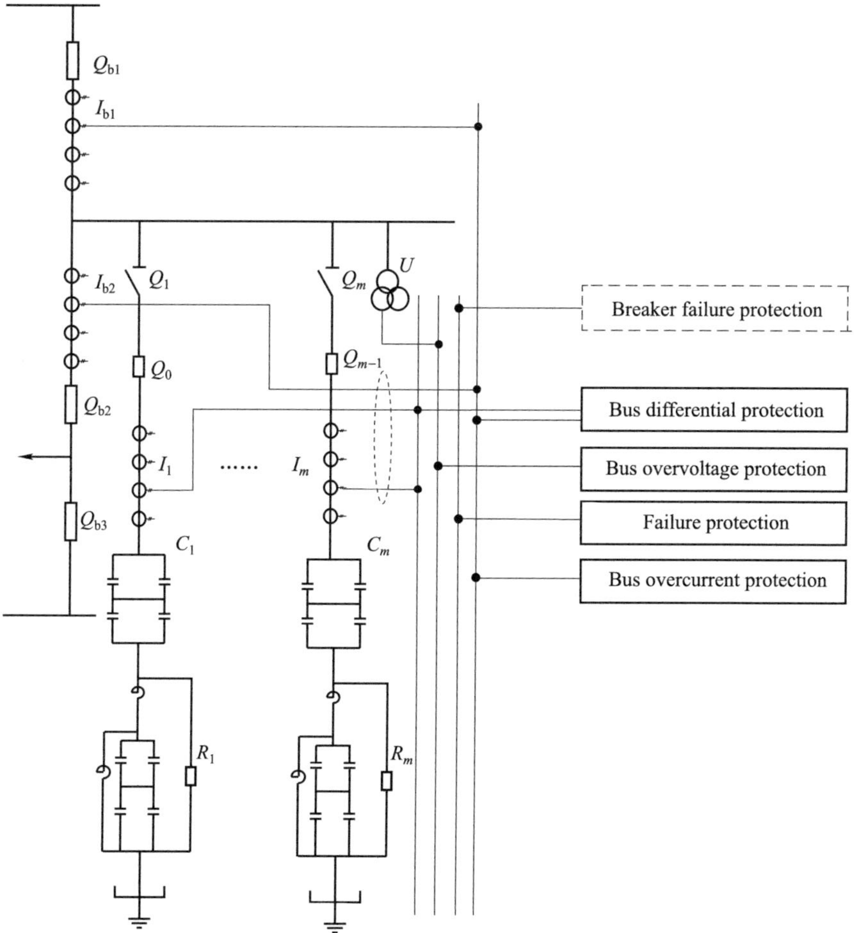

Fig. 45 Protection measuring points and configuration in the AC filter bus

The principle of converter DC overvoltage protection is basically the same as that of pole DC overvoltage protection, but the protected zones are different. The converter DC overvoltage protection protects the equipment in the converter zone, while the pole layer DC overvoltage protection protects all the equipment in the pole zone.

Consequence of protective action: request control system to switch; phase-I protection activated, inverter Y blocked; phase-II protection activated, inverter X blocked; AC circuit breaker tripped; start failure; AC circuit breaker locked; fault recording triggered.

Table 48 DC overvoltage protection of converter

Targeted failure	Protect all devices in the converter zone from overvoltage caused by various reasons
Protection principle	HV converter $\|U_{DL} - U_{DM}\| > U_{D\_set}$ LV converter $\|U_{DM} - U_{DN}\| > U_{D\_set}$
Protection zone	2
Protection coordination	Insulation coordination with primary equipment. The measuring point U_{DM} is added between the HV and LV converters. The converter DC overvoltage protection and the pole DC overvoltage protection are basically the same in principle, but the protected zones are different. The DC overvoltage protection protects the equipment in the converter zone, and the pole DC overvoltage protection protects the equipment in the entire pole zone
Backup protection	Voltage control function in the control system
Communication dependent	No
Recorded data	All voltages and protection actions
Protection action consequence	(1) Request control system switchover (2) Converter blocking (3) AC circuit breaker tripping (4) Start failure (5) Lock AC circuit breaker (6) Triggering fault recording

2. Harmonic protection

Considering that harmonic protection is configured according to the withstand capacity of equipment, in the case of strong coupling of the two AC grids on the layered accessed convert station, it is impossible to accurately locate the harmonic source position relying on the AC voltage characteristic quantity. In order to minimize the risk of single pole power loss, the 50 Hz harmonic protection and 100 Hz harmonic protection on the layered access side are assigned to the converter protection host.

The harmonic protection setting is based on the principle of preventing current interruption when the load is small and matching the harmonic overload capacity of the device when the load is large. The principle of converter harmonic protection is shown in Table 49.

(III) Strategy for Measurement Fault

The voltage balance control of the UHV DC system with layered access at the receiving end needs to monitor the voltage of the HV and LV converters in real time, and the converter voltage is calculated from the DC pole line voltage U_{DL}, neutral line voltage U_{DN}, and the newly introduced midpoint divider voltage U_{DM}, as follows.

Voltage of HV converter

$$U_{conv11} = U_{DL} - U_{DM} \tag{13}$$

Table 49 Harmonic protection of converter

Targeted failure	Mainly used as a backup protection when the AC system cannot be disconnected due to an asymmetrical fault
Protection principle	HV and LV converter: $I_{DCxN\_100Hz} > I_{set}$ and, $U_{ac} < U_{ac\_set}$ The 100 Hz harmonic protection of the converter layer is only enabled in the layered accessed converter station, and the low signal for determining the AC system voltage is added according to the different accessed AC system
Protection zone	2
Protection coordination	It is coordinated with the tripping time of the AC system and the overstress capacity of the valve. It prevents current interruption when the system is lightly loaded and is coordinated with the harmonic overload capacity of the device when it is heavily loaded
Backup protection	It acts as backup protection
Communication dependent	No
Recorded data	I_{DCxN_100Hz} and protection actions
Protection action consequence	(1) Request control system switchover (2) Converter blocking (3) AC circuit breaker tripping (4) Start failure (5) Lock AC circuit breaker (6) Triggering fault recording

Voltage of LV converter

$$U_{conv12} = U_{DM} - U_{DN} \tag{14}$$

where U_{DL} is DC pole line voltage, kV;

U_{DM} is midpoint divider voltage, kV;

U_{DN} is neutral line voltage, kV.

From formula (13) and formula (14), it can be seen that once U_{DM} fails or the output measurement value is incorrect, it will directly affect the voltage balance control of the HV and LV converter, and then affect the stable operation of the DC system. Therefore, in order to ensure that the DC system will not be blocked by the converter due to the control system action when the midpoint voltage divider fails, the converter voltage calculation value is introduced into the control.

During normal operation, the converter voltage on the inverter side can be calculated by the following formula

$$U_{conv} = U_{di0} \cos\gamma - (d_x - d_r) \cdot \frac{U_{di0N}}{I_{dcN}} \cdot I_{dc} \tag{15}$$

where U_{conv} is converter voltage, kV; U_{di0} is ideal no-load DC voltage, kV; d_x is relative inductive voltage drop; d_r is relative resistive voltage drop; I_{dc} is direct current, kA; γ is extinction angle, (°); U_{di0N} is nominal ideal no-load DC voltage, kV; I_{dcN} is nominal value of direct current, kA.

When the control system detects an abnormal measurement value of U_{DM}, it issues an alarm and switches the system. If the abnormality still exists, the control system uses the calculated value of U_{DM} to replace the measured value for control, achieving voltage balance control of the HV and LV converter, which can avoid system operation problems caused by U_{DM} measurement faults.

IV. Overall Design Scheme

(I) DC Control System Design Scheme

1. Control system function hierarchy and configuration

(1) Design principles of function hierarchy. For the connection method of double 12-pulse converter series structure in each pole of UHV DC projects, to improve the reliability and availability of the DC system, the design of the DC control and protection system meets the following principles.

   1) The control and protection is configured based on each 12-pulse converter as a basic unit. The implementation of the control functions and protection configuration of each 12-pulse converter are independent of each other, which is conducive to the independent disconnection of a single 12-pulse converter without affecting the normal operation of others. At the same time, the physical connections between each 12-pulse control and protection system should be simplified as much as possible.
   2) The failure of a single component in the control and protection system cannot cause any 12-pulse converter in the DC system to stop operating.
   3) In the event of a fault in a higher-level control unit, the 12-pulse control unit still has the ability to maintain the running state of the DC system and continuing to run or exiting operation according to the operator's instructions.
   4) Any circuit fault or measurement device fault of a pole or converter should not affect another pole or converter on the same pole through the signal exchange interface between converters, the signal exchange interface with other control levels, or the device power supply. When a pole or converter device is under maintenance (including withdrawal from operation, maintenance, and re-entry stages), it will not impose any restrictions on the operating mode of the other pole or converter on the same pole, nor will it cause any control mode or function of the other pole or converter on the same pole to fail and let alone cause the other pole or converter on the same pole to stop operating.

(2) Hierarchy structure and function configuration. For converter station whose HV and LV converters are connected to the same AC system, its control and protection system is configured in accordance with the bipolar, pole and converter

level of the conventional UHV DC system, and is functionally divided into the station control/bipolar control level, pole control level and converter control level. For the UHV DC converter station with layered access to two AC grids, since its HV and LV converters are also within the same station, it still adopts the structure of bipolar level control, pole level control, and converter level control. However, due to the layered access, there is a significant difference within each control level compared to conventional UHV DC project, especially in the station control/bipolar control level, where it is necessary to add separate control of the reactive power of the two AC systems.

Figure 46 shows the schematic diagram of the hierarchy structure of the DC control system. The system level controller implements control functions for the AC and DC systems, the regional level implements control functions for more than one station, such as the master/slave station selection function. And the control functions of the station control/bipolar control level, pole control level, and converter control level are as follows.

Station/Bipolar control level: The level controls two poles simultaneously in a DC transmission system, which implements the calculation and distribution of power/current commands, the coordination of current commands between stations, the

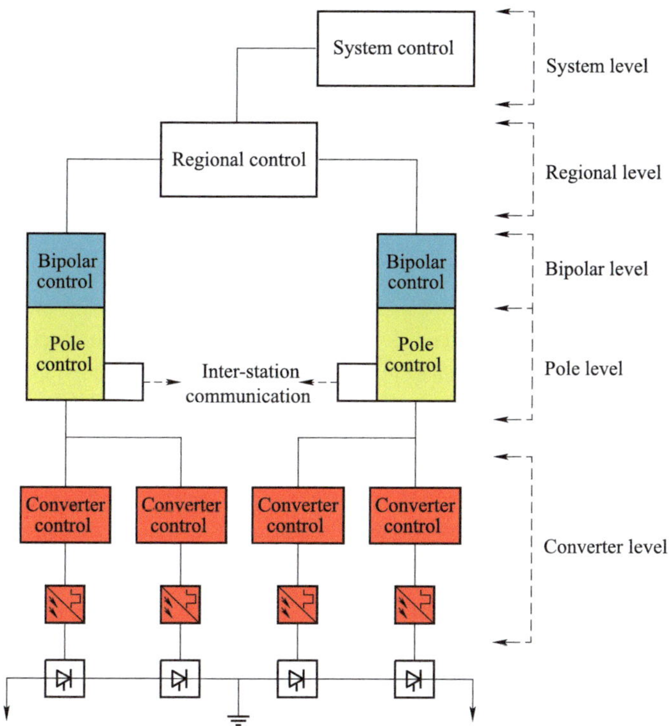

Fig. 46 Schematic diagram of the hierarchy structure of the DC control system

switching control of station reactive equipment, and the station DC sequence control function. Functions related to bipolar control are implemented as much as possible at the pole level to ensure that any single-circuit fault will not disturb both poles. The main functions include: ① setting the power set value of the bipolar; ② bipolar current balance control; ③ inter-pole power transfer control; ④ independent control of reactive power and AC bus voltage of the two AC systems; ⑤ interface of safety and stability control equipment, etc.

Pole control level: The main functions of the pole control level include: ① providing the current set value to the converter level after calculation, controlling the current of the DC transmission system. The current set value of the master control station is given by the power control module or manually set, and transmitted to the slave control station through communication device. ② DC transmission power control. ③ Pole start and stop control. ④ Fault handing and control, including retard and block and automatic restart control, low-voltage current limiting control, etc. ⑤ Remote control and communication between the same pole of each converter station, including the transmission of current set value and other continuous control information, the transmission of AC and DC equipment operation status information and measurement values, etc.

Converter control level: The level in the DC implements the necessary control functions and valve triggering functions for the operation of the converter, mainly including closed-loop control of DC current, DC voltage, converter voltage, converter voltage balance, extinction angle, as well as deblocking and blocking functions of the converter. The converter control also has a manual function of current rise and fall, which serves as a backup function in the event of a fault in the bipolar/pole control host. The main control functions include converter trigger control, constant current control, constant extinction angle control, DC voltage control, converter voltage control, converter voltage balance control, limit control of maximum and minimum value of firing angle, DC voltage, DC current, as well as converter blocking and deblocking sequence control, etc. For converter stations that are not layered accessed, as the HV and LV converters of the same pole are connected to the same AC grid, the closed-loop control functions of DC current, DC voltage, and extinction angle in the converter control level can also be configured in the pole control level, and the converter control level only receives the firing angle signal issued by the upper level. For layered accessed converter stations, the firing angle of each 12-pulse converter is independently controlled, and the closed-loop control of DC current, DC voltage, and extinction angle is generally located in the converter control, while configuring the converter voltage balance control module to achieve the goal of balanced operation of the converter voltage, without bringing over-stress to the main equipment.

2. Redundancy of the control system

The control system is designed according to the principle of complete dualization at all levels, divided into system A and system B. The scope of dualization starts from the secondary winding of the measurement and includes the complete measurement circuit, signal input/output circuit, communication circuit, host, and all related DC control devices.

The pole control system implements the functions of the system, region, bipolar, and pole control levels, while the converter control system implements the functions of the converter control level. Additionally, a separate station control system can be added to implement the functions of the system, region, and station/bipolar control levels. To meet the requirements for high reliability, both the pole control system and the converter control system adopt redundant configurations. The redundant pole control system and converter control system achieve complete control functions through cross-connection. Only the system currently in the active state communicates effectively with the control system of another converter or pole control system, ensuring that in the event of a failure in any one control system, there is no restriction or unavailability of the function of the other level of control system (pole control system or converter control system) or of the two control systems of the other converter that are not faulty. During system switching, the pole control system and the converter control system can switch from system A to system B and vice versa.

(II) DC Protection System Design Scheme

1. Function hierarchy of DC protection system

For the connection mode of double 12-pulse converter in series, in order to eliminate the connection between the converters, avoid the impact of the maintenance of the single 12-pulse converter on the other operating converter, and improve the reliability of the entire system, it is necessary to ensure the independence of the converter level protection, i.e., each 12-pulse converter uses a separate protection device. Therefore, the DC protection system is divided into pole protection system and converter protection system, which respectively implement pole/bipolar protection and converter protection functions, as shown in Fig. 47. The hierarchical structure is described as follows.

(1) The pole protection system configures an independent pole protection host (PPR) to complete the pole and bipolar protection functions. DC filter protection is integrated into pole protection and is not configured independently.
(2) The converter protection system configures an independent converter protection host (CPR) to complete all protection functions of the converter. The electrical protection of the converter transformer is integrated into the converter protection and is not configured independently.
(3) I/O units are configured according to poles or converters. When a pole or converter is out of operation, the corresponding protection host and I/O device can be operated to the maintenance state, and any operation can be performed on this converter without affecting the system operation.

2. Protection redundancy scheme

   (1) "Two out of three" implementation scheme. The UHV DC system with layered access adopts a triple redundant protection configuration, and the logic of "two out of three" is used at the outlet. This logic is also configured in the independent "two out of three" host and control host (pole control host and converter control host).

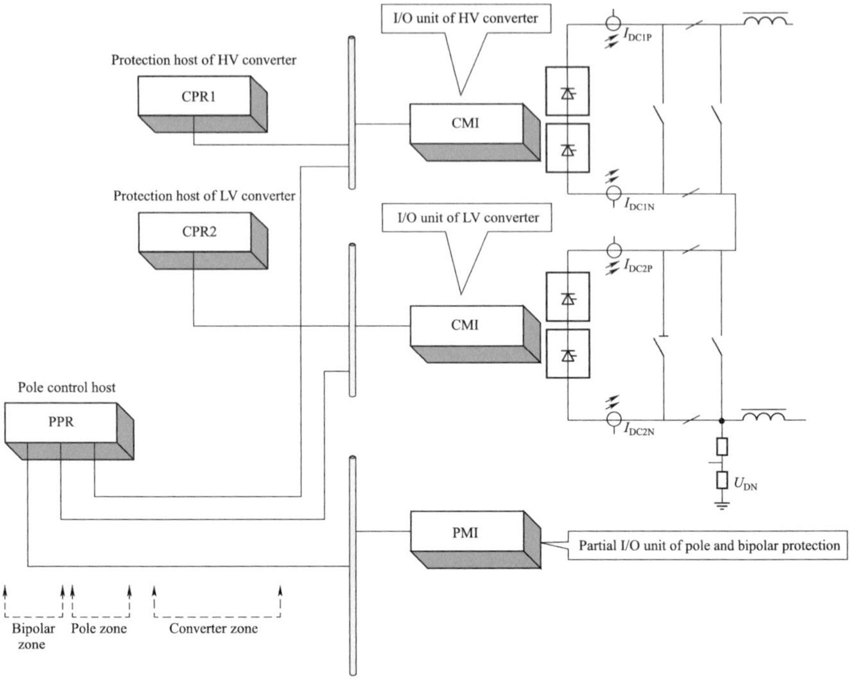

Fig. 47 Function Hierarchy of DC protection system

1) "Two out of three" host. The redundant configuration of the pole level and the converter level is implemented by the pole "two out of three" host (P2F) and the converter "two out of three" host (C2F), respectively, using a separate DC protection device, which is the same hardware platform as the protection host. The "two out of three" host receives information about the action of each set of protection, and its "two out of three" logic output realizes functions such as tripping the transformer incoming feeder breaker and starting switch failure protection.
2) "Two out of three" control host logic. The same "two out of three" logic is also configured in each level of control system hosts. Each control host also receives information about the action of each set of protections, and implements functions such as blocking and tripping the AC breaker through the same "two out of three" protection logic.

Three sets of pole protection hosts and three sets of converter protection hosts, all communicate with the "two out of three" host and the control host on the same level via fiber optics, transmitting digitally verified signals. The triple protection and the "two out of three" logic form a whole, and only when two of the same type of protection actions are judged to be correct in the three sets of protection hosts (extreme protection hosts or converter protection hosts), the output is allowed to be blocked or tripped to ensure reliability and safety. In addition, when one of the three sets of protection hosts is out of service for some reason, the "one out of two" protection logic is adopted. When two of the three sets of protection hosts are out of service, the "one out of one" protection logic is adopted. When all three sets of protection hosts are out of service, the control system issues a blocking command. The "two out of three" functions in the protection system of PCS9550 are shown in Fig. 48.

(2) Features of the "two out of three" scheme.

1) The "two out of three" function is implemented separately in the independent "two out of three" host and control host. The "two out of three" host outlet realizes the switching function of the converter transformer, and the host "two out of three" logic realizes the DC blocking and switching function of the converter transformer.

Once the protection activated, if the redundant control system fails to complete the blocking in extreme cases, after the "two out of three" host outlet disconnects the converter transformer incoming feeder breaker, the circuit breaker's pre-trip signal (early make signal) notifies the pole controller to block.

2) The protection host is connected to the "two out of three" host and control host through optical fiber, transmitting verified digital signals, which improves the reliability and anti-interference ability of signal transmission.
3) The "two out of three" function is implemented according to the protection classification, not simply the "or" of the tripping outlet, which improves the accuracy and reliability of the "two out of three" logic. Since all the protection hosts send digital signals to the "two out of three" host and control host, the "two out of three", "one out of two", "one out of one" and other logics can be

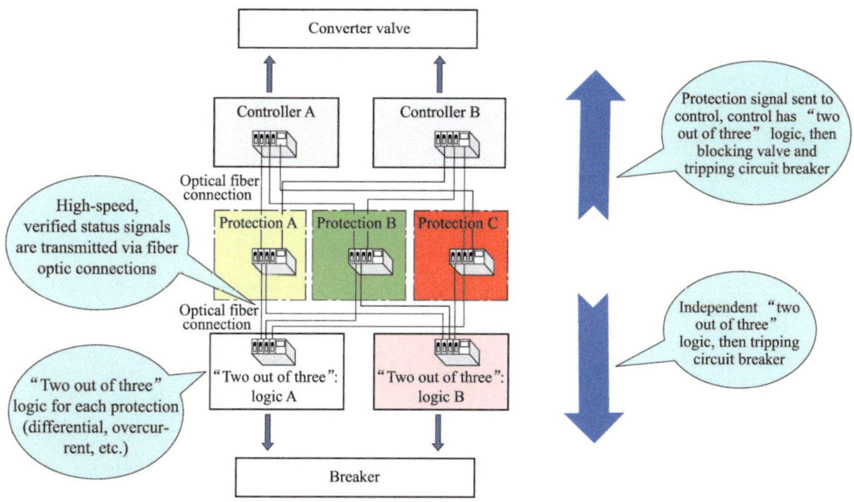

Fig. 48 "Two out of three" functions in the protection system of PCS9550

selected according to the type of protection. For example, in the case of three sets of protection functions, only when two or more protections have the same type of protection action, the "two out of three" logic will be allowed. Because the protection is judged according to the specific type of protection, rather than simply taking the tripping contact for "or" logic processing, the accuracy and reliability of the protection action logic is greatly improved.

4) The "two out of three" configuration has independent hosts, and their working status can be monitored at the operator's workstation.

3. Non-electrical protection of converter transformers

The non-electrical protection of the converter transformer in the layered accessed UHV DC system does not set up independent protection devices, adopts the "two out of three" output logic. The non-electrical tripping signal is collected through the non-electrical power interface device (NEPA/B/C), and sent to the converter control host (CCP) through a fiber-optic Ethernet point-to-point connection to implement the "two out of three" logic. The non-electrical output logic of the converter transformer is shown in Fig. 49.

4. AC filter protection

The AC filter protection is configured by banks. Each bank of AC filters is equipped with 2 sets of protection, including bus protection and AC filter protection, providing fully redundant protection. In addition, a separate operation box needs to be configured to realize the control interface with the AC filter sub-bank breaker.

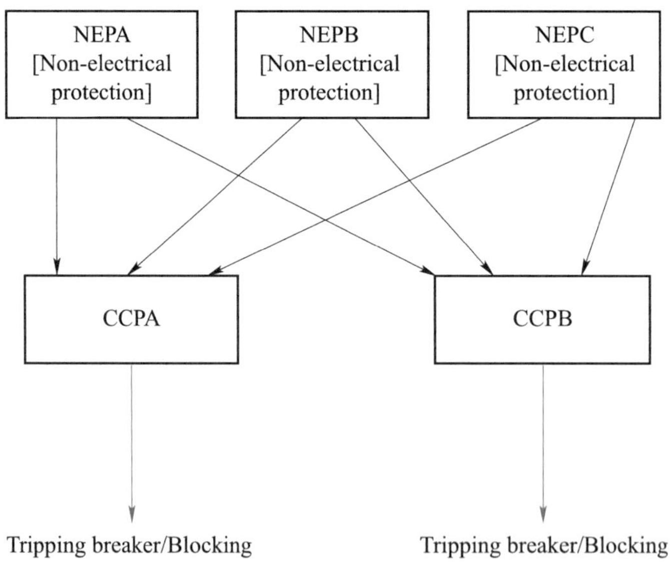

Fig. 49 Non-electrical output logic of the converter transformer

The AC filter protection should be able to interface with the electronic TA in any of the following types:

(1) Digital interface type of IEC standard protocol (IEC 60044-8 or IEC 61850).
(2) Digital interface type of TDM protocol.
(3) Analog interface type.

The tripping circuit of AC filter protection configured by banks is shown in Fig. 50.

The tripping circuit of the bus-related protection, except for the AC filter bus incoming feeder breaker, is similar to that in Fig. 50.

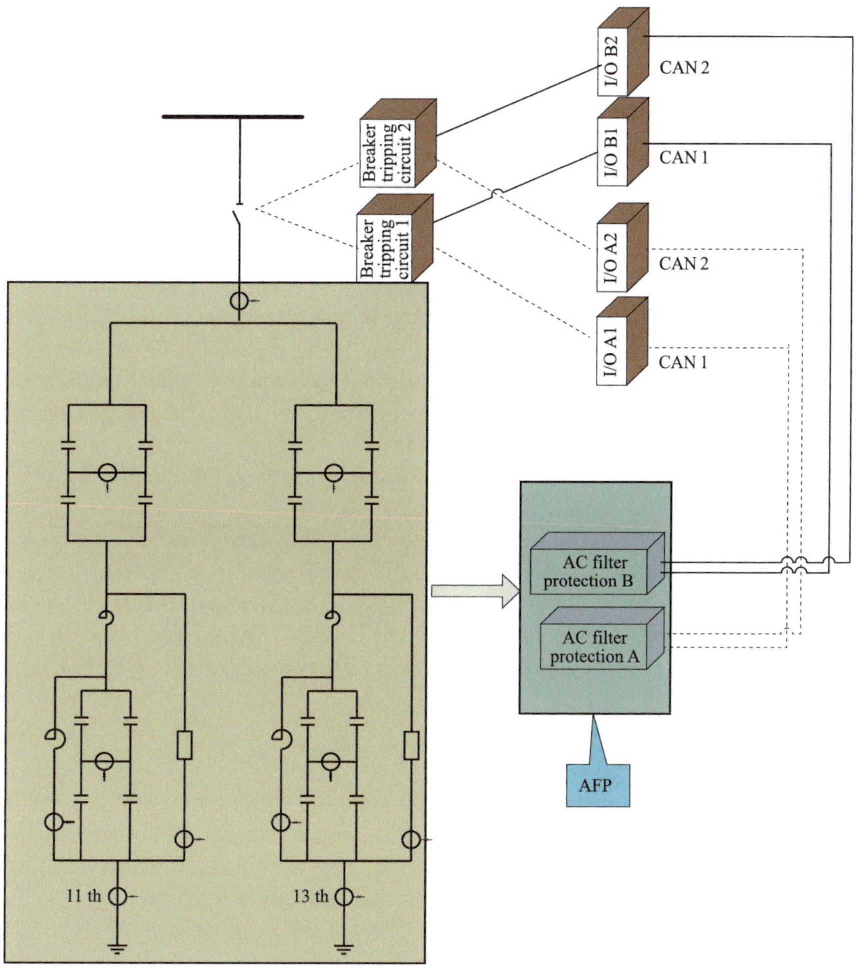

Fig. 50 Tripping circuit of AC filter protection configured by banks

(III) DC Control Protection System Interface

The DC control protection system reserves interfaces with the converter valve control system, converter valve cooling system, converter transformer, and safety and stability control system in the converter station, and adopts the corresponding UHV DC converter station general interface technical specifications.

1. Interface with the converter valve control system

The converter control protection (CCP) and valve base electronics (VBE) both adopt a dual redundant configuration scheme, with a "one-to-one" connection between the CCP and VBE, and a "one main and one backup" mode in normal operation.

The CCP and VBE systems in the main (ACTIVE) state are actually responsible for controlling the converter valve and issuing blocking commands. The CCP and VBE systems in the backup (STANDBY) state must be in hot standby mode unless they are unavailable. In other words, except for not sending firing pulses to the valve tower, the other functions of firing pulse generation, return pulse generation, protection, alarm, blocking, monitoring, etc. are the same as those of the main system. The VBE in standby mode detects a tripping signal (VBE_Trip) and sends it to the CCP, but the corresponding CCP does not send it. In the event of a failure in the main system, the standby system is automatically switched on. Events and alarms generated by the VBE are sent directly to the operator workstation (OWS) via the field bus.

All digital signals between the VBE and CCP systems use optical modulation signals, and the carrier frequency error must not exceed 10%. The signal channel uses a multimode optical fiber with a wavelength of 820nm, the control protection input is not less than − 25 dBm, and the control protection output is not less than - 15 dBm.

The control signals from CCP to VBE include main signal (ACTIVE), normal/abnormal voltage signal (VOLTAGE), control pulse signal (CP), deblock signal (DEBLOCK), bypass pair operation signal (BPPO), inverter operation status signal (INV_Ind) and waveform recording signal (REC_Trig). The signals from VBE to CCP include VBE available signal (VBE_OK), VBE tripping signal (VBE_Trip) and feedback pulse signal (FP).

2. Interface with the converter valve cooling system

The valve cooling control protection (VCCP) and the converter control protection (CCP) are both dual redundant configurations, using a "one main and one backup" mode. The dual redundant VCCP and CCP are "cross-connected", that is, each set of VCCP exchanges signals with two sets of CCP systems in real time.

The CCP system in active status is actually responsible for the control of the DC system and the issuing of blocking commands, while the CCP in standby status is not allowed to issue blocking commands from the VCCP. The VCCP in either active or standby status only uses information from the active CCP to complete related controls. Unless the system is unavailable, both sets of VCCP need to be in working status. For control information, both the active and standby CCP only use signals from the active VCCP for related DC system controls. For protection action signals, both the active and standby CCP simultaneously use signals from both the active and standby VCCP.

All digital signals between the VCCP and CCP systems use optical modulation signals, the duty cycle of carrier signal is 50%, and the frequency error should not be greater than 10%. Analog quantities all use optical signals, the communication protocol is IEC 60044-8. The alarm, event, and other information generated by the VCCP are transmitted to the OWS via optical fiber, and the communication protocol is IEC 61850 or Profibus.

The signals from CCP to VCCP include active signal (ACTIVE), remote switch valve cooling main pump command (switch pump), and deblock signal (DEBLOCK). The modulated signals from VCCP to CCP include tripping command (VCCP_ TRIP), power runback command (RUNBACK), availability signal (VCCP_OK), ready for operation signal (VCCP_ RFO), redundant cooling capacity (REDUNDANT), active signal (VCCP_ACTIVE), and temperature signal.

3. Interface with converter transformer

The schematic diagram of the secondary system structure of the converter transformer is shown in Fig. 51. The converter transformer control system is used to control the operation of the UHV DC converter transformer cooling system. After calculating and processing the operating temperature, load, and cooler status of the converter transformer, it combines with the DC control system command to control the start and stop of the converter transformer cooler, and sends the cooler operating status to the DC control protection system. The technology route of transformer electronic control (TEC) or programmable logic controller (PLC) can be adopted. The control cabinet is used to collect the electrical and non-electrical signals of the UHV DC converter transformer, and pass the signals to the converter transformer control system, intelligent component cabinet, DC control protection system, etc. It accepts control commands issued by the converter transformer control system and operators, and controls the cooler and on-load tap changer. It supplies power to the converter transformer cooler, converter transformer control system, intelligent component cabinet, and other converter transformer components. The intelligent component cabinet is used to collect operating parameters such as gas in the oil of the converter transformer, core clamp grounding current, SF_6 pressure, oil temperature, winding temperature, on-load tap position, etc., and sends the data to the integrated online monitoring platform. The integrated online monitoring platform monitors the information of equipment of the converter station for data recording, statistics, and analysis, without performing any control.

The secondary signal of converter transformer is hard-wired to the control cabinet, and the control cabinet is hard-wired to the converter transformer control system. The control cabinet is hard-wired to the intelligent component cabinet, and the communication signals include bushing pressure, tap changer position, converter transformer winding temperature, oil temperature, and oil level. The control cabinet is also hard-wired to the DC control protection system. The communication between the converter

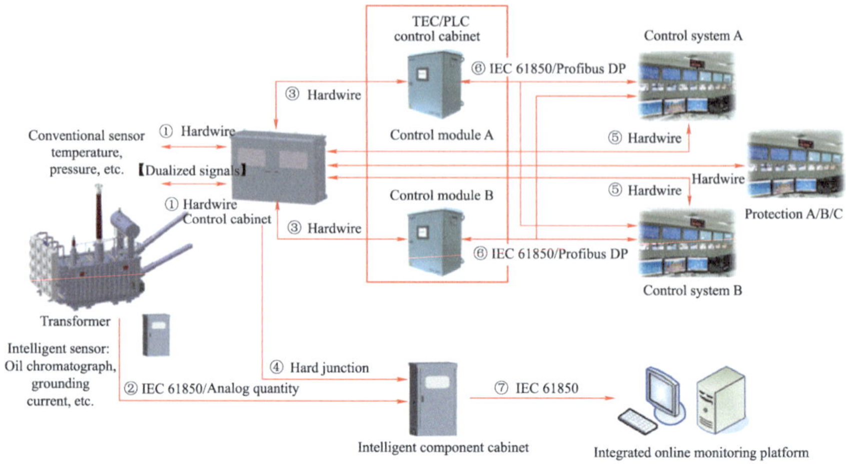

Fig. 51 Schematic diagram of the secondary system structure of the converter transformer

transformer control system and the DC control and protection system is carried out using IEC 61850 or Profibus DP.

The control signals from the DC control and protection system to the converter transformer control system include remote forced input/withdrawal signals (including reset signals), converter transformer differential signals, heavy gas protection action of complete cut-out cooler (including pump and fan) and converter transformer charging signal. The signals from the converter transformer control system to the DC control protection system include cooler operating status (each group's switching status and fault information), overload capacity of converter transformer, tap changer position, winding temperature, and oil temperature.

4. Interface with the safety and stability control system

The pole control protection (PCP) and safety and stability control device (SSC) both adopt a dual configuration (triple protection) scheme, with PCP and SSC cross-connected. During normal operation, PCP adopts a "one main one standby" mode, with two sets of SSC operating independently, without a main or standby distinction.

Both the active and standby PCP devices send signals to the safety and stability control device, which executes the commands of the active PCP system. If the active PCP system fails, it automatically switches to the standby system.

The protocol IEC 60044-8 is used for communication between the PCP and the SSC, with the communication rate of 5 Mbit/s. The communication fiber belongs to the scope of the DC control and protection system. The transmission medium is multimode fiberglass with ST or LC interfaces.

The signals from the DC control system to the safety and stability control system include DC power drop signal, DC power drop amount, DC pole control mode, DC pole control active status, DC pole control standby status, converter maximum deliverable power, converter power delivery value, converter abnormal shutdown signal, converter operating status, converter normal shutdown signal, DC power drop flag, and DC commutation failure, etc. The signals from the safety and stability control system to the DC control system include increasing DC power signal, increasing DC power capacity, increasing DC power flag, decreasing DC power signal, decreasing DC power capacity, decreasing DC power flag, DC bipolar blocking signal, and DC island operation mode, etc.

6 Electromagnetic Environment

The electromagnetic environment study of the ± 1100 kV converter station mainly focuses on the total electric field at the ground level and the power frequency electric field. The DC ground total electric field of DC yard is focused, which is affected by the diameter and height of the tubular busbar. The power frequency electric field of AC filter yard is focused, which is affected by the height of the filter capacitor tower and the tubular busbar. Therefore, the electromagnetic environment of the converter station not only relates to the health of the station's operation and inspection personnel but also determines the project investment. It is necessary to accurately predict the level and distribution of ground total electric field and power frequency electric field by combining field tests and theoretical calculations.

I. Experiments on the Electromagnetic Environment of the DC Side of the ± 1100 kV Converter Station

The ground total electric field of UHV DC converter station directly affects the health of the operating personnel in the station and must be properly addressed. The main facility that generates the total electric field in the DC yard area is the overhead bare tubular bus. The DC total electric field at the ground level under the tubular busbar consists of two parts: ① The nominal electric field generated by the charge on the tubular busbar; ② When the electric field strength on the surface of the tubular busbar exceeds the corona inception field strength, the free charge generated by the corona discharge on the surface of the tubular busbar forms an ion flux field in space. The former is mainly related to the ground voltage and height of the tubular busbar, while the latter is also related to the diameter of the tubular busbar in addition to these two factors. Therefore, whether the selection of the tubular busbar and the determination of the ground height are reasonable not only determines the strength of the total electric field at the ground level of the converter station, but also affects the construction cost.

China has conducted a large amount of research on the selection and ground height optimization of the tubular busbar in the DC yard during the preliminary feasibility study of the ± 800 kV UHV DC transmission project. However, the ± 1100 kV UHV DC transmission project has no precedent in the world. The large increase in voltage will cause the corona discharge characteristics of the tubular busbar surface to change, and it cannot be simply extrapolated from the design experience of the ± 800 kV UHV DC converter station. A combination of theoretical calculations and full-scale tests should be used to determine the economical and reasonable tubular busbar type and ground height for the ± 1100 kV converter station.

The research object is the selection of tubular busbars for the ± 1100 kV converter station. First, through the simulation calculation of the ground total electric field of different types of tubular busbars, combined with the control value of the ground total electric field, a preliminary tubular busbar type suitable for the ± 1100 kV converter station is proposed. Then, a ± 1100 kV DC yard tubular busbar prototype test circuit is set up at the Beijing UHV DC Test Base to carry out ground total electric field tests of two different diameter tubular busbars to verify the effectiveness of the calculation method. Finally, the recommendations for the economical and reasonable tubular busbar type and ground height for the ± 1100 kV converter station are given based on technical and economic requirements.

(I) Control value of ground total electric field

1. Control value of the ground total electric field for the ± 1100 kV converter station

Currently there is no standard for the control value of the total electric field for the ± 1100 kV converter station. However, from the perspective of control mechanisms, the control indicators of the ground total electric field should be basically consistent whether it is an EHV or UHV DC converter station. Therefore, the control value of the ground composite total field for the ± 1100 kV converter station can be determined by referring to the relevant standards for ± 500 and ± 800 kV converter stations that are already in operation and have been accepted. For the ± 1100 kV converter station, it is recommended that the ground total electric field in most areas should be controlled to not exceed 30 kV/m, except for a few locations. Under this electric field limit, the human body has no obvious direct perception of the DC electric field, and relatively obvious transient electric shocks can be largely avoided.

2. Control value of the ground total electric field when considering only a single tubular busbar

The ground total electric field in the converter station is not only generated by the tubular busbar, but also by a large number of various types of flexible conductors and fittings. Therefore, the actual ground total electric field under the tubular busbar is greater than that generated by the tubular busbar alone. If the ground total electric field in most areas of the converter station is controlled below 30 kV/m, then that generated by a single tubular busbar should be less than 30 kV/m, but there is currently no conclusion on the specific control value. Considering the variety of conductor types and the complex layout in the DC yard, it is currently difficult to give specific values through simulation calculations. To solve this problem, a special measurement of the ground total electric field of multiple ± 500, ± 660 and ± 800 kV converter stations in China was carried out, and the distribution characteristics and maximum value of the ground total electric field of different voltage levels were obtained. At the same time, the ground total electric field generated only by a single tubular busbar was calculated, and the difference between the calculation value of a single tubular busbar and the maximum measured value of the actual converter stations was obtained. The results are listed in Table 50.

According to the comparison between calculation and measurement, it can be known that the maximum measured value of the ground total electric field of the converter station is basically larger than the calculated value of a single tubular busbar, and the difference is 5–10 kV/m. Therefore, in order to control the ground total electric field within a large range of the converter station to less than 30 kV/m, it is appropriate to control the ground total electric field generated by a single tubular busbar within 20–25 kV/m.

(II) Calculation Method of Ground Total Electric Field

The tubular busbar is regarded as an infinite straight conductor, and the ground total electric field under the tubular busbar is solved according to its calculation method for the DC transmission line.

At the ground level below the DC tubular busbar, the constraint equations of the total electric field $\boldsymbol{E}_S$, space charge density ρ and ion flux density $\boldsymbol{J}$ is represented as

$$\nabla \cdot \boldsymbol{E}_S = \rho/\varepsilon_0 \tag{16}$$

$$\boldsymbol{J} = K\rho\boldsymbol{E}_S \tag{17}$$

$$\nabla \cdot \boldsymbol{J} = 0 \tag{18}$$

where ε_0 is the air permittivity, K is the ion mobility.

Table 50 Maximum measured value in the converter stations and the calculated value for a single tubular busbar of the ground total electric field

Voltage level (kV)	Converter station	Maximum value of ground total electric field (kV/m)		Difference between calculated value and measured value (kV/m)
		Calculated value for a single tubular busbar	Measured value in the converter stations	
±800	Fengxian	19.4	25.2–28.1	5.8–8.7
	Zhongzhou	18.3	22.8–29.4	4.5–11.1
±660	Yinchuandong	18.2	21.5–30.7	3.3–12.5
±500	Jiangling	15.9	18.5–28.6	2.6–12.7
	Yidu	15.9	21.0–24.4	5.1–8.5
Average				4.3–10.7

The flux tracing method is used to calculate the ground total electric field of the DC tubular busbar, which simplifies the solution of the ground total electric field of the DC transmission line. The basic assumptions of the flux tracing method include:

(1) The space charge only affects the magnitude of the electric field, not its direction, that is, the Deutsch assumption, expressed in mathematical form as

$$\boldsymbol{E}_S = A\boldsymbol{E} \tag{19}$$

where A is scalar function, $\boldsymbol{E}$ is nominal electric field.

(2) When ionization occurs near the surface of the wire, the field strength on the surface of the wire remains at the corona inception field strength.
(3) The area between the positive wire and the ground only contains positive ions, and the area between the negative wire and the ground only contains negative ions.
(4) The threshold corona of the positive and negative wires are equal.
(5) The diffusion effect of ions is not considered.
(6) The mobility of positive and negative ions are the same and are constants unrelated to the electric field strength.

Under normal altitude conditions, the corona inception field strength of tubular bushar at the ± 1100 kV converter station can be selected as follows: 18.4 kV/cm for dry tubular busbar, 11.3 kV/cm for wet tubular busbar, and 9.2 kV/cm for tubular busbar with a certain amount of dirt on the surface.

(III) Calculation Results of the Maximum Ground Total Electric Field for Tubular Busbars of Various Diameters

The relationship curves between the maximum ground total electric field and the height above ground for five common diameters of tubular busbars are calculated, taking into account the dry, wet and specific pollution conditions of the tubular busbars. The diameters of the tubular busbar are set to be φ 300, φ 350, φ 400, φ 450 and φ 500 mm. The calculation results do not consider altitude and are only applicable to low-altitude areas. Figures 52, 53, 54, 55, and 56 show the results of the maximum strength of ground total electric field as a function of the tubular busbar height in the dry, wet and polluted states for the five common diameters.

As can be seen from Figs. 52, 53, 54, 55, and 56:

(1) φ 300 mm tubular busbar. When the height above the ground is between 16.5 and 19.6 m in dry conditions or between 24 and 27.3 m in wet conditions, it can meet the requirement that the maximum ground total electric field is below 20–25 kV/m; but in polluted conditions, to meet the requirement that the maximum ground total electric field is less than 25 kV/m, the height above the ground will exceed 30 m.
(2) φ 350 mm tubular busbar. When the height above the ground is between 16.8 and 20 m in dry conditions, between 19.5 and 22 m in wet conditions, or between

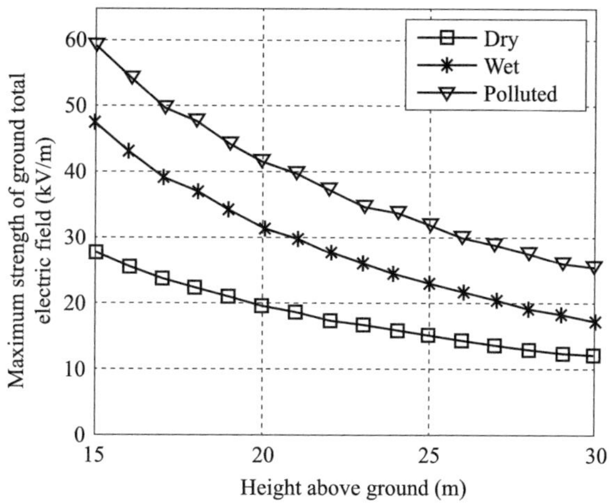

Fig. 52 Relation between maximum ground total electric field for 300 mm tubular busbars and height above ground

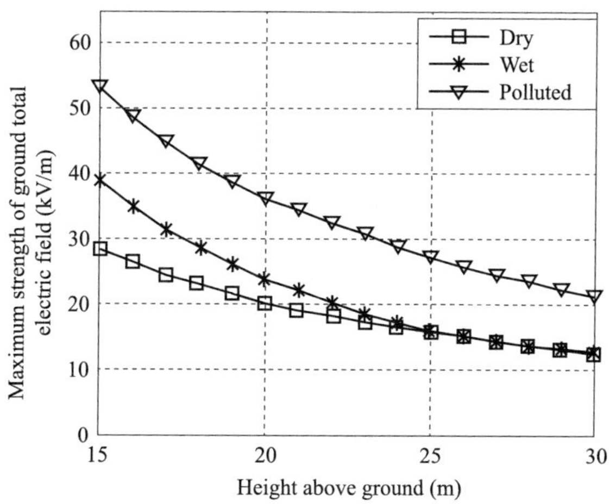

Fig. 53 Relation between maximum ground total electric field for 350 mm tubular busbars and height above ground

26.7 and 31.3 m in polluted conditions, it can all meet the requirement that the maximum ground total electric field is below 20–25 kV/m.

(3) φ 400 mm tubular busbar. When the height above ground is between 17.2 and 21 m in dry and wet conditions, and between 23.5 and 27 m in polluted conditions, it can meet the requirement that the maximum ground total electric field is below 20–25 kV/m.

(4) φ 450 mm tubular busbar. When the height above ground is between 17.5 and 21 m in dry and wet conditions, and between 20.2 and 23 m in polluted conditions, it can meet the requirement that the maximum ground total electric field is below 20–25 kV/m.

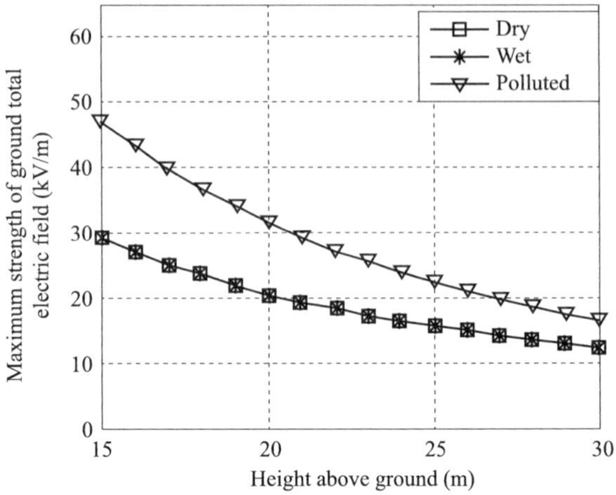

Fig. 54 Relation between maximum ground total electric field for 400 mm tubular busbars and height above ground

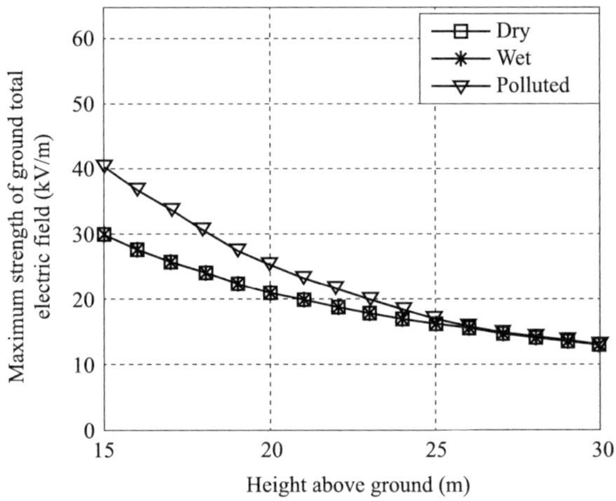

Fig. 55 Relation between maximum ground total electric field for 450 mm tubular busbars and height above ground

(5) φ 500 mm tubular busbar. It can meet the requirement that the maximum ground total electric field is below 20–25 kV/m when the height above ground is between 18 and 21.4 m in dry, wet, and polluted conditions.

Based on the above calculation and analysis results, the minimum heights of the tubular busbar above the ground under dry, wet and polluted conditions are given in Table 51, with the maximum value of the ground total electric field controlled at 20–25 kV/m.

As can be seen from Table 51, for tubular busbars with diameters of 300–500 mm, as long as the height above ground is reasonably selected, the ground total electric

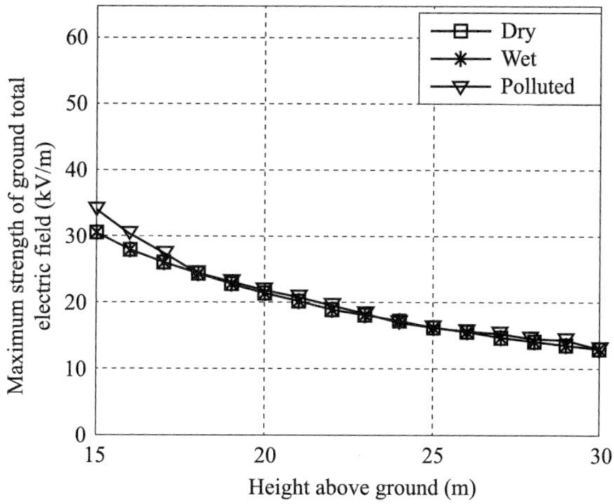

Fig. 56 Relation between maximum ground total electric field for 500 mm tubular busbars and height above ground

Table 51 Minimum height above the ground for tubular busbars with different diameters (ground total electric field strength controlled at 20–25 kV/m)

Diameter of tubular busbar (mm)	Minimum height in dry state (m)	Minimum height in wet state (m)	Minimum height in polluted state (m)
300	16.5–19.6	24.0–27.3	> 30.0
350	16.8–20.0	19.5–22.0	26.7–31.3
400	17.2–21.0	17.2–21.0	23.5–27.0
450	17.5–21.0	17.5–21.0	20.2–23.0
500	18.0–21.4	18.0–21.4	18.0–21.4

field level can meet the requirements. However, whether the calculation method chosen in this book is reasonable needs to be further verified by experiments.

(IV) Ground Total Electric Field Test of Tubular Busbar

Based on the design experience of the tubular busbar of the ± 800 kV UHV DC transmission project, tests are conducted on tubular busbars with diameters of 300 and 400 mm.

1. Test layout scheme

The ground total electric field test for the ± 1100 kV tubular busbar is carried out on the testing line in the SGCC's UHV DC Test Base. According to the measuring experience, the maximum value of the ground total electric field usually appears under the "T" or "L" shaped tubular busbar. In order to simulate an environment close to the actual layout of the converter station, a partial modification was made to the UHV DC testing line, and an "L" shaped test model was built, as shown in Fig. 57. In Fig. 57a, the lengths of the two tubular busbars arranged at right angles are 24 m and 16 m, respectively, and the heights of the centers of the tubular busbars above the ground are both 16.2 m. The diameters of the tubular busbars used for the test are 300 and 400 mm, respectively, and the materials are aluminum. The ground total electric field tests were conducted under voltages of both + 1100 and − 1100 kV for each type of tubular busbar.

On the ground under the north–south oriented tubular busbar (24 m in length), two survey lines are arranged perpendicular to the direction of the tubular busbar. Survey line *a* is located halfway from the end of the tubular busbar, and survey line *b* is located at 3/4 of the tubular busbar end. According to the on-site test conditions, the measurement area of ground total electric field is mainly concentrated under the tubular busbar and on the west side.

2. Test results for φ 400 mm tubular busbar

Under sunny weather and clean surface conditions, by applying + 1100 and − 1100 kV, the measured results of total electric field at two survey lines a and b below the tubular busbar with 400 mm diameter and the calculated results of the nominal electric field were obtained, as shown in Fig. 58.

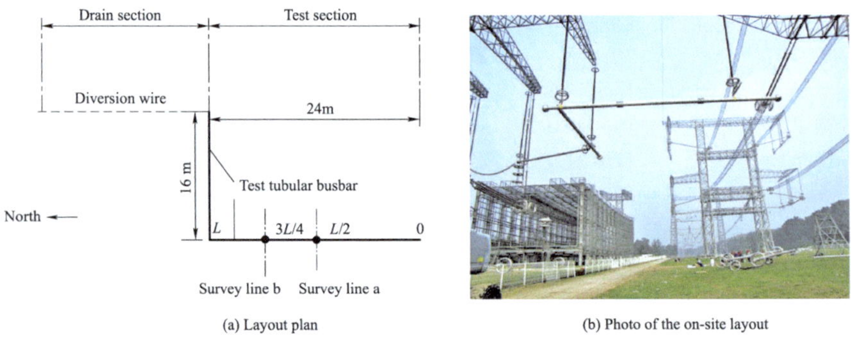

Fig. 57 Layout for tubular busbar test

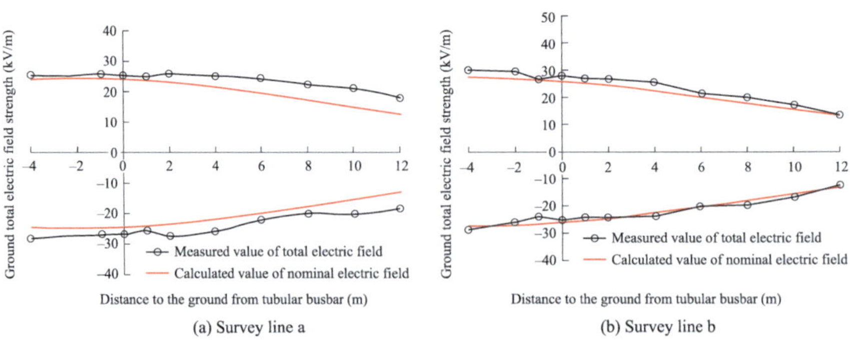

Fig. 58 Measured and calculated results of the ground total electric field for 400 mm diameter tubular busbar

As can be seen from Fig. 58, the trend with distance for the measured results of the total electric field and the calculated results of the nominal electric field is basically the same. On the survey line b, which is 3/4 the length of the tubular busbar, the measured value is in very good agreement with the calculated value of the nominal electric field. On the survey line a, which is in the middle of the tubular busbar, the measured value is slightly different from the calculated value in the area far from the tubular busbar due to the close proximity to the tower and the uneven ground. However, in general, the measured values of the ground total electric field under the tubular busbar are basically consistent with the calculated results of the nominal electric field, indicating that under the test voltage of ± 1100 kV, the surface of the dry and clean 400 mm diameter tubular busbar is not coronaed, and the ground total electric field can be directly analyzed and predicted using the three-dimensional electrostatic field calculation method.

3. Test results for φ 300 mm tubular busbar

Under the condition of clean surface, the measurement results of total electric field of the 300 mm diameter tubular busbar on two survey lines on the ground are shown in Fig. 59. The maximum ground total electric field is 34–37 kV/m, which is over 30% higher than the measurement results and nominal calculation values of the 400 mm diameter tubular busbar. Since the nominal electric fields under the 400 and 300 mm diameter tubular busbars are basically equal, the ground total electric field measurement result of the 300 mm diameter tubular busbar is significantly higher than the nominal electric field calculation value, indicating that under the ± 1100 kV DC test voltage, even if the tubular busbar is in a clean surface state, the surface of the 300 mm diameter tubular busbar has already occurred a certain degree of corona discharge, and the nominal electric field distribution obtained by the three-dimensional electrostatic field calculation method cannot reflect the total electric field distribution at this time. Considering the wet and dry conditions of the tubular busbar separately, use the three-dimensional total electric field calculation method to solve the distribution curve of the DC total electric field on the ground, and the calculation results are shown in Fig. 59. As can be seen from the figure, the measurement results and calculation results of the total electric field are basically consistent with the trend of distance change, and all measurement results are between the calculation results of the wet and dry conditions. As long as the appropriate corona inception field strength of the tubular busbar is selected, the calculation results will be very consistent with the measurement results, proving the effectiveness of the three-dimensional total electric field calculation method of the tubular busbar.

(V) Determination of ± 1100 kV Tubular Busbar Type and Ground Clearance

For indoor tubular busbar, the probability of dampness and surface roughness is very small, it is recommended to determine the ground clearance under dry conditions. For outdoor tubular busbar, the ground clearance can be determined under wet conditions, and if surface contamination occurs during long-term operation, regular cleaning and maintenance should be carried out.

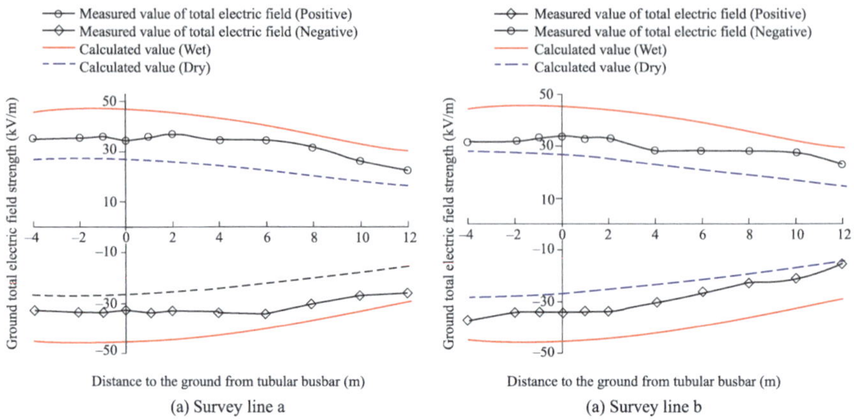

Fig. 59 Measured and calculated results of the ground total electric field for 300 mm diameter tubular busbar

Combining the field test results of the true tubular busbar and results in Table 52, it is suggested:

(1) For low-altitude indoor ± 1100 kV converter stations, it is recommended to use a φ 300 mm tubular busbar, and the minimum ground clearance can be 16.5–19.6 m.
(2) For low-altitude outdoor ± 1100 kV converter stations, it is recommended to use a φ 400 mm tubular busbar, and the minimum ground clearance can be 17.2–21.0 m. At the recommended height, it is advisable to clean the tubular busbar when the total electric field is found to be high for a long period of time. If the ground clearance is increased to 23.5–27 m, it is not necessary to clean and maintain the tubular busbar.

II. Study on the Electromagnetic Environment of the AC Side of the ± 1100 kV Converter Station

(I) Control Value of Ground Power Frequency Electric Field

At present, there is no standard for the control value of the power frequency electric field of the ± 1100 kV converter station. However, from the perspective of control mechanism, the control index of the power frequency electric field is consistent

Table 52 Maximum value of the power frequency electric field strength on the patrol corridor in front of the AC filter yard at different heights of the tubular busbar

Voltage level (kV)	1000			750		
Height of the tubular busbar (m)	16	16.5	17	13	13.65	14.5
Maximum value of the power frequency electric field strength (kV/m)	12.1	11.8	11.5	13.4	12.2	11.6

whether it is a UHV DC converter station or AC substation. The design specifications of the substation can be referred to. The requirement in Section 6.0.4 of GB 50697—2011 "Code for design of 1000 kV substation" is that the electrostatic induction field strength (field strength at 1.5 m above ground) inside the site of the 1000 kV outdoor distribution device should not exceed 10 kV/m, but may be allowed to reach 15 kV/m in a small number of areas.

(II) Calculation Method of Power Frequency Electric Field

The common methods for three-dimensional numerical calculation of power frequency electric field include charge simulation method, boundary element method, finite element method, and the combination of two methods. These calculation methods have been proven to be able to solve the three-dimensional electric field problem of complex models. The overall structure of the capacitor bank in the AC filter yard of the converter station is relatively complex, and there is no relevant literature to model the entire AC filter yard. In order to accurately assess the electromagnetic environment inside the AC filter yard, it is necessary to carry out numerical simulation of the power frequency electric field inside the AC filter yard considering the entire field. Due to the large number of devices and the complex layout, if the charge simulation method is used, it is difficult to set the simulated charge, and if it is not set properly, it will bring a large error. Therefore, the finite element method is generally used.

For the three-phase AC yard, since the voltage is a phasor, the voltage of each phase of the tubular busbar is often expressed in complex form for calculations, and for electric field calculations, the phasor value of the phase voltage can be expressed as

$$\begin{cases} \dot{U}_{\mathrm{a}} = U(\cos\theta + \mathrm{j}\sin\theta) \\ \dot{U}_{\mathrm{b}} = U[\cos(\theta - 120^\circ) + \mathrm{j}\sin(\theta - 120^\circ)] \\ \dot{U}_{\mathrm{c}} = U[\cos(\theta + 120^\circ) + \mathrm{j}\sin(\theta + 120^\circ)] \end{cases} \tag{20}$$

where θ is initial phase angle (°), U is peak value of line voltage.

$\theta = 0^\circ$ is usually taken as the calculation condition for simulation.

The electric potential of any node n within the field can be written as a formula composed of real and imaginary parts, that is,

$$\dot{U}_n = U_{\mathrm{real}} + \mathrm{j}U_{\mathrm{image}} \tag{21}$$

where U_{real} is the real part of the electric potential at node n, U_{image} is the imaginary part of the electric potential at node n.

By loading the potential as an excitation condition, divided into real and imaginary parts separately, while keeping the model and boundary conditions unchanged, the individual components of the field strength at a point in the field can also be expressed in complex form.

$$\begin{cases} \boldsymbol{E}_{nx} = E_{nx,\mathrm{r}} + \mathrm{j}E_{nx,\mathrm{j}} \\ \boldsymbol{E}_{ny} = E_{ny,\mathrm{r}} + \mathrm{j}E_{ny,\mathrm{j}} \\ \boldsymbol{E}_{nz} = E_{nz,\mathrm{r}} + \mathrm{j}E_{nz,\mathrm{j}} \end{cases} \tag{22}$$

The vector value of total field strength at point n can be represented as

$$\boldsymbol{E}_n = (E_{nx,\mathrm{r}} + \mathrm{j}E_{nx,\mathrm{j}})\boldsymbol{e}_x + (E_{ny,\mathrm{r}} + \mathrm{j}E_{ny,\mathrm{j}})\boldsymbol{e}_y + (E_{nz,\mathrm{r}} + \mathrm{j}E_{nz,\mathrm{j}})\boldsymbol{e}_z \tag{23}$$

where $\boldsymbol{e}_x$, $\boldsymbol{e}_y$, $\boldsymbol{e}_z$ are the unit vectors of the field strength in the x, y, z directions at node n.

The maximum amplitude of the field strength at node n is

$$\boldsymbol{E}_n = \sqrt{E_{nx,\mathrm{r}}^2 + E_{ny,\mathrm{r}}^2 + E_{nz,\mathrm{r}}^2 + E_{nx,\mathrm{j}}^2 + E_{ny,\mathrm{j}}^2 + E_{nz,\mathrm{j}}^2} \tag{24}$$

Using the above method, the problem of solving the time-harmonic field of the power frequency electric field can be converted into the solution of the electrostatic field. The values of power frequency potential and electric field at each point in the field are calculated by formula (24). There are many charged conductors in the AC side of the UHV DC converter station, and its power frequency electric field conforms to the quasi-static field model, which can be solved by the electrostatic field solver in the finite element analysis software.

(III) Analysis of the Impact of Equipment in the AC Filter Yard on the Power Frequency Electric Field

Figure 60 shows a schematic layout of the whole AC side area. There are 8 banks of AC filters in the converter station, which are arranged symmetrically on both sides, and only one of them is selected for simulation analysis. The black boxes in the figure show the AC filter yards, and the ABC three-phase filter towers are arranged in each bank of AC filter yards. The red dashed line is the centre line of the patrol corridor around the AC filter yard, and the research concern is also the distribution of the power frequency electric field at a height of 1.5 m above the ground on this centre line. The centerline of the patrol corridor on the incoming line side of the AC filter yard is 7.5 m away from the fence of the filter field, and the centerline of the patrol corridor on both sides parallel to the bus incoming line is 13.5 m away from the fence. The side view of the AC filter yard is given in Fig. 61, in which the tubular busbar is at a distance of 16.5 m from the ground.

For a 1000 kV AC filter yard, the height of the filter capacitor tower is about 14 m. Under the requirement of satisfying the air insulation gap distance, the power frequency electric field strength on the ground will also determine the distance between the charged equipment in the AC filter yard and the ground, as well as the distance between the surrounding fences and the equipment body, thus determining the size of the entire AC filter yard. This will largely affect the economy of the AC filter yard.

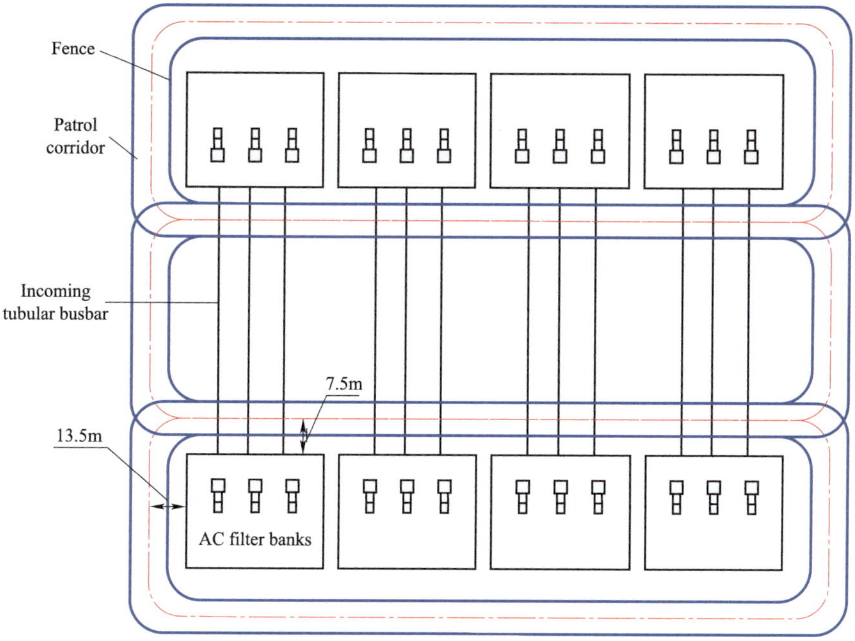

Fig. 60 Schematic diagram of the simplified layout of the 1000 kV AC side area

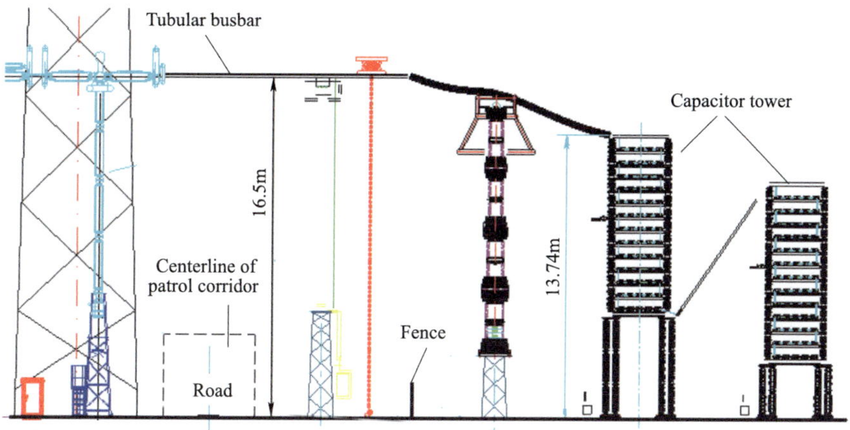

Fig. 61 Side view of the AC filter yard

In order to separately study the contributions of the incoming tubular busbar and the capacitor tower inside the AC filter yard to the power frequency electric field on the patrol corridor around the AC filter yard, simulation models were established without considering the incoming tubular busbar and with considering partial influence of the incoming tubular busbar, as shown in Fig. 62a, b. A schematic of the location of the three patrol corridors is also labelled in the figure. It is assumed that the filters operate at the rated voltage during calculation, and the real and imaginary parts are calculated. The applied real voltage is: Phase A is 857.3 kV; Phase B is − 428.6 kV; Phase C is − 428.6 kV; the imaginary voltage is: Phase A is 0 kV; Phase B is 742.4 kV; Phase C is − 742.4 kV. The real voltage and imaginary voltage are loaded onto the corresponding capacitor banks, and the ground is at 0V. Since the power frequency electric field decays quickly with distance, the size of the solution area (external air) is set to 200% of the basic model to meet the calculation requirements.

The simulation analyzed the distribution of the power frequency electric field strength on the patrol corridor around the AC filter yard when considering part of the tubular busbar and when not considering the tubular busbar, as shown in Fig. 63. It can be seen that without considering the tubular busbar, the power frequency electric field generated by the AC filter yard at the center of the road in front is lower than 1.8 kV/

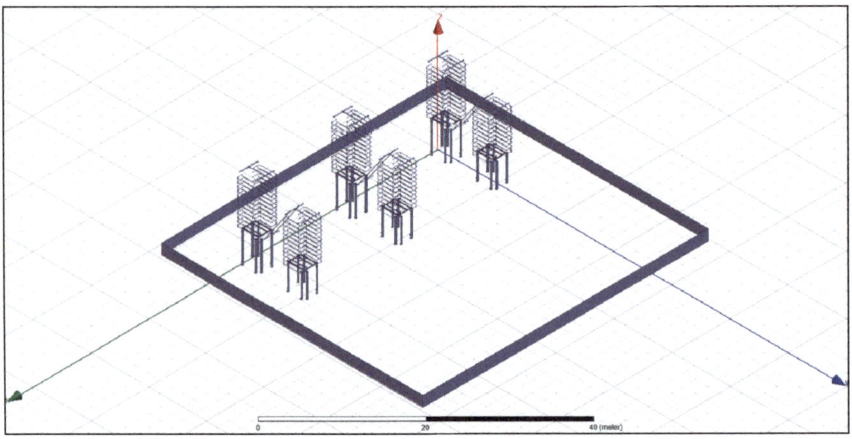

(a) Without considering incoming tubular busbar

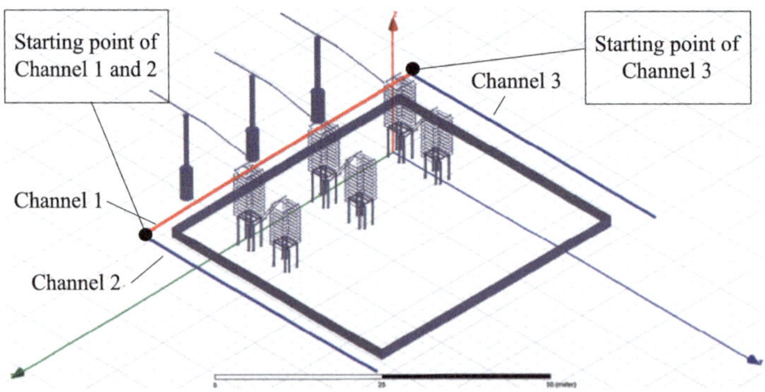

(b) Considering incoming tubular busbar

Fig. 62 Three-dimensional simulation modelling of AC filter yard m; when considering the influence of part of tubular busbar, the power frequency electric field strength is up to 9.5 kV/m. It can be seen that the contribution of the capacitor tower in the AC filter yard to the electric field on the patrol corridor is only 18.95%. When the height of the capacitor tower is appropriately reduced, assuming that the power frequency electric field caused by the capacitor tower increases by 20%, the contribution to the electric field is only 3.79%, which is a very small change. Therefore, the power frequency electric field on the patrol corridor 1 in front of the filter field is mainly affected by the voltage on the tubular busbar, and the equipment such as the capacitor tower inside the AC filter yard has a small impact. In order to reduce the strength of the power frequency electric field on the patrol corridor 1, the height of the tubular busbar can be appropriately increased, and the height of the capacitor tower of the AC filter can be appropriately reduced under the condition of meeting the insulation requirements. In addition, in order to simplify the calculation, the capacitor tower can be omitted from the model, which greatly reduces the amount of calculation. Thus, the long calculation duration and low calculation accuracy of the power frequency electric field under complex equipment structures can be overcome.

(IV) Power Frequency Electric Field in the AC Side Area under Different Tubular Busbar Heights and Phase-to-Phase Spacings

A simplified three-dimensional simulation model of the AC side area is established, as shown in Fig. 64. The location of the patrol corridor is consistent with that in Fig. 65. Based on this calculation model, the influence of the tubular busbar height and the phase-to-phase spacing on the distribution of the power frequency electric field on the patrol corridor 1 was obtained separately.

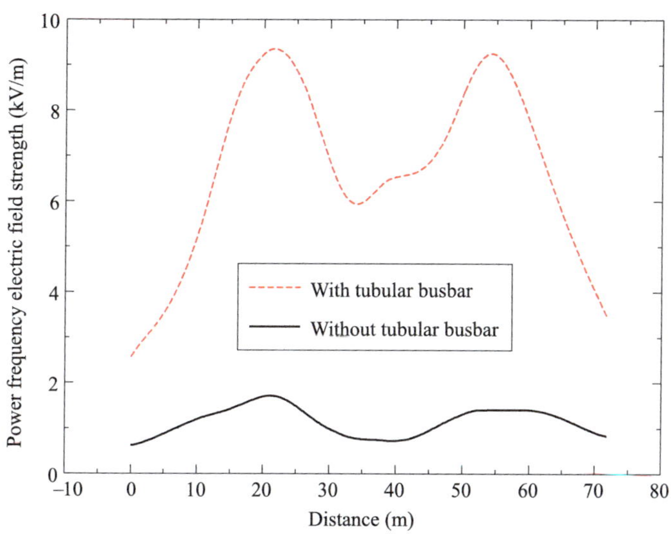

Fig. 63 Electric field distribution on the patrol corridor 1 in front of the AC filter yard

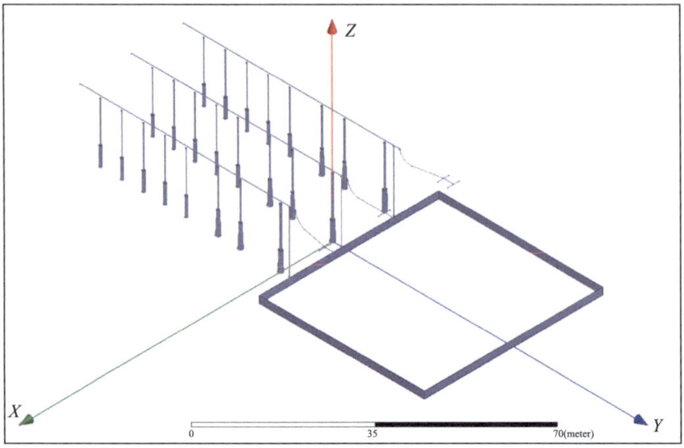

Fig. 64 Simplified calculation model for the 1000 kV AC yard

The power frequency electric fields under different tubular busbar heights and phase-to-phase spacings are shown in Figs. 65 and 66, and the maximum power frequency electric field strengths are shown in Tables 52 and 53. The following conclusions can be drawn from the figures: ① The power frequency electric field strength increases as the height of the tubular busbar decreases; ② The power frequency electric field strength decreases as the phase-to-phase spacing decreases. The main reason is that the voltage phase angles between the three phases differ by 120°, and the electric field strength formed between the three phases also differ by 120°. When the distance between the three phases is closer, the electric field is more likely to offset each other. When the distance between the three phases is too far, the counteraction effect between the three phases weakens, and the electric field increases instead. Therefore, in the selection of phase-to-phase spacing, while ensuring the requirements of insulation strength, the phase-to-phase spacing should be minimized as much as possible.

(V) Determination of the Tubular Busbar Height and Phase-to-Phase Spacing within the AC Side Area

According to the above calculation results and the requirements of GB 50697—2011 6.0.4: the level of electrostatic induction field strength (space field strength at a distance of 1.5 m above the ground) in the outdoor distribution equipment site should not exceed 10 kV/m, but in some areas it can be allowed to reach 15 kV/m. Considering the simulation error and a certain margin, the recommended height and phase-to-phase spacing of the tubular busbar are as follows:

(1) The height of the tubular busbar for the 1000 kV AC filter yard is 16.5 m, and the phase-to-phase spacing is 14.2 m.
(2) The height of the tubular busbar for the 750 kV AC filter yard is 13.65 m, and the phase-to-phase spacing is 10.75 m.

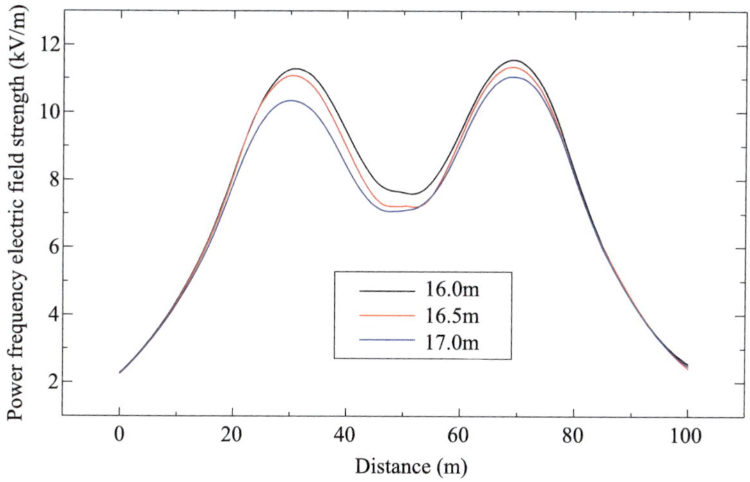

Fig. 65 Influence of the busbar height on the distribution of the power frequency electric field in the patrol corridor in front of the filter field

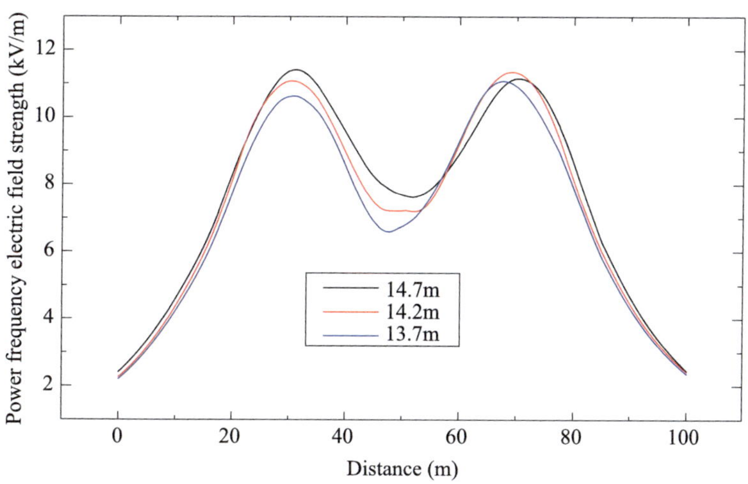

Fig. 66 Influence of the phase-to-phase spacing on the distribution of the power frequency electric field in the patrol corridor in front of the filter field

Table 53 Maximum value of the power frequency electric field strength on the patrol corridor in front of the AC filter yard at different phase-to-phase spacings

Voltage level (kV)	1000			750		
Phase-to-phase spacing (m)	13.7	14.2	14.7	10	10.75	11.5
Maximum value of the power frequency electric field strength (kV/m)	6.7	7.2	7.7	6.5	7	8.4

Chapter 3
± 1100 kV Converter Station Apparatus

The ± 1100 kV UHV DC transmission project is a brand new topic with no existing experience for reference. In the preliminary research of the CG project, China has summarized experience from the construction and operation of several ± 800 kV UHVDC transmission projects. Based on the current level of DC equipment manufacture, a ± 1100 kV UHVDC transmission system with a rated transmission capacity of 12 GW, the highest voltage level and the largest transmission power in the world, has been planned.

The research and development of ± 1100 kV, 12 GW equipment are crucial for the CG project. The UHV equipment has the following main characteristics.

(1) It has a large number for UHV engineering equipment, and each equipment has its own characteristics. It has no universal equipment specifications, and also the participation of experts from various specialties in the research work is needed.
(2) The equipment has a high insulation level, large single equipment capacity, and high difficulty in manufacture. The development of equipment needs to determine the reasonable insulation level as the first step. The over designed insulation level will increase the project investment and difficulty of manufacture; The less designed insulation level may cause safety problems for the operation of equipment.
(3) The equipment is complex, and multiple suppliers are involved in the development of new equipment. New equipment includes converter transformers, DC wall bushings, AC/DC filters, smoothing reactors, DC circuit breakers, disconnectors and fast grounding switches, surge arresters, control protection and measurement equipment, etc. The schematic DC system diagram of the CG project is shown in Fig. 1.

Z. Liu, *±1100kV UHV DC Power Transmission Technology*,
https://doi.org/10.1007/978-981-96-3712-6_3

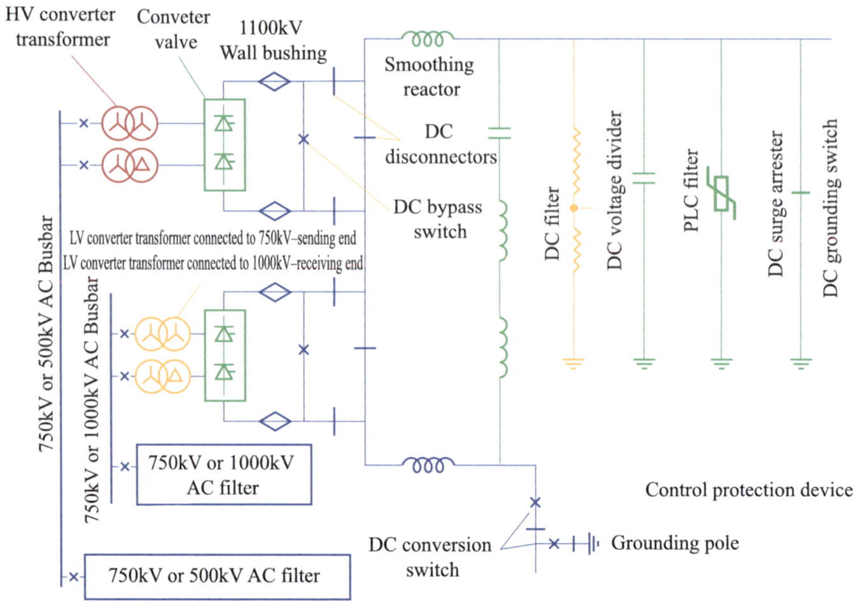

Fig. 1 Schemitic configuration of CG project

The most difficult equipment on development (the red part in Fig. 1) mainly includes the HV converter transformer at sending/receiving end, 1100 kV DC wall bushings, etc.; the equipment with relatively large development difficulty (the yellow part in Fig. 1) mainly includes the 750/1000 kV LV converter transformer, converter valve, 1100 kV DC bypass switch/DC voltage divider, 1100 kV ACF group circuit breaker, etc.; the equipment that needs new development and has relatively small difficulty (the green part in Fig. 1) mainly includes 1100 kV smoothing reactor, DC filter, PLC filter, surge arrester, disconnectors, etc.

1 Converter Transformer

As the world's first research and application of 1100 kV converter transformer, it has no relevant experience for reference. The equipment has a high insulation level, large single capacity, and high difficulty for manufacture. Its internal structure is complex, with characteristics such as electric field concentration. Due to its size and weight transportation restrictions, the control of its external dimensions and weight is important. To ensure the safe and reliable installation and operation of the equipment, it is necessary to optimize and demonstrate the design schemes for the winding, main insulation, grid-side outlet device, valve-side lead, oil tank mechanical strength, etc.

I. Technical Requirements

(I) Environmental Conditions

The highest temperature of the Changji converter station is 41.6 °C, and the lowest temperature is − 42 °C, designed for 5 mm ice coverage, pollution level E, altitude 510–520 m, seismic intensity level 7, dynamic acceleration peak 0.2*g*. The highest temperature at Guquan converter station is 40.7 °C, the lowest temperature is − 13.8 °C, designed for 10 mm ice coverage, pollution level D1, altitude 57.6–89.1 m, seismic intensity level 6, dynamic acceleration peak 0.2*g*. The environmental conditions of the converter station are shown in Table 1.

(II) AC System Conditions

At the 750 kV side of Changji converter station, the lowest steady-state voltage on extreme condition is 713 kV, the lowest steady-state voltage is 750 kV, the steady-state voltage on normal condition is 775 kV, the highest steady-state voltage is 800 kV, and the highest steady-state voltage on extreme condition is 800 kV. At the 500 kV side of Guquan converter station, the lowest steady-state voltage on extreme condition is 475 kV, the lowest steady-state voltage is 490 kV, the normal steady-state voltage is 510 kV, the highest steady-state voltage is 525 kV, and the highest steady-state voltage on extreme condition is 550 kV; At the 1000 kV side of Guquan converter station, the lowest steady-state voltage on extreme condition is 950 kV, the lowest steady-state voltage is 1000 kV, the normal steady-state voltage is 1050 kV, the highest steady-state voltage is 1070 kV, and the highest steady-state voltage on extreme condition is 1100 kV. The AC system conditions and frequency of the converter station are shown in Tables 2 and 3.

(III) Transportation on Large Equipment

The transportation requirements for the converter transformers at Changji converter station: the HV side is assembled in a factory with production and testing conditions, where is nearby the sending end of converter station; the LV side can be transported by

Table 1 Environmental conditions in converter station

Parameters	Changji converter station	Guquan converter station
Maximum temperature in extreme condition (°C)	41.6	40.7
Minimum temperature in extreme condition (°C)	− 42	− 13.8
Designed ice-coating (mm)	5	10
Pollution class	e	D1
Altitude (m)	510–520	57.6–89.1
Earthquake intensity	7	6
Dynamic peak acceleration	0.2*g*	0.2*g*

Table 2 AC system conditions in converter station (kV)

Parameters	Changji converter station	Guquan converter station	
		500 kV Side	1000 kV Side
Minimum steady state voltage in extreme condition	713	475	950
Minimum steady state voltage	750	490	1000
Nominal steady state voltage	775	510	1050
Maximum steady state voltage	800	525	1070
Maximum steady state voltage in extreme condition	800	550	1100

Table 3 Frequency in AC system (Hz)

Converter station	Changji	500 kV side on Guquan	1000 kV side on Guquan
Nominal frequency	50	50	50
Dynamic frequency deviation	± 0.2	± 0.1	± 0.1
Frequency deviation after clearance of fault in 10 minutes	± 0.5	± 0.5	± 0.5
Frequency deviation under contingency	± 1.0	± 1.0	± 1.0

railway + road. The limit of transportation size for the HV side converter transformer ($L \times W \times H$): 15,000 mm × 5,500 mm × 6,500 mm, transportation weight limit: 600 t; the limit of railway transportation size for the LV side converter transformer ($L \times W \times H$): 13,000 mm × 3,500 mm × 4,850 mm, railway transportation weight limit: 360 t.

The transportation requirements for the converter transformer at Guquan converter station: waterway + road, the limit of transportation size ($L \times W \times H$): 15,500 mm × 5500 mm × 6500 mm, transport weight limit: 600 t.

(IV) Main Parameters

(1) Changji Converter Station. The converter transformer is single-phase, double-winding, on-load tap changer, oil-immersed transformer, with ODAF cooling method, and the neutral point of the grid side is directly grounded; the number of voltage regulation levels of the on-load tap changer installed at the neutral point of the grid side: (− 5, + 25), each level voltage: 0.86%U_N (U_N is the rated value of the grid side voltage, kV). When the average temperature rise of the winding is not exceeded 55 K, the capacity of both the grid side and valve side windings is 607.5 MVA.

The maximum steady-state voltages at the winding on grid side, Y-connected valve side and D-connected valve side are 461.9, 140.6 and 243.6 kV respectively; the rated voltages at the winding on grid side, Y-connected valve side and D-connected valve side are 447.5, 140.6 and 243.6 kV respectively. The rated frequency is 50 Hz.

The rated continuous current of the main tap at the winding on grid side is 1358 A, at the winding on Y-connected valve side is 4454 A, and at the winding on the D-connected valve side is 2572 A; under the most negative tap 1.05 (per unit value) for 2 h without spare conditions, the continuous current of the grid side winding is 1518 A, the Y-connected valve side winding is 4454 A, and the D-connected valve side winding is 2572 A. The current density of the winding under the rated tap does not exceed 3.2 A/mm^2, and the magnetic flux density does not exceed 1.75 T.

The impedance voltage and allowable variation range between the grid side and the valve side (YNyn0, YNd11): the rated tap (main tap) is (20 ± 0.9)%, the minimum tap (− 5 taps) is (20 ± 0.9)%, the maximum tap (+ 25 taps) is (20 ± 0.9)%. The impedance deviation between all converter transformers is less than 2%; the DC bias current is 10 A.

See Table 4 for the insulation level.

Table 4 Insulation level

Item		Winding on grid side (kV)	Winding on valve side (kV)			
			High voltage		Low voltage	
			Y1	D1	Y2	D2
Lightning impulse *LI*	Terminal 1	1950	2300	1980	1350	1240
	Terminal 2	185	2300	1980	1350	1240
Chopped lightning impulse *LIC*	Terminal 1	2100	2530	2175	1485	1360
	Terminal 2	–	2530	2175	1485	1360
Switching impulse *SI*	Terminal 1	1550	–	–	–	–
	Terminal 2	–	–	–	–	–
	Terminal 1 + Terminal 2	–	2100	1840	1250	1175
AC apply in short term (neutral point)	Terminal 1 + Terminal 2	95	–	–	–	–
AC induction in short term	Terminal 1	900	–	–	–	–
AC induction in long term + partial discharge	Terminal 1 (U1)	800	239	414	239	414
	Terminal 1 (U2)	693	211	365	211	365
AC apply in long term + partial discharge	Terminal 1 + Terminal 2	–	1297	987	676	366
DC apply in long term + partial discharge	Terminal 1 + Terminal 2	–	1791	1353	914	475
DC polarity reversal + partial discharge	Terminal 1 + Terminal 2	–	1386	1021	655	289

(2) Guquan Converter Station. The converter transformer of Guquan Converter Station is a single-phase, double-winding, on-load tap changer, oil-immersed transformer, with ODAF cooling method, and the neutral point of the grid side is directly grounded; the on-load tap changer installed at the neutral point of the grid side, the number of high voltage regulation levels: (– 5, + 25), each level voltage 1.25% U_N, the number of low voltage regulation levels (– 10, + 20), each level voltage 0.65% U_N. When the average temperature rise of the winding is $\leq$ 55 K, the capacity of both the grid side and valve side windings is 587.1 MVA.

The highest steady-state voltage of the grid side winding of the HV converter transformer is 303.1 kV, and the highest steady-state voltage of the grid side winding of the LV converter transformer is 617.8 kV; the Y-connected valve side winding is 138.1 kV, and the D-connected valve side winding is 239.2 kV. The rated voltage of the grid side winding is 294.5 kV, the rated voltage of the LV converter transformer at the grid side winding is 606.2 kV; the Y-connected valve side winding is 131.8 kV, and the D-connected valve side winding is 228.3 kV, with a rated frequency of 50 Hz.

The main tap rated continuous current of the HV converter transformer at the grid side winding is 1993.8 A, and the main tap rated continuous current of the LV converter transformer at the grid side winding is 968.4 A; the Y-connected valve side winding is 4454 A, and the D-connected valve side winding is 2572 A. The continuous current without backup for 2 h under the most negative tap 1.05 (per unit value) of the HV converter transformer at the grid side winding is 2276 A, and the continuous current without backup for 2 h under the most negative tap 1.05 (per unit value) of the LV converter transformer at the grid side winding is 1109 A; the Y-connected valve side winding is 4454 A, and the D-connected valve side winding is 2572 A. The current density of the winding under the rated tap should not exceed 3.2 A/mm^2, and the magnetic flux density should not exceed 1.75 T.

The impedance voltage and allowable variation range between the grid side and the valve side (YNyn0, YNd11): the rated tap (main tap) is (22 ± 1.0)%, the minimum tap (– 5 tap) is (22 ± 1.0)%, the maximum tap (+25 tap) is (22 ± 1.0)%. The impedance deviation between all converter transformers should not exceed 2%; the DC bias current is 10 A.

See Tables 5 and 6 for insulation level.

(V) Performance Requirements

(1) The temperature rise limit at the rated capacity, the top oil temperature rise requirement is $\leq$ 45 K, the average winding temperature rise requirement is $\leq$ 53 K, the hot spot temperature rise of the winding requirement is $\leq$ 66 K, the oil tank, iron core and structural parts temperature rise requirement is $\leq$ 73 K, the requirement of winding hot spot temperature during short-term overload is $\leq$ 120 °C.
(2) Under the system's highest voltage and frequency, and when the valve side is at the highest steady-state voltage, the converter transformer should

Table 5 Insulation level on HV side

Item		Winding on grid side (kV)	Winding on valve side (kV)	
			Y1	D1
Lightning impulse *LI*	Terminal 1	1550	2300	1930
	Terminal 2	185	2300	1930
Chopped lightning impulse *LIC*	Terminal 1	1705	2530	2125
	Terminal 2	–	2530	2125
Switching impulse *SI*	Terminal 1	1175	–	–
	Terminal 2	–	–	–
	Terminal 1 + Terminal 2	–	2100	1785
AC apply in short term (neutral point)	Terminal 1 + Terminal 2	95	–	–
AC induction in short term	Terminal 1	680	–	–
AC induction in long term + partial discharge	Terminal 1 (U1)	550	234	405
	Terminal 1 (U2)	476	206	358
AC apply in long term + partial discharge	Terminal 1 + Terminal 2	–	1292	982
DC apply in long term + partial discharge	Terminal 1 + Terminal 2	–	1786	1347
DC polarity reversal + partial discharge	Terminal 1 + Terminal 2	–	1384	1018

be able to operate normally; the converter transformer should be able to operate continuously under 110% of the rated voltage when unloaded.

(3) The bushing outlet terminal on grid side should be designed according to anti-corona requirements.

(4) The minimum air gap requirement of bushing to ground/wall on converter transformer valve side, see Table 7.

II. Technical Key Points

(I) Technical Features

The transmission capacity of the ± 800 kV UHVDC transmission project has been increased from 8, 10 to 12 GW, and the voltage level has been increased to ± 1100 kV. The receiving end is layered into the 500/1000 kV AC grid. To achieve this goal, the converter transformer of large capacity and layered access technology is the key to the main equipment research and development. In this case, the capacity of the converter transformer has increased by 20%, and the voltage has increased by 37.5%, resulting in a significant increase in the temperature rise and insulation design difficulty of the converter transformer. The developed 1100 kV converter transformer is shown in Figs. 2 and 3.

Table 6 Insulation level on LV side

Item		Winding on grid side (kV)	Winding on valve side (kV)	
			Y2	D2
Lightning impulse *LI*	Terminal 1	2250	1350	1235
	Terminal 2	185	1350	1235
Chopped lightning impulse *LIC*	Terminal 1	2400	1485	1355
	Terminal 2	–	1485	1355
Switching impulse *SI*	Terminal 1	1800	–	–
	Terminal 2	–	–	–
	Terminal 1 + Terminal 2	–	1250	1165
AC apply in short term (neutral point)	Terminal 1 + Terminal 2	95	–	–
AC induction in long term + partial discharge	Terminal 1 (U1)	1100	234	405
	Terminal 1 (U2)	953	206	358
AC apply in long term + partial discharge	Terminal 1 + Terminal 2	–	672	362
DC apply in long term + partial discharge	Terminal 1 + Terminal 2	–	908	470
DC polarity reversal + partial discharge	Terminal 1 + Terminal 2	–	653	287

Table 7 Minimum air gap requirement of bushing to ground/wall on converter transformer valve side

	Type of converter transformer	Terminal	Changji	Guquan
Air gap of bushing to ground/wall on valve side (m)	Y1	Valve a (ground/wall)	11.1	11.1
		Valve b (ground/wall)	10.7	10.7
	D1	Valve a and b (ground/wall)	10.5	10.5
	Y2	Valve a (ground/wall)	4.6	4.6
		Valve b (ground/wall)	4.6	4.6
	D2	Valve a and b (ground/wall)	4.5	4.5

Fig. 2 1100 kV converter transformer in Changji converter station

Fig. 3 1100 kV converter transformer in Guquan converter station

1. High short-circuit impedance

The impedance of the converter transformer is usually higher than that of the AC transformer, which is not only to limit the short-circuit current according to the short-circuit capacity of the converter valve, but also to limit the rise rate of the valve current during commutation. However, the high short-circuit impedance will increase reactive power loss and reactive power compensation equipment quantity, and cause excessive voltage drop of commutation. With the increase of DC transmission voltage, the capacity of a single converter transformer further increases. Due to manufacture reasons and large equipment transportation restrictions, the impedance of converter transformer at Changji converter station reaches 20%, and that of the converter transformer at Guquan converter station reaches 22%. In addition, the difference between the impedances of each phase of the converter transformer must be kept to a minimum (generally not more than 2%), otherwise it will cause an increase in the non-characteristic harmonic component of the converter transformer current.

2. Large rated capacity

The DC transmission capacity of the ± 1100 kV UHVDC transmission project reaches 12 GW, and the capacity of the converter transformer also increases accordingly. The rated capacity of the converter transformer at Changji converter station reaches 607.5 MVA, and the rated capacity of the converter transformer at Guquan converter station reaches 587.1 MVA.

3. High short-circuit current withstand capability

Due to the presence of a DC component in the fault current, the maximum asymmetric short-circuit current decay time that the converter transformer can withstand is longer, and it will remain at a relatively high level until the protection action. The short-circuit electrodynamic force is directly proportional to the square of the short-circuit current amplitude. The short-circuit electrodynamic force is applied to the winding and lead support structure. The converter transformer should be able to withstand larger short-circuit stresses. The commutation failure of the converter valve will also cause the converter transformer to sustain more electrodynamic force impacts.

4. Wide range of on-load tap changer

The converter transformer has a wide range of on-load voltage regulation to ensure that the voltage variation and trigger angle operate within an appropriate range. Especially during operation of reduced DC voltage, the number of positive taps can reach more than 20. The voltage regulation range of the converter transformer at Changji converter station is $(-5, +25) \times 0.86\%$, and the voltage regulation range of the converter transformer at Guquan converter station is $(-5, +25) \times 1.25\%$ and $(-10, +20) \times 0.65\%$.

(II) Key Technologies of Railway Transportation of Converter Transformer with a Large Capacity

1. Structure of converter transformer

The structure of the single-phase 4-column converter transformer at the Changji converter station is shown in Fig. 4. Magnetic shunts are set in the insulating terminal rings at the upper and lower terminlas of the winding, which are of a split-core structure, with 4 pieces on each column, connected to the grounding lead and the clamp.

Fig. 4 Main structure of single-phase 4-column converter transformer at the low voltage side of Changji converter station

The insulating platen of upper and lower terminals have specially shaped electrostatic shielding tubes at the magnetic shunt locations, connected to the clamps, improving the local electric field distribution. The insulation structure is complex, using a large number of molded insulation parts to ensure electrical strength.

The lower support uses an oil-guided pad structure, with the net and valve side windings independently oiling, providing good heat dissipation, strength, and easy assembly. The insulating end uses a combination of end rings, corner rings, and sealing rings, with a reasonable oil circuit structure, ensuring even oil distribution.

The main structure is compressed by pressure blocks, avoiding the impact of the pin structure on the end wiring, making the structure compact and reliable.

All parts on the structure are rounded to reduce the probability of partial discharge. The valve side windings between each column are connected by a "hand-in-hand" connection, with insulation paper wrapped around the outside of the shielding tube.

There are equipotential connection lines inside the shielding tube. The "hand-in-hand" connection and outgoing wires at the valve side windings leave through the winding side, the outgoing wires leave at the upper end on grid side windings, make the spatial layout compact. A large diameter metal tube is used for shielding of lead wires. One end of the shielding tube extending into the winding body and the other end directly extending into the equalizing ball at the end of the tube, making the shielding

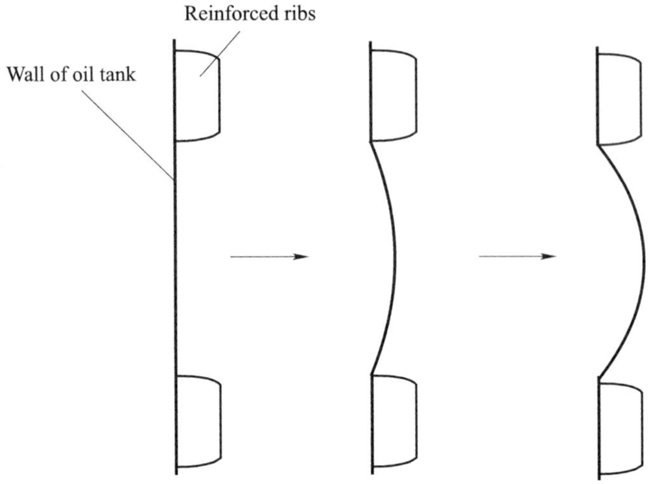

Fig. 5 Optimization of oil tank structure

structure reasonable and reliable. The outside of the shielding tube is wrapped with paper insulation, meeting the insulation withstand voltage requirements of AC and DC.

2. Special structure of arch-shaped oil tank

The converter transformer with a large-capacity for railway transportation uses a special arch-shaped oil tank structure, ensuring that the winding insulation distance, magnetic flux density (1.78 T), current density (regulation 2.9 A/mm^2, grid side 3.5 A/mm^2, valve side 3.5 A/mm^2) and other control indicators do not increase compared to conventional converter transformers, and the total loss (1400 kW) is at the same level. The optimization of the oil tank structure is shown in Fig. 5.

The converter transformer oil tank at the Changji converter station is reinforced with a groove-shaped reinforcing iron. The materials for the oil tank walls, bottom, cover, and reinforcing iron are all high-strength structural steel. The bottom of the tank uses a whole steel plate, each of which is ultrasonically inspected before welding to ensure the quality of the steel plate. The welding seams of the oil tank assembly and the important reinforcing rib are also ultrasonically inspected. The welding of the oil tank uses high-quality welding rods, advanced equipment such as submerged arc welding machines, and gas shielded welding machines to ensure the quality of the welding. The seals use molded rubber materials. The inner wall of the oil tank is welded with a copper shield. The long positioning groove along the flange of the tank is matched with the cover for positioning, the sealing surface is flat, and it is a welded structure to ensure sealing performance.

3. Forced-directed oil cooling method

Figure 6 shows the forced-directed oil cooling method for converter transformers with large-capacity in railway transportation. The cold oil mainly enters the winding (winding loss) directly through the pipeline, and part of it enters the oil tank (iron core loss and other stray losses) through the bypass or opening pipe. In the forced oil non-directed cooling method, the oil cooled by the cooler mainly enters the oil tank directly through the pipeline, and the flow of oil is mainly caused by the temperature difference of the oil, as shown in Fig. 7.

The temperature rise test adopted a more stringent temperature stability standard, which was raised from a temperature rise of less than 1 K within 1 h to a temperature rise of less than 1 K within 3 h, and the overall temperature rise process exceeded 30 h.

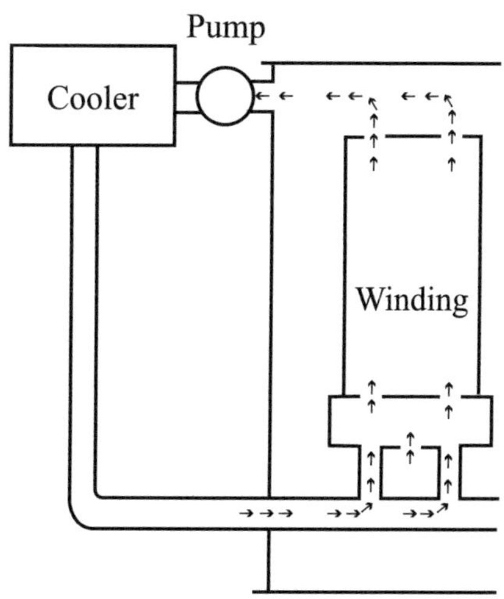

Fig. 6 Forced-directed oil cooling method

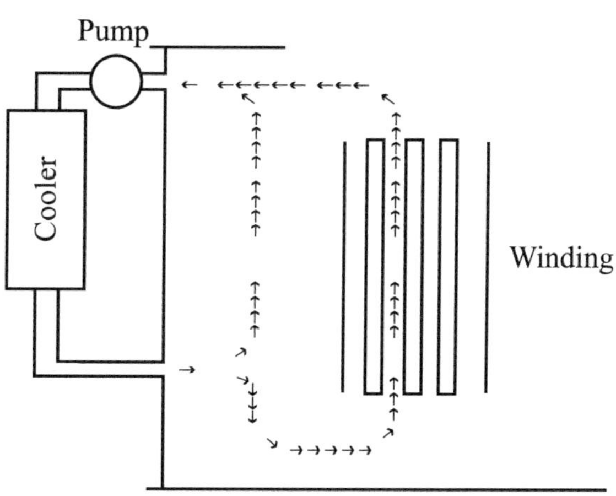

Fig. 7 Forced non-directed oil cooling method

(III) Key Technologies of 1000 kV Converter Transformer Connected to Grid Side

The grid side access of the 1000 kV converter transformer faces the "double difficulty" of insulation design and leakage magnetic control, and the grid side winding adopts a simple structure and does not require a split terminal-out structure of winding, which further increases the difficulty of insulation design.

1. Winding structure

The winding structure of the 1000 kV converter transformer connected to the grid side is sorted as iron core—valve side—grid side—regulation. The regulation winding adopts a single-layer cylindrical structure, the grid side winding adopts a continuous structure, and the upper and lower ends of the valve side winding adopt an internal screen continuous structure. The longitudinal capacitance is adjusted by changing the number of turns of the internal shield of the winding end wire group, and it leads to obtain a good lightning impulse voltage waveform distribution. This structure facilitates the 1000 kV end-out of the grid side and can better control the short-circuit impedance size deviation.

2. Outgoing line structure

In order to adapt to the characteristics of layered access at 1000 kV grid side, and considering the insulation distance and mechanical strength, a design of platen shown in Fig. 8 was developed for winding opening, which resolved the conflict between insulation and mechanical requirements for winding opening.

In order to strictly test the insulation performance of outgoing line structure at 1000 kV grid side, during the long-duration induction AC voltage withstand test (ACLD) with partial discharge measurement, an excitation of 1100 kV voltage was

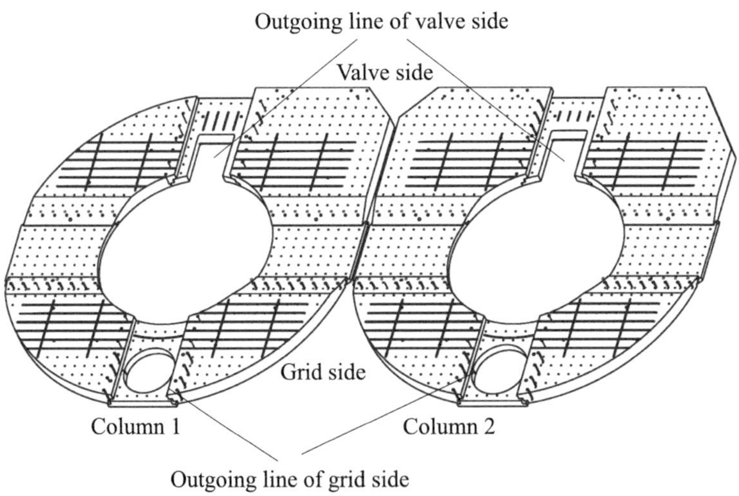

Fig. 8 Platen structure of winding

applied for 5 min (without frequency correction), achieving the design goal of no partial discharge.

3. Shielding structure

Since the grid side voltage of converter transformer reaches 1000 kV, its magnetic leakage and temperature rise control become problems that must be solved. For this reason, a new structure of large-volume composite shielding is adopted to solve the problem of shielding of magnetic leakage under large current. Aluminum shielding is laid on the inner walls of both sides of the oil tank, and magnetic shielding is laid on the upper and lower inner walls of the oil tank (that is, the top and bottom of the tank).

In addition, the grid side winding of 1000 kV converter transformer adopts a combination of self-adhesive flat wires, which effectively controls the winding loss through reasonable design. The loss calculation results are shown in Table 8. Table 8 shows that the calculation results of total load loss are all less than the guaranteed value of 1150 kW.

In summary, the temperature rise test results of 1000 kV converter transformer at grid side are shown in Table 9. As shown in Table 9, it is clear that the temperature rise margin is sufficient (about 15 K lower than the required value).

(IV) Key Technology of 1100 kV Converter Transformer Connected to Valve Side

The arrangement of valve side bushing of ± 1100/825 kV converter transformer differs from that of ± 800 kV level converter transformer. It is not placed on the short-axis box wall, but on the large turning elevation seat on the box cover, ensuring the corona ring at the top of the valve side bushing's insulation distance from the ground. The appearance and body schematic diagrams of HV converter transformers at Changji and Guquan converter stations are shown in Figs. 9 and 10.

(V) Large equipment Transportation

The transportation parameters of the HV converter transformer are large, and the road transportation distance is long. To meet the project needs, a new 650 t bridge frame car and a 600 t converter transformer on-site transfer cart have been developed, both of them are currently the world's largest, as shown in Figs. 11 and 12.

Table 8 Calculation of load losses (kW)

Item	Basic resistance loss	Eddy current loss	Stray loss	Total loss of load	Harmonic loss
Y-connected winding on grid side	856	158	272	1286	1551
D-connected winding on grid side	855	148	271	1274	1529

Table 9 Measured results (averaged) of temperature rise of 12 GW, 1000 kV converter transformer at grid connected side (K)

Winding	Average temperature rise of winding		Temperature rise of hot spot on winding		Temperature rise in top layer of oil		Average temperature rise of oil
	Measured value	Guaranteed value	Measured value	Guaranteed value	Measured value	Guaranteed value	Measured value
Winding on grid side	33	53	40	66	20	50	17
Y-connected winding on valve side	36	53	49	66			
D-connected winding on valve side	36	53	47	66			

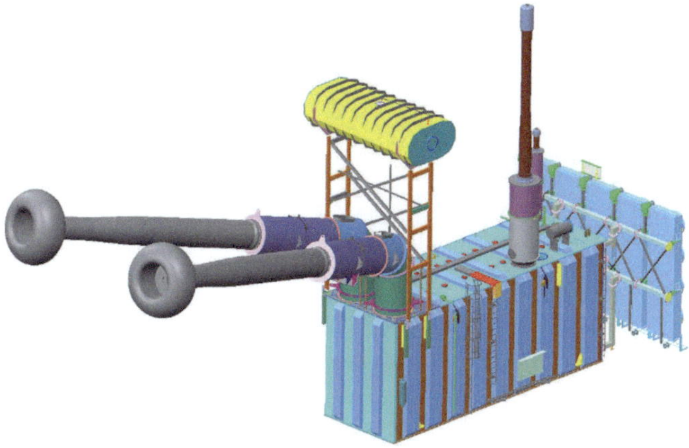

Fig. 9 External configuration of converter transformer at HV side of Changji station

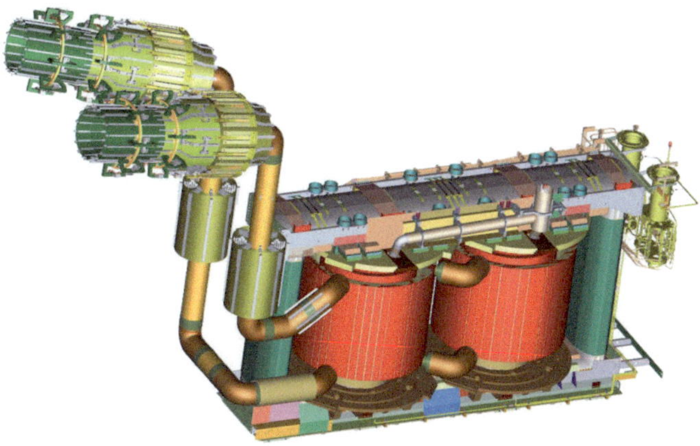

Fig. 10 External configuration of converter transformer at HV side of Guquan station

(VI) Equipment Upgrade and Renovation

Due to the increase in size and weight of 1100 kV converter transformer, and the thickening of insulation materials, the corresponding winding, iron core processing, structure assembly, gas phase drying tank, lifting and transportation, processing lathe and other equipment all need to be upgraded and transformed. The requirements for upgrading and renovation equipment are shown in Table 10.

Fig. 11 650 t bridge-type frame vehicle

Fig. 12 600 t on-site transfer vehicle of converter transformer

Table 10 Requirement of equipment upgrade and renovation

Item	Requirement of parameter	Item	Requirement of parameter
Vertical coil winding machine	Radius 4.2/3.5 m, weight of manufactured winding 20 t	Kerosene vapor-phase drying tank	15 m × 6 m × 7 m, minimum Power 600 kW
Winding press	Compressible radius 4.5/3.8 m, pressure 200–450 kn	Environmental requirements for "desert houses"	24 m × 7 m × 7.5 m, Humidity < 30%
Air ride truck	Maximum load weight 850 t	Pump for vacuum and hot oil circulation	Single truck power ≥ 6000 m^3/h
Hoisting and lifting facility	Gantry crane of final assembly shop 450 t	Sand blasting and spraying shop for oil tank	15 m × 6 m × 6 m

The 1100 kV converter transformer, DC withstand voltage of DC bushing, AC withstand voltage, impact withstand voltage test device and test hall and other test facilities need to be fully upgraded. The effects of the newly built assembly workshop and test hall are shown in Fig. 13. The upgrading requirements of test equipment are shown in Table 11, and the upgrade of main equipment on each factory has been shown in Table 12.

(a) Assembly workshop

(b) Test hall

Fig. 13 Newly established assembly workshop and test hall

Table 11 Requirement of testing facility upgrade and renovation

Item	Parameters
Series resonance device	Rated Voltage: 1800 kV, Rated Capacity: 25 200 kVA
DC voltage generator	Rated Voltage: 2250 kV, Rated Current: 35 mA
Impulse voltage generator	Rated Voltage: 6000 kV, Energy Requirement: 900 kJ
Grounding resistance	< 0.5 Ω

Table 12 Situation of equipment upgrade and renovation for different manufacturers

Manufacturers	Manufacture of converter transformer	Production process	Renovation project
Manufacturer 1 of converter transformer	LV side 275 kV 1 unit, 550 kV 1 unit, HV side 825 kV 1 unit, 1100 kV 1 unit at Changji converter station	Laminated iron core	Iron core shearing device
			Iron core stacking equipment
			Work platform
		Winding wiring	Vertical coil winding machine
			Horizontal coil winding machine
			Adjustable mold
			Winding press
			Winding lifting bracket
			Winding transfer system
		Production of oil tank	Automatic sand blasting machine
		Desiccation	Vapor-phase drying tank
		Transportation tool	High tonnage air cushion vehicle
			Large scale loading vehicle
	LV side 275 kV 1 unit, 550 kV 1 unit, HV side 825 kV 1 unit, 1100 kV 1 unit at Changji converter station	Rountine test	AC withstand voltage test device
			DC system withstand voltage test
			Partial discharge monitoring equipment
			Ultrasonic detector
			UV imager
Manufacturer 2 of converter transformer	LV side 275 kV 1 unit, 550 kV 1 unit, HV side 825 kV 1 unit, 1100 kV 1 unit at Guquan converter station	Laminated iron core	Iron core stacking platform
		Winding wiring	Winding hoisting system
		Active part assembly	Body lifting system
		Transportation tool	Air cushion car

(continued)

Table 12 (continued)

Manufacturers	Manufacture of converter transformer	Production process	Renovation project
		Test	Impulse generator
			Layout of the experimental hall
		Bushing workshop	Newly established plant
			Renovation of connection between existing factory building and the testing hall
			Upgrade core body bracket
			Upgrade processing and forming equipment
			Upgrade transportation tool for entire transformer
Manufacturer 3 of converter transformer	LV side 275 kV 1 unit, 550 kV 1 unit, HV side 825 kV 2 units, 1100 kV 2 units at Changji converter station	Winding	Winding set table
			Winding press
		Iron core	Iron core stacking platform
		Desiccation	Vapor-phase drying tank
		Transportation tool	Air cushion transport vehicle
			Transformer lifting beam and lifting rope
		General assembly	Oil filter machine
			Assembly work frame
		Rountine test	Impulse voltage generator
			Chopping device
			Capacitive voltage divider
			Compensation capacitor tower
			Series resonant device
			DC voltage generator
			Variable frequency power supply
			Shielding Hall

(continued)

Table 12 (continued)

Manufacturers	Manufacture of converter transformer	Production process	Renovation project
Manufacturer 4 of converter transformer	HV side 825 kV 4 units, 1100 kV 4 units at Guquan converter station	Production of oil tank	Lifting equipment
			Submerged arc welding
			Welding robot
		Active part assembly	Lifting equipment for entire set of windings
			Body lifting equipment
		Desiccation	Vapor-phase drying tank
			Foundation of vapor-phase drying tank
			Kerosene moisture content test
			Purchase 2 air cushion vehicles (with a load capacity of 400 t)
		General assembly	Low frequency heating device
			Assembly plant
			Air cushion car
			1 newly established mobile vacuum unit
			1 newly established vacuum oil filter
			General assembly bridge crane
		Routine test	Experimental Hall
			Power frequency and series resonances
			DC withstand voltage device
			Impulse voltage generator
Manufacturer 5 of converter transformer	LV side 275 kV 5 units, 550 kV 5 units at Guquan converter station	Winding	Upgrade 1 horizontal winding machine
			Add 16 dedicated adjustable molds (maximum diameter 3300 mm)
			Upgrade coil press

(continued)

Table 12 (continued)

Manufacturers	Manufacture of converter transformer	Production process	Renovation project
		Transportation tool	Purchase 2 air cushion vehicles (with a load capacity of 280 t)
		Production of oil tank	New spray drying integrated room (20 m × 9.5 m x 10 m)
Manufacturer 6 of converter transformer	HV side 825 kV 6 units, 1100 kV 6 units at Changji converter station	Winding	Add 1 set of winding lifting equipment
			Renovation of sand blasting room
			Transformation of winding machine
		Routine test	Series resonant device
			DC voltage generator
			Impulse voltage generator
			Grounding resistance

(VII) Solutions for Key Manufacturing Problems

The iron core stacking platform and DC generator are shown in Figs. 14 and 15 respectively.

Fig. 14 Iron core stacking platform

Fig. 15 DC generator

Table 13 Hotspot temperature rise margin of converter transformer at LV side of Changji converter station (K)

Optimization method	Original method, change coefficient of hotspot		Implementation of blocking holes etc.				Change of wire gauge
No. of transformer	Y1	Y2	Y3	Y4	Y5	Y6	Y7
Hotspot temperature rise at grid side	66.0	60.0	65.1	64.2	63.7	59.0	58.0
Optimization method	Original method, change coefficient of hotspot		Implementation of blocking holes etc.				Change of wire gauge
No. of transformer	D1	D2	D3	D4	D5	D6	D7
Hotspot temperature rise at grid side	63.4	62.1	57.7	58.7	62.1	58.6	56.7

(1) Tooling, test facility modification and verification: Problems such as damage to test equipment like lightning impact and DC generator, and the angle issue of the iron core standing tooling have been resolved.
(2) Design issues: Two successive design optimizations have been made to increase the hotspot temperature rise margin of the LV converter transformer at Changji converter station. For details, see Table 13.
(3) Design issues: The "upside-down" error inside the outgoing line device of the 1100 kV converter transformer at Changji converter station has been corrected.

(a) 1100kV Insulation fall off of outgoing feeder of converter transformer

(b) HD original outgoing feeder

(c) Optimized structure

Fig. 16 Insulation fall off of outgoing feeder, HD original outgoing feeder and optimized strcuture of 1100 kV converter transformer

Fig. 17 Creepage of clamp iron yoke cushion block

An external suspension frame has been added to the outgoing line device of the 825 kV converter transformer to ensure reliable support and suspension of the overweight insulator. The 1100 kV converter transformer outgoing line insulation falling off, the 825 kV converter transformer original outgoing line structure schematic and the improved structure are shown in Fig. 16.

(4) Design issues: Additional copper shielding insulation has been installed on the network side of the LV converter transformer at Guquan converter station to solve the "ground potential, high field strength" design weakness of the iron core clamp, improving insulation margin. All subsequent product tests have passed at once. The creepage of the clamp iron yoke pad and the creepage of the network side outgoing line device are shown in Figs. 17 and 18 respectively.

Fig. 18 Creepage of outgoing feeder at grid side

Fig. 19 Installation of protective screen to keep low humidiy in hall

(5) Test issues: Through multiple technical measures (such as corona ring replacement, etc.), problems such as poor conditions in the test halls of multiple converter transformer manufacturers and twists and turns in the test process have been solved. A test summary has been compiled to pre-control test risks. The installation of barriers to ensure low humidity in the hall and the high field strength position of the larger corona ring shielding casing are shown from Figs. 19, 20 and 21.

Fig. 20 Large corona ring to shield high field intensity area of bushing

Fig. 21 Implementation of hard pipe to decrease interference

(6) Design issues: For the first time, a tail ring that reduces the local electric field was applied at the root of the 1100 kV converter transformer bushing at the Changji converter station. The internal insulation of the valve-side outgoing device (elevated seat) is tightly fitted, with a gap $\leq$ 16 mm from the end of the bushing, and the bushing fixing point is nearly 3000 mm from the end. The self-weight of the bushing is large. When the front end of the bushing is hanging down and the tail end is upturned, mechanical damage "indentations"

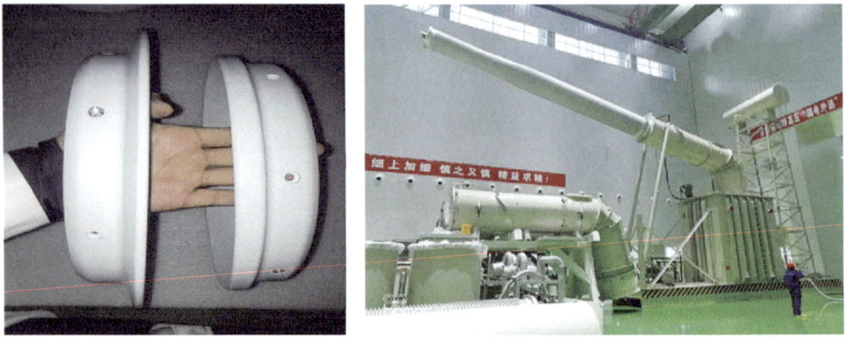

(a) Comparison of tailed ring before and after optimization

(b) HY independent test facility

Fig. 22 Comparison of tailed ring before and after optimization, HY independent test device of 1100 kV converter transofmer bushing

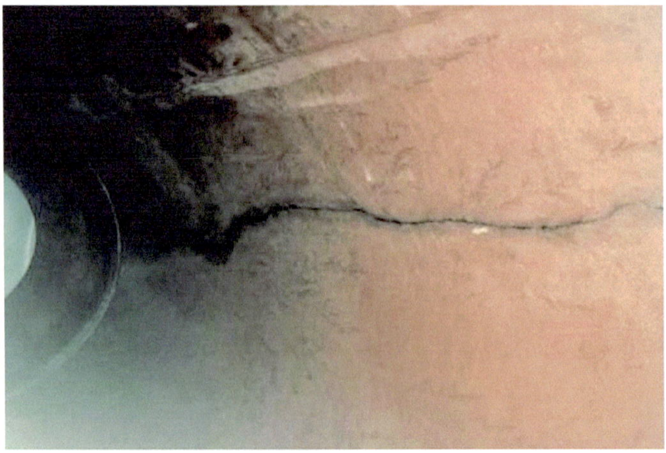

Fig. 23 Creepage of hoist seat at 825 kV converter transformer

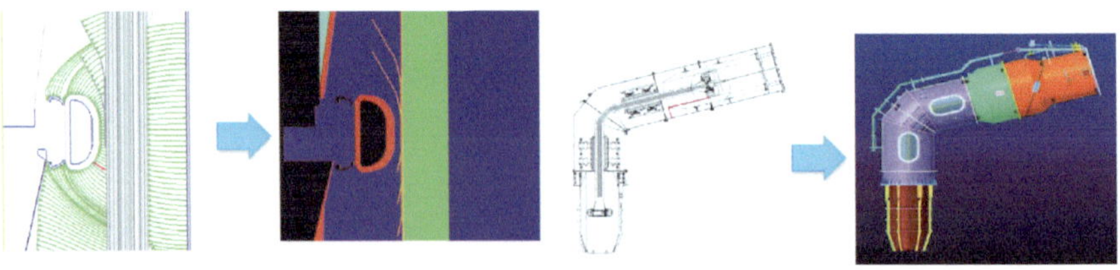

(a) add angle ring

(b) expanding

Fig. 24 Improvement of design (add angle ring + expanding)

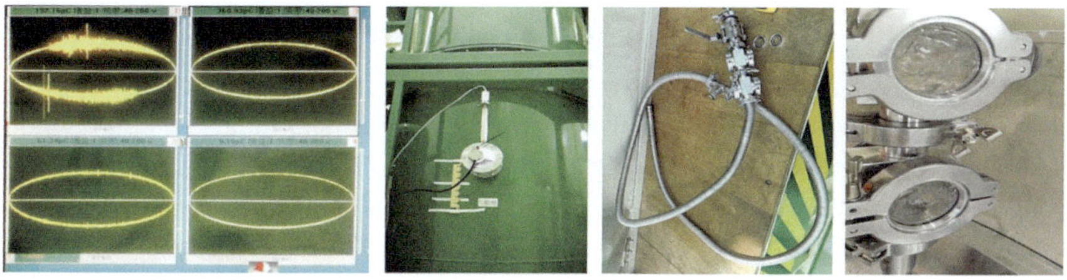

(a) Discharge waveform of bubble (b) Vacuum display of oil filter machine (c) High sealing metal hose (d) Monitoring of oiling bubble

Fig. 25 Oiling method improvement

appear on multiple devices. The radial size of the bushing tail ring is improved, the insulation gap is increased under the premise of ensuring the balance of the AC and DC electric fields, and the mechanical strength is enhanced by using bonded cardboard. The above measures have been successfully verified on two converter transformer bodies through endoscope inspection and insulation testing, and a separate test device has been developed to further strengthen the assessment. A comparison before and after the improvement of the 1100 kV converter transformer bushing tail ring, and the 1100 kV converter transformer independent test device are shown in Fig. 22.

(7) Design issues: The insulation design margin of the 825 kV converter transformer outgoing device is tight under the AC electric field, the tolerance for installation process is low, and the extreme test conditions and installation errors are not fully considered. About 50% of the converter transformers discharge during the AC external test on the valve side. Increasing the formed angle ring to increase the creepage distance, expanding the diameter of the elevated seat shell and strictly controlling the eccentricity of the equalizing ball to increase the distance to the ground, the subsequent tests passed at one time. The creepage of the 825 kV converter transformer elevated seat, and the design improvement (adding angle ring + expanding diameter) are illustrated in Figs. 23 and 24 respectively.

(8) Process issues: The valve-side bushing of the HV converter transformer at the Changji converter station faces a significant increase in voltage level, an increase in the oil cavity of the capacitor core, and a small tilt angle of the installation position. The oil injection requirements of the converter transformer adopt a new high vacuum low gas content index control, which is about one order of magnitude higher than the commonly used gas content index in China. The insufficient original process level leads to partial discharge exceeding the standard during multiple bushing tests. The valve-side bushing uses a high-sealing metal hose for vacuuming, and the pressure of the last stage oil filter vacuum degassing tank is controlled to $\leq$ 40 Pa, and bubbles are monitored in real time throughout the process. After the process adjustment, all tests passed at one time. See Fig. 25 for the improvement of the oil injection process. This process has been promoted and used in field assembly and multiple transformer factories.

2 Valve-Side Bushings and Wall Bushings

The valve-side bushing of UHV converter transformer is one of the key components of the converter transformer, playing the roles of electrical connection, insulation isolation, and mechanical connection. It is the necessary channel for the UHV substation and converter station to send out electrical energy. The wall bushing is the only electrical connection device connecting the high-voltage large-capacity electrical equipment inside and outside the valve hall of the DC transmission project converter station. It is in the "neck" position of the DC transmission system, carrying the voltage and current of the entire system, and playing the role of insulation and mechanical support. The wall bushing has been subjected to high voltage and large current for a long time, the operating conditions are complex, and it is difficult to control the internal and external insulation field strength of the bushing and the heating of the center conductor.

I. Technical Requirements

(I) Valve Side Bushing

(1) The types of valve side bushings for various models of converter transformers are shown in Table 14.
(2) If SF_6 gas insulation is used, an observable density (pressure) gauge or density relay should be installed, and the analog signal should be transmitted remotely.
(3) The requirements for bushing current are shown in Table 15.
(4) The test of temperature rise uses the method of point distribution along the guide rod. The two ends of the capacitor core must be distributed with points. The distance between any two temperature measurement points cannot be greater than 300 mm, and the applied current must not be less than the value specified in Table 16. The temperature rise test of the valve side bushing needs to consider the influence of transformer oil temperature (90 °C) and valve hall temperature (50 °C).
(5) The voltage for measuring the dielectric loss and capacitance of the bushing is shown in Table 16.
(6) The insulation level of the valve side bushing of the converter transformer is increased by different coefficients compared to the insulation level of the converter transformer winding, where the DC voltage is 1.15, DC polarity reversal is 1.15, externally applied AC voltage is 1.10, lightning impulse is 1.10 (routine test can be 1.05), operation impulse is 1.10 (routine test can be 1.05). The voltage for the type test of the valve side bushing must be carried out according to the above requirements, and the specific values are shown in Table 17.
(7) The valve side bushing of HV converter transformer does not carry out the 1 min short-time withstand test at power frequency, and uses the externally applied AC long-time withstand test instead.

(8) The AC 1 min withstand voltage of the valve side bushing capacitor tap is 2.0 kV.
(9) The minimum creepage distance of the valve side bushing is 14.0 mm/kV.

(II) Wall Bushing

(1) The natural environmental conditions are as required in the first section of this chapter.
(2) Valve hall environmental conditions: ① Fully enclosed indoors, slightly positive pressure, with ventilation and air conditioning; ② Maximum temperature + 50 °C; ③ Minimum temperature + 10 °C; ④ Maximum humidity 60%RH.
(3) Indoor DC yard environmental conditions: ① Fully enclosed indoors, with ventilation; ② Maximum temperature + 50 °C; ③ Minimum temperature + 10 °C; ④ Maximum humidity 60%RH.

Table 14 Bushing style of converter transformer at valve side

Bushing Type	Y1	D1	Y2	D2
Type	Filled SF_6	Filled SF_6	Dry type or filled SF_6	Dry type or filled SF_6

Table 15 Requirement of bushing current (A)

Converter station location	Item	Bushing at valve side			
		Y1	D1	Y2	D2
Changji	Rated current	5033	2906	5033	2906
	Testing current for temperature rise	5879	3395	5879	3395
Guquan	Rated current	5032	2906	5032	2906
	Testing current for temperature rise	5879	3395	5879	3395

Table 16 Dielectric losses and capacitive measured voltage on bushing (kV)

Converter station location	Voltage	Winding on valve side			
		Y1	D1	Y2	D2
Changji	$0.5U_r/\sqrt{3}$	440	334	229	123
	$1.05U_r/\sqrt{3}$	923	702	480	259
	$1.5U_r/\sqrt{3}$	1319	1003	686	369
Guquan	$0.5U_r/\sqrt{3}$	438	333	227	122
	$1.05U_r/\sqrt{3}$	920	699	477	255
	$1.5U_r/\sqrt{3}$	1315	998	681	365

Note U_r is RMS value of rated voltage on bushing, kV

Table 17 Type test voltage of bushing at valve side of converter transformer (kV)

Converter station location	Testing item	Winding on valve side			
		Y1	D1	Y2	D2
Changji	Lightning impulse *LI*	2530	2178	1485	1364
	Switching impulse *SI*	2310	2024	1375	1293
	AC apply in long term + partial discharge	1427	1086	744	403
	DC apply in long term + partial discharge	2060	1556	1051	546
	DC polarity reversal + partial discharge	1594	1174	753	332
Guquan	Lightning impulse *LI*	2530	2123	1485	1359
	Switching impulse *SI*	2310	1964	1375	1282
	AC apply in long term + partial discharge	1421	1080	739	398
	DC apply in long term + partial discharge	2054	1549	1044	541
	DC polarity reversal + partial discharge	1592	1171	751	330

(4) Direct current system conditions: ① The project's rated voltage is 1100 kV; ② Continuous operating voltage: 1122 kV (DC) + harmonic 75 kV (root mean square value); ③ Installation environment: indoor field, outdoor field, valve hall.
(5) Bushing leakage rate: Guaranteed not to reach the alarm value for 10 years, and the leakage rate is no more than 0.5% per year.
(6) The bushing's terminal is a high conductivity flat plate terminal, with the terminal surface silver-plated.
(7) The pulling force that the bushing terminal should withstand is not less than: 3000 N in the horizontal longitudinal direction, 2000 N in the horizontal transverse direction, and 2000 N in the vertical direction.
(8) The terminal can withstand at least 400 N m of torque without deformation.
(9) The bushing temperature rise test is conducted under simulated conditions of the maximum operating temperature. Under the maximum operating temperature, a current of temperature rise test is applied to the bushing. After the bushing temperature rise stabilizes and lasts for 2 h, the temperature rise data is recorded. The temperature rise must not exceed the limit specified by the standard, then the bushing is considered to have passed the temperature rise test.
(10) The umbrella type of the wall bushing is a non-spiral structure of large and small umbrellas, and the umbrella shape coefficient is not more than

4. The outdoor dry arc distance of the 1100 kV wall bushing is not less than 12.6 m.

(11) See Table 18 for the bushing rating requirements.

II. Key Technologies

The bushing has both internal and external insulation, the electric field is complex, and the requirements of its structure and size are strict. In actual design, the main problems to be solved are heating, dielectric loss, thermal breakdown, and sealing. In UHV systems, due to the high electric field, the requirements for its length and diameter are very demanding, often becoming a constraint in equipment manufacturing.

In the ± 1100 kV, 12 GW project, the main insulation of the valve side bushing is oil-impregnated or rubber-impregnated paper capacitor core, the capacitor core is filled with SF_6 gas between the bushing sheath, the wall bushing uses pure SF_6 gas insulation and rubber-impregnated paper core plus SF_6 gas insulation type, whether it is the valve side bushing or the wall bushing, they all use a composite insulation sheath structure. But no matter what technical route, it is necessary to solve and balance the problems brought by the insulation performance, temperature rise performance, and mechanical performance of the bushing. The HV converter transformer valve side bushing and DC wall bushing in the ± 1100 kV, 12 GW project usually adopt a small angle horizontal arrangement. Due to the large size, the requirements for its bending load resistance are very high; the large load current plus the heat island effect in the converter station area makes the hot spot temperature rise problem of the bushing prominent; in order to have both impact, power frequency, DC, polarity reversal voltage withstand ability, it inevitably poses a challenge to insulation design.

After the voltage and capacity are increased, the design of the converter transformer valve side bushing and DC wall bushing mainly faces the problems of equipment insulation and heat dissipation capacity, and the further increase in size and weight puts higher requirements on mechanical strength.

(I) Comparison on Technical Parameters

1. Main parameter comparison of valve side bushing

The rated current of the converter transformer valve side bushing is directly related to the rated current of the DC system. Tables 19, 20 and 21 respectively provide the parameters of the valve side bushing current for ± 1100 kV, 12 GW and ± 800 kV, 10 GW and 8 GW projects. By comparison, it can be seen that the rated current and temperature rise test current of the converter transformer valve side bushing of the 12 GW project, and the multiple of the increase in DC capacity are similar. Considering harmonics and a certain margin, the multiple has increased. Because the requirements for the oil temperature of the converter transformer and the temperature of the valve hall are more stringent after the capacity is increased, it is necessary to clearly limit the oil temperature of the converter transformer and the temperature of the valve hall when conducting temperature rise tests on the valve side bushing.

The insulation level of the converter transformer valve side bushing will increase by different factors compared to the insulation level of the converter transformer

Table 18 Requirement of rated parameters of bushing

Item		1100 kV Busbar
Position of bushing voltage		Changji/Guquan
Long term DC current (A)		5523
Test current of temperature rise (A)		5839
Short-time withstand current (kA, RMS) 1 s		16
Peak withstand current (kA, peak value)		40
Rated DC voltage, to ground U_{dN} (kV)		1100
Maximum continuous DC voltage, to ground U_{dmax} (kV)		1122, supposed by harmonics 75 kV (RMS)
Dielectric loss factor, capacitive measured voltage	Level 1 test voltage (kV, RMS)	424
	Level 2 test voltage (kV, RMS)	891
	Level 3 test voltage (kV, RMS)	1273
	Lightning impulse test voltage (kV, peak value)	2420
	Switching impulse test voltage (kV, peak value)	2100
	Power frequency 1 min test voltage (kV, RMS)	1190
Dielectric loss factor, capacitive measured voltage	Power frequency 1 h test voltage (kV, RMS) + partial discharge	1190
	DC 2 h test voltage (kV) + partial discharge	1683
	DC polarity reversal test (kV) + partial discharge (90/90/45 min)	− 1403/+ 1403/− 1403
	DC wet-withstand voltage (kV)	1403
	Maximum test voltage of partial discharge test	
	Level 1 test voltage (kV, RMS)	424
	Level 2 test voltage (kV, RMS)	891
	Level 3 test voltage (kV, RMS)	1273
	Test voltage due to radio interference, terminal to ground (kV, RMS)	849
Minimum creepage	Outdoor (kV/mm)	45
	DC yard indoor (kV/mm)	25
	Valve hall (kV/mm)	14

Table 19 Current parameters of bushing at valve side of ± 1100 kV, 12 GW project (A)

Item	Bushing at valve side			
	High voltage		Low voltage	
	Y1	D1	Y2	D2
Rated current	5033	2906	5033	2906
Test current of temperature rise	5879	3395	5879	3395

Table 20 Current parameters of bushing at valve side of ± 800 kV, 10 GW project (A)

Item	Bushing at valve side			
	High voltage		Low voltage	
	Y1	D1	Y2	D2
Rated current	5766	3329	5766	3329
Test current of temperature rise	6500	3900	6500	3900

Table 21 Current parameters of bushing at valve side of ± 800 kV, 8 GW project (A)

Item	Bushing at valve side			
	High voltage		Low voltage	
	Y1	D1	Y2	D2
Rated current	4596	2653	4596	2653
Test current of temperature rise	5390	3120	5390	3120

winding, and the routine test voltage of the valve side bushing must be conducted according to the requirements of insulation level. Tables 22, 23 and 24 respectively provide the insulation level and test voltage parameters of the valve side winding of the 12, 10 and 8 GW projects. It can be seen that the insulation level of the converter transformer valve side bushing of the 12 GW project is similar to the multiple of the increase in voltage level.

2. Comparison of main parameters of wall bushing

Tables 25 and 26 respectively provide the main parameters of the 1100, 800, 550 and 400 kV wall bushings used in the 12, 10 and 8 GW projects. By comparing the data, it can be seen that before and after the capacity of the wall bushing is increased, the rated voltage and related test voltage of the bushing remain unchanged, the rated current and overload current increase, and the short-time withstand current remains unchanged. After increasing the transmission capacity, the design of the wall bushing is largely similar to the design of the 8 GW project.

Through Tables 25 and 26, it can be seen that compared with the ± 800 kV, 8 GW, 10 GW projects, the rated voltage and external insulation level of the 550, 1100 kV wall bushings of the ± 1100 kV, 12 GW project have significantly increased, and the short-time withstand current remains unchanged. In addition, the 12 GW project

Table 22 Insulation level of 12 GW project (kV)

Item		Winding on grid side	Winding on valve side			
			High voltage		Low voltage	
			Y1	D1	Y2	D2
Lightning impulse *LI*	Terminal 1	1950	2300	1980	1350	1240
	Terminal 2	185	2300	1980	1350	1240
Chopped lightning impulse (*LIC*)	Terminal 1	2100	2530	2175	1485	1360
	Terminal 2	–	2530	2175	1485	1360
Switching impulse *SI*	Terminal 1	1550	–	–	–	–
	Terminal 2	–	–	–	–	–
	Terminal 1 + Terminal 2	–	2100	1840	1250	1175
AC apply in short term (neutral point)	Terminal 1 + Terminal 2	95	–	–	–	–
AC induction in short term	Terminal 1	900	–	–	–	–
AC induction in long term + partial discharge	Terminal 1 (U1)	800	239	414	239	414
	Terminal 1 (U2)	693	211	365	211	365
AC apply in long term + partial discharge	Terminal 1 + Terminal 2	–	1297	987	676	366
DC apply in long term + partial discharge	Terminal 1 + Terminal 2	–	1791	1353	914	475
DC polarity reversal + partial discharge	Terminal 1 + Terminal 2	–	1386	1021	655	289

needs to pay more attention to the heat generation problem brought by large current. Compared with previous projects, the design is more detailed in specifying the current density of the equipment terminal board. For the wall bushing, it is stipulated that the overlap length of the terminal rectangular conductor joint should not be less than the width of the conductor, and the current density is clearly specified according to Table 27.

(II) Technical Route 1 of Implementation Method on Bushing

Compared to the ± 800 kV, 8 GW project, the changes required in bushing technology on the valve side are mainly reflected in: the rated current and temperature rise test current increase, while the insulation level also increases simultaneously; for the DC wall bushings, only the rated current and overload current increase slightly. The enhancement of current carrying capacity inevitably requires temperature rise control, while ensuring its mechanical characteristics, the insulation level is slightly increased or remains unchanged as the size of the equipment structure changes. In the actual application of the 12 GW project, there are mainly two technical routes that

Table 23 Insulation level of 10 GW project (kV)

Item		Winding on grid side	Winding on valve side			
			High voltage		Low voltage	
			Y1	D1	Y2	D2
Lightning impulse *LI*	Terminal 1	1550	1870	1600	1300	1175
	Terminal 2	185	1870	1600	1300	1175
Chopped lightning impulse (*LIC*)	Terminal 1	1705	2060	1760	1430	1293
	Terminal 2	–	2060	1760	1430	1293
Switching impulse *SI*	Terminal 1	1175	–	–	–	–
	Terminal 2	–	–	–	–	–
	Terminal 1 + Terminal 2	–	1675	1360	1175	1050
AC apply in short term (neutral point)	Terminal 1 + Terminal 2	95	–	–	–	–
AC induction in short term	Terminal 1	680	–	–	–	–
AC induction in long term + partial discharge	Terminal 1 (U1)	550	178	307	178	307
	Terminal 1 (U2)	476	154	265	154	265
AC apply in long term + partial discharge	Terminal 1 + Terminal 2	–	941	724	481	264
DC apply in long term + partial discharge	Terminal 1 + Terminal 2	–	1298	992	648	342
DC polarity reversal + partial discharge	Terminal 1 + Terminal 2	–	1004	749	462	207

can meet its technical requirements, and this section introduces their implementation methods respectively.

1. Valve side bushing

The valve side bushing uses an oil-impregnated paper capacitor core, the external insulation on the air side uses a silicone rubber composite insulator sheath, the capacitor core inside the sheath is equipped with a fiberglass internal insulation sleeve, and SF_6 gas is filled between the internal insulation sleeve and the external insulation sleeve. The capacitor core without external insulation sleeve is directly immersed in transformer oil. When this design is used, the capacitor core has good heat dissipation performance, and the converter transformer oil is doubly isolated from the valve hall through the internal insulation sleeve and the insulator outer sheath. Regardless of the type of bushing, its heat dissipation process mainly relies on the thermal conductivity of its own material, which is a passive heat dissipation process. In response to the temperature rise problem brought about by the increase in current for the valve side bushing, common measures include reducing the loss of the conductive

Table 24 Insulation level of 8 GW project (kV)

Item		Winding on grid side	Winding on valve side			
			High voltage		Low voltage	
			Y1	D1	Y2	D2
Lightning impulse *LI*	Terminal 1	1550	1800	1550	1300	1175
	Terminal 2	185	1800	1550	1300	1175
Chopped lightning impulse (*LIC*)	Terminal 1	1705	1980	1705	1430	1293
	Terminal 2	–	1980	1705	1430	1293
Switching impulse *SI*	Terminal 1	1175	–	–	–	–
	Terminal 2	–	–	–	–	–
	Terminal 1 + Terminal 2	–	1620	1315	1175	1050
AC apply in short term (neutral point)	Terminal 1 + Terminal 2	95	–	–	–	–
AC induction in short term	Terminal 1	680	–	–	–	–
AC induction in long term + partial discharge	Terminal 1 (U1)	550	180	312	180	312
	Terminal 1 (U2)	476	156	270	156	270
AC apply in long term + partial discharge	Terminal 1 + Terminal 2	–	914	697	481	265
DC apply in long term + partial discharge	Terminal 1 + Terminal 2	–	1260	954	648	342
DC polarity reversal + partial discharge	Terminal 1 + Terminal 2	–	972	717	462	207

rod, increasing the cooling effect, and evenly distributing the heat. A heat pipe is a passive heat transfer component that uses phase change principles and capillary action. Its superconducting heat and isothermal properties make it a good temperature control tool, and its heat transfer efficiency is much higher than that of pure copper of the same material.

The use of heat pipe technology to change the structure of the current-carrying conductive rod can significantly improve the heat dissipation efficiency. Although there is no need to change the design of the bushing's internal and external structure, the structure of the current-carrying and heat pipe is complex and requires high standard of sealing. In addition, by increasing the diameter of the conductive rod, the current-carrying area can be increased to control the heat generation of the bushing. This method is simple in structure, but requires redesigning the internal and external insulation of the bushing.

Figure 26 shows an improved schematic diagram of outer sheath design of composite insulation. It can be seen that for the composite insulation outer sheath of the bushing, in addition to adopting the outer shape of the bushing at a higher

Table 25 Comparison of main parameters of 1100, 800 kV bushing

Item	12 GW project	10 GW project	8 GW project
Rated current (A, DC)	5523	6328	5046
2 h overload current (A, DC)	5839 (test current of temperature rise)	6693	5335
Short-time withstand current (kA, RMS) 1 s	16	16	16
Peak withstand current (kA, peak value)	40	40	40
Rated DC voltage, to ground U_{dN} (kV)	1100	800	800
Maximum continuous DC voltage, to ground U_{dmax} (kV)	1122	816	816
Maximum voltage of equipment U_m, to ground (kV, RMS)	–	577	577
Lightning impulse test voltage (kV, peak value)	2420	1870	1800
Switching impulse test voltage (kV, peak value)	2100	1675	1620
Power frequency 1 min test voltage (kV, RMS)	1190	865	865
DC polarity reversal test (kV) + partial discharge (90/90/45 min)	– 1403/+ 1403/– 1403	– 1020/ + 1020/– 1020	– 1020/ + 1020/– 1020
DC wet-withstand voltage (kV)	1403	1020	1020

voltage level, the radius of the umbrella skirt tip has also been optimized to reduce the risk of end corona discharge. At the same time, the diameter-changing area of the insulation outer sheath has also been lengthened.

2. DC wall bushing

The DC wall bushing of the 12 GW project Changji converter station adopts a pure gas insulation type, and the air side uses a silicone rubber composite insulator outer sheath, completely abandoning the capacitor core. This design is oil-free, has no risk of explosion and combustion, and has good heat dissipation performance. The current-carrying conductor is designed in one piece, and the conductor rod has no intermediate connection.

Compared with the ± 800 kV, 10 GW project, the total length of the 1100 kV bushing is 5.58 m longer than the 800 kV bushing, reaching 26.2 m. The diameter

Table 26 Comparison of main parameters of 550 kV, 400 kV bushing

Item	12 GW project	10 GW project	8 GW project
Rated current (A, DC)	5523	6328	5046
2 h overload current (A, DC)	5839	6693	5335
Short-time withstand current (kA, RMS) 1 s	16	16	16
Peak withstand current (kA, peak value)	40	40	40
Rated DC voltage, to ground U_{dN} (kV)	550	400	400
Maximum continuous DC voltage, to ground U_{dmax} (kV)	585	408	408
Maximum voltage of equipment U_m, to ground (kV, RMS)	–	289	289
Lightning impulse test voltage (kV, peak value)	1350	980	903
Switching impulse test voltage (kV, peak value)	1200	880	825
Power frequency 1 min test voltage (kV, RMS)	688	500	500
DC polarity reversal test (kV) + partial discharge (90/90/45 min)	− 713/ + 713/− 713	− 510/ + 510/− 510	− 510/ + 510/− 510
DC wet-withstand voltage (kV)	713	510	510

Table 27 Current density of uncoated joints (A/mm^2)

Rated current (A)	Current density of copper joints	Current density of aluminum joints
< 200	0.258	0.78 times current density of copper joints
200–2000	$0.258 - 0.875\ (I_N - 200) \times 10^{-4}$	
> 2000	0.1	

Note I_N is rated current of wall bushing, A

of the insulation outer sheath remains unchanged; the total length of the 550 kV bushing is 2.96 m longer than the 400 kV bushing, reaching 14.4 m. The diameter of the insulation outer sheath has increased from 786 to 902 mm, thereby significantly improving the heat dissipation capacity of the bushing, and the temperature difference required for the same loss is smaller. The schematic diagrams of the 550, 1100 kV bushing shapes are shown in Figs. 27 and 28.

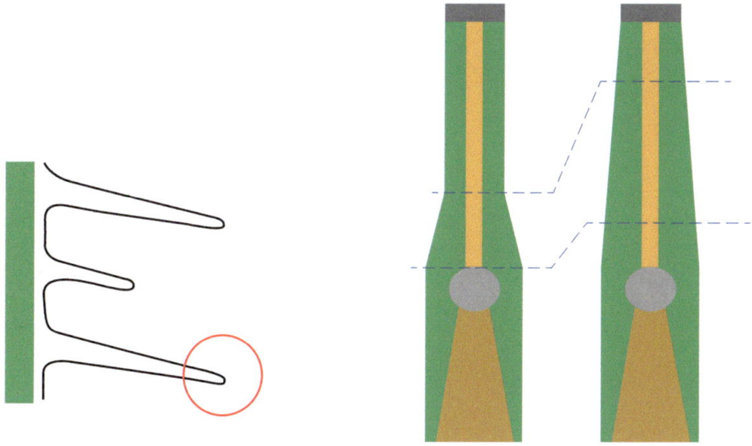

(a) Optimization of umbrella petticoats (b) Variable diameter area of extended insulation jacket

Fig. 26 Schematic configuration and design of polymer-housed insulation jacket

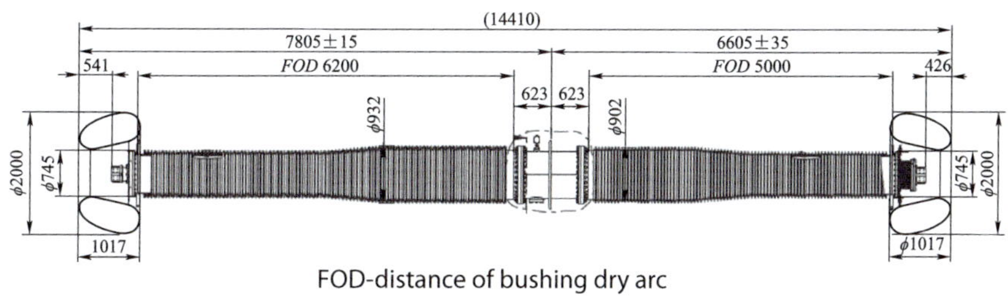

Fig. 27 Schematic diagram of 550 kV bushing

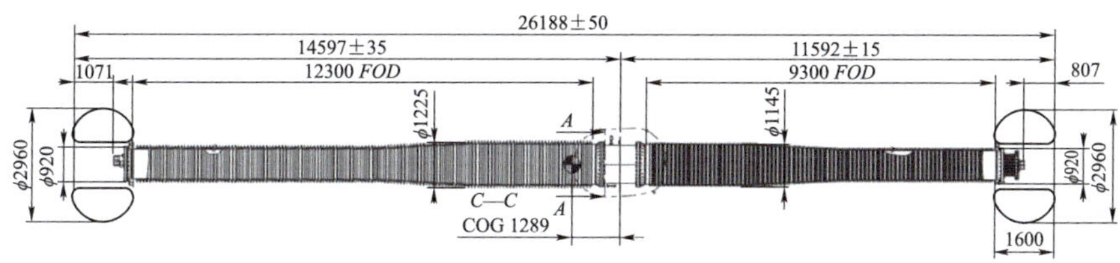

Fig. 28 Schematic diagram of 1100 kV bushing

Compared with the ± 800 kV, 10 GW project, the 1100 and 550 kV bushings adopt an optimized umbrella structure, which improves the height and spacing of the insulation umbrella group, and the performance against pollution, rain flash, and snow flash is even better. The umbrella structure is shown in Figs. 29 and 30.

The outdoor environment of the ± 1100 kV, 12 GW project can reach a minimum temperature of − 42 °C, therefore, a special cold start state evaluation was carried out

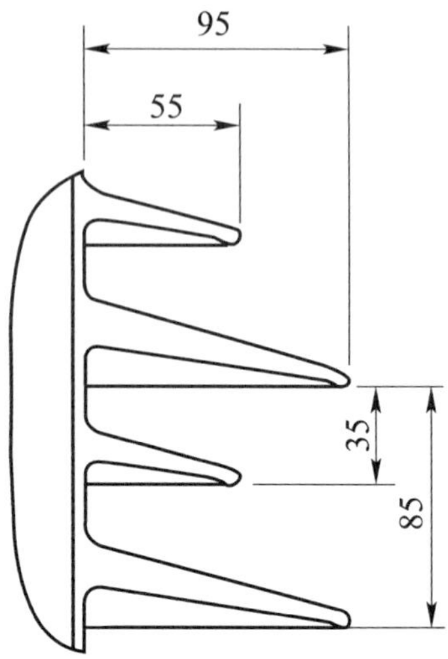

Fig. 29 Umbrella structure of 1100 kV bushing in outdoor side

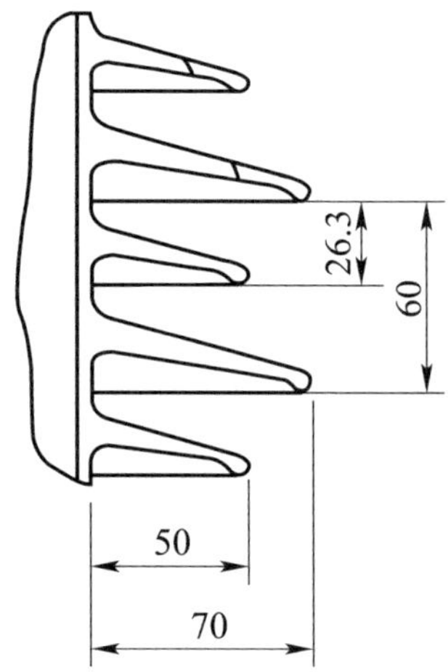

Fig. 30 Umbrella structure of 550 kV bushing in outdoor side

for the DC wall bushing. Based on the thermal conduction and flow characteristics of SF_6 gas, after professional simulation analysis, it was determined that all wall bushings used in outdoor environments need to be equipped with heating devices to avoid the risk of liquefaction of the internal SF_6 gas during cold start. A heating belt is installed on the indoor test flange of the bushing to heat the internal SF_6 gas.

(III) Technical Route 2 of Implementation Method on Bushing

The valve side bushing of the 12 GW project receiving end transformer adopts Technical Route 2. Its main insulation uses epoxy resin impregnated paper capacitor core, and the outer sheath uses silicone rubber hollow composite insulator. SF_6 gas is filled between the capacitor core and the outer sheath. Its heat dissipation performance is worse than that of oil-impregnated paper bushing, but the risk of explosion and combustion is lower due to the oil-free design.

In terms of insulation structure design, the matured main insulation design scheme of Technical Route 2 is adopted, that is, using epoxy resin impregnated paper capacitor core, and using SF_6 gas as auxiliary insulation. The capacitor core is made of insulating paper and aluminum foil wound into a concentric cylindrical capacitor, which is solidified by vacuum drying and impregnating epoxy resin. Its internal structure is shown in Fig. 31.

The valve side bushing of Technical Route 2 adopts capacitive equalization. Through reasonable arrangement of aluminum foil and built-in equalization method, the axial and radial electric field can be accurately controlled to achieve the purpose of uniform electric field.

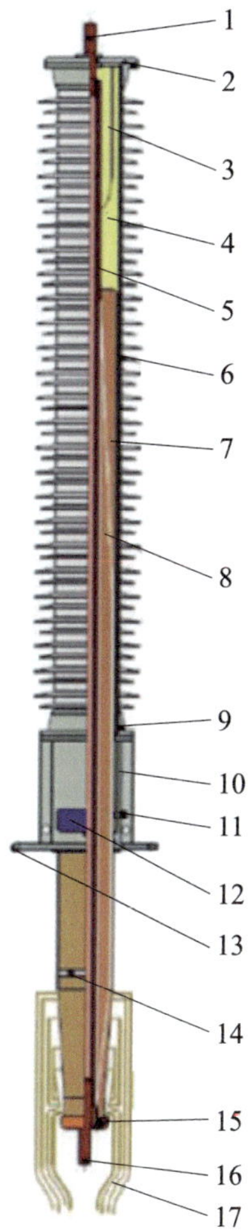

Fig. 31 Internal structure of valve side bushing of Technical Route 2. 1—Terminal block; 2—Head terminal cover; 3—Grading cover; 4—Filling medium; 5—Conductive rod; 6—Polymer-housed jacket; 7—Capacitor core; 8—Aluminum foil screen; 9—Outer jacket lower flange; 10—Bushing flange; 11—DILO valve (if any); 12—Voltage divider box; 13—Air release plug; 14—Grounding strip; 15—Bottom terminal plate; 16—Wiring terminals in oil; 17—Formed parts

(IV) Domestic Research and Development of Wall Bushing

Relying on the National 863 Project, two different technical routes of 1100 kV DC wall bushings in China have passed type tests at the China Electric Power Research Institute and are arranged to grid-connected commissioning at Changji and Guquan converter stations.

1. 1100 kV wall bushing developed by Chinese manufacturer 1

The overall structure is mainly divided into five parts: hollow composite insulator, wall bushing cylinder, equalizing system, central conductor and support structure, and internal shielding system. The overall dimensions of the 1100 kV wall bushing at Changji converter station are shown in Fig. 32.

(1) Hollow composite insulator. The hollow composite insulator is mainly composed of a glass fiber winding tube, a silicone rubber umbrella skirt, and upper and lower flanges. The total length of the outdoor bushing is 14 m, the total designed creepage distance is 58,300 mm, the dry arc distance is 13,180 mm, and the creepage ratio is 52 mm/kV. The total length of the indoor bushing is 13 m. In order to consider the universality of the umbrella group mold, the total designed creepage distance is 53,800 mm, the dry arc distance is 12,180 mm, and the creepage ratio is 48 mm/kV. In terms of umbrella structure, for horizontally used bushings, the possibility of rainwater bridging and short-circuiting between umbrellas is small. Theoretically, the following three umbrella shapes

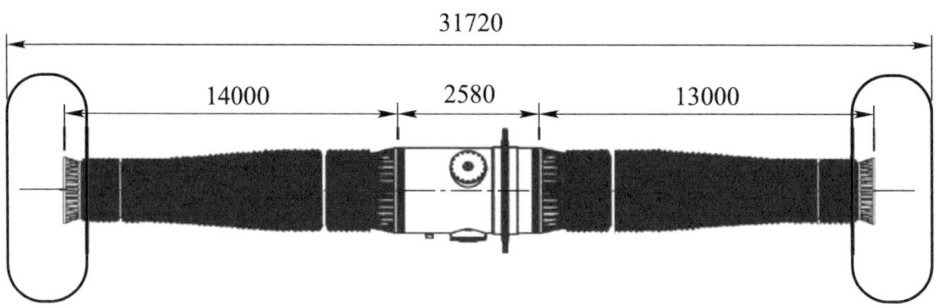

Fig. 32 General size of 1100 kV wall bushing in Changji converter station

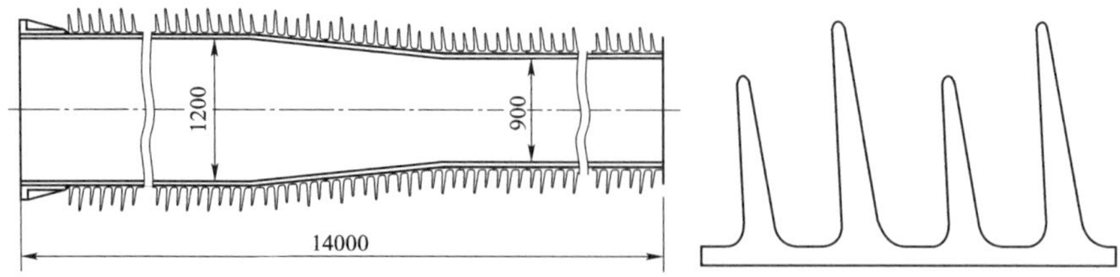

(a) Hollow composite insulator (b) Umbrella structure

Fig. 33 Hollow composite insulator and umbrella structure

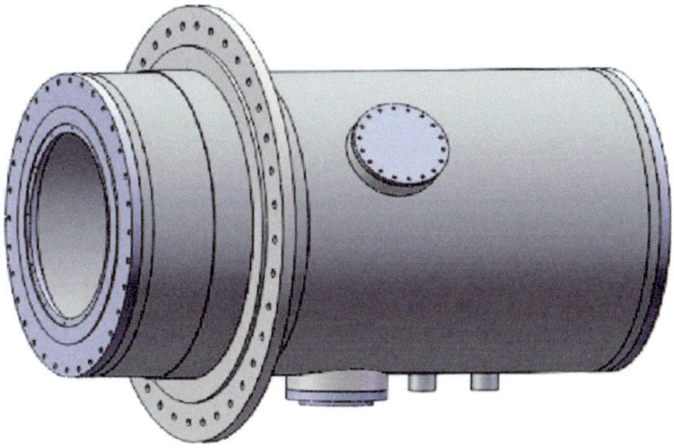

Fig. 34 Wall bushing cylinder

Fig. 35 Structure of conductive rod

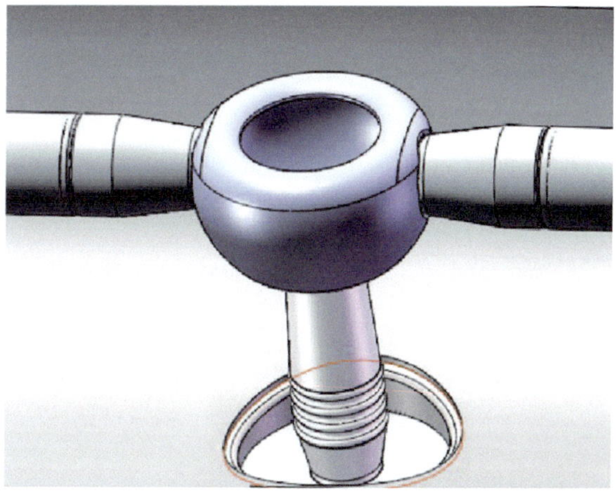

Fig. 36 Connection structure in middle part of conductive rod

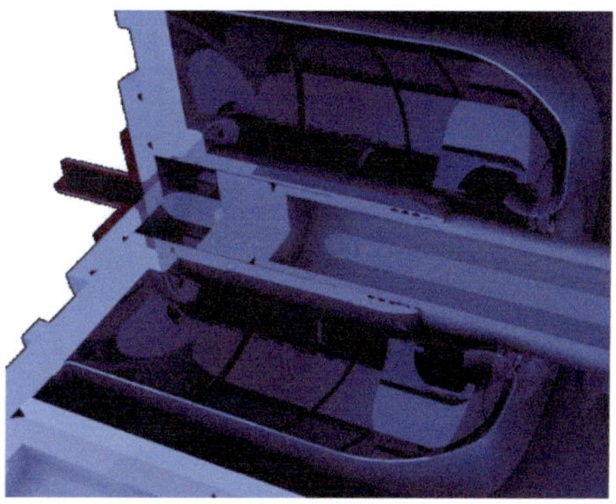

Fig. 37 Connection structure in end sliding part

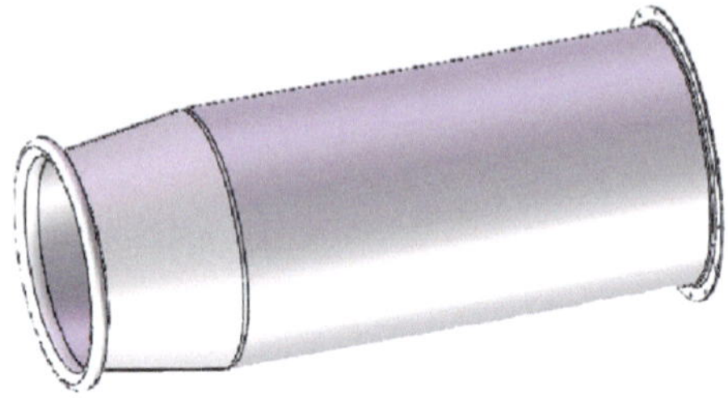

Fig. 38 Structure of internal shield

Fig. 39 Configuration of corona ring

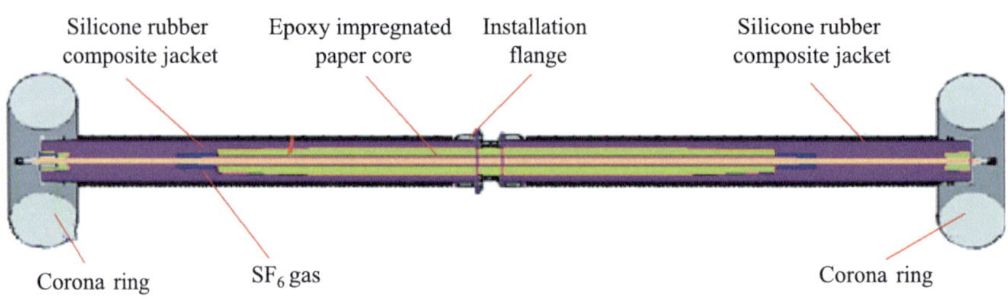

Fig. 40 Structure of bushing

can be used. However, under the same dry arc distance, the "one large and one small umbrella" can achieve a larger creepage distance than the "one large and two small umbrellas", so the umbrella structure with one large and one small arrangement is chosen. The hollow composite insulator and umbrella structure are shown in Fig. 33.

(2) Wall bushing cylinder. As one of the key components of the bushing, the wall bushing cylinder is a pressure vessel component, as shown in Fig. 34. During the construction, installation, and operation process, the material and welding process of the tube body are crucial for the safe operation of the wall bushing. According to the requirements of performance parameter, the cylinder material is chosen as Q345R pressure vessel steel plate.

(3) Central conductor and support structure. The conductive rod adopts a design scheme with a fixed middle and sliding ends. The sliding part is connected with spring fingers, which can allow the conductive rod to slide during thermal expansion and contraction, and maintain good flow capacity, as shown in Fig. 35. The design against foreign objects is mainly reflected in two aspects: one is the design of the end shield, the shield is turned inward, which can play a role in equalizing the electric field and effectively control the metal powder that may be generated by the sliding part; the second is the middle support insulator is installed in a recessed socket, which can make the impurities of the middle bushing cylinder enter the recessed area and not spread outwards. Figures 36 and 37 show the middle connection structure and end sliding connection structure of the conductive rod.

(4) Internal shielding structure. After a large amount of simulation calculations by the project team, a single-layer shielding structure was determined as shown in Fig. 38. The lower end flange of the shield is connected to the wall bushing cylinder, serving as a grounding shield. The material used is aluminum alloy, which is lightweight and has a small deformation amount. It can play a role in uniform axial and radial electric fields and is one of the commonly used shield structures in bushings.

(5) End corona ring. The end equalizing system mainly makes the end electric field uniform through the corona ring. In structural design, it is necessary to ensure that the outer surface electric field distribution of the corona ring is minimized by changing the arc shape of the outer part of the corona ring without changing the outer diameter, as shown in Fig. 39.

2. 1100 kV wall bushing developed by Chinese manufacturer 2

    (1) Overall structure. The main insulation of the bushing uses a whole epoxy resin impregnated paper capacitor core, the outer insulation uses a hollow composite insulator, and SF_6 gas is filled between the capacitor core and the hollow composite insulator, which can meet the insulation strength and partial discharge requirements of the bushing under various environments. The bushing structure is shown in Fig. 40.

A strong spring compensation structure is used to apply assembly force to the whole bushing, reducing the squeeze on the core body caused by the deflection of the central conductor tube; improving the rigidity of the outer insulating hollow composite insulator, reducing its end droop offset.

The SF_6 gas filled between the capacitor core and the Polymer-housed forms an epoxy core—SF_6 composite insulation, which has high insulation strength. The middle part uses an integral conduit for current carrying, and both ends use a strap finger for current carrying, which has high current carrying reliability.

The head design uses a strong spring compensation structure to ensure that the pressure on each sealing surface of the bushing end is not less than 2 MPa when the temperature changes from − 40 to + 50 °C under large bending moment conditions,

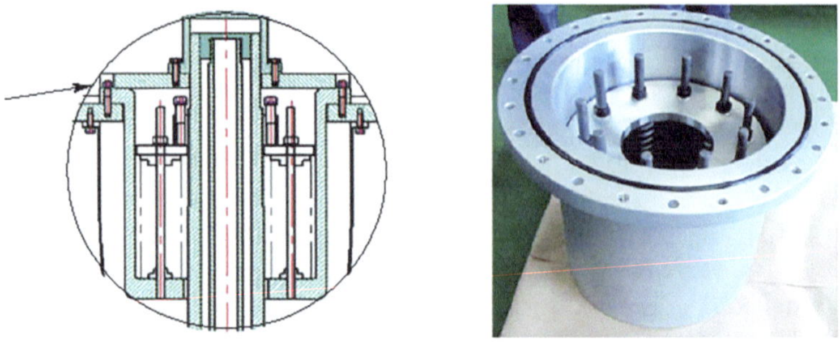

(a) Internal terminal structure (b) Spring assembly structure

Fig. 41 Internal terminal structure and spring assembly structure of bushing

and the overall sealing reliability of the bushing is high. The indoor end structure and spring assembly structure of the bushing are shown in Fig. 41.

The middle part uses a flange butt joint structure, and the butt joint is tightened with high-strength bolts; the middle part of the core body uses two supports to prevent eccentricity, and buffer material is filled between the support point and the core body to prevent the local stress of the core body from being too high and damaged. The middle structure and outdoor end structure of the bushing are shown in Fig. 42.

The current-carrying tube of the DC wall bushing adopts an integral structure, and the conductor tube directly extends out of the two end covers. The strap is in the high-voltage area outside the bushing insulation. By tightening the conductor tube with a strong spring, when the temperature of the conductor tube rises, the elongation of the conductor tube decreases, and the tension decreases, which can effectively reduce the elongation of the conductor tube.

(2) Current carrying capacity design. The bushing adopts a double conduit structure, the core winding conduit does not carry current, and the conductive tube is made of purple copper material. There is a gap between the winding tube and

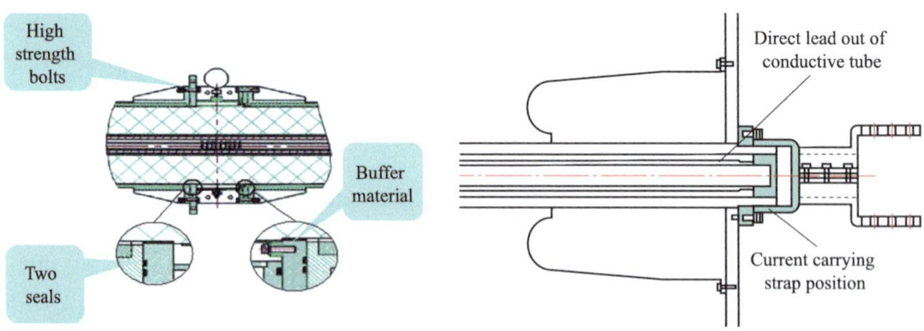

(a) Middle part (b) External terminal structure

Fig. 42 External terminal structure and middle part of bushing the conductive tube, which suppresses the temperature exceeding the standard caused by current carrying heating and the impact on the main insulation, effectively conducts and dissipates heat, solves the problem of heat accumulation, achieves uniform heating, and reduces the impact of heat release on the main insulation.

The current density of the conductive tube section of the ± 1100 kV DC wall bushing is 0.73 A/mm^2. Based on the successful development of the ± 1100 kV DC wall bushing, the ± 800 kV and ± 400 kV DC wall bushings designed by similar technology have a current density of 0.54 A/mm^2 according to the long-term temperature rise test current of 6644 A; according to the 2 h temperature rise test current of 7025 A, the current density of the conductive tube section is 0.57 A/mm^2, which meets the current carrying and thermal performance requirements of the bushing. At the same time, the use of a chamber communication method allows SF_6 to circulate better, enhancing the cooling effect.

3. Test situation

Chinese Manufacturer 1 has UHV insulation test rooms, high voltage test rooms, and mechanical test rooms, which can complete all type tests and routine tests of ± 1100 kV DC wall bushings. The routine test of the ± 1100 kV DC wall bushing is shown in Fig. 43.

The main equipment includes ZDFI-2000 kV/30 mA DC voltage generator (see Fig. 44a), 1800 kV/5000 kVA power frequency test system, 4000 kV/400 kJ air cushion mobile impulse voltage generator, LDS-6 partial discharge measurement system, WRV1.5/540G variable frequency resonance test system, 1600 kV/1000 kVA power frequency test system, 3600 kV/270 kJ impulse voltage generator complete test device (see Fig. 44b), temperature rise test equipment, characteristic tester, temperature rise automatic tester, etc., which can carry out DC withstand voltage, power frequency withstand voltage test, lightning impulse, operation wave impulse withstand voltage test, partial discharge measurement, long-term current heating, high

Fig. 43 Routine test of ± 1100 kV DC wall bushing

(a) ZDF1-2000kV DC voltage generator

(b) 3600kV/270kJ Impulse voltage generator and testing devices

Fig. 44 Testing equipment of Chinese manufacturer 1

voltage insulation test, mechanical test, sealing test, moisture measurement, temperature rise test, power frequency withstand voltage test, rainproof test, mechanical performance, mechanical durability and other type, research tests for switchgear of 1100 kV and below voltage levels.

According to the provisions of GB/T 22674—2008 "Bushings for DC Systems" and other standards, considering the latest engineering parameters, the type test plan for ± 1100 kV DC pure SF_6 gas insulated wall bushings was determined. The product passed the full set of type tests at one time, and completed more than 30 UHV test items.

(3) 1100 kV wall bushing test of Chinese Manufacturer 2 (see Fig. 45). According to the provisions of standards such as GB/T 22674—2008, and considering the latest engineering parameters, the type test scheme for ± 1100 kV DC pure SF_6 gas-insulated wall bushings has been determined. The product passed the full set of type tests at the Power Industry Power Equipment and Instrument Quality Inspection and Testing Center in one go, completing a total of 22 UHV test items.

1) Type test items include power frequency dry withstand voltage test and partial discharge measurement, lightning impulse dry withstand voltage test, operating impulse wet withstand voltage test, temperature rise test, cantilever load test, internal pressure test of gas-insulated and gas-impregnated bushings, and size check.
2) Routine test items include dielectric loss factor and capacitance measurement, lightning impulse dry withstand voltage test, power frequency dry

(a) Example I (b) Example II

Fig. 45 Test of wall bushing of Chinese manufacturer 2

withstand voltage test, repeated measurement of dielectric loss factor and capacitance, DC withstand voltage test and partial discharge measurement, repeated measurement of dielectric loss factor and capacitance, tap insulation test, internal pressure test of gas-insulated and gas-impregnated bushings, seal test of gas-insulated and gas-impregnated bushings, and appearance and size check.

3) Special tests include uniform rain DC voltage test, non-uniform rain DC voltage test.
4) Additional tests include radio interference voltage and visible corona test, thermal short-time current withstand test.

(V) Solving the Problem of 1100 kV Valve Side Bushing Made by a Foreign Manufacturer

At the beginning of development, the 1100 kV valve side bushing made by a foreign manufacturer had serious problems with core casting cracking (see Fig. 46), with 6 out of 8 damaged by the old process.

Fundamental cause: The process control of the epoxy curing process was not in place, the heat reaction was unbalanced, the internal reaction was fast, the external reaction was slow, and the epoxy was not fully cured, leading to thermal stress breakdown.

Solution: Embed temperature sensors, establish process control speed and standards; use heat pipe technology internally to further release excess heat, etc. The bushing passed the type test successfully with the new process.

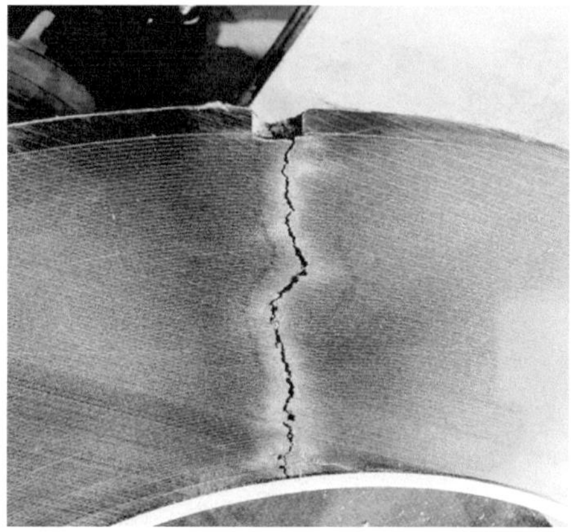

Fig. 46 Core cracking

3 12 GW/1100 kV Converter Valve

The ± 1100 kV, 12 GW project has increased the external insulation requirements for the core equipment converter valve, bringing new challenges to the design and operation of the converter valve. It is necessary to carry out research and tackle key problems on the basis of previous projects. This section mainly introduces the development work and corresponding results of the 1100 kV UHV DC transmission converter valve.

I. Technical Requirements

(1) For environmental conditions, see the relevant requirements in the first section of this chapter.
(2) The converter valve should be an indoor type double crystal thyristor converter valve with air insulation and water cooling. The creepage distance of the external insulation should not be less than 14 mm/kV (calculated by the highest DC voltage).
(3) The converter valve should not only have the ability to withstand normal operating voltage and current, but also have the ability to withstand the impact voltage and current caused by the maloperation of the valve's triggering system, faults in various parts of the station, or faults in the AC system.
(4) The converter valve must be designed to be fault-tolerant. During the operating period between two scheduled maintenance, the failure or damage of valve components will not cause more damage to the thyristor level, and the valve still has the ability to operate.
(5) The converter valve should use low-noise components to reduce the noise level of the valve during operation.
(6) The configuration of the surge arrester is shown in Fig. 47.

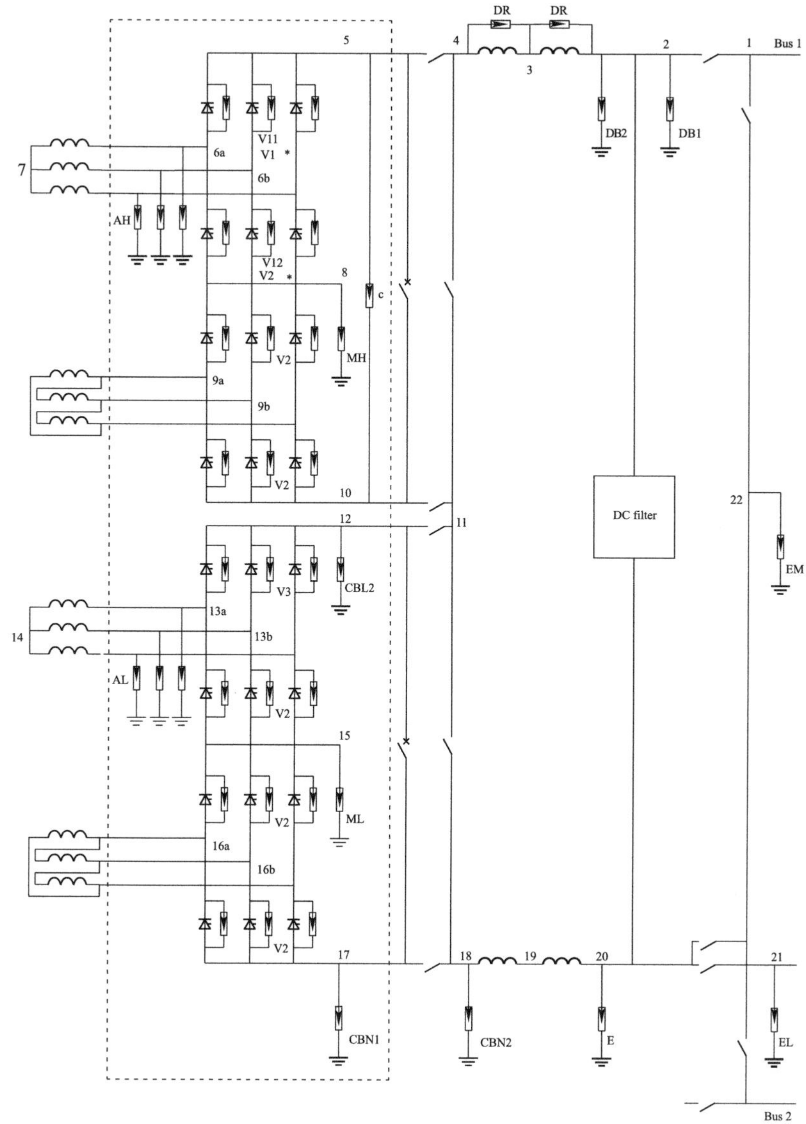

V1 (V11), V2 (V12), V3—Valve surge arrester; ML—Surge arrester of lower 12 pulse converter unit and 6 pulse converter bridge; MH—Surge arrester of upper 12 pulse converter unit and 6 pulse converter bridge; CBL2—Surge arrester at DC busbar of middle point of upper and lower 12 pulse converter unit (to ground); DB1—Surge arrester on DC line; DB2—Surge arrester on DC busbar; CBN1, CBN2, E, EL, EM—Surge arrester on neutral busbar; DR—Smoothing reactor with its parallel surge arrester; AH—Surge arrester on bushing at valve side of upper 12 pulse high voltage YY converter transformer; AL—Surge arrester on bushing at valve side of lower 12 pulse high voltage YY converter transformer

Fig. 47 Configuration of surge arrester

(7) For thyristor components, the converter valve uses 6 inch thyristor components. The thyristors of the same single valve should come from the same manufacturer and should not be mixed. Each thyristor component should have the ability to independently bear the rated current, overload current, and various transient surge currents. The main circuit should not adopt a design with thyristor components in parallel.

(8) Redundancy, during the 12-month operating period between two scheduled maintenance, if there are no damaged thyristor components at the beginning of this operating period, and no thyristor components are replaced during the operating period, for the 48 single valves of the whole station bipolar, the single valve with all redundant thyristor levels damaged should not exceed one. The number of redundant thyristor levels in each valve should not be less than 2.5 times the expected value of the number of thyristor levels damaged during the 12-month operating period, or should not be less than 3 levels (whichever is larger).

(9) Mechanical performance, the converter valve can adopt a suspended structure, should adopt a modular design, components can be replaced; the trigger system layout should facilitate the opening and replacement of optical fibers, and should also avoid mechanical damage to the optical fibers during installation; the structure of the converter valve should ensure that the leaked cooling fluid automatically flows out along the groove, leaves the live parts, flows to a detector and alarms, and does not cause any damage to any components. If a water sprinkler fire extinguishing system is installed in the valve hall, it should be ensured that the operation of the sprinkler system will not cause any damage to any components.

(10) Voltage withstand capability, the converter valve should be able to withstand normal operating voltage and various overvoltages. The converter valve can obtain sufficient voltage tolerance by using thyristors in series; the dispersion of overvoltage protection level and other non-linear factors inside the valve affecting the valve's voltage tolerance should be considered. Under the condition that all redundant thyristor levels are damaged, the insulation of single and multiple valves should have the following safety factors: ① For operating surge voltage, 10% above the surge arrester protection level; ② For lightning surge voltage, 10% above the surge arrester protection level; ③ For steep wavefront surge voltage, 15% above the surge arrester protection level. Under the maximum design junction temperature condition, when the inverter side converter valve is in the recovery period after commutation, the thyristor should be able to withstand a forward transient voltage peak equivalent to the protection trigger voltage level.

(11) Current withstand capability, the converter valve should have the ability to bear the rated current, overload current, and various transient surge currents. After the converter valve operates at any power level between the minimum power and 2 h overload, it should have at least the 3 s temporary overload capacity required by the technical specifications without engaging the standby cooling. The main circuit should not adopt a design

with thyristor components in parallel. For transient overcurrents caused by faults, the converter valve should have the following withstand capabilities: ① Withstand capability of short-circuit current with subsequent lockout; ② Withstand capability of short-circuit current without subsequent lockout; ③ Withstand capability of additional short-circuit current.

(12) The operating capability under AC system failure, when the AC system failure causes the three-phase average rectified voltage measured at the AC bus of the converter station to be greater than 30% of the normal voltage, but less than the extreme minimum continuous operating voltage and lasts for 1 s, the DC system should be able to operate continuously and stably; in the event of a severe AC system failure, causing the three-phase average rectified voltage measured at the AC bus of the converter station to be 30% or less of the normal value, if possible, the DC current should be maintained by continuing to trigger the valve group to operate at a certain value, thereby improving the recovery performance of the HVDC system.

(13) The converter valve uses a photoelectric conversion triggering system. Optical isolation is used between high and low voltage circuits. When the AC system failure causes the AC bus voltage at the converter station to drop to the following values and lasts for the corresponding periods, all energy storage devices in the thyristor level triggering circuits should have enough energy to continuously provide triggering pulses to the thyristor components, so that the converter valve can be safely turned on: ① AC system single-phase ground fault, the fault phase voltage drops to 0, lasting at least 0.7 s; ② AC system three-phase ground short-circuit fault, voltage drops to 30% of normal voltage, lasting at least 0.7 s; ③ When the AC system three-phase ground metal short-circuit fault occurs, the voltage drops to 0, lasting at least 0.7 s, followed by the clearing of this type of fault and the recovery of commutation voltage, there should be enough energy storage in the valve triggering circuit to safely trigger the thyristor components.

(14) Control system, the control system of the converter valve should ensure the correct operation of the converter valve under normal or fault conditions of the primary system. Under no circumstances should the improper operation of the control system cause damage to the converter valve. The control system should be fully redundant and should have comprehensive self-checking and alarm functions.

(15) Thyristor monitoring system, remote monitoring is carried out in the control room of the converter station to confirm the status of each thyristor level and correctly indicate any abnormalities or damage to any thyristor or other related electronic devices. After the redundant thyristor level is damaged, the monitoring device should issue an alarm. When the damaged thyristor level exceeds the redundancy number, causing the operating thyristor converter valve to face more serious damage, information should be sent to the monitoring system or other protection systems to lock the converter.

(16) Each thyristor level in the valve should have a protective triggering system to provide overvoltage protection triggering for the thyristor level. The design should allow the thyristor pole to operate under the condition of continuous action of protective triggering. Under the maximum swing load power frequency overvoltage, such as the swing load power frequency overvoltage after the AC system fault, the protective triggering of the valve should not act due to the transient overshoot of the inverter commutation, and should not affect the recovery of the DC system afterwards. In addition, rapid changes in the triggering angle during normal control should not cause protective triggering action.
(17) The cooling system must have sufficient cooling capacity to effectively cool the converter valve under various operating conditions. Each valve hall sets up an independent closed-loop water cooling system for a set of converter valves, and the main equipment of the internal cooling system of the valve (but not limited to) includes circulating water pumps, deionization devices, degassing tanks (if needed), expansion pressure tanks (or high-position expansion pressure water tanks), mechanical filters, supplementary water pumps, electric heating devices, distribution and control protection equipment.
(18) The thyristor valve should be able to eliminate any cause of fire in its design, manufacture, and installation, as well as the possibility of fire spreading within the valve. The non-metallic materials inside the valve should be flame-retardant and have self-extinguishing properties, and the materials should comply with UL94. V0 Material Standard. All plastics should contain a sufficient amount of flame retardant, such as aluminum trihydride (AlH_3), but it should not reduce other essential physical properties of the material, such as mechanical strength and electrical insulation characteristics. Due to the high corrosiveness and toxicity of the substances produced by the combustion of brominated flame retardants, the use of these substances as fillers is not allowed. The design of the converter valve should be oil-free. The design of the thyristor electronic device unit should be reasonable, without the risk of overheating and arcing.

II. Key Technologies

(1) Research and Development of Thyristor

1. Optimization of layout design on cathode

The world's largest single-chip capacity new 6 inch thyristor (8.5 kV/5500 A) has been successfully developed, which optimizes the layout design of the thyristor (see Fig. 48), improves the turn-on characteristics and dv/dt performance of the thyristor, and conducts a comparison test of thyristor characteristics and valve assembly with reference to the 800 kV/10 GW project.

Looking down at the core of the thyristor from the cathode direction, its cathode graphic structure is shown in Fig. 49. As can be seen from Fig. 49, the effective

Fig. 48 Design and optimization of internal layout of thyristor

–Short circuit point, gate and amplification gate –Active conduction area of cathode; –Junction termination

Fig. 49 Schematic diagram of cathode of thyristor

cathode area is the total area of the chip minus the area occupied by the short circuit point, gate, enlarged gate, and junction terminal.

In Fig. 49, from the center radially outward, the yellow areas are the gate, enlarged gate, and short circuit point, with a total area of S_p; the red area is the effective cathode, with an area of S_e; the purple area is the junction terminal, with an area of S_t. In terms of current carrying capacity, the effective conduction area ratio of the thyristor R_s is defined as the effective cathode area S_e to the total chip area ($S_p + S_e + S_t$), that is, $R_s = S_e/(S_p + S_e + S_t)$.

From the perspective of increasing the flow capacity, it is necessary to increase the red area, namely S_e. Under the condition that the total area of the chip remains unchanged, it is necessary to reduce the area of the purple terminal S_t and the yellow gate, the enlarged gate, and the short-circuit point S_p. From the perspective of reducing the shutdown time, it is necessary to increase the number of yellow short-circuit points, refine the enlarged gate branches, and reduce the radius of the circular gate and the enlarged gate. Through simulation and experimental verification, a small

Fig. 50 Schematic diagram of layout of compact structure on cathode of thyristors

diameter densely distributed short-circuit point; fine branches, small radius enlarged gate layout, referred to as dense and small layout, as shown in Fig. 50. Quantitative calculations show that this dense and small layout increases the effective conduction area of the cathode by 7.8%.

Refining the enlarged gate branches and reducing the radius of the circular enlarged gate are beneficial to reducing the shutdown time and increasing the effective conduction area of the cathode, becoming a compatible and most worthy of exploration highlight. Whether this optimization direction is established is the key to the success of R&D. According to the thyristor turn-on mechanism, refining the enlarged gate branches reduces the ability and rate of the enlarged gate strong trigger current to transport to the deep cathode, thereby reducing the rise rate of the turn-on threshold current. The rise rate of the turn-on threshold current d*i*/d*t* has been difficult to accurately calculate and measure so far. After in-depth research, a "thyristor branch full of $N+$ enlarged gate" structure was invented, which has the powerful function of increasing the conduction ability and speed of the strong trigger current, enabling the refinement of the enlarged gate branches and the reduction of the circular enlarged gate radius to be smoothly implemented. This new type of enlarged gate structure introduces a reduction in shutdown time t_q of about 100 μs, while the forward voltage drop V_{TM} and recovery charge Q_{rr} remain basically unchanged, and the measured data are shown in Figs. 51 and 52.

The ability of d*i*/d*t* has also been improved, and the lower limit specification of d*i*/d*t* value is *S* increased to *T*, and no abnormalities were found in the tightened test, the results are shown in Fig. 53 shown.

As mentioned above, by introducing a novel gate structure that enhances the triggering current conduction capability and speed, the reliable application of the delicate new layout is ensured, while reducing the turn-off time and increasing the

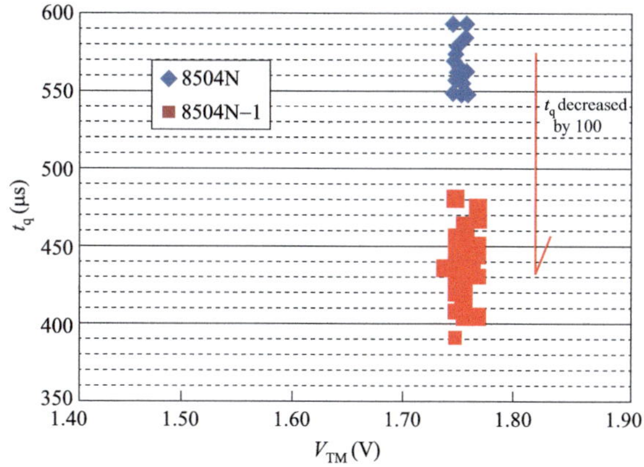

Fig. 51 Test results of unchanged reduce off-time V_{TM}

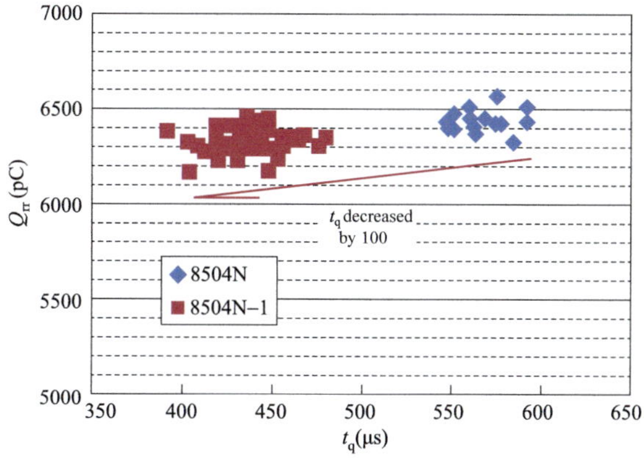

Fig. 52 Test results of unchanged reduce off-time Qrr

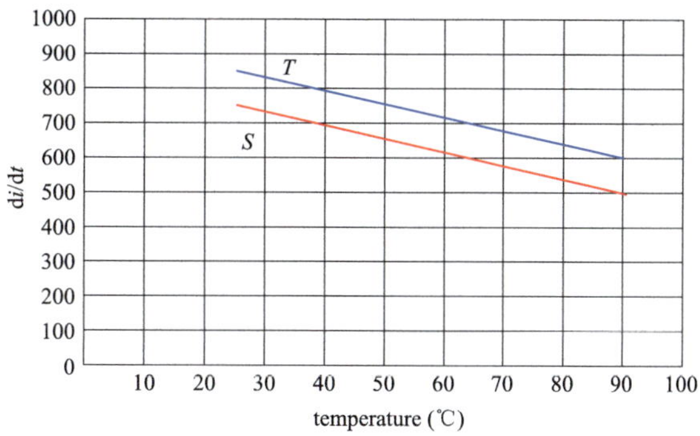

Fig. 53 d*i*/d*t* ightened test data

effective conduction area. This solves the technical problems encountered in the development of the 5500 A, 8500 V thyristor in the CG project.

2. P-type doping on radial variation

For the thyristor body, a high concentration P-type layer is still needed as the substrate of emitter junction to optimize high temperature voltage resistance, d*v*/d*t*, turn-off time, etc., after the symmetrical PNP structure is turned off. This is generally formed by the second P-type doping. However, this P-type layer is not desired in the gate amplification and negative angle junction terminal area. Due to the difficulty in making the P-type deep diffusion impurity mask layer, this P-type layer is not masked during production. As a result, to ensure the terminal breakdown voltage required by the first low concentration P-type doping, a thickness that is submerged by the second P-type doping must be reserved, and this thickness is greater than 30 μm. From the perspective of wafer thickness, considering the double-sided effect, there is a 60 μm wafer thickness problem. From a process perspective, the depth of the diffusion junction is directly proportional to the square root of the diffusion time, and the time required for diffusion increases geometrically as the increase in junction depth. The depth of the first low concentration P-type doping junction is the deepest, and continuing to deepen it takes too much time, while reducing the depth saves significant time. If the depth of aluminum diffusion junction is 100 μm and takes 40 h, if it is increased by 30 μm, it will take 67.6 h, and if it is reduced by 30 μm, it will only take 19.6 h. It can be seen that reducing the depth of the first low concentration P-type doping junction is also significant for shortening the process cycle, improving process stability, and saving energy.

As a deep optimization of design and process, the patented technology of deep junction P-type doping on radial variation has been developed and applied to achieve the selectivity of the second P-type doping, reducing the depth of the first low concentration P-type doping junction by more than 30 μm. It also makes the low concentration area where the terminal space charge region of the blocking junction can be expanded more abundant. After optimization by P-type radial doping variation, the thickness of the short base region is also reduced to the extreme, and the voltage-thickness ratio reaches an unprecedented height.

The 5500 A, 8500 V UHV thyristor center double amplification gate P-type doping variation structure is characterized by extremely low P-type doping concentration in the lateral base region below the 1AG, 2AG N + area, which increases the lateral base resistance, reduces the area occupied by the amplification gate, and improves the trigger sensitivity. At this time, the area occupied by the two-stage amplification gate is even smaller than that occupied by the conventional single-stage amplification gate, but the d*i*/d*t* tolerance is greatly improved.

3. Positive and negative symmetrical mesa terminal

As early as the late 1960s, Kohl studied the junction termination of mechanical processing-type table-top thyristors. Due to various constraints at the time, on one hand, the processing was complex and inefficient, on the other hand, the level of raw silicon material was low, the thyristor had a small diameter, low voltage, thin chip, and the processing was difficult with a low yield rate. Nowadays, thyristor chips have a large diameter, high voltage, and sufficient thickness, and the processing methods are diverse, which can be waterjet or laser. Its efficiency, yield rate, and control precision have met the requirements for mass production; at the same time, it is obvious that using a table-top type to replace the general negative angle contributes to increasing the voltage and enlarging the effective cathode area. A special topic was carried out to develop this into a patented technology and standardize this structure and processing technology. The table-top type significantly increases the effective cathode area, with a quantitative calculation result of increasing the effective cathode area by 6.1%, and forming a smaller negative angle at the leading edge of the blocking junction is more conducive to improving the P–N junction breakdown voltage. Test statistics show that the blocking voltage is increased by 8%.

4. Summary of optimization analysis

Based on the theory, characteristics and process practice of power semiconductor device, matching the new forward and reverse symmetric table-top junction termination will make the long base area of thyristor extremely thin. Combined with the patented technology "forward and reverse symmetric P-type doping on radial variation", the limit thinning short base area improves the blocking voltage, achieving an unprecedented voltage-thickness ratio. If the voltage-thickness ratio of the 4000 A, 8500 V 6 inch UHV thyristor chip from Xiangjiaba is set as 1, then the voltage-thickness ratio of the 8500 V thyristor chip used in other typical projects is shown in Table 28.

With the effective support of the patented technology "thyristor branch full of N + amplifying gate", a small diameter densely distributed short circuit point is optimized and designed; fine branches, small radius amplifying gate layout. Not only has it significantly increased the effective conduction area ratio, improved the on-state current, but also unconditionally reduced the turn-off time. This dual optimization effect has solved the key technical problem. At the same time, the use of forward and reverse symmetric table-top modeling has reduced the area occupied by the end terminal itself, increased the effective conduction area ratio, and raised the effective conduction area ratio to the extreme. Also taking the effective conduction area ratio of the 4000 A, 8500 V 6 inch UHV thyristor chip from Xiangjiaba as the benchmark

Table 28 Ratio of normalized voltage and thickness of 6 in, 8500 V thyristor in typical projects

Specifications	XS project	HZ project, JH project	JCG project
4000 A, 8000 V	1		
5000 A, 8500 V		1.06	
5500 A, 8500 V			1.125

set as 1, the effective conduction area ratio of the 8500 V thyristor chip used in other typical projects is shown in Table 29. The changes in the technical parameter specifications of the 6 in, 8500 V thyristor used in various typical projects are shown in Table 30. The 5500 A, 8500 V thyristor used in the CG project is the thyristor with the largest effective conduction area ratio in the world so far.

As shown in Table 30, the 6 inch, 8500 V thyristor has been significantly optimized in terms of current and turn-off time parameters from the XS project to the CG project. The main limiting parameters, avalanche voltage and non-repetitive critical current rise rate, have also been appropriately increased, with no compromises on other parameters. This is sufficient to demonstrate that the comprehensive parameter level of the thyristor has been effectively improved, thus it can fully and reliably meet the application requirements of the CG project's 5500 A, 1100 kV UHV DC transmission project.

(II) Comparative Experiment of Key Parameters of Thyristor

The test items for the 8.5 kV, 5500 A thyristor mainly include on-state voltage drop, turn-off time, reverse recovery charge, gate trigger voltage/current, reverse avalanche capability, and high-temperature turn-off test. The specific test parameters are shown in Table 31.

The test parameter requirements for the 7.2 kV, 6250 A thyristor used by a certain manufacturer in the CG project are consistent with the comparative test parameters of the thyristor used in the ± 800 kV, 10 GW project. The specific test items and parameters are shown in Table 32.

Table 29 Ratio of normalized active conduction area of 6 in, 8500 V thyristor in typical projects

Specifications	XS project	HZ project, JH project	CG project
4000 A, 8000 V	1		
5000 A, 8500 V		1.07	
5500 A, 8500 V			1.135

Table 30 Change of technical parameters of 6 in, 8500 V thyristor in typical projects

Parameter specifications	XS project 4000 A, 8000 V	HZ project, JH project 5000 A, 8500 V	CG project 5500 A, 8500 V
Gate voltage V_{T0} (V)	1.25	1.25	1.2
Slope resistance ∂_{T} (mΩ)	0.18	0.124	0.118
Non-repeated critical rate of rise of current di/dt (A/μs)	3200	3200	3500
Surge current *ITSM* (kA)	48	51	58
Turn off-time t_q (μs)	780	700	500
Avalanche voltage V (V)	9100	9300	9500

Table 31 Comparative test and parameters of 8.5 kV, 5500 A thyristor

No.	Name of parameters	Limit of parameters	Testing conditions
1	On-state voltage drop V_{TM} (V)	≤ 1.81	Junction temperature $T_{vj} = 90$ °C, Average current in conduction $I_T =$ 5500 A
2	Turn off-time t_q (μs)	≤ 500	$T_{vj} = 90$ °C, $I_T = 2000$ A, $V_R = 200$ V, d*i*/d*t* has – 1.5 A/μs, – 4.0 A/μs 2 types
3	Reverse recovery charge Q_r (μC)	5400–55000	$T_{vj} = 90$ °C, $I_T = 2000$ A, d*i*/dt has – 1.5 A/μs, – 4.0 A/μs
4	Gate trigger voltage V_{GT} (V)	≤ 2.6	$T_{Vj} = 25$ °C, $V_{AK} = 12$ V
5	Gate trigger current I_{GT} (mA)	≤ 400	$T_{Vj} = 25$ °C, $V_{AK} = 12$ V
6	Reverse avalanche ability (long term avalanche) I_{RAL} (A)	≤ 20	$T_{vj} = 90$ °C, $V_{RAL} = 9100$ V, $t_p = 250 +$ 2500 μs, 3 pulses
7	High temperature reverse bias		$T_{vj} = 90$ °C, ① AC full wave, $V_{peak} \geq$ 6000 V, 8 h ② Power frequency binormal half wave $V_{peak} \geq 6400$ V, 3 h

Table 32 Comparative test and parameters of 7.2 kV, 6200 A thyristor

No.	Name of parameters	Limit of parameters	Testing conditions
1	On-state voltage drop V_{TM} (V)	≤ 1.68	$T_{vj} = 90$ °C, $I_T = 5500$ A
2	Turn off-time t_q (μs)	≤ 500	$T_{vj} = 90$ °C, $I_T = 2000$ A, $V_R =$ 200 V, d*i*/d*t* = – 1.5 A/μs, – 4.0 A/μs 2 types
3	Reverse recovery charge Q_r (μC)	5400–55 000	$T_{vj} = 90$°C, $I_T = 2000$ A, d*i*/d*t* has – 1.5 A/μs, – 4.0 A/μs
4	Gate trigger voltage V_{GT} (V)	≤ 2.6	$T_{Vj} = 25$ °C, $V_{AK} = 12$ V
5	Gate trigger current I_{GT} (mA)	≤ 400	$T_{Vj} = 25$ °C, $V_{AK} = 12$ V
6	Reverse avalanche ability (long term avalanche) I_{RAL} (A)	≤ 20	$T_{vj} = 90$ °C, $V_{RAL} = 8100$ V, $t_p =$ 250 + 2500 μs, 3 pulses
7	High temperature reverse bias		Power frequency binormal half wave $V_{peak} \geq 5700$ V, 3 h

1. Comparison of two types of 8.5 kV, 5500 A thyristors on the same platform
   (1) On the Chinese test platform A, two types of 8.5 kV, 5500 A thyristors are measured, and the test data is shown in Table 33.
   (2) On the Chinese test platform B, two types of 8.5 kV, 5500 A thyristors are measured, and the test data is shown in Table 34.

Table 33 Comparison of measured results of Chinese testing platform A

No. of Thyristor	Q_{rr} (μC)		t_q (μs)		V_{TM} (V)		I_{AL} (A)	
	– 1.5 A/μs, 90 °C		– 1.5 A/μs, 90 °C		5500 A, 90 °C		8100 V, 90 °C	
	Platform A	Platform B	Platform A	Platform B	Platform A	Platform B	Platform A	Platform B
1	5071	5358	424.1	383.2	1.76	1.79	0.8	0.6
2	5076	5296	399.7	367.4	1.77	1.79	0.6	0.7
3	5115	5375	370.2	372.5	1.76	1.78	0.9	0.7
4	5088	5366	409.0	382.6	1.76	1.77	0.7	0.8
5	5055	5385	406.8	383.3	1.75	1.79	0.6	0.7
6	5086	5391	409.4	377.4	1.75	1.78	0.5	0.7
7	5027	5380	388.6	393.4	1.76	1.77	1.1	0.6
8	4988	5349	408.5	393.7	1.76	1.78	1.2	0.7
9	5067	5388	401.9	376.0	1.76	1.78	0.5	0.8
10	5058	5359	431.1	398.5	1.73	1.78	0.7	1.2
Average	5063.1	5364.7	404.93	382.8	1.756	1.781	0.76	0.75

Table 34 Comparison of measured results of Chinese testing platform B

No. of thyristor	Q_{rr} (μC)		t_q (μs)		V_{TM} (V)		I_{AL} (A)	
	– 1.5 A/μs, 90 °C		– 1.5 A/μs, 90 °C		5500 A, 90 °C		8100 V, 90 °C	
	Platform A	Platform B	Platform A	Platform B	Platform A	Platform B	Platform A	Platform B
1	5356.5	5788.3	414.0	407.0	1.78	1.79	0.42	0.34
2	5164.4	5689.5	430.0	356.0	1.75	1.81	0.42	0.42
3	5172.1	5843.6	413.0	389.0	1.78	1.79	0.40	0.40
4	5203.2	5750.6	436.0	379.0	1.78	1.79	0.42	0.42
5	5352.1	6042.1	404.0	360.0	1.77	1.79	0.40	0.38
6	5244.2	5943.0	427.0	387.0	1.78	1.80	0.48	0.42
7	5139.2	5862.6	409.0	363.0	1.78	1.80	0.48	0.36
8	5088.1	5805.9	389.0	366.0	1.77	1.80	0.42	0.38
9	5063.0	5853.3	419.0	369.0	1.78	1.79	0.46	0.38
10	5151.5	5813.3	392.0	386.0	1.78	1.80	0.46	0.38
Average	5193.4	5839.2	413.3	376.2	1.78	1.80	0.436	0.368

(3) Summary of comparison of test data. Through the analysis of the above data, it can be seen that although there are certain differences in the measurement results on the two platforms, the overall trend is consistent:

1) At d*i*/d*t* = – 1.5 A/μs, under the condition of thyristor junction temperature of 90 °C, the turn-off time t_q of both types of 8.5 kV/5500 A thyristors is far less than 500 μs, meeting the technical specification

requirements. Relatively speaking, the thyristor t_q on the Chinese test platform A is generally less than that on the Chinese test platform B.

2) The on-state voltage drop of both types of 8.5 kV, 5500 A thyristors is less than 1.81 V, meeting the technical specification requirements. Relatively speaking, the on-state voltage drop of the thyristor on the Chinese test platform A is generally less than that on the Chinese test platform B.
3) The avalanche leakage current of both types of 8.5 kV, 5500 A thyristors is far less than the standard value of 20 A, meeting the technical specification requirements. Relatively speaking, the avalanche leakage current of the thyristor on the Chinese test platform A is generally less than that on the Chinese test platform B.
4) As di/dt increases, Q_{rr} and t_q correspondingly increase, but the increasing of t_q is relatively small.

2. Comparison of the same 8.5 kV, 5500 A thyristor on different platforms

(1) Comparison of measurement data of 8.5 kV, 5500 A thyristor of Company A on two platforms. Table 35 shows the test data of Company A's thyristor on two platforms.

(2) Comparison of measurement data of 8.5 kV, 5500 A thyristor of Company B on two platforms. Table 36 shows the test data of Company B's thyristor on two platforms.

(3) Summary of comparison of test data. From the above data, there is a certain deviation in the data measured by the same thyristor on the platforms of Company A and Company B.

1) The thyristor of Company A Q_{rr} has a smaller measurement result on the test platform of Company A, and the stability of the test platform of Company A is better; The thyristor of Company A's t_q has very close test results on both platforms, with t_q slightly larger on the platform of Company A; The thyristor of Company A's V_{TM} has a smaller measurement result on the platform of Company A, but the deviation is relatively stable, generally 0.1–0.2 V; The avalanche current of the thyristor of Company A is smaller on the test platform of Company B.
2) The thyristor of Company B Q_{rr} has a smaller measurement result on the test platform of Company B, and the stability of the test platform of Company A is better, which is consistent with the comparison test results of the thyristor of Company A on two test platforms; The thyristor of Company B's tq has a smaller measurement result on the Xi'an test platform, which is opposite to the comparison test results of the thyristor of Company A on the two test platforms, in general, the deviation measured on the two platforms is small, the average deviation of t_q is within 10 μs.
3) The thyristor of Company B VTM has a larger measurement result on the test platform of Company A, which is consistent with the comparison test results of the thyristor of Company A on two test platforms.

Table 35 Comparison of measured results of thyristor of company A in two testing Platforms

No	No. of device	Q_{rr} (μC)		t_q (μs)		V_{TM} (V)		I_{AL} (A)	
		− 1.5 A/μs, 90 °C		90 °C		5500 A, 90 °C		9100 V, 90 °C	
		Platform A	Platform B	Platform A − 1.5 A/μs	Platform B − 1.5 A/μs	Platform A	Platform B	Platform A	Platform B
1	316H0855 − 20	5788.3	5359	407.0	398.5	1.79	1.78	0.38	1.2
2	316H0855 − 3	5689.5	5296	356.0	367.4	1.81	1.79	0.38	0.7
3	316H0855 − 13	5843.6	5380	389.0	393.4	1.79	1.77	0.38	0.6
4	316H0855 − 15	5750.6	5388	379.0	376.0	1.79	1.78	0.40	0.8
5	316H0855 − 10	6042.1	5391	360.0	377.4	1.79	1.78	0.42	0.7
6	316H0855 − 8	5943.0	5385	387.0	383.3	1.80	1.79	0.38	0.7
7	316H0855 − 5	5862.6	5375	363.0	372.5	1.80	1.78	0.34	0.7
8	316H0855 − 2	5805.9	5358	366.0	383.2	1.80	1.79	0.42	0.6
9	316H0855 − 6	5853.3	5366	369.0	382.6	1.79	1.77	0.36	0.8
10	316H0855 − 14	5813.3	5349	386.0	393.7	1.80	1.78	0.42	0.7
Average		5839.2	5364.7	376.2	382.8	1.80	1.781	0.368	0.75

Table 36 Comparison of measured results of thyristor of Company B in two testing platforms

No.	No. of device	Q_{rr} (μC)		t_q (μs)		V_{TM} (V)		I_{AL} (A)	
		− 1.5 A/μs, 90 °C		90 °C		5500 A, 90 °C		9100 V, 90 °C	
		Platform A	Platform B	Platform A − 1.5 A/μs	Platform B − 1.5 A/μs	Platform A	Platform B	Platform A	Platform B
1	65.2.16470.36	5356.5	5071	414.0	424.1	1.78	1.76	0.48	0.8
2	65.2.16474.19	5164.4	5058	430.0	431.1	1.75	1.73	0.40	0.7
3	65.2.16472.29	5172.1	5067	413.0	401.9	1.78	1.76	0.46	0.5
4	65.2.16472.20	5203.2	4988	436.0	408.5	1.78	1.76	0.48	1.2
5	65.2.16471.31	5352.1	5086	404.0	409.4	1.77	1.75	0.42	0.5
6	65.2.16470.43	5244.2	5076	427.0	399.7	1.78	1.77	0.46	0.6
7	65.2.16471.02	5139.2	5115	409.0	370.2	1.78	1.76	0.40	0.9
8	65.2.16471.13	5088.1	5055	389.0	406.8	1.77	1.75	0.42	0.6
9	65.2.16472.14	5063.0	5027	419.0	388.6	1.78	1.76	0.42	1.1
10	65.2.16471.12	5151.5	5088	392.0	409	1.78	1.76	0.42	0.7
Average		5193.43	5063.1	413.3	404.93	1.775	1.756	0.436	0.76

4) The avalanche current of the thyristor of Company B is smaller on the test platform of Company B, and the stability of the test platform of Company B is better, which is consistent with the comparison test results of the thyristor of Company A on two test platforms.

3. Comparison of two 7.2 kV, 6250 A thyristors on the same platform

(1) Test data of 7.2 kV, 6250 A thyristor of Company C on the test platform of Company A is shown in Table 37.

(2) Summary of test data comparison. Through data analysis, it can be seen that:

1) The reverse recovery charge Q_{rr} of the 6250 A thyristors provided by Company C and Company B both meet the technical specification requirements of 5000–5400 μC. Relatively speaking, the Q_{rr} of the thyristor from Company C is larger.

2) The turn-off time t_q of the 6250 A thyristors provided by Company C and Company B are both far less than the technical specification requirement of 500 μs. Relatively speaking, the t_q of the thyristor from Company C is larger.

3) The 6250 A thyristors provided by Company C and Company B both meet the technical requirement of having an on-state voltage drop of less than 1.68 V. The on-state voltage drops of the thyristors from Company C and Company B are very close, and their stability is good.

4) The 6250 A thyristors provided by Company C and Company B both meet the technical requirement of having a reverse avalanche leakage current of less than 20 A. The reverse avalanche capabilities of the

Table 37 Comparison of measured results of Platform A

No. of Thyristor	Q_{rr} (μC)		t_q (μs)		V_{TM} (V)		I_{AL} (A)	
	– 1.5 A/μs, 90 °C		– 1.5 A/μs, 90 °C		5500 A, 90 °C		8100 V, 90 °C	
	Platform A	Platform C	Platform A	Platform C	Platform A	Platform C	Platform A	Platform C
1	5083.6	5308.6	245.0	322.0	1.61	1.61	0.4	0.4
2	5060.2	5226.2	234.0	277.0	1.60	1.62	0.5	0.4
3	5050.0	5196.1	305.0	289.0	1.60	1.63	0.5	0.4
4	5051.9	5363.1	342.0	308.0	1.61	1.62	0.4	0.4
5	5091.2	5180.6	264.0	298.0	1.61	1.60	0.4	0.4
6	5193.2	5359.3	275.0	372.0	1.61	1.61	0.5	0.4
7	5143.9	5184.8	265.0	289.0	1.61	1.62	0.5	0.4
8	5004.8	5294.9	214.0	299.0	1.60	1.61	0.5	0.4
9	4997.4	5274.2	248.0	291.0	1.61	1.61	0.5	0.4
10	5009.3	5227.7	228.0	299.0	1.62	1.62	0.4	0.4
Average	5068.6	5261.6	262.0	304.4	1.61	1.62	0.46	0.4

thyristors from Company C and Company B are very close, and their stability is good.

(III) Component Special Tests

Special test items have been developed according to the characteristics of the CG Project. The test items include maximum continuous overload test, protective triggering test, minimum turn-off angle test, single-cycle fault current test, and 3-cycle fault current test. The test parameters are set according to the most severe operating parameters of the project. These tests can prove that the converter valve has complete functions and sufficient design margin.

(1) The results of the maximum continuous overload test show:

   1) Converted to a single thyristor level, the turn-on and turn-off voltage stress is 3.0–3.4 kV, and the on-state current is about 6160 A. Under the most severe and recurring conditions, the voltage and current at turn-on and turn-off do not exceed the bearing capacity of the thyristor components and other circuit components in the valve.
   2) The valve losses have not exceeded the specified limits. Under the operating conditions of maximum continuous operating current, minimum cooling flow, and highest inlet water temperature, the valve module can operate safely for a long time without any component overheating.
   3) The reactor can suppress the discharge rate of stray capacitance at the moment of thyristor turn-on, effectively protecting the thyristor. The non-linear characteristics of the saturated reactor can suppress the rise rate of the turn-on current.

(2) The results of the protective triggering test show:

   1) The thyristor level has the correct forward overvoltage protection function. When the normal triggering function of a certain level of thyristor is lost, the gate triggering monitoring circuit board will detect that the forward voltage of this level of thyristor exceeds a certain limit, and will send a triggering signal to the thyristor to avoid the thyristor from being damaged by overvoltage.
   2) The thyristors, gate circuits, and damping resistors of each valve factory have slightly increased losses during continuous forward overvoltage protection, but the temperature is within safe limits and no overheating has occurred.
   3) The minimum turn-off angle test shows: at the most stringent shutdown test conditions with minimum turn-off angle (12°), the highest thyristor junction temperature and $\mathrm{d}i/\mathrm{d}t = -3$ A/μs, the valve components are fully functional, there will be no phase change failure, they have the correct voltage and current waveforms, and all the internal circuits of the valve operate correctly.
   4) The results of single-cycle fault current test show that: under the most severe conditions where the junction temperature of the thyristor is 90°, the single-cycle fault current is at its maximum, and the overvoltage during the recovery

period occurs simultaneously, under the conditions of the most unfavorable block recovery time, the valve components still have sufficient blocking ability after the occurrence of the single-cycle fault current, and will not re-conduct.

5) The results of 3-cycle fault current test show that: under the most severe conditions where the junction temperature of the thyristor is 90° and the 3-cycle fault current is at its maximum, under the conditions of the most unfavorable reverse recovery time, the valve components are fully functional after the occurrence of the 3-cycle fault current, and no thyristor is damaged.

In summary: The various indicators of the converter valve components of each valve factory in CG project meet the technical specification requirements, the functions of the valve are complete, no components are overheating, and the valve components can reliably shut down under the steady-state or transient $\gamma_{\min}$ value conditions.

(IV) Research on the External Insulation Performance of 1100 kV Converter Valves

The converter valve is the main equipment in the valve hall. As the voltage level is raised from 800 to 1100 kV, the withstand level of the DC busbar's switching impulse voltage is raised from 1675 to 2100 kV. The primary issue is to study the external insulation design of the converter valve and improve its electrical strength.

There are usually two methods to improve the electrical strength of the external insulation of the converter valve: Method 1 increases the air distance of the converter valve to surrounding objects; Method 2 optimizes the structure of the converter valve shield. For Method 1, due to the saturation characteristics of the air gap under operating shock, this method requires a significant increase in the gap distance, which will cause a significant increase in the size of the valve hall; for Method 2, increasing the curvature radius of the shield can significantly increase the withstand level of the air gap's switching impulse voltage, reduce the requirement for the air distance of the converter valve to surrounding objects, and improve the electrical strength of the external insulation of the converter valve.

The external insulation design of the converter valve in CG project adopts Method 2. First, the simulation of finite element method is used to study the electric field distribution on the surface of the 1100 kV UHV DC converter valve shield, and structural optimization is carried out based on the simulation results; then, the valve tower after the shield structure optimization is subjected to withstand and $U_{50\%}$ tests of switching impulse voltage, to verify the design of the converter valve shield and the minimum air distance from the converter valve to the surrounding walls and ground in the actual valve hall.

The simulation model includes the valve tower shield and the arrester corona ring, considering the actual operating conditions of the converter valve, assigning voltage to the converter valve shield, and setting ground potential around the valve tower model according to the actual ground, wall, and top of the valve hall, to calculate the electric field strength on the surface of the shield. The distribution of the electric field on the surface of the shield under switching overvoltage is shown in Fig. 54.

Fig. 54 Field distribution on surface of shielding cover under switching overvoltage

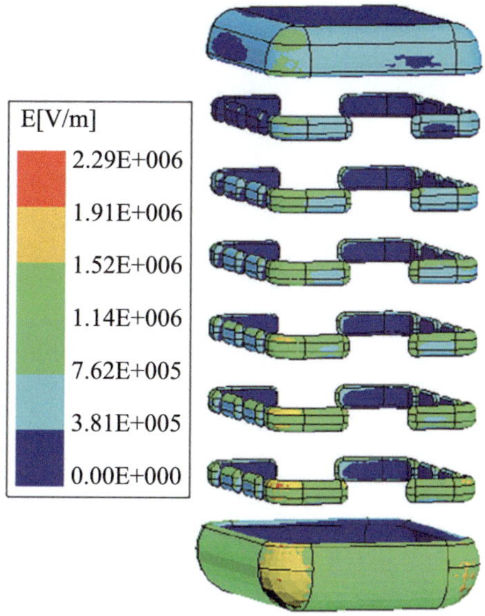

Based on past engineering design experience and considering a certain design margin, the side shield, bottom shield, and the corona ring at the bottom of the surge arrester exceed the control limit of the electric field strength. Therefore, the 1100 kV converter valve shield was optimized based on the structure of the 800 kV converter valve shield and the corona ring at the bottom of the surge arrester. By increasing the radius of curvature of the shield's ground-facing surface, the improvement effect of the electric field strength on the surface of the 1100 kV converter valve shield is significant, meeting the requirements of the control range of the electric field strength on the shield surface, with the maximum electric field reduction reaching about 35%.

(V) $U_{50\%}$ Test Verification on External Insulation of the Converter Valve

A switching impulse voltage test on external insulation was conducted on the HV valve tower of the 1100 kV converter valve after the shield was improved. The test needs to simulate the walls and ground around the converter valve.

The switching impulse $U_{50\%}$ test was conducted using the lifting method specified in Appendix A of GB/T 16927.1—2011 "High Voltage Test Techniques Part 1: General Definitions and Test Requirements". When the distance between the valve tower and the side wall is 5 m, the positive polarity $U_{50\%}$ voltage is 2498 kV. Under negative polarity switching impulse voltage, a voltage of − 2550 kV was applied by 10 times, and no discharge occurred. The multiple valve operation impulse voltage is 2155 kV, the minimum air distance between the actual valve tower in valve hall and the wall is 13 m, and the test results show that the design of the converter valve shield is reasonable and has a large safety margin.

(VI) Measures to Improve the Cooling Capacity of the Converter Valve

1. Overall design scheme of valve cooling

The cooling capacity of the 12 GW/1100 kV converter valve should ensure that there is at least 10% redundancy in the internal cooling system under non-fault operating conditions. For the external cooling system, an air cooler + closed cooling tower method is used. Each set of heat exchanger bundles in air cooler has a 30% redundancy. The redundancy capacity of the closed cooling tower in series with the air cooler, and it should meet the operation needs of the rated load under the highest environmental temperature when a cooling tower is lost (mixed water occurs and the operation personnel cannot immediately shut down the valve after losing the cooling tower).

2. Measures to improve the internal cooling system compared to the 5000 A project

   (1) Design margin. To ensure that the valve cooling system meets the cooling requirements of the converter valve under various operating conditions, the design of the cooling capacity of the 12 GW/1100 kV converter valve internal cooling system is consistent with the current 5000 A converter valve, that is,
      1) Under the premise of knowing the alarm value of the converter valve inlet temperature, take 2 °C lower than the alarm value of the converter valve inlet temperature as the rated inlet water temperature of the converter valve.
      2) The flow calculation of the internal cooling system meets the requirement of the maximum heat exchange required by the converter valve under 2 h overload operation (including the highest outdoor ambient temperature), and leaves more than 10% redundancy according to the valve cooling flow protection set value requirement.

   Through the combined effect of the above two aspects, the design of the internal cooling system can meet the cooling requirements of the converter valve under various operating conditions, and the above design margin indicators.

   (2) Pipeline optimization. The following design methods and technical measures are used to ensure the uniform flow of parallel and series branches in the diverter valve distribution pipe, thereby ensuring sufficient cooling for each heating element.
      1) The module distribution pipe adopts a diagonal water inlet and outlet method, which improves the uniformity of the water distribution in each branch of the module.
      2) The pressure-flow of components such as saturated reactors, water resistors, and thyristor radiators are matched and combined in design, striving to keep the flow resistance of each branch consistent.

3) For branches that cannot meet the consistent flow resistance, the method of increasing the pipe length or pipe resistance is used to meet the basic consistent requirement of flow resistance.
4) The simulation software PIPEFLOW is used for calculation to achieve balanced flow in each branch.
5) Through the overall flow pressure test of the valve module and the ultrasonic flow test of each branch, the rationality of the water distribution design has been verified.
6) The water pressure of branch pipe of each valve component is not less than 1.6 MPa.

The thyristor radiator, damping resistor, and saturated reactor are connected by small-diameter PVDF pipes. The joints of the PVDF pipes are equipped with O-rings made of Ethylene Propylene Diene Terpolymer (EPDM) rubber material. The pipe joints are connected to the radiator, damping resistor, and saturated reactor using quick thread joints. The connection with plastic components (damping resistors) uses PVDF nuts, and the connection with metal components (radiators, saturated reactors) uses metal nuts. This design makes the thermal expansion coefficient of the nut consistent with the cooled components, which is not prone to loosening and leakage due to temperature changes, improving the reliability of the connection and reducing the probability of leakage.

(3) Measures to prevent electrode corrosion. The cooling water needs to flow over metal parts in different positions and at different potentials, and a weak leakage current will be generated in the waterway between metal parts at different potentials. Therefore, these metal parts may be subject to electrolytic corrosion. The leakage current density generated by the pressure difference in the water pipe is controlled at the microampere level per square centimeter. However, even with such a low current density, if protective measures are not taken, electrolytic corrosion of aluminum radiators and saturated reactors may still occur. To solve this problem, the following methods are adopted:

1) The conductivity of the coolant in the cooling system is controlled within 0.3 μS/cm, and the oxygen content is within $(2 \times 10^{-5})\%$ by nitrogen pressure regulation and deoxygenation device, thereby controlling the leakage current at a lower level and slowing down the corrosion rate of the pipeline.
2) Corrosion-resistant materials such as 316 L and aluminum alloy are used for metals in contact with the cooling medium.
3) All stainless steel equipment and pipe welding adopt protective gas blowing on the back of the weld, and the inside of the pipe is strictly acid-washed, degreased, cleaned, and rinsed to keep the inside of the pipe clean.
4) The interlayer of the valve module adopts a spiral water pipe, the waterway resistance is high, which greatly reduces the stray current in the main water pipe and reduces the amount of electrical corrosion.
5) The fixed potential electrode configured in the main water pipe of the valve module adopts an integrated design of the equipotential electrode and the water pipe sealing element, a conical equipotential electrode design

arranged in the axial direction facing the current and water flow, and a process method combining stainless steel material and nano coating. This effectively avoids the chlorine current corrosion of the metal component body and the O-ring, and improves the sealing reliability of the waterway.

6) Each radiator, damping resistor, and saturable reactor inlet and outlet are equipped with anti-corrosion electrodes, which can prevent electrolytic corrosion of the contact surface of the aluminum radiator, damping resistor, and saturable reactor body with the coolant. After calculating with Faraday's electrolysis theorem and combining with practical engineering experience, the design of anti-electrolytic corrosion electrodes can meet the technical requirements of at least 40 years of service life. Anti-corrosion tests have shown that after using the above methods, the converter valve and its components still have extremely high corrosion resistance even if they work in high-temperature cooling media for a long time with the leakage current exceeds the normal working value.

(4) Main pump reliability.

1) The base of the main circulation pump is equipped with a vibration damping device. When selecting the main circulation pump, choose a rust-proof model or undergo rust-proof treatment.
2) The main circulation pump operates in the high-efficiency area at the rated flow rate. The flow, head of delivery, and efficiency of the main circulation pump at the working point do not produce negative deviations, and the positive deviation does not exceed 3%. The maximum amplitude limit of the main circulation pump group is within the range stipulated by GB/T 29531—2013 "Contract Vibration Measurement and Evaluation Method", and the amplitude of operating vibration limit of the main circulation pump group does not exceed 0.05 mm. When the main circulation pump group is reversed due to a sudden power outage, its reverse speed does not exceed 120% of the rated speed of the main circulation pump, and it will not cause any damage to the equipment. The main circulation pump can run continuously for more than 25,000 h, the mechanical seal can run continuously for 8000 h, and the design life of the main circulation water pump is more than 131,000 h.
3) The main circulation pump is redundantly configured, automatically switched regularly, and the switching cycle is no longer than one week (can be set); it can automatically switch back when the switch is unsuccessful. The switching of the main circulation pump does not affect the normal operation of the soft starter. The main circulation pump switching has a manual switching function.
4) The main circulation pump uses AC power, and the pump can work normally within the range of power fluctuation − 10% to + 15%. When an AC system fault causes the average rectified voltage measured on the AC bus in the converter station to be 30% higher than the normal voltage, but less than

the extreme minimum continuous operating voltage and lasts for 1 s, the valve cooling system can run continuously and stably. It can withstand an overvoltage level of RMS phase voltage not less than 1.5 (per unit value), with a tolerance time of 300 ms; It can withstand an overvoltage level of RMS phase to phase voltage not less than 1.3 (per unit value), with a tolerance time of 300 ms.

5) The total delay of the unsuccessful switching judgment of the main circulation pump and the return switching time is less than the low flow protection tripping time. The low flow protection tripping time is not less than 10 s.
6) At the moment of shutdown of the main circulation pump, considering the principle of flow inertia and in conjunction with the requirements of the converter valve, the delay of action of ultra-low cooling water protection is generally set to 15 s, and the action delay setting time is longer than the 400 V standby self-casting delay. Considering the time for the main circulation pump of the valve cooling system to restart and establish pressure and flow after power loss, the maximum allowable power loss time of the main circulation pump is no more than 6 s, and the total delay of the unsuccessful switching criterion of the main circulation pump is 5 s. The flow alarm and trip protection settings of each level of the converter valve cooling system can meet various operating requirements including the switching of the main circulation pump and the standby self-casting of the distribution device.

3. Measures for improvement of external cooling system compared to the ± 800 kV, 8 GW project

(1) Protection measures with extremely low temperature.

1) Set up electric heaters. The temperature in the converter station is low in winter. In order to prevent the water in the outdoor heat exchange equipment (air cooler and closed cooling tower) and pipes from freezing or the water temperature being too low when the converter valve is shut down in winter, electric heating devices are set up on the circulating water pipes between the valve cooling equipment, and it is designed to operate the pump without stopping during the shutdown of the converter valve (unless the valve cooling equipment fails), in order to prevent the water in the outdoor equipment and pipes from icing. The electric heating equipment is set up in the inlet pipe of the air cooler. When the outdoor square temperature is extremely low in winter, it can prevent the cooling medium entering the room from being too low to damage the converter valve, and it can also prevent the outdoor air cooler and closed cooling tower pipes from icing when the converter valve is shut down.
2) Set up louvers. If the valve cooling system is completely shut down, measures such as closing the louvers are taken to reduce natural radiation cooling and ensure that the internal pipes will not ice within 4 h, so as to have enough time for anti-icing treatment, such as installing anti-icing equipment.

3) Valve, motor, steel structure design. The motor uses SKF/NSK bearings and is filled with low-temperature resistant grease, which can operate smoothly and prevent icing in an environment of – 40 °C for a long time. The valve meets the low-temperature use requirements as low as – 40 °C, and the steel frame uses D-grade product, which is resistant to low temperatures.

(2) Optimization of external cooling layout to improve heat dissipation efficiency in response to the heat island effect. Since the heat island effect mainly causes a decrease in the cooling capacity of the air cooler, measures have been taken to optimize the external wind cooling layout in high-temperature and arid areas to improve the efficiency of local heat dissipation:

1) Increase the air inlet height of the air cooler to improve the heat dissipation efficiency;
2) Heat-proof wind return sealing plates are set up around the air cooler inspection channel;
3) The air cooler adopts a wind-drawing design to prevent hot wind from returning.

(3) Reliability of valve cooling power supply and control/protection.

1) The main circulation pump equipment of the internal cooling of the valve is powered separately by two power sources. When one power source fails, it automatically switches to another power source for input. At this time, the main circulation pump does not switch. The main circulation pump is powered by an independent dual power source after switching, and does not share power with other equipment, which can prevent the main circulation pump from stopping due to short circuits and other faults caused by other equipment. The external cooling equipment of the valve is powered by at least two AC power sources, which are divided into multiple independent dual power sources through wiring terminals, and are separately powered to the external cooling equipment.
2) The valve cooling control/protection system is powered by a switching resource from DC 220 V/DC 110 V to DC 24 V, with the A control system connected to two DC power supplies, and the B control system connected to two DC power supplies. For non-redundant instruments and single signals, they are powered by a dedicated C route power supply. For signals under control protection, DC 110 V voltage is used. Since there are four downstream signals, four DC power supplies are used respectively. Any power supply failure does not affect the correct transmission of downstream signals to the two PLC control systems. The control circuits of the main circulating pump and electric heater do not use DC power, but use the same AC power as the main circuit of the equipment.
3) The valve cooling control protection system is redundantly set up, with two sets of valve cooling control protection systems sampling and working at the same time, but only one is in the activated state. The DC is locked only when both hosts fail, and the tripping circuit

with dynamic breaking contacts has contact monitoring function. Valve water cooling protection is configured in a redundant manner, each set of complete, independent water-cooling protection devices can handle all types of valve cooling system failures that may occur. Under normal circumstances, the redundant water-cooling protection systems are in operation, allowing a short-term exit of one protection system.

4) To ensure the safe and stable operation of the system and prevent the valve from stopping due to instrument failure, redundant instruments are set up to monitor important parameters of the converter valve cooling system (inlet valve temperature of cooling water, conductivity of cooling water, flow rate of cooling water, etc.), which are hot standby and have reliable anti-refusal and anti-maloperation measures. According to the "three out of two" principle, three sets of independent input and output units are configured; redundant optical fiber interfaces are configured to realize communication between the DC control protection system and the external valve cooling monitoring system.

III. Design of Technical Scheme

(I) Overall Design Scheme of Converter Valve

The converter valve of Changji Converter Station adopts air insulation, deionized pure water cooling, and a suspended double valve structure. It adopts an integrated valve tower design, which extends the assembly, testing, and transportation unit of the converter valve to the valve module, not only improving the production testing efficiency, but also facilitating the on-site installation and construction. The design comprehensively considers the AC, DC, and impulse voltage stress of the valve tower during operation, selects a reasonable air and creepage distance, ensures that the converter valve does not break down or flashover during operation, and will not cause surface discharge of insulation material due to aging or contamination of insulation material. The internal air distance of the converter valve is selected according to the strictest working conditions, and the creepage distance is designed to be not less than 14 mm/kV.

Table 38 shows the main technical differences between the 800 kV converter valve and the 1100 kV converter valve. A comprehensive comparison of the specification parameters of the 1100 kV/5455 A and 800 kV/5000 A converter valves in Table 38 shows that the differences between the two types of projects are mainly reflected in the rated voltage, overload, and DC voltage tolerance. Based on the clarification of these differences, the converter valve factory has carried out the design of converter valves with higher voltage tolerance.

(II) Structural Design of Converter Valve

After the UHV DC transmission project is upgraded from 800 to 1100 kV, the main parameters of the converter valve equipment are shown in Table 39. The 1100 kV converter valve uses 8.5 kV/5500 A thyristors, and the number of thyristors in series

Table 38 Major technical difference between 800 and 1100 kV converter valves

Parameters		800 kV/5000 A	1100 kV/5455 A	Note
Current	Rated DC current (A)	5000	5455	
	2 h overload current (A)	5335	5523	
Voltage	Rated unloaded DC voltage (kV)	231.5	319.04	
	Maximum unloaded DC voltage (kV)	240	329	
	Load shielding coefficient of temporary overvoltage (per unit value)	1.3	1.3	
Wide angle operation	Firing angle of reduced voltage operation with rated current (°)	15.0	20.9	Rectifier
	Extinction angle of reduced voltage operation with rated current (°)	35.4	31.4	Inverter
	Inductive voltage drop	10.5	10.0	
Short circuit current	Short circuit peak current of single cycle, with following blocking (kA)	50	55	
	Short circuit peak current of three cycles, without following blocking (kA)	50	55	
	Peak blocking voltage with following blocking (kV)	292	379	
Insulation level	*SIWL* (kV, peak value)	451	627	
	LIWL (kV, peak value)	445	606	

in a single valve is greater than the number of thyristors in series in a 5000 A converter valve.

The converter valves all adopt air insulation, deionized water cooling, indoor installation of suspended double valve structure, a total of 6 layers; each single valve includes 6 valve modules, each double-valve has 12 valve modules, as shown in Fig. 55. The converter valve mainly consists of suspension parts, valve frame, busbar, thyristor module, PVDF water pipe, layer shielding, optical cable trough, layer assembly, valve surge arrester, etc.

Table 39 Rectifier side converter structure and parameters

Parameters	800 kV/5000 A	1100 kV/5455 A
SIWL (kV)	451	627
Valve base type (suspended or supported)	suspended	suspended
Multiple valve type (double valve or quadruple valve), quantity	double valve	double valve
Thyristor type	8.5/5000	8.5/5500
Series connection number of single valve thyristor	57 + 3	81 + 3
Number of single valve modules	4	6
Number of valve reactors	8	12
Damping capacitance (μF)	2.0	2.0
Damping resistance (mΩ)	40	40
DC voltage grading resistance (mΩ)	88	88
FOP protection level (V)	8.1	8.1

In order to meet the different technical requirements of different projects, the converter valve adopts standardized, modular design. Project operation shows that modular design has good availability, high reliability and the most economical project cost, which is the best way to achieve standardization. The 1100 kV/5455 A converter valve continues to use the standard converter valve modular design, each single valve consists of 6 modules, each module contains 2 components, each component consists of 7 thyristor levels, 1 valve reactor, etc., as shown in Fig. 56.

The 1100 kV/5455 A converter valve adopts a small component structure, which is compact, flexible, easy to install, and meets fire and seismic performance requirements, mainly with the following technical advantages:

(1) Standardized component design.
(2) Deionized water-cooling thyristors and valve reactors.
(3) Use of indirectly cooled damping resistors through radiators, with high power and strong heat dissipation capacity.
(4) The cooling circuit adopts a reasonable layout and uses proven fireproof materials.
(5) The number of auxiliary parts is reduced, which lowers the failure rate and simplifies maintenance.
(6) The entire valve tower adopts a suspension structure design, increasing the seismic resistance of the converter valve.

(III) Electrical Design of Converter Valve

The electrical design of the converter valve should consider the voltage stress and current stress that may be caused under various circumstances. The electrical design mainly includes the following aspects: ① Thyristor selection; ② Series level design; ③ Forward overvoltage protection; ④ Design of damping and voltage equalizing circuit; ⑤ Valve reactor design; ⑥ Design of converter valve insulation coordination.

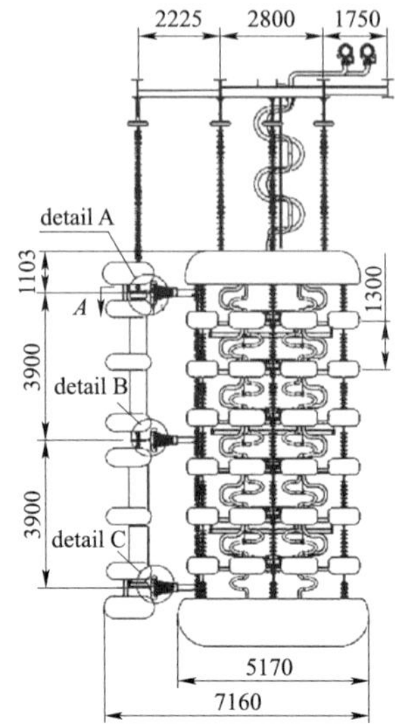

(a) front view of double valve tower in low voltage valve hall

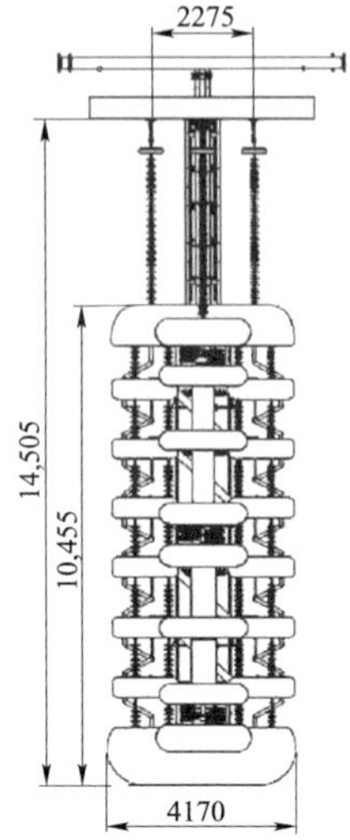

(b) side view of double valve tower in low voltage valve hall

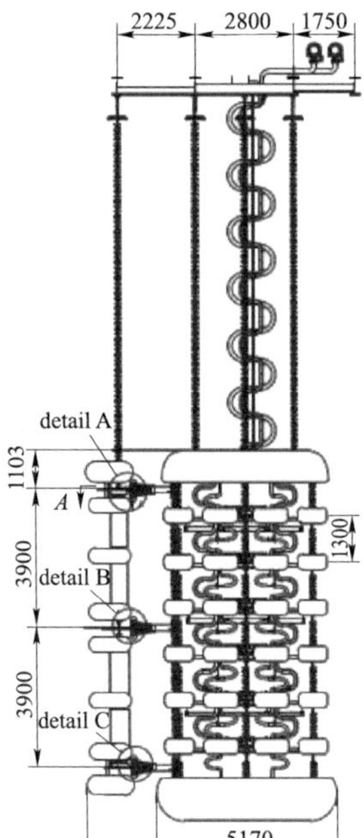

(c) front view of double valve tower in high voltage valve hall

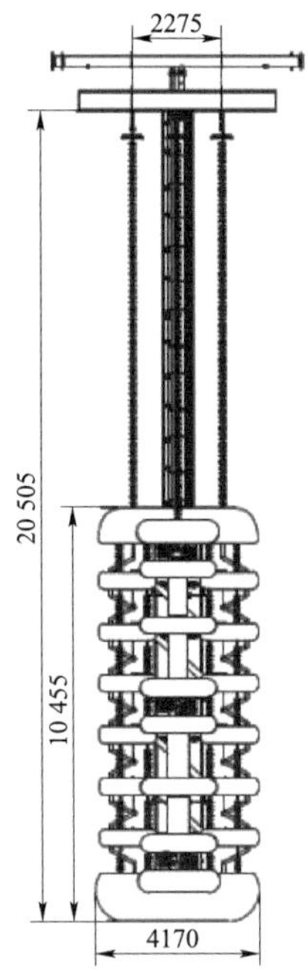

(d) side view of double valve tower in high voltage valve hall

Fig. 55 Configuration of double valve tower

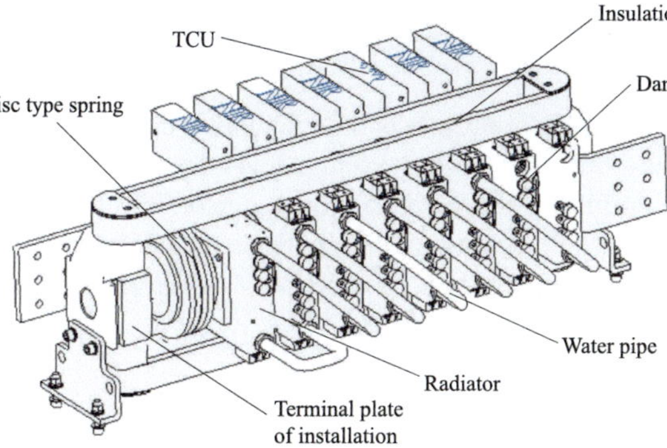

Fig. 56 Module of 1100 kV converter valve

According to the parameter requirements of the 1100 kV/5455 A UHV DC transmission system, the main electrical parameters of the determined 1100 kV/5455 A converter valve are shown in Table 40.

Table 40 Major design parameters of 1100 kV, 5455 A single valve

Item	Sending station	Receiving station	Item	Sending station	Receiving station
Thyristor type	6 in	6 in	Damping resistance R_{sn} (Ω)	3360	3108
Series connection number of single valve thyristor n_t	84	84	Valve section impulse grading capacitor (nF)	–	–
Redundancy of single valve n_{red}	3	3	Number of saturated reactors n_{rea}	12	12
Module number of single valve n_{mod}	12	12	Firing type	Electric trigger	Electric trigger
Damping capacitance C_{sn} (μF)	0.0238	0.0214	Structure of valve tower	Double valve tower with suspended structure	Double valve tower with suspended structure

4 Smoothing Reactor

The smoothing reactor of the CG project is independently developed and manufactured in China, and it is the world's highest voltage level dry-type air-core smoothing reactor. The single smoothing reactor has an inductance of 75 mH, and the winding weight reaches 98–106 t, which is a significant increase in quality and size compared to previous project smoothing reactors, posing new requirements for production manufacturing technology.

I. Technical Requirements

Table 41 Requirement of specific creepage distance of external insulation on pole line and neutral point equipment in Changji converter station (mm/kV)

Location	Insulation type	Umbrella	Vertical bushing and pillar insulator			
		Average diameter (mm)	250–300	400	500	600
Pole line and neutral point equipment	Composite insulator	Size of umbrella (C type)	47	49	51	53

Table 42 Requirement of specific creepage distance of external insulation on neutral line in Changji converter station (mm/kV)

Insulation type	Umbrella	Pillar insulator	Vertical bushing
	Average diameter (mm)	250–300	
Composite insulator	Size of umbrella (C type)	46	45

Table 43 Requirement of specific creepage distance of external insulation on pole line and neutral point equipment in Guquan converter station (mm/kV)

Location	Insulation type	Umbrella	Vertical bushing and pillar insulator			
		Average diameter (mm)	250–300	400	500	600
Pole line and neutral point equipment	Composite insulator	Size of umbrella (C type)	52	53	55	57

Table 44 Requirement of specific creepage distance of external insulation on neutral line in Guquan converter station (mm/kV)

Insulation type	Umbrella	Pillar insulator	Vertical bushing
	Average diameter (mm)	250–300	
Composite insulator	Size of umbrella (C type)	45	44.6

Table 45 Rated current (maximum environmental temperature)

DC current at rated power under ideal conditions I_{dN} (A)	5455
Maximum continuous DC current I_{mcc} (A)	5523
2 h overload current I (A)	5839
Transient fault current (kA, peak value) (waveform shown in Fig. 57)	40

Table 46 Rated voltage (kV)

Under ideal conditions, rated power, DC voltage, to ground U_{dN}	Pole line	1100
	Neutral line	150
Maximum continuous DC voltage, to ground U_{dmax}	Pole line	1122, supposed simultaneously with harmonics 75 kV
	Neutral line	200

Table 47 Insulation level and testing voltage (testing level of total series condition) (kV)

Item		Insulation level	
		Pole line	Neutral line
Switching impulse withstand level (peak value)	Between Terminals	2100	2100
	Terminal to Ground	2100	2100
Lightning full impulse withstand level (peak value)	Between Terminals	2600	2600
	Terminal to Ground	2580	600
Lightning chopped impulse withstand level (peak value, between terminals)		2860	2860
DC withstand (120 min, to ground)		1755	225

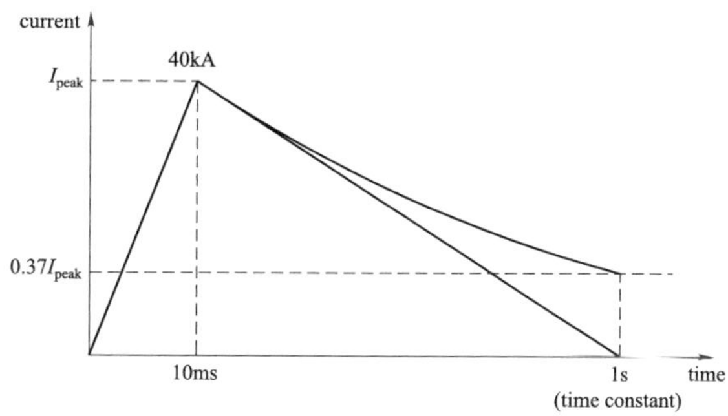

Fig. 57 Transient fault current

(1) The environmental conditions are detailed in the first section of this chapter, and the indoor field conditions are discussed in the second section.
(2) The supports for the DC field pole line and neutral line smoothing reactor are required to use composite insulators, and the creepage distance requirements are as follows:
   1) The creepage distance required for the external insulation of the 1100 and 550 kV equipment at the Changji converter station should not be less than the values in Table 41.
   2) The creepage distance of the neutral line at the Changji converter station should not be less than the values in Table 42.
   3) The creepage distance required for the external insulation of the 1100 and 550 kV equipment at the Guquan converter station should not be less than the values in Table 43.
   4) The creepage distance of the neutral line at the Guquan station should not be less than the values in Table 44.
(3) Transportation conditions, the transportation restrictions for the Changji and Guquan converter stations are outer diameter (width) × transport height: 5.5 m × 4.5 m, with a weight limit of 100 t.
(4) The dry smoothing reactors on the pole line and neutral bus of the Changji converter station are all composed of 2 series-connected 75 mH coils. The dry-type smoothing reactors on the pole line and neutral bus of the Guquan converter station are also composed of 2 series-connected 75 mH coils.
(5) Dry-type smoothing reactor: air core, outdoor type, natural cooling; rated inductance: 2 × 75 mH, 0%– to 5%.
(6) The current rating (at the highest environmental temperature) is shown in Table 45.
(7) The voltage rating is shown in Table 46.
(8) Insulation level and test voltage (total test level after series connection) are shown in Table 47.
(9) The steady-state temperature rise limit under maximum continuous current: When used outdoors, the temperature rise requirement after adding the harmonic equivalent current to the maximum continuous current (5523 A) is: the hotspot temperature rise should not exceed 105 K, and the average temperature rise should not exceed 80 K (long-term operating current 5523 A plus equivalent harmonic current).
(10) After the smoothing reactor is put into operation, noise measurements should be conducted at a location 5 m away from the contour of the smoothing reactor and 2 m above the ground. The measured noise (sound pressure level) should not exceed 75 dB (A).
(11) Insulation heat resistance grade: It is required to use multi-strand type displacement wires for winding, with an H-class insulation layer between the strands. The heat resistance grade of other insulation materials should not be lower than F-class.

II. Key Technologies

The rated DC voltage of the smoothing reactor in the CG Project has been increased from ± 800 kV to ± 1100 kV, and the rated inductance has been increased from 50 to 75 mH. The equipment manufacturer overcame difficulties and finally completed the design and manufacture of the equipment. When the rated current and capacity of a single UHV smoothing reactor increase, the design difficulty, volume, and weight also increase. From a design perspective, dry-type hollow smoothing reactors need to face new problems such as an increase in the outer diameter of the winding, an increase in the number of encapsulation layers, and the need to increase the current-carrying area of the wire and reduce the current density of the wire to achieve the required temperature rise limit. Some auxiliary devices such as noise reduction devices are not suitable for smoothing reactors with increased capacity, and the structure of some parts of the support system also needs to be reinforced.

During the development process of the ± 1100 kV/5455 A smoothing reactor, technical problems were tackled, in-depth research was conducted, simulation modeling, prototype testing were carried out, and the smoothing reactor was designed and manufactured. After testing by authoritative institutions, the test values all meet the technical requirements, laying a solid foundation for the smooth manufacture of the equipment. Figures 58 and 59 show the actual equipment.

Fig. 58 Smooting reactor in Changji converter stations

Fig. 59 Smooting reactor in Guquan converter stations

(I) Characteristics of the 12 GW Smoothing Reactor

1. Increase in coil inductance

The single pole of the polar line is composed of two 75 mH windings in series, with a total inductance of 150 mH. Compared with the past, the total inductance remains unchanged, but the inductance of a single winding has increased by 50% (from 50 to 75 mH) compared with previous projects. Inductance is a key parameter of the smoothing reactor, and an increase in inductance means that the weight of the reactor will increase significantly under the same conditions.

2. Increase in coil impulse level

Since the single pole of the polar line is connected in series by two 75 mH smoothing reactors, after the rated voltage is increased from ± 800 kV to ± 1100 kV, the insulation level of the line is increased to 2600 kV for the first time, making the insulation level of a single winding completely inconsistent with that at ± 800 kV. Through simulation software, the peak voltage values of lightning impact at the beginning and end of the winding when a 2600 kV level lightning wave invades were calculated. According to the simulation results, the lightning impulse voltage of each winding was determined. Figures 60 and 61 show the simulation results of the lightning impulse of the ± 1100 kV series winding of smoothing reactor.

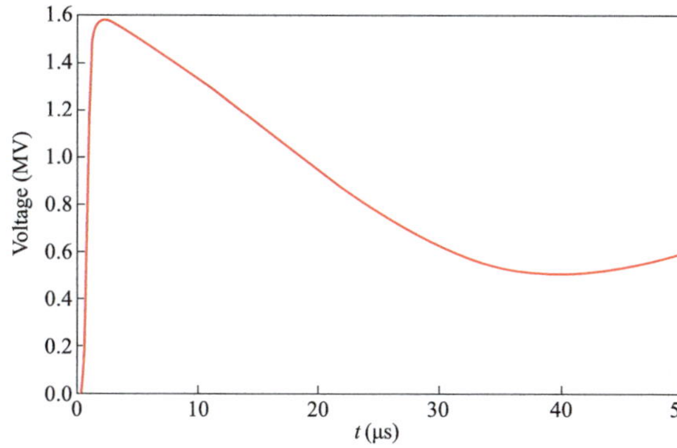

Fig. 60 Lightning impulse waveform on the first winding

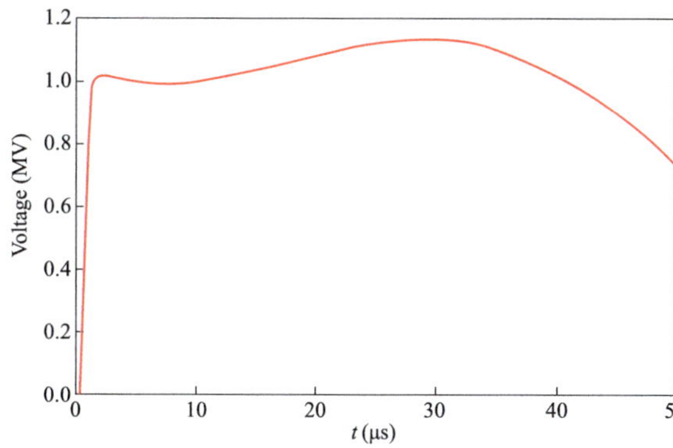

Fig. 61 Lightning impulse waveform on the second winding

3. Increase in insulation level of terminal-to-ground

After the rated DC voltage is increased from ± 800 kV to ± 1100 kV, the lightning impulse full wave level from terminal to ground is increased from 1950 to 2580 kV, the switching impulse level from terminal to ground is increased from 1675 to 2100 kV, the DC withstand level (referred to ground) is increased from 1236 to 1755 kV. The increase in insulation level makes the pillar insulator completely inconsistent with that at 800 kV. Therefore, the equipment factory has designed and manufactured special pillar insulators for the CG project.

(II) Research on the Mutual Inductance of Series Windings

The single pole of the polarity line is composed of two 5455 A/75 mH windings in series, with a total inductance of 150 mH. Such a geometrically large hollow winding, the corresponding magnetic field is also a problem that must be considered. The magnetic field of the hollow reactor is divergent. This divergent magnetic field affects the self-inductance of the two windings themselves, causing the total inductance to be far greater than the design value, thereby affecting the entire line. On the other hand, due to the existence of leakage magnetic field, it also affects other equipment near the reactor. Figures 62 give the values of the magnetic induction intensity at different

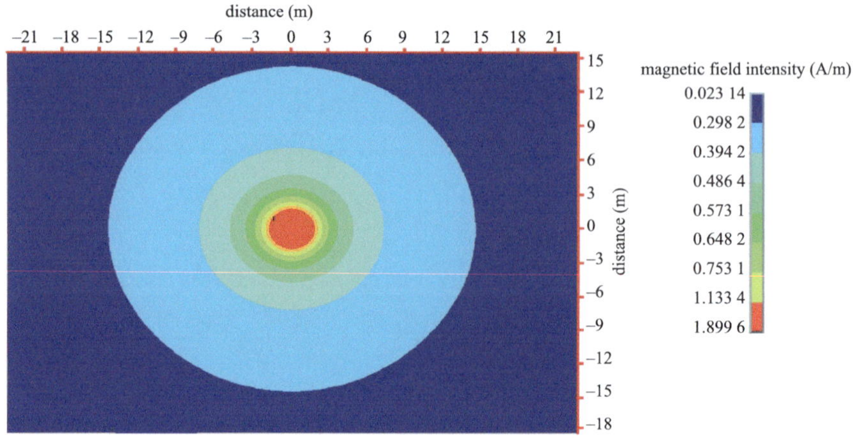

Fig. 62 Magnetic simulation nephogram of smoothing reactor

distances. Combining the calculation method and simulation results, determine the distance between the two windings, ensure that the coefficient of mutual inductance between the two windings in series meets the requirements, and ensure that the winding meets the specification requirements for other equipment.

(III) Research on Shielding Structure

After the rated DC voltage is increased from ± 800 kV to ± 1100 kV, the impulse levels of the winding and pillar insulator are increased from 2100 and 1950 to 2600 and 2580 kV respectively, which are about 23 and 32% higher than the previous 8 GW project. First, through simulation calculations, a shielding structure that meets the corona and radio interference requirements of the ± 1100 kV smoothing reactor was determined, and through a series of research and experiments, the discharge characteristics of the shielding structure under different electrode forms were studied. Using calculation and experimental verification, a scheme of shielding structure layout that meets the terminal-to-terminal and terminal-to-ground insulation requirements of the ± 1100 kV smoothing reactor was determined, as shown in Fig. 63.

After the rated DC voltage is increased from ± 800 kV to ± 1100 kV, the switching impulse from terminal to ground is increased from 1675 to 2100 kV, which requires the equipment to withstand an switching voltage of 2100 kV to the surrounding grounded body. According to the true type test and the overall potential and surface electric field distribution nephogram (see Fig. 64), it was finally determined that the shielding device of the polarity line reactor consists of 7 layers of corona rings, of which the winding body has 4 layers of corona rings, the support structure has 3 layers of corona rings, and the diameter of the corona ring is increased from 200 to 300 mm.

Fig. 63 Configuration and diagram of shielding structure of ± 1100 kV smoothing reactor

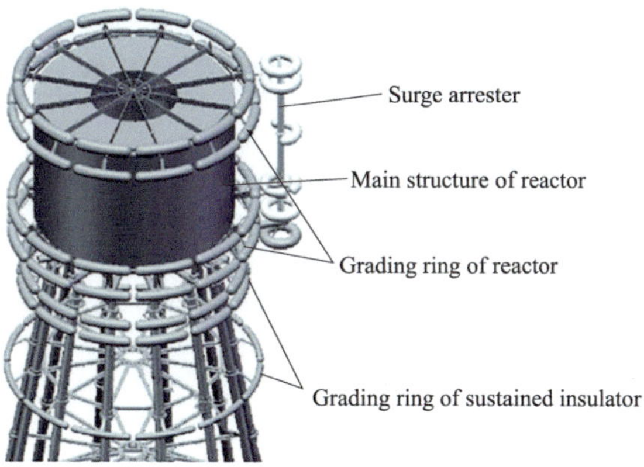

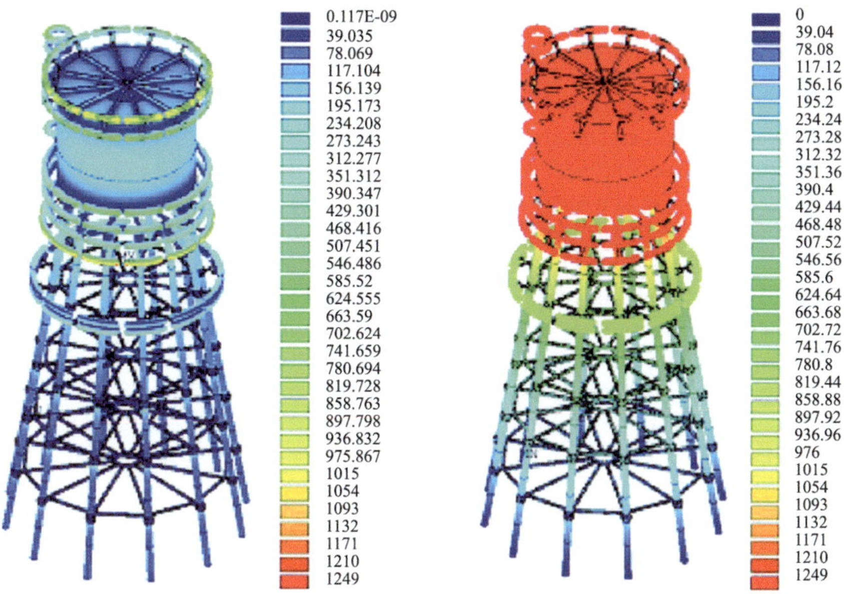

(a) Distribution of entire electric field (kV/m) (b) Distribution of entire potential (kV)

Fig. 64 Overall potential and surface electric field distribution nephogram

After testing by authoritative institutions, the smoothing reactor $U_{50\%}$ test meets the requirements, the safety margin is 1.7% when the switching impulse waveform parameter is 250/2500 μs, and the safety margin is 2.0% when the switching impulse waveform parameter is 500/5000 μs.

The short-circuit resistance of the smoothing reactor is mainly determined by the performance of the encapsulation material. Through physical and chemical tests, experiments were conducted on different formulas and curing temperatures. After research, the impulse resistance of the encapsulation material was increased to 6.55 kJ/m^2, the compressive strength increased to 147.6 MPa, improving the safety factor of radial and axial stress.

(IV) Study on the Distribution of Direct and Harmonic Currents

The direct current of the smoothing reactor is distributed according to the conductivity in each encapsulation layer, and the harmonic current is distributed according to the reactance. Compared with the previous ± 800 kV, 8 GW project, the rated direct current of the ± 1100 kV, 12 GW project has increased by 9% (from 5000 to 5455 A), making the distribution of direct and harmonic currents in each encapsulation layer more complex. For this reason, the equipment factory has developed a current distribution calculation program which is suitable for the ± 1100 kV/5455 A/ 75 mH smoothing reactor, and it can accurately calculate the distribution of direct and harmonic currents in each encapsulation layer of the smoothing reactor.

(V) Study on Distribution of Temperature Rise

The rated direct current of the smoothing reactor has increased from 5000 to 5455 A, making the calculation of distribution of temperature rise in each layer more complex. The thermal convection, thermal conduction, and thermal radiation of each encapsulation layer are inconsistent with the current 5000 A/75 mH reactor. Through simulation software, the smoothing reactor is simulated under the condition of rated direct current plus harmonics, and Fig. 65 shows the thermodynamic field simulation results of the 12 GW smoothing reactor. From Fig. 65, the temperature rise distribution of each encapsulation layer can be known, and the simulation results can be compared and adjusted with the design scheme to finally determine the scheme. The temperature rise of the smoothing reactor has been optimized from 60 K of the ± 800 kV, 8 GW DC transmission project to 47.2 K, and the hotspot temperature rise has been optimized from 76.8 to 73.3 K.

(VI) Control of Heating of the Wiring Terminal Board

When the rated direct current is increased to 5455 A, optimization should be started from improving the process level of machining, welding, installation, etc., and at the same time, the wiring terminal board and metal parts of tubular busbar should be changed from single-sided connection to double-sided connection, so that the contact current density of the wiring terminal board of the ± 1100 kV, 12 GW project is significantly reduced compared with the ± 800 kV, 8 GW project, avoiding the occurrence of abnormal overheating of the terminal board. Figure 66 shows a three-dimensional schematic diagram of the double-sided connection of the 12 GW/5455 A smoothing reactor wiring terminal board.

Figures 67 and 68 respectively show the size diagrams of the wiring terminal boards of the 8 and 12 GW projects. It can be seen from the figures, the size of the wiring terminal board of the smoothing reactor of the 8 GW UHV DC transmission project is 200 mm × 250 mm, and the size of the wiring terminal board of the smoothing reactor of the ± 1100 kV, 12 GW project has increased to 280 mm ×

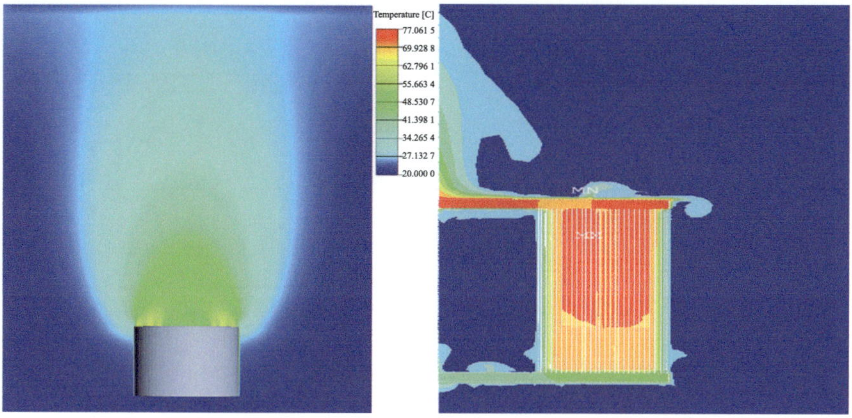

Fig. 65 Simulation results of thermodynamic field of 12 GW smoothing reactor

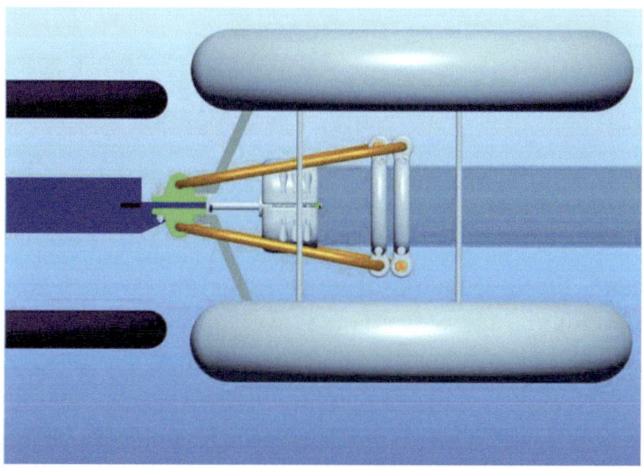

Fig. 66 3-dimensional diagram of double connection of wiring terminal board on 12 GW/5455 A smoothing reactor

280 mm. The contact current density of the wiring terminal board has been optimized from 0.063 to 0.043 A/mm^2, and it significantly reduces the occurrence of abnormal overheating of the terminal board.

(VII) Measures to Reduce Noise

In order to reduce the noise of smoothing reactors under high current, it is necessary to consider to avoid the inherent vibration frequency of the equipment, optimizing the structure of the noise reduction device, selecting a reasonable wire insulation structure, and adopting conductor impregnation and optimized winding processes. On one hand, during the winding process, each wire passes through a conduit of epoxy resin, which allows the epoxy resin to maximumly fill the gaps between the wires, and between the wires and the encapsulating yarn, forming the winding as a whole body, reducing the noise generated by wire vibration, and reducing the noise from the source, which is the body of the smoothing reactor winding. On the other

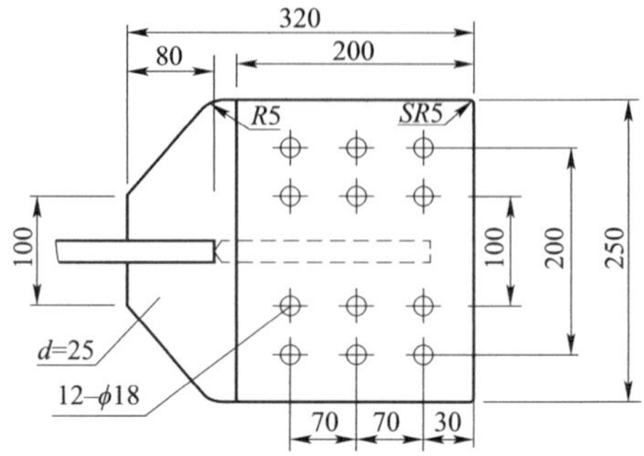

Fig. 67 Size of wiring terminal board on 8 GW project

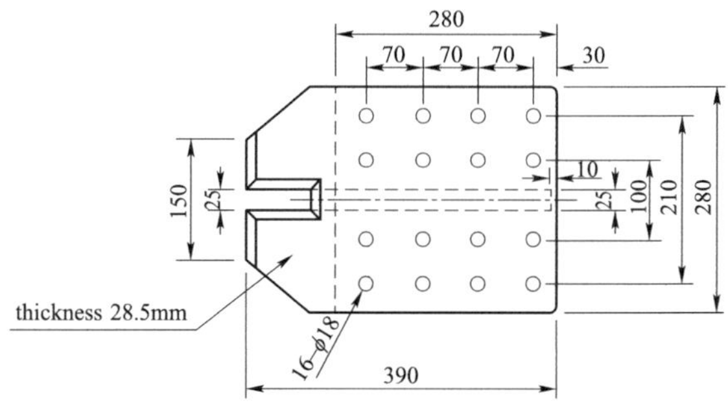

Fig. 68 Size of wiring terminal board on 12 GW project

hand, the installation of noise reduction devices absorbs the noise emitted by the smoothing reactor from the noise propagation path.

The noise reduction device can reduce the noise of the main body of smoothing reactor, but it also has a certain impact on the heat dissipation of the main body of smoothing reactor. It is necessary for the main body's ventilation area to meet the requirements of air flow heat dissipation. Therefore, the equipment factory optimizes the structure of the noise reduction device while ensuring noise reduction. Compared with the noise reduction device of the previous 5000 A smoothing reactor, the heat dissipation ratio of the optimized noise reduction device has increased from 1.11 to 1.14, and the air flow heat dissipation effect is better.

(VIII) Anti-seismic Research

The BKK-80000/110 high anti-seismic parallel reactor has passed the seismic prototype test with horizontal design accelerations of 0.4 and 0.5 g respectively at professional research institutions. This anti-seismic test has the strongest earthquake intensity with reactor protype ratio 1:1 level in China so far, and it has accumulated rich design experience for smoothing reactors in terms of anti-seismic resistance.

Based on the measured data of the mechanical performance parameters of the supporting structure components during the anti-seismic prototype test of BKK-80000/110, the smoothing reactor equipment factory has modeled and simulated the support scheme of the ± 1100 kV/5455 A/75 mH smoothing reactor, and Fig. 69

shows the corresponding anti-seismic simulation nephogram. To ensure that the support system of the smoothing reactor meets the anti-seismic requirements under a horizontal acceleration of 0.2 g, the smoothing reactor equipment factory has conducted research and analysis on the support system and determined the final design scheme, so that the smoothing reactor with increased mass can meet the requirements of stable operation.

III. Design Technical Scheme

(I) Overall Structure

The winding of the smoothing reactor at the Changji converter station has a total of 26 encapsulations, and the winding of the smoothing reactor at the Guquan converter station has a total of 31 encapsulations. Taking the Changji converter station as an example, the overall installation structure of the smoothing reactor on the side of pole bus is shown in Fig. 70. Its support part adopts a solid composite insulator

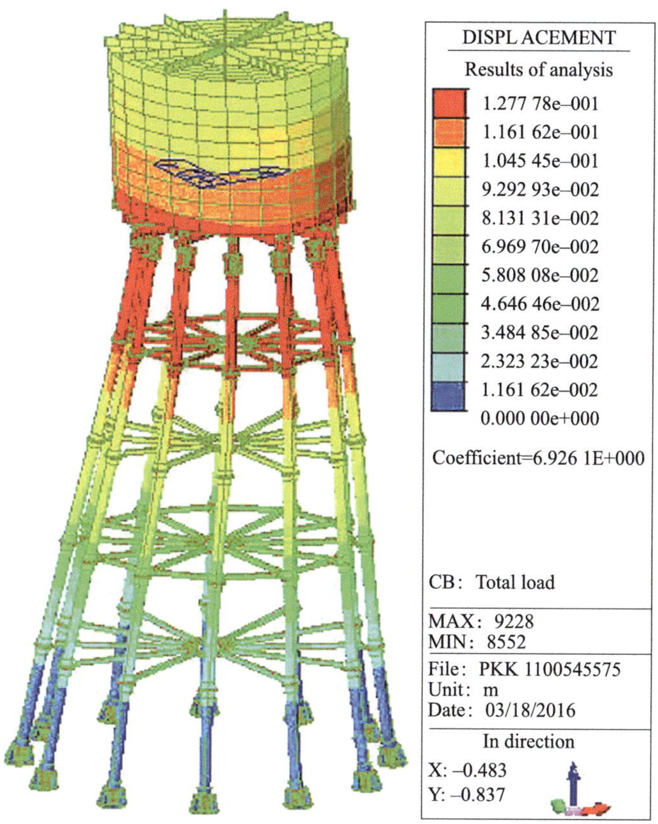

Fig. 69 Simulation nephogram of anti-seismic performance for smoothing reactor

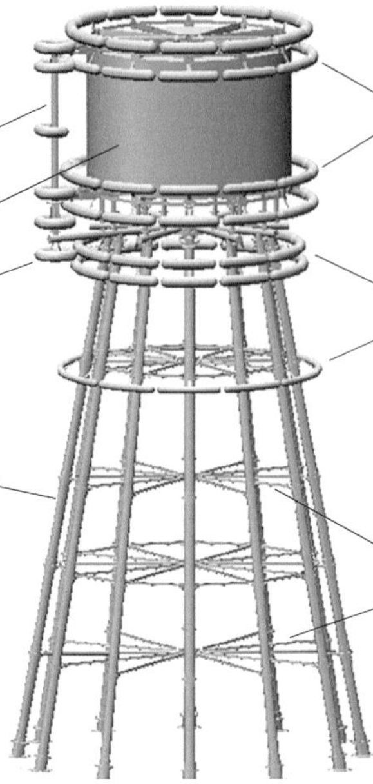

Fig. 70 Entire installation structure of smoothing reactor on pole bus side

with 12-column, and it is inclined at 10° for support, each column is composed of 5 insulators with various heights, and metal tie rods are used between each insulator to fix insulator in each column. The smoothing reactor on the side of pole bus is arranged indoors, and there is no need to install a noise reduction device. Brackets are provided at the upper and lower terminals of the winding for the installation of surge arresters. The smoothing reactor operates on a UHV DC line, so shielding devices are installed at all high field strength positions of the smoothing reactor to prevent corona discharge.

The overall installation structure of the smoothing reactor on the neutral line side is shown in Fig. 71. The support part adopts a 12-column solid composite insulator for vertical support, the smoothing reactor on the neutral line side is arranged outdoors, and noise reduction devices are designed above, in the middle, below, and inside the reactor main body, which completely wraps the reactor inside, of that can reduce noise and prevent rain water invasion. The smoothing reactor on the neutral line side does not need to install a surge arrester. Also, because the smoothing reactor operates on a UHV DC line, shielding devices are installed at all high electric field strength positions of the smoothing reactor to prevent corona discharge.

(II) Structure of Noise Reduction Device

When the smoothing reactor is operating, the rated current is composed of several harmonic and direct currents, which generate vibrational magnetic field forces of various frequencies and a stable static magnetic field force. The former generates

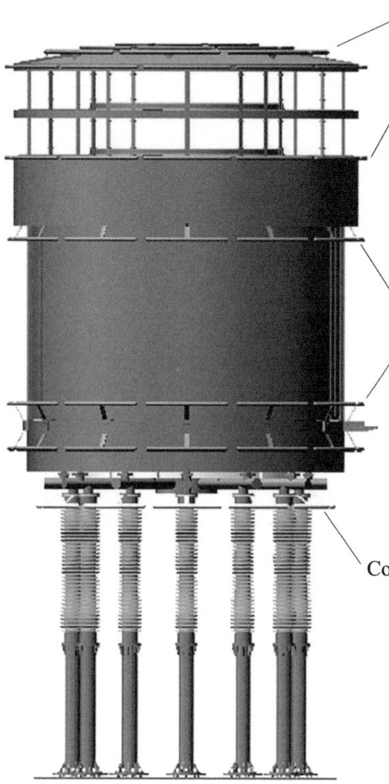

Fig. 71 Entire installation structure of smoothing reactor on neutral line side

electromagnetic noise, while the latter does not contribute to the noise. When the reactor winding carries the current, the current flowing through the winding will generate a magnetic field inside and outside the reactor, and the magnetic field in turn acts on the current-carrying winding, thus generating a magnetic field force on the winding. When the current in the winding changes with time, the size and direction of the magnetic field change accordingly, so the magnetic field force received by the winding wire changes in size, causing the winding to vibrate. As a result, the interaction of multiple currents with different frequencies causes vibrations with different frequencies.

To reduce the noise, the smoothing reactor has set up a noise reduction device, which is basically the same as previous projects. The noise reduction device consists of an upper muffler, an external cylindrical sound-absorbing cover, an internal sound-absorbing tube, a bottom cylindrical sound-absorbing cover, and a bottom grille muffler. Among them, the internal sound-absorbing tube has been installed inside the reactor before leaving the factory; the external cylindrical sound-absorbing cover is a cylinder composed of dozens of middle sound cavities spliced together, which needs to be installed on the periphery of the reactor body on site to reduce the sound radiation to the outside space; to prevent rainwater from entering the sound cavity, an upper muffler is installed on the top of the reactor to shield the reactor body from

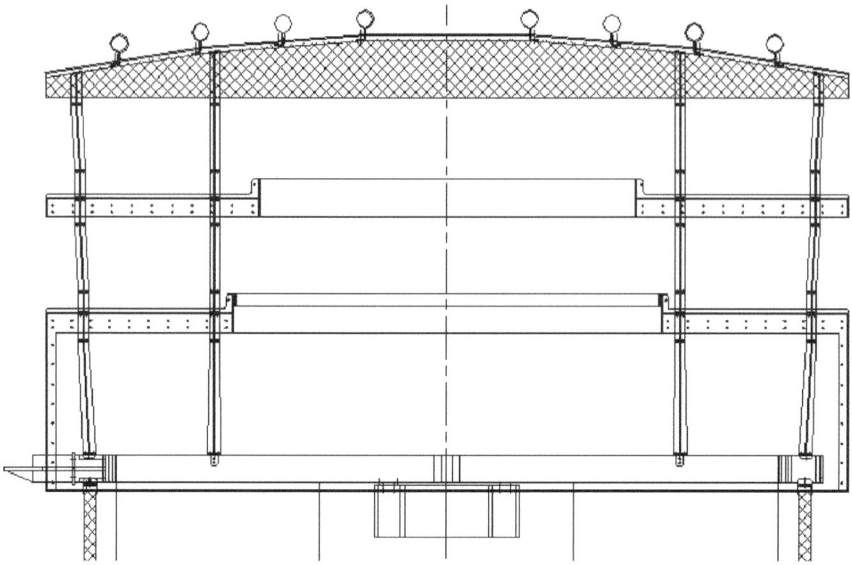

Fig. 72 Upper muffler structure

rainwater. The upper muffler mainly consists of a top cover, a partition, and a main body. The body of muffler and the reactor body, and the muffler body and the muffler top cover are all supported by two circles of metal rods both inside and outside, as shown in Fig. 72.

(III) Grille Structure of Bird-proof

The grille structure of bird-proof on smoothing reactor is basically the same as previous projects. Due to some gaps and vacancies between the center of the silencer body and the center of the lower terminal of the smoothing reactor body, birds can enter the silencer body through these gaps and nest on the reactor. When the nesting position is on the upper terminal of the seal, it will have a certain impact on the operation of the reactor. The smoothing reactor in 12 GW Changji converter station has set up a bird-proof grille, the schematic diagram of the structure is shown in Figs. 73 and 74. The material of bird-proof grille has a high heat resistance level and a certain hardness and toughness, which meets the requirements of installation and implementation. The bottom of bird-proof grid has been installed inside the reactor when the smoothing reactor leaves the factory. After the silencer body is assembled, the upper bird-proof net is fixed at the installation position indicated in the instruction manual on site.

(IV) Conductor

To ensure that all technical indicators of the smoothing reactor can meet the requirements, the equipment factory has specially designed the most critical material of

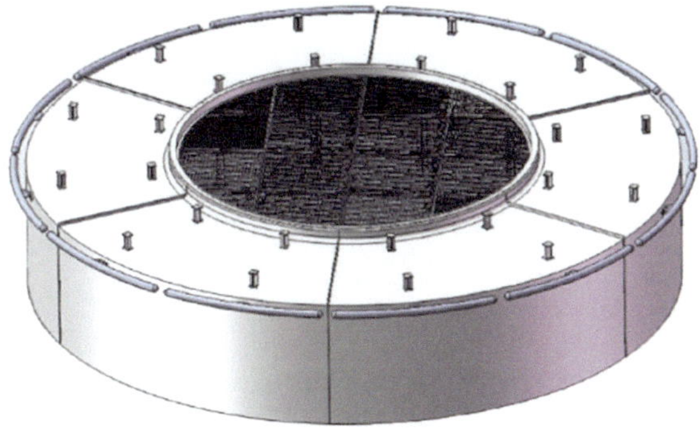
Fig. 73 Diagram of upper grille structure of bird-proof

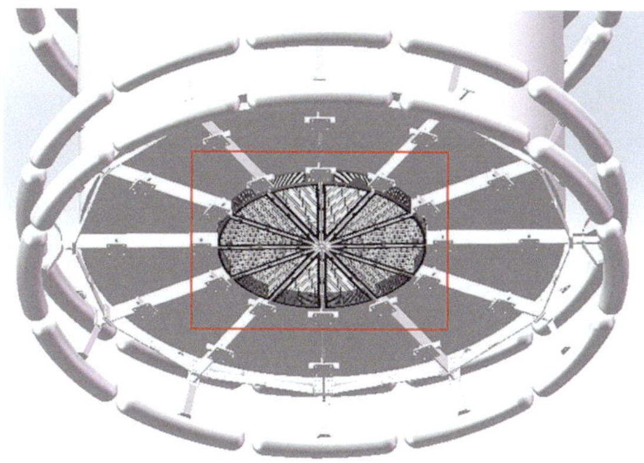
Fig. 74 Diagram of bottom grille structure of bird-proof

the reactor, the electromagnetic wire, and the diameter, cross-section, and size of the electromagnetic wire are designed according to the results of repeated iterative calculations, thereby ensuring the optimal performance of the reactor product. To ensure the insulation tightness of the electromagnetic wire, the equipment factory has wound an epoxy impregnated glass fiber mixed woven tape on the surface of the electromagnetic wire. An example of the conductor structure is shown in Fig. 75.

5 DC Switchgears

DC switchgears is a key device in the current loop of the UHV DC transmission system. The further increase in the capacity of UHV DC transmission, especially the increase of system voltage from 800 to 1100 kV, puts higher technical performance requirements on switching apparatus.

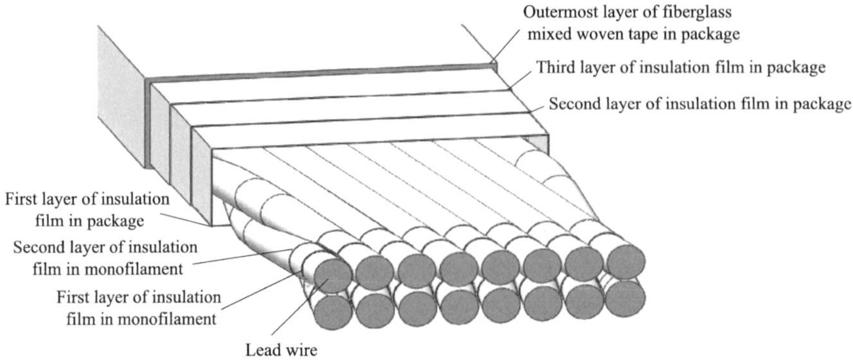

Fig. 75 Example of conductor structure

The DC switchgears in the UHV converter station mainly includes DC disconnectors and grounding switches, DC conversion switches, and DC bypass switches. The technical difficulties in the development of DC switching gears mainly correspond on the current increase issues related to conductor heating, external insulation, and equipment mechanical stability problems. The type of large-capacity DC switching gears inherits the design of the ± 800 kV UHV project DC switching gears, and has been improved or redesigned for product heating and other issues after the current increases. This section focuses on the technical requirements of DC switching gears, especially the requirements for rated current and overload current, as well as the development results of switching apparatus.

Table 48 Configuration and parameters of disconnector

No.	Item	Standard parameters
1	Structural type or model	Double column horizontal telescopic type
	Grounding switch	Not grounded
2	Electric or manual	Electric and manual
	Motor voltage (V)	AC 380/220
	Control voltage (V)	DC 220/110

Table 49 Parameters of operating mechanism of grounding switch

No.	Item		Standard parameters
1	Electric or manual		Electric and manual
	Motor voltage (V)		AC 380/220
	Control voltage (V)		DC 220/110
2	Backup auxiliary contact	Disconnector (pair)	10
		Grounding switch (pair)	8

Table 50 Rated parameters of disconnector

No.	Item		Standard parameters
1	Rated DC voltage (kV)		1122
2	Rated DC current (A)		6000
3	2 h overload DC current (A)		6600
4	Rated lightning impulse withstand peak voltage (1.2/50 μs) (kV)	Open contacts	2550
		To ground	2550
5	Rated switching impulse withstand peak voltage (250/2500 μs) (kV)	Open contacts	2100
		To ground	2100
6	DC withstand voltage, 1 h (kV, 直流)	Open contacts	1683
		To ground	1683
7	Rated short time withstand current and time of duration (kA/s)		2/14
8	Rated withstand pead current (kA)		35
9	Mechanical stability (time)		≥ 1000
10	Short time power frequency withstand voltage of auxiliary and control circuit (kV)		2
11	Radio interference voltage ($1.1U_d$) (μV)		≤ 2000
12	Static mechanical load of wiring terminals (N)	Horizontal Lengthways	3500
		Horizontal Transverse	3500
		Vertical	3500
		Safe coefficient	Static 2.75, Dynamic 1.7
13	Ability of harmonic current interruption of high voltage isolation switch at DC filter	Interruption of current (A, RMS)	300
		Recovery voltage (kV)	82.5

Table 51 Rated parameters of grounding switch

No.	Item		Standard parameters
1	Rated voltage (kV, DC)		1122
2	Rated lightning impulse withstand peak voltage (kV)	Open Contacts	–
	Value (1.2/50 μs) (kV)	To Ground	2550
	Rated switching impulse withstand pcak voltage (kV)	Open Contacts	N.A
	Value (250/2500 μs) (kV)	To Ground	2100
3	DC withstand voltage, 1 h (kV, DC)	Open Contacts	N.A
		To Ground	1683
4	Rated short time withstand current and time of duration (kA/s)		2/14
5	Rated withstand peak current (kA)		35
6	Radio interference voltage ($1.1U_d$) (μV)		≤ 2000

I. Technical Requirements

(I) DC Disconnectors and Grounding Switch

(1) The environmental conditions are as required in the first section of this chapter.

(2) The disconnector of high-voltage DC filter should have the ability to switch the operated DC filter at 50% DC rated voltage and below.

(3) The dry arc distance of the 1100 kV pole line disconnector should not be less than 13 m, and the break distance should be considered as not less than 7.3 m.

(4) The pole line disconnector and grounding switch use silicone rubber composite insulators.

(5) The structure and type parameters of the disconnector are shown in Table 48.

(6) The switching mechanism and parameters of grounding switch are shown in Table 49.

(7) The rated parameters of disconnector are shown in Table 50.

(8) The rated parameters of grounding switch are shown in Table 51.

(II) DC Conversion Switch

(1) The environmental conditions are as required in the first section of this chapter.

(2) The DC conversion switch in the UHV converter station includes Metallic Return Transfer Breaker (MRTB), Ground Return Transfer Switch (GRTS), Neutral Bus Switch (NBS), and Neutral Bus Grounding Switch (NBGS). The schematic diagram of DC conversion switch is shown in Fig. 76.

Reactors, capacitors, and non-linear resistors are installed on the insulating platform. The capacitors, non-linear resistors, and SF_6 interrupters are electrically connected to the insulating platform, while the reactors are insulated from the insulating platform.

(3) MRTB: The metallic return circuit of converter station at sending end is equipped with 1 MRTB. The MRTB consists of three parallel branches, namely the SF_6 interrupter, the oscillating circuit, and the non-linear resistor. To achieve the conversion requirements, oscillations must be generated internally to extinguish the arc in the SF_6 interrupter when the current passes through zero. The general requirements for MRTB are shown in Table 52.

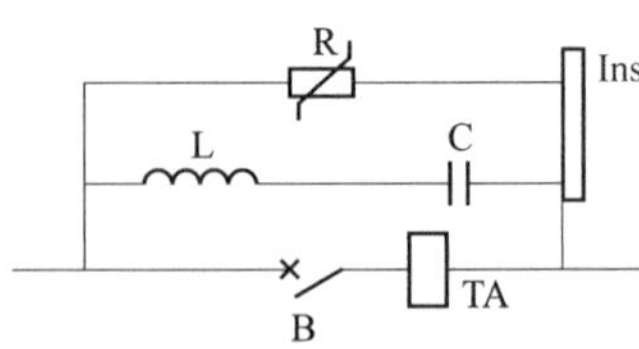

Fig. 76 Theoretical diagram of DC conversion switch B—SF_6Chopper; L—Reactor; C—Capacitor; R—Varistor; TA—Current Transformer

Table 52 General requirement of MRTB

No.	Parameters	Required value
1	Rated current (A, DC)	6000
2	Overload current (A)	6600
3	Rated voltage (kV, DC)	150
4	Maximum conversion current (A, DC)	5059

(4) GRTS: The ground return circuit of converter station at sending end is equipped with 1 GRTS. The three parallel paths that make up it are the SF_6 interrupter, the oscillating circuit, and the non-linear resistor. The general requirements are shown in Table 53.

(5) NBS and NBGS: Both the converter stations at sending and receiving ends are equipped with 2 NBS and 1 NBGS. The NBS is composed of SF_6 interrupter, oscillating circuit, and non-linear resistor in three parallel paths. The NBGS should have a fast disconnector and must achieve rapid grounding operation, otherwise, the voltage of the neutral bus will lose control. The NBGS should have the ability to transfer DC current, the general requirements are shown in Table 54.

(6) The terminal board should be a double-sided flat plate. The allowable static mechanical load of the main terminal board and wire clamp should not be less than the following values: ① Its static safety factor is 2.75; ② Longitudinal in the horizontal direction is 3000 N; ③ Horizontal in the horizontal direction is 2000 N; ④ The vertical direction is 2000 N; ⑤ The

Table 53 General requirements of GRTS

No.	Parameters	Required value
1	Rated current (A, DC)	6000
2	Overload current (A)	6600
3	Rated voltage (kV, DC)	150
4	Maximum conversion current (A, DC)	1647

Table 54 General requirements of NBGS

Type	No.	Parameters	Required value
NBS	1	Rated current (A, DC)	6000
	2	Overload current (A)	6600
	3	Rated voltage (kV, DC)	150
	4	Maximum conversion current (A, DC)	5839
NBGS	1	Rated current (A, DC)	–
	2	Overload current (A)	–
	3	Rated voltage (kV, DC)	150
	4	Maximum conversion current (A, DC)	5839

main terminal board and wire clamp should be able to withstand a torque of 1000 N m without deformation.

(7) SF_6 The interrupter generally uses an AC SF_6 circuit breaker, the interrupter should be an outdoor single-phase type, with spring or hydraulic actuation mechanism. The actuation mechanism must be reliable, and remote/local/off selection switch can realize remote, local off, and on operations. The switch of local operation selection should be placed within the switch actuation mechanism. Without mechanical adjustment, maintenance, or replacement of parts, the number of no-load operations of the SF_6 interrupter should not be less than 10,000 times.

(8) Operation sequence, the normal operation of the DC conversion switch is O–C–O cycle. In reverse operation, the action of the circuit breaker is C–O–C; the interrupter in NBS must realize the C–O–C operation cycle. In case of conversion failure or motor power loss, this function can ensure that the interrupter reaches the closed position; the interrupters in MRTB, GRTS, and NBGS must realize the O–C operation cycle, in case of conversion failure or motor power loss, this function can ensure that the interrupter reaches the closed position; the high-speed disconnector must realize the C–O operation cycle.

(9) All equipment materials (including insulators) should consider the impact of sand and wind. The wind speed that the equipment body can withstand should be converted to the wind speed at the highest point of the equipment, and the amplification effect of sandstorms on wind pressure should be fully considered. Measures such as thickening the wall of the equipment shell should be taken; the protection level of meters such as density meters is IP55, and measures such as installing protective covers should be taken to prevent sand and dust from entering. Outdoor terminal boxes and all other boxes should use thickened using stainless steel shells, choose double or triple door sealing structures, and use wind and sand resistant relays.

(10) The switch control cabinet and terminal box should have good sealing performance, have a double insulation structure, come with heating function and temperature control device inside the cabinet, automatically start when the temperature is below zero, ensure that the temperature inside the cabinet is not below zero, and the heater power should be able to meet the operating requirements at extremely low temperatures. The terminal box cable should be entered from the bottom.

(11) The DC conversion switch uses ceramic insulation or silicone rubber composite insulation, and according to the requirements of the operation management unit, if the equipment uses ceramic insulation, it needs to be coated with RTV-II type coating (factory coated, if damaged, repaired on site), the RTV-II coated by the manufacturer has a three-year warranty. If the hydrophobicity is lower than HC6 during the warranty period, the manufacturer is responsible for supplementary coating.

(12) The temperature rise test method is "after the equipment reaches a stable temperature at the rated current, the current is increased to the overload current for 2 h".

(III) DC Bypass Switch

(1) See the relevant requirements in the first section of this chapter for environmental conditions.
(2) See the relevant requirements in the fifth section of this chapter for wind and sand low temperature requirements.
(3) See Table 55 for the technical parameters of the pole line bypass switch.

II. Key Technologies

(I) Technical Features of DC Switch

1. Technical parameters of DC disconnector and grounding switch

In the case of a system with 1100 kV/5455 A, the rated current of the DC disconnector is set to 8000 A, and the temperature rise test current is 8800 A, which can meet the needs of the 1100 kV/12 GW UHV DC transmission project.

2. Technical parameters of DC conversion switch and bypass switch

The DC conversion switch and bypass switch are important equipment in the DC field of the converter station, mainly used for the conversion of various operating modes of the DC transmission system, grounding system conversion, etc. The DC conversion switches (MRTB, GRTS, NBGS, NBS) are all installed on the neutral line, completing the conversion of DC current. The DC voltage sustained by the switches is not high, but the conversion current after the DC current is increased will be larger than the previous UHV DC transmission project, especially after the DC line is lengthened, the conversion energy is larger and the conversion circuit needs to be redesigned. The DC bypass switch is divided into pole line bypass switch and midpoint bypass switch, mainly used for isolating the converter valve, and the current requirement is not high.

Table 56 provide the main technical parameters of the ± 1100 kV, 12 GW project DC conversion switch, Table 57 provides the main technical parameters of the ± 800 kV, 10 GW project DC conversion switch, and Table 58 provides the requirements of the DC conversion switch in the previous 7.2 GW UHV DC transmission project. From Tables 56, 57, and 58, it can be seen that after the transmission capacity is increased, the conversion currents of MRTB, GRTS, and NBS have increased significantly, and the energy absorbed by the surge arrester in the auxiliary circuit during the conversion process has increased, so it is necessary to redevelop the auxiliary circuit of the DC circuit breaker. Therefore, the DC circuit breaker in the previous UHV DC transmission project cannot be directly used in the 10 GW UHV DC transmission project. It is necessary to conduct in-depth research on the original basis, redesign the auxiliary circuit, and meet the requirements of larger conversion current. The DC conversion switch of the CG project has cancelled the

Table 55 Technical parameters of pole line bypass switch

No.	Item			Required value
1	Open contacts (individual)			2
2	Rated voltage (kV, DC)		To ground	1122
			Open contacts	561
3	Current stress		Maximum operating current (A, DC)	5838
			Rate short time withstand current (2 s) (kA, DC)	14
			Rate withstand peak current (kV, peak value)	35
4	Testing voltage of external insulation			
4.1	Lightning impulse withstand	To ground (kV, peak value)		2550
		Open contacts (kV, peak value)		1313
4.2	Switching impulse withstand	To ground (kV, peak value)		2100
		Open contacts (kV, peak value)		1180
4.3	DC voltage withstand, 1 min	To ground (kV, peak value)		1683
		Open contacts (kV, peak value)		815
5	Maximum closing time (ms)			≤ 90
6	Mechanical stability (time)			≥ 3000
7	Rated switching sequence			CO–15 s–CO
8	Testing voltage of radio interference (μV)			≤ 2000
9	SF_6 gas humidity (μL/L)	Handing over and acceptance check		≤ 150
		Long term operation allowable value		≤ 300
10	SF_6 gas leakage rate (%/year)			≤ 0.5
11	SF_6 gas purity (%)			99.8
12	Operating mechanism or type			Hydraulic or spring
	Motor voltage (V)			AC 380/220, DC 220/110
	Power of closing operation	Rated operating voltage (V)		DC 220, DC 110
		Allowable range of operating voltage		85–110%, 30% not operating
		Number of windings (set)		2

(continued)

Table 55 (continued)

No.	Item		Required value
		Steady state current of each winding	DC 220 V, 2.5[a] or DC 110 V, 5 A
		Rated operating voltage (V)	DC 220, DC 110
		Allowable range of operating voltage	65–110%, 30% not operating
		Number of windings (set)	2
		Surge current of each winding (A)	Bidders provide
		Steady state current of each winding	DC 220 V, 2.5 A or DC 110 V, 5 A
	Heater	Voltage (V)	AC 220
		Power of each phase (W)	Bidders provide
	Backup auxiliary contacts	Number (pair)	10
		Interrupting ability	DC 220 V, 2.5 A or DC 110 V, 5 A
13	Period of maintenance (Year)		≥ 20
	Hydraulic structure	24 h pressured times (times)	≤ 2
	Spring structure	Energy storage time (s)	≤ 20
14	Static load of terminal	Horizontal lengthways (N)	3500
		Horizontal transverse (N)	2000
		Vertical (N)	2000
		Safe coefficient	Static 2.75, Dynamic 1.7

parallel branch of the shunt, adopted a new type of current-carrying break, with a rated current-carrying capacity of 6000 A, and the overload current can reach 6600 A for 2 h.

The main technical parameters of the bypass switch for the ± 1100 kV, 12 GW project and the ± 800 kV, 10 GW project are shown in Table 59. It can be seen that the main technical differences of the bypass switch for the ± 1100 kV, 12 GW project are the significant increase in voltage level and insulation withstand level, the need to increase the length of the break and pillar insulator, and pass the corresponding type test.

Table 56 Major technical parameters of DC conversion switch in ± 1100 kV, 12 GW project

Parameters	MRTB	GRTS	NBS	NBGS
Rated current (A, DC)	6000	6000	6000	–
Withstand peak voltage (kA, peak value)	35	35	35	35
Short time withstand current, 2 s (kA, peak value)	14	14	14	14
Maximum conversion current (A, DC)	5059	1647	5839	5839

Table 57 Major technical parameters of DC conversion switch in ± 800 kV, 10 GW project

Parameters	MRTB	GRTS	NBS	NBGS
Rated current (A, DC)	6600	6600	6600	–
Withstand peak voltage (kA, peak value)	45	45	45	45
Short time withstand current, 2 s (kA, peak value)	18	18	18	18
Maximum conversion current (A, DC)	5900	2500	6400	6400

Table 58 Major technical parameters of DC conversion switch in ± 800 kV, 7.2 GW project

Parameters	MRTB	GRTS	NBS	NBGS
Rated current (A, DC)	5017	5017	5017	–
Withstand peak voltage (kA, peak value)	30	30	30	30
Short time withstand current, 2 s (kA, peak value)	12	12	12	12
Maximum conversion current (A, DC)	4463	972	5017	–

Table 59 Major technical parameters of DC bypass switch

No.	Item	± 1100 kV, 12 GW project	± 800 kV, 10 GW project
1	Maximum operation current (A, DC)	5839	6600
2	Rated short time withstand current (2 s, min)	14	18
3	Rated withstand peak current (kA, peak value)	35	45
4	Testing current of temperature rise (kA, peak value)	–	
5	Rated voltage (kV, DC)	550	408
	Maximum voltage (kV, DC)	561	408
	Maximum sustained voltage on open of contacts (kV, DC)	561	408
6	Insulation Level		
	Lightning withstand voltage (open of contacts, kV)	1313	950
	Lightning withstand voltage (to ground, kV)	1313	950
	Switching withstand voltage (open of contacts, kV)	1180	850
	Switching withstand voltage (to ground, kV)	1180	850
	DC withstand voltage, 1 min (to ground, kV)	815	612
	DC withstand voltage, 1 min (open of contacts, kV)	815	612

(II) DC Switch Temperature Rise Test

Through optimized design, the DC conversion switch used in the ± 1100 kV, 12 GW project can reach a current-carrying capacity of 6000 A, and through the corresponding temperature rise test, the temperature rise results of key parts are shown in Table 60. Table 60 shows that the rated current and overload current capacity of the new type of DC conversion switch meet the requirements of relevant specifications. The temperature rise values of key parts are shown in Table 60.

(III) External Insulation Test

The insulation level of the DC pole line area equipment is high, the discharge curve enters the saturation area, and the shape of the electrode of the equipment voltage equalization shielding device is very sensitive, requiring detailed design, establishing the selection of the indoor DC field equipment voltage equalization shielding device, and greatly improving the reliability of the project.

Combining the characteristics of pole line equipment, we have developed new types of voltage equalization shielding devices such as "apple-shaped" grading covers, carried out full three-dimensional electric field simulation calculations, repeatedly iterated to establish device selection, and conducted $U_{50\%}$ prototype test verification, as shown in Figs. 77, 78 and 79.

The surface field strength of 1100 kV pole line equipment is less than 15 kV/cm, and it passed the $U_{50\%}$ test at one time, completely solving the problem of the sensitivity of the electrode shape of indoor DC field equipment.

III. Technical of Scheme DC Switch

(I) Technical Scheme of DC Disconnector

The conventional ± 800 kV disconnector is a mature product, with a double-column horizontal telescopic structure, and the pillar insulator adopts a tripod structure to increase stability (see Fig. 80). Under the voltage of ± 1100 kV, the creepage distance and overvoltage level of the equipment increase significantly, especially the requirement for insulation level increases greatly, which puts extremely high requirements on the insulation level of the equipment.

The finally adopted scheme is: ± 1100 kV double-column horizontal telescopic outdoor UHV DC disconnector, and it mainly consists of a base, a pillar insulator,

Table 60 Temperature rise on key parts

Product	Testing current of temperature rise (A)	Temperature rise of wiring terminals (K)	Contacts (K)	Limit of temperature rise (K)
DC conversion switch	6000	28.1	55.4	65
DC conversion switch	6600 (2 h overload current)	30.3	62.9	65

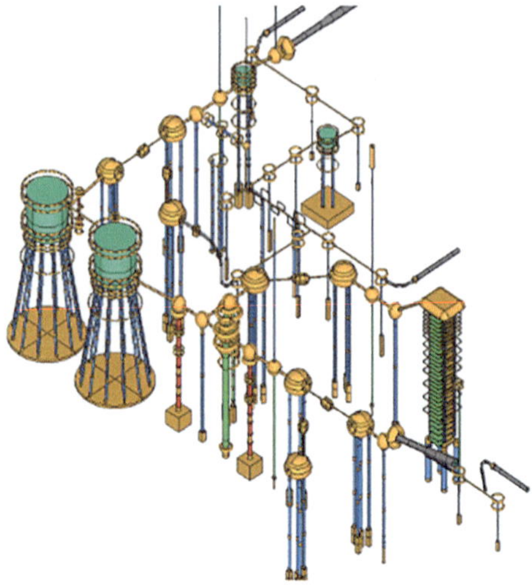

Fig. 77 3-D electric field simulation of DC indoor yard

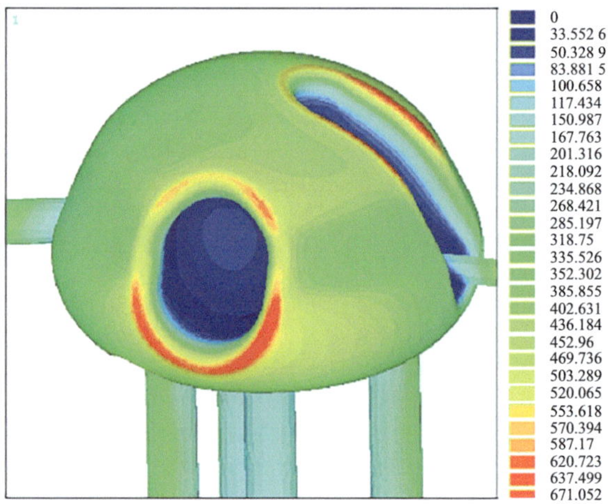

Fig. 78 Electric field simulation on "apple-shaped" grading covers

an operating insulator, a main gate blade, a static contact, an equalization cover, an operating mechanism, etc. The pillar insulator stands on the base, the operating insulator is installed in the center of the moving tripod, the main gate blade and the moving equalization cover are installed on the top of the moving pillar insulator, the static contact and the static equalization cover are installed on the top of the static pillar insulator, and the pillar insulator adopts a tripod structure to increase stability.

The main technical measures are as follows:

(1) The moving and static sides adopt a new type of bun-shaped hemispherical grading cover (see Fig. 81), the elbow adopts a new type of hemispherical shell grading cover, which can effectively reduce the unevenness of electric field

Fig. 79 Real type discharge test on "apple-shaped" grading covers

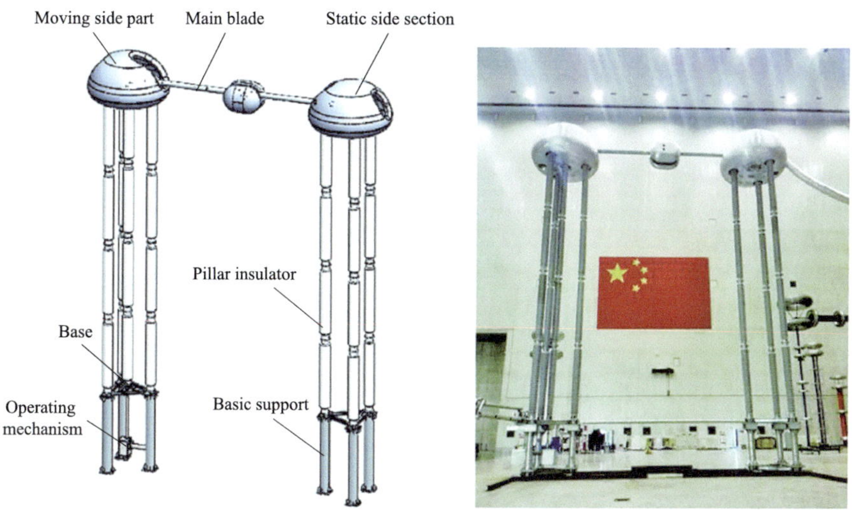

Fig. 80 DC disconnector

on main gate blade, eliminate edge effects, and greatly improve the insulation performance of the disconnector.

(2) A new type of composite pillar insulator is used, with a height of 14 m and a creepage distance of 52,100 mm.

(3) The moving contact adopts a ring-shaped stylus structure (see Fig. 82). The ring-shaped contact generates contact pressure through the ring-shaped springs on the outer edge of the stylus, with multiple meshing points. The stylus are

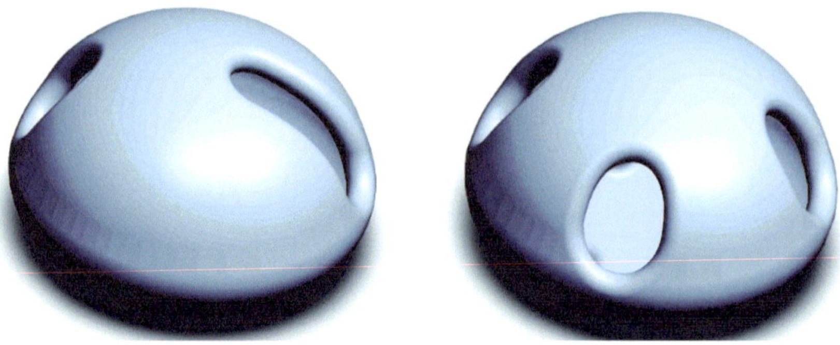

Fig. 81 Bun-shaped hemispherical grading cover

Fig. 82 Ring-shaped stylus structure

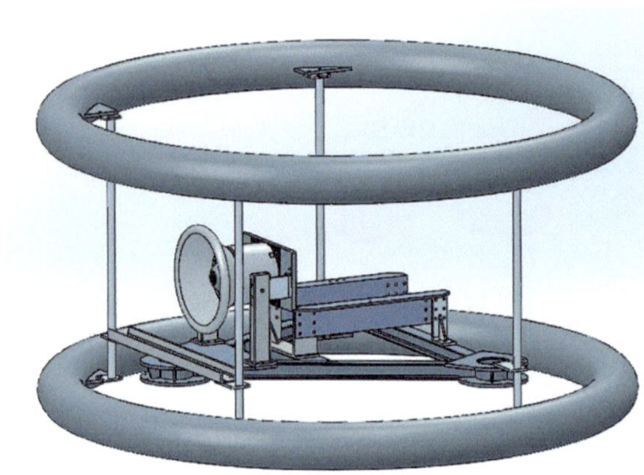

directly connected to the outside of the conductive tube, reducing fixed connection points, making it easier to control assembly quality, more reliable than internal connections, and ensuring the reliability of current flow.

(II) Technical Solution of DC Conversion Switch

The current conversion process of the DC conversion switch is complex, involving the coordination of oscillation circuit parameters, the energy absorption of nonlinear resistors, and the self-excited oscillation of the arc. The development of a DC conversion switch after capacity increase has a high degree of difficulty. According to the current carrying and current conversion capabilities of existing equipment, the development of a DC conversion switch with a rated current of 6600 A requires special attention to the following areas of research.

1. Current carrying capacity

The ± 1100 kV, 12 GW project uses a new type of HPL245B1 breaker, with a rated current carrying capacity of 6000 A, and an overload current of up to 6600 A for 2 h.

The maximum current carrying capacity of traditional AC circuit breakers is 5000 A (RMS value). To meet the current carrying requirement of 6600 A (DC), a new type of AC circuit breaker needs to be designed or a parallel structure of circuit breakers needs to be adopted. The development cycle of a new type of AC circuit breaker is long, the technical difficulty is high, and the arc characteristics of the new circuit breaker are uncertain, posing a great challenge to the design of the DC conversion switch. However, the structure of parallel 5000 A circuit breakers can meet the current carrying requirements without changing the arc characteristics of the circuit breaker, which can meet the construction requirements of the project.

2. Current conversion capability of the DC conversion switch

The DC conversion switch in the 7.2 GW UHV DC transmission project already has a conversion current capacity of 5000 A, but it cannot meet the conversion current requirement of 6400 A for the new project. It is necessary to further enhance the transfer current capability of the DC conversion switch and verify through transfer current tests that the equipment meets project requirements.

3. Parameter and its configuration of the conversion circuit

The passive-type conversion switch mainly uses the negative resistance characteristic of the arc, but the larger the current, the more unfavorable the self-excited oscillation, and the more difficulty encounters for the conversion. Therefore, it is necessary to well coordinate the parameters of conversion circuit and breaking device. Through experiments and simulation calculations, refine the arc simulation model, carry out simulation calculations on the breaking process of the DC conversion switch, and study the coordination of the conversion circuit and the breaking device under different parameters and configurations.

4. Design of energy absorption device and insulation platform on DC conversion switch

There is a large inductance between the line and the ground, storing a huge amount of energy. After current conversion, this part of the energy needs to be absorbed by the nonlinear resistor of the DC conversion switch, among which the MRTB sustains the most severe operating conditions. Moreover, after the rated operating current is increased from 5000 to 6250 A, the energy absorbed by the nonlinear resistor of the DC conversion switch increases to 1.56 times than the previous amount. The increase in energy absorption leads to an increase in the number of nonlinear resistors, an increase in the load on the insulation platform, and the demand for optimized design of the structural strength of the insulation platform.

(III) Technical Solution of DC Bypass Switch

In UHV DC transmission projects, each 12-pulse converter has a parallel bypass switch. The main function of the bypass switch is to bypass the converter valve that is out of operation, allowing the system to operate in an incomplete manner, thereby reducing power loss and the impact on the AC systems on both sides, and improving system availability. Under normal operating conditions, the bypass switch only participates in the current loop for a short time, and the impact of the current on the bypass switch is not significant.

Due to the significant increase in rated voltage and insulation withstand level, the structure of the HPL1100B2 type pole bypass switch in the 12 GW UHV DC transmission project has changed significantly. Compared with the 8, 10 GW UHV DC transmission project, the gap length has increased from 5.86 to 6.99 m; the height has increased by 6 m compared to the bypass switch of the 8, 10 GW UHV DC transmission project, reaching 18.9 m.

The midpoint bypass switch, based on the HPL800B2 type bypass switch of the original 8, 10 GW UHV DC transmission project, has been optimized. The insulation gap length has increased from 5.86 to 6.99 m, and the switch height remains unchanged.

All new bypass switches have undergone various type tests again, and the product design and test results all meet the requirements of technical specifications.

6 DC Voltage Divider

The DC voltage divider of the CG project uses a resistive-capacitive voltage divider, which is converted into a serial digital optical signal on-site through a remote module. The insulation level of this device has been improved in comparison to the ± 800 kV DC voltage divider. While meeting the insulation level requirements, it has higher requirements for its anti-interference ability, measurement accuracy, reliability, etc.

I. Technical Requirements

(1) See the relevant requirements in the first section of this chapter for environmental conditions, and the second section of this chapter for indoor field conditions.
(2) The DC voltage divider is a resistive voltage divider with capacitive compensation, including HV arm, LV arm, control room electronic measuring equipment and connecting cables. The body of the voltage divider is installed within the insulator (insulation tube).
(3) Use analog signals, each channel should have an independent low-pass filter, secondary voltage divider with high-stability resistance, and isolation amplifier. Any output channel failure should not cause abnormal signals in other output channels.

(4) Each DC voltage divider is equipped with a set of IED online monitoring system at the device main structure to monitor the SF_6 gas pressure. The power supply and signal transmission optical fiber are provided by other equipment suppliers.
(5) See Table 61 for technical parameters.
(6) See Table 62 for insulation level.

II. Key Technologies

The design calculations for key components of the 1100 kV DC voltage divider are given below.

(I) Calculation of 1100 kV DC Voltage Divider Bushing

1. Creepage distance of external insulation

Considering the highest operating voltage of the system and various correction factors, the required external insulation creepage distance L of the hollow composite bushing is calculated by

Table 61 Technical parameters

Item	Parameters
Maximum sustained DC voltage of pole line equipment (kV)	1122, including 60 kV (RMS) harmonics
Maximum sustained DC voltage of 550 kV pole line equipment (kV)	561 (Guquan)
Maximum sustained DC voltage of neutral busbar equipment (kV)	150/40 (Changji/Guquan)
Rated primary DC voltage of pole line equipment (kV)	1100
Rated DC voltage of 550 kV pole line equipment (kV)	550 (Guquan)
Neutral busbar equipment rated primary DC voltage (kV)	60/20 (Changji/Guquan)
Measurement range of pole line voltage (kV)	± 1800
Measurement range of 550 kV pole line voltage (kV)	± 850
Measurement range of neutral busbar equipment voltage (kV)	± 200
Transient voltage measurement range of neutral busbar equipment (kV)	± 400
Isolated A/D system output (per unit value)	1.0
Pole busbar voltage (kV)	1100
550 kV pole busbar voltage (kV)	550
Neutral busbar voltage (kV)	100

Table 62 Insulation level

Item	Parameters
Rated lightning impulse withstand voltage of pole line equipment (1.2/50 ms) (kV, peak value)	2600
Rated lightning impulse withstand voltage of 550 kV pole line equipment (1.2/50 ms) (kV, peak value)	1450
Rated lightning impulse withstand voltage of neutral busbar equipment (1.2/50 ms) (kV, peak value)	600
Rated switching impulse withstand voltage of pole line equipment (kV, peak value)	2100
Rated switching impulse withstand voltage of 550 kV pole line equipment (kV, peak value)	1250
Rated switching impulse withstand voltage of neutral busbar equipment (kV, peak value)	550
Pole line wet state dc withstand voltage and partial discharge measurement test (60 min, wet) (kV)	1683
550 kV pole line wet state DC withstand voltage and partial discharge measurement test (60 min, wet) (kV)	842
Neutral busbar dc withstand voltage and partial discharge measurement test (60 min, dry) (kV)	225
Power frequency withstand voltage test on pole line (1 min) (kV, RMS)	1264
550 kV power frequency withstand voltage test on pole line (1 min) (kV, RMS)	632
Power frequency withstand voltage test on neutral busbar (1 min) (kV, RMS)	275

$$L = RUK_1K_2K_3 \tag{1}$$

where

R Creepage distance ratio, generally taken as 52 mm/kV;
U Maximum voltage of the equipment, kV;
K_1 Altitude correction factor, calculated based on an altitude of 1000 m;
K_2 Reduction factor of silicone rubber pollution resistance, taken as 0.9;
K_3 Diameter correction factor, taken as 1.2.

According to formula (1), $L = 52 \times 1100 \times 1 \times 0.9 \times 1.2 = 61{,}776$ (mm).

2. Dry arc distance of bushing

The dry arc distance of bushing S is calculated based on the creepage distance, the calculation formula is given below

$$S = L/\alpha_{\mathrm{CF}} \tag{2}$$

where α_{CF}— Creepage factor of the composite bushing, generally taken as 4.5 for DC.

According to formula (2), $S = 61{,}776/4.5 = 13{,}728$ (mm).

The actual creepage distance of bushing $L_1 = 63{,}594 > L$; actual arc distance $S_1 = 14{,}400$ mm $> S$, meets the requirements.

(II) External Electric Field Simulation Analysis of 1100 kV DC Voltage Divider

1. Design of corona ring

The HV terminal of the ± 1100 kV DC voltage divider uses two double-layer corona rings, with the corona ring diameter φ 600 mm, the large ring center diameter φ 3000 mm, the small ring center diameter φ 1800 mm, and the distance between the large and small rings is 700 mm (see Fig. 83). Under the action of the electromagnetic field, the electromagnetic field distribution around the corona ring is similar to a sphere, improving the shielding effect and the electric field distribution is uniform. The LV terminal uses a single ring with a diameter φ 400 mm, and a center diameter φ 1700 mm (see Fig. 84).

2. Establishment of a three-dimensional model

The DC yard equipment of this project is an indoor station. According to the installation position of the ± 1100 kV DC voltage divider, a three-dimensional model is established. A model is established for the inner wall and the top roof of the DC field, as shown in Fig. 85.

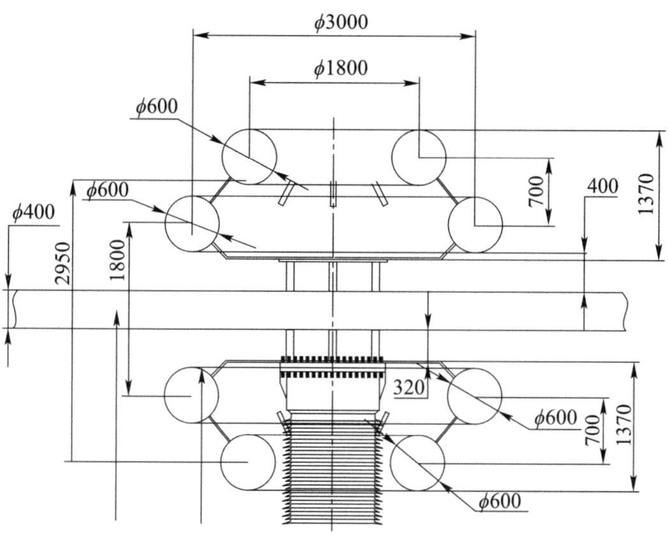

Fig. 83 Installation diagram of HV side corona ring

Fig. 84 Installation diagram of LV side corona ring

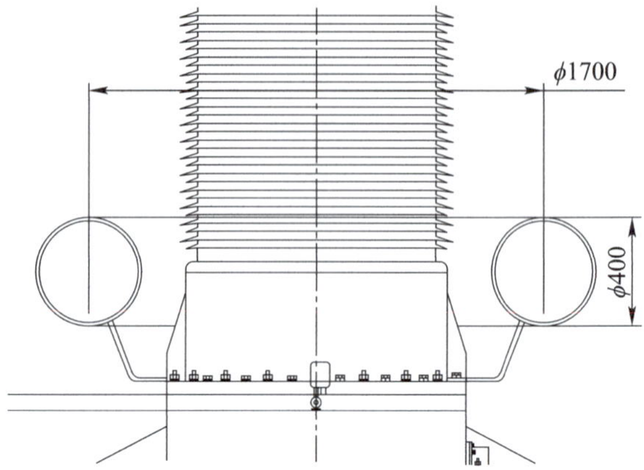

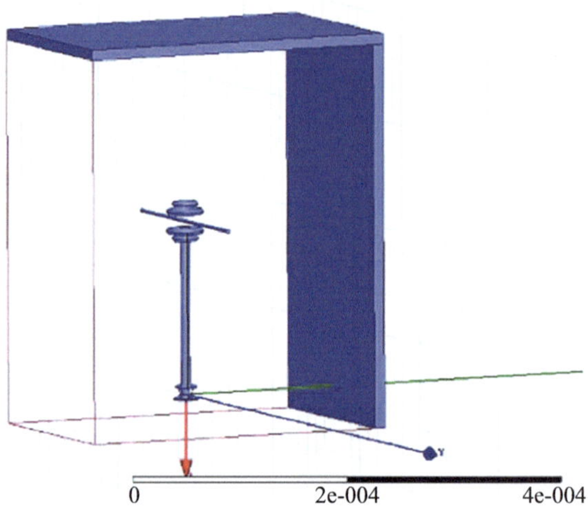

Fig. 85 A 3-D model of 1100 kV DC voltage divider

3. Simulation analysis results

The results of electric field simulation of ± 1100 kV DC voltage divider under DC withstand voltage condition are given below.

(1) The overall electric field distribution cloud diagram and voltage distribution cloud diagram of ± 1100 kV DC voltage divider are shown in Fig. 86, where the maximum electric field strength is 1.4230 kV/mm, concentrated in the outer edge area of the HV terminal of corona ring.
(2) The maximum electric field strength of the insulating bushing is 0.387 8 kV/mm, concentrated in the lower area of the HV terminal of corona ring. The electric field strength of the LV terminal of the insulating bushing is about 0.09–0.12 kV/mm, as shown in Fig. 87.

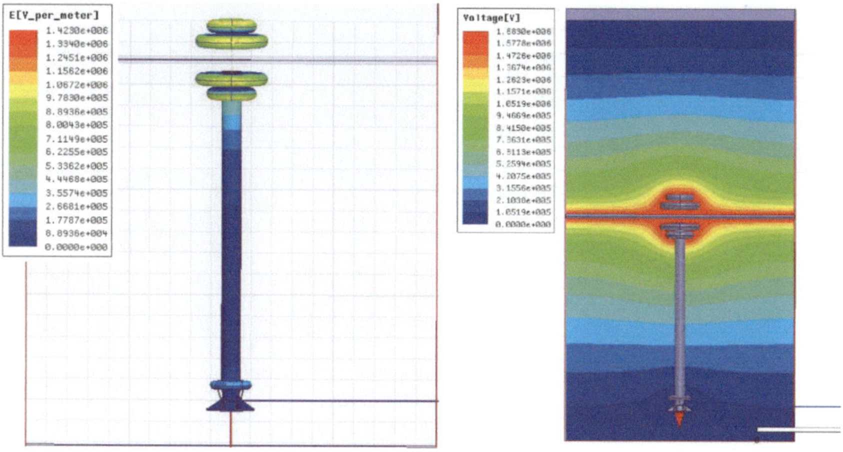

(a) Nephogram of electric field distribution (b) Nephogram of voltage distribution

Fig. 86 Nephogram of distribution of electric field and voltage on ± 1100 kV DC voltage divider

Fig. 87 Electric field distribution of insulator bushing

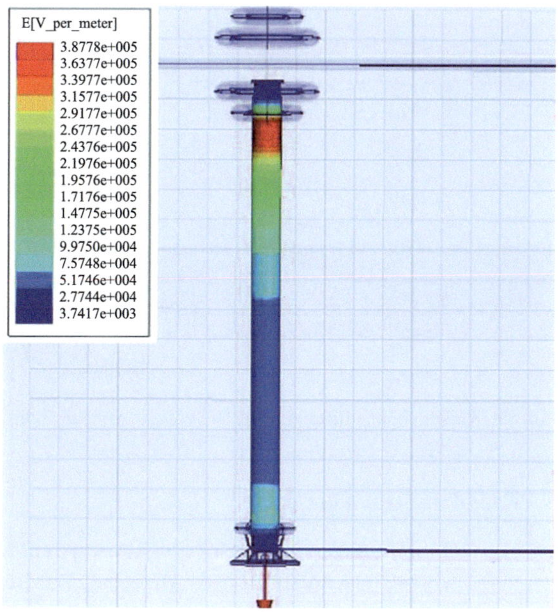

(3) The maximum electric field strength of the HV terminal of corona ring is 1.4230 kV/mm, concentrated in the outer edge area of the HV terminal of corona ring. The maximum electric field strength of the LV terminal of grading ring is 0.27 kV/mm, concentrated in the outer edge area of the LV terminal of corona ring, as shown in Fig. 88.

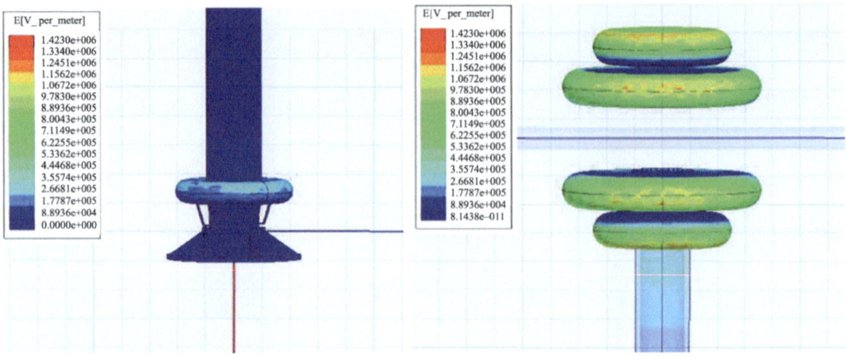

Fig. 88 Electric field distribution of HV side corona ring

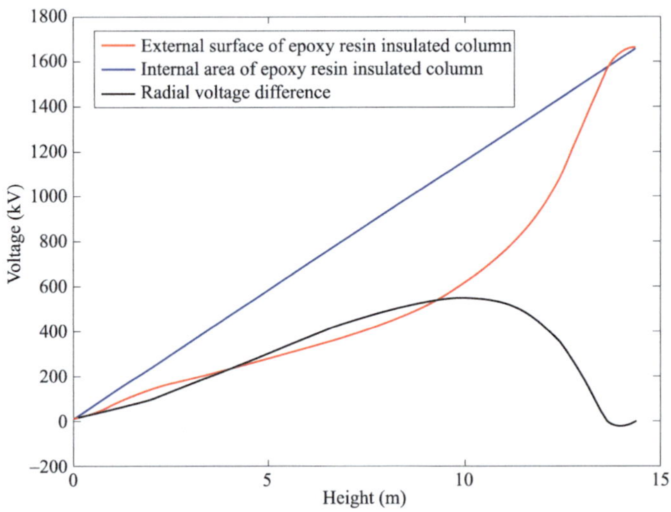

Fig. 89 Radial voltage difference of DC voltage divider at different heights

4. Radial insulation check

   (1) Simulation Analysis. In the resistor–capacitor unit of the DC voltage divider, the resistors are uniformly distributed from top to bottom, so it can be assumed that the internal electric field caused by the resistors is uniformly distributed from the HV terminal to the LV terminal. There are radial voltage differences between the external electric field at different heights of the voltage divider. Figure 89 illustrates the radial voltage difference between the internal and external electric fields of the DC voltage divider exposed to 1683 kV. At a height of 10 m in the voltage divider, the maximum internal radial voltage difference is 546 kV.

(2) Theoretical calculation. Through theoretical calculation, the radial insulation strength of ± 1100 kV DC voltage divider U is given by:

$$U = LE_{\mathrm{O}} \tag{3}$$

where

L the maximum radial insulation distance, in mm;
E_{O} the critical field strength of SF_6 gas under a standard atmospheric pressure, generally 8 kV/mm.

The maximum radial insulation distance of ± 1100 kV DC voltage divider is 262 mm, and according to formula (3), the maximum radial insulation strength is greater than 2096 kV.

In summary, the design strength of radial insulation of ± 1100 kV DC voltage divider is much greater than the insulation strength under actual operating conditions, meeting product performance requirements.

(III) Strength check of 1100 kV DC voltage divider

The wind load calculation formula is given by

$$P_{\mathrm{w}} = Ck_{\mathrm{n}}qA \tag{4}$$

where

P_{w}: the wind load acting on the object, in N;
C: the wind force coefficient, taken as 1.3;
k_{n}: the height variation coefficient of wind pressure, taken as 1;
q: the basic wind pressure, calculated based on a sustained wind speed of 34 m/s;
A: the horizontal windward area, A.

The basic wind pressure calculation formula has

$$q = \frac{1}{2}\rho v^2 = 0.646\,5 \times 34^2 = 747.5\ (\mathrm{N/m^2})$$

The horizontal windward area has

$$A = 3.5 \times 2.64 + 15.5 \times 1 = 24.74\,(\mathrm{m^2})$$

Therefore, the wind load has

$$P_{\mathrm{w}} = 1.3 \times 1 \times 747.5 \times 24.74 = 24\,(\mathrm{kN})$$

Using ANSYS software for simulation calculation, applying a force of 24 kN to the HV terminal of 1100 kV DC voltage divider, the simulation results are shown from Figs. 90, 91, and 92, the maximum displacement is 64.655 mm, and the maximum stress on base screw is 150.13 MPa.

The pre-tightening force of the M20 hot-dip galvanized 8.8 grade screw under the action torque of 220 N m is $F = C/[0.16P + U_0\ (0.58\ d_2 + d_m/2)]$. The pitch of M20 screw is 2, so the minor diameter of thread is 17.835 mm.

The stress generated by the pre-tightening force is 43,400/(8.92 × 8.92 × 3.14) = 173.7 (MPa).

In summary, the maximum stress acting on the screw is 323.7 MPa, which is less than the nominal tensile strength of 800 MPa of M20 8.8 grade hot-dip galvanized bolt, meeting the requirements.

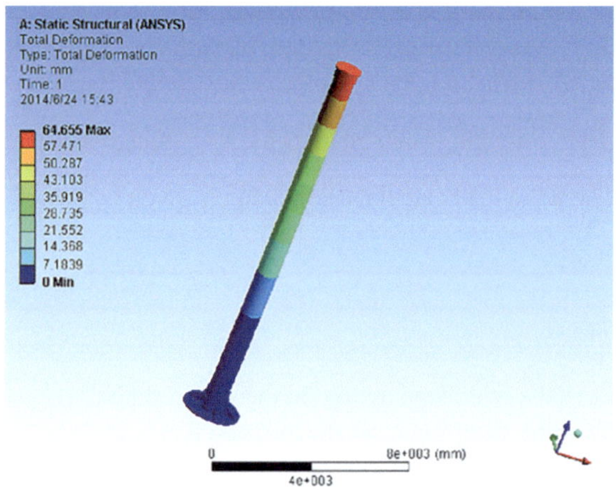

Fig. 90 Maximum offset under wind load

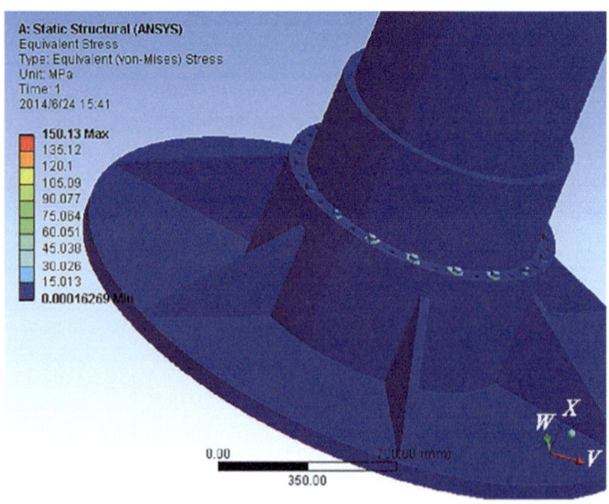

Fig. 91 Maximum stress under wind load

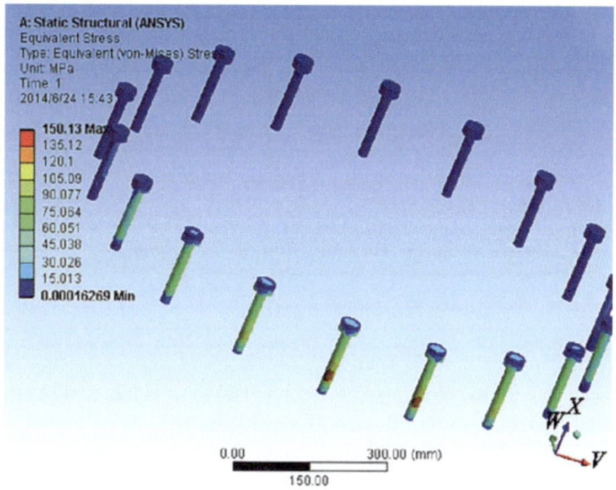

Fig. 92 Maximum stress of screws under wind load

(IV) Anti-seismic Analysis of 1100 kV DC Voltage Divider

According to the requirements of GB/T 26217—2019 "DC Voltage Measurement Device for High Voltage Direct Current Transmission System" and GB 50260—2013 "Anti-Seismic Design Code for Power Facilities", the DC voltage divider product should be checked for anti-seismic capability or related tests should be conducted.

According to GB 50260—2013, the anti-seismic capability of 1100 kV DC voltage divider is checked. GB 50260—2013 points out that electrical facilities can use static method, equivalent base shear method, mode-superposition response spectrum method or time history analysis method for anti-seismic analysis. According to design method requirements of GB 50260—2013, DC voltage divider should adopt the mode-superposition response spectrum method.

1. Static pre-stress analysis

    (1) Anti-seismic analysis modeling and description for transformer. The anti-seismic analysis of transformer uses the general commercial analysis software Ansys Workbench 15.0. The geometric parameters, material properties, and connection structures used in the modeling and analysis process are set according to the actual transformer product and design parameters. To ensure the accuracy and appropriate efficiency of the calculation, the model has been appropriately simplified.

The main structure of 1100 kV DC voltage divider is a hollow composite bushing, which accounts for more than 60% of the weight of transformer. The stiffness and strength of transformer are mainly guaranteed by composite bushing, and the inner diameter of its epoxy glass fiber tube is 800 mm.

The main material parameters of transformer are shown in Table 63.

The finite element method model of transformer is initially established using the three-dimensional software Creo, imported into the general finite element analysis software Workbench for mesh division, material property setting, and solution. The

Table 63 Performance of major material parameters of transformer

Material	Elastic modulus (GPa)	Failure stress (MPa)	Allowable stress (MPa)	Density (kg/m3)	Poisson's Ratio
Epoxy fiberglass winding pipe	21	225	90	2000	0.3
ZL104-T6	70	257	103	2800	0.33
5A02	70	240	120	2850	0.33
Q235	210	235	120	7850	0.3

Note The allowable stress of the material here is the design allowable value given considering the load characteristics of the transformer product

model uses solid elements (solid22), and the connections between the epoxy tube, aluminum flange, and base are rigid. The Bonded connection method is used in the Workbench program.

(2) Static pre-stress analysis. Establishing the voltage transformer calculation model, as shown in Fig. 93, and apply wind load, self-weight, and tensile load for static strength calculation.

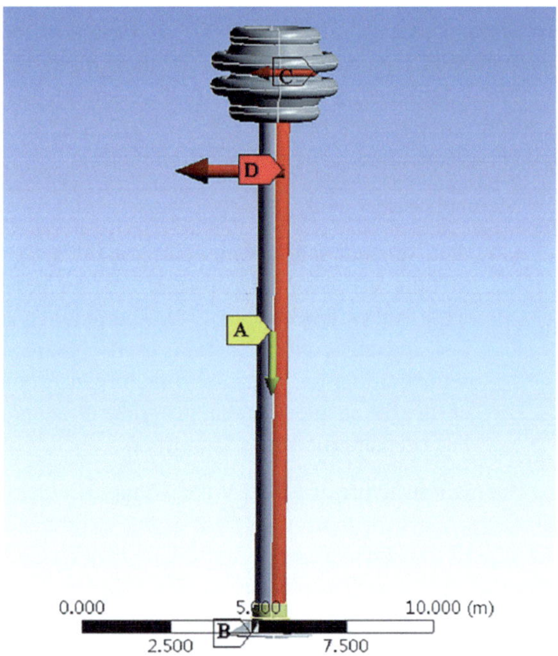

Fig. 93 Model of transformer

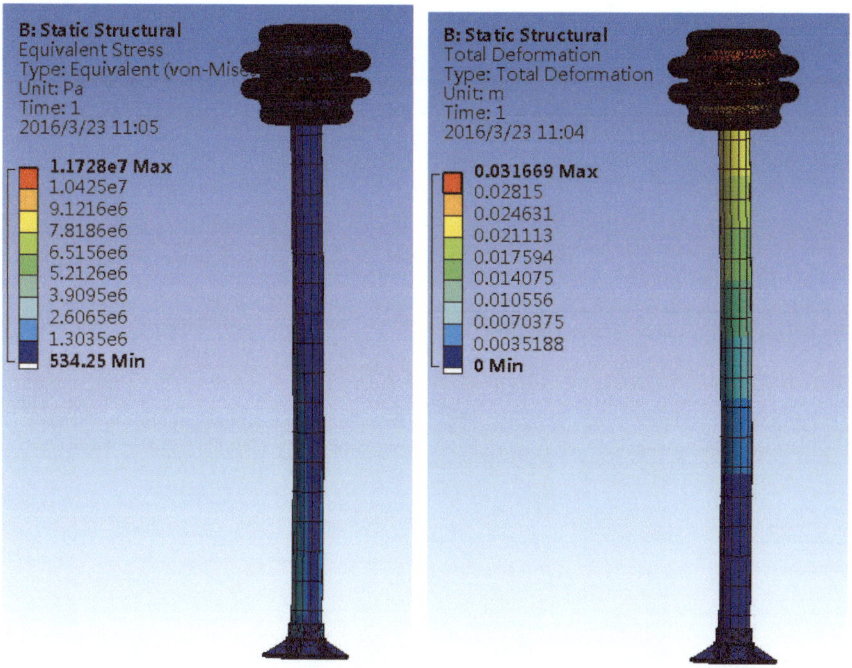

(a) stress nephogram (b) displacement nephogram

Fig. 94 Stress/strain simulation analysis diagram

According to Fig. 94, the results of the static pre-stress analysis show that under static stress conditions, the maximum stress of transformer occurs at the root of flange under the insulator, which is far lower than the allowable stress of the material 11.7 MPa.

2. Modal analysis

In the anti-seismic analysis of electrical equipment, the first few modes often play a major role. Therefore, only the first 6 modes are extracted in the calculation and the results are superimposed. The transformer modal shapes and natural frequencies obtained are shown in Table 64 and Fig. 95. The first and second order, third and fourth order modal frequencies and shapes are the same, which is determined by the symmetrical rotary structure of transformer.

3. Acceleration spectrum analysis (Anti-seismic)

Based on the modal shape data of transformer, considering factors such as the dynamic amplification factor and damping ratio, the curve of earthquake acceleration standard spectrum is drawn as the input condition for the spectrum analysis, and the horizontal and vertical two-way spectrum analysis is carried out, and the strength safety factor is checked.

Table 64 Calculated modal vibration and natural frequency of transformer

Mode No.	Frequency (Hz)	Vibration type
1	1.34	*X*-axis transverse bending
2	1.3401	*Y*-axis transverse bending
3	7.6369	*X*-axis transverse bending
4	7.7534	*Y*-axis transverse bending
5	9.5506	*Z*-axis longitudinal stretching
6	11.849	*X*-axis transverse bending

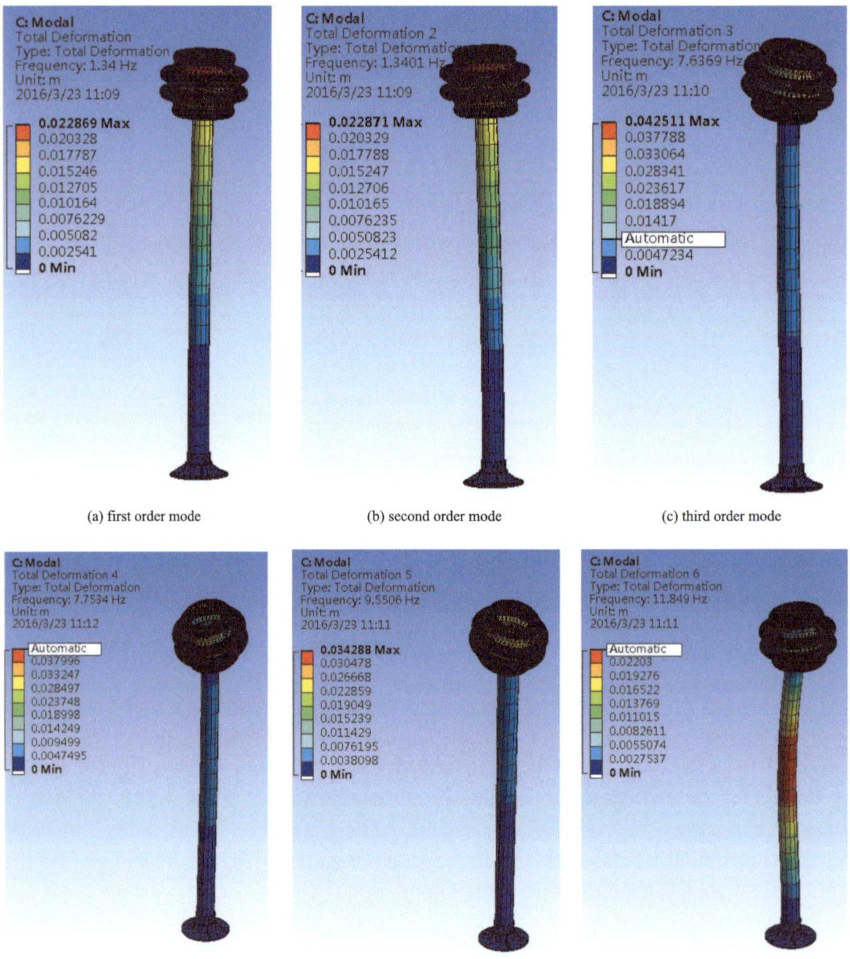

(a) first order mode (b) second order mode (c) third order mode

(d) fouth order mode (e) fifth order mode (f) sixth order mode

Fig. 95 Vibration diagram of DC voltage divider

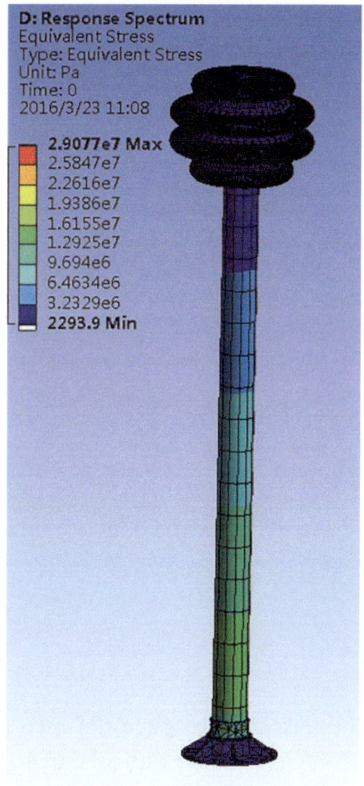

Fig. 96 Maximum stress diagram of transformer under earthquake action

Since the structure is basically centrally symmetric, only the load on $X + Z$ direction is considered in the analysis. The analysis results show that under the above-mentioned two-way earthquake response, the dangerous section appears in the lower flange of the bushing and the root area of the glass fiber tube, and the calculated stress value is 29 MPa, as shown in Figs. 96 and 97.

The maximum stress occurs in the root area of composite bushing, with a value of 29 MPa, and the maximum deformation occurs at the top of transformer, with a value of 197 mm.

In summary, the 1100 kV DC divider is a symmetrical structure, with a uniform internal mass distribution and a low center of gravity. The outer bushing has good rigidity, and its destructiveness under earthquake excitation is not strong. The calculated stress results of the simulation analysis also conform to the above analysis, and the structure can withstand an earthquake magnitude of 8.

III. Technical Scheme of Design

(I) Changji Converter Station DC Divider

The DC divider mainly relies on the insulation of the composite hollow insulator. To ensure the insulation level of the DC divider, it is necessary to calculate

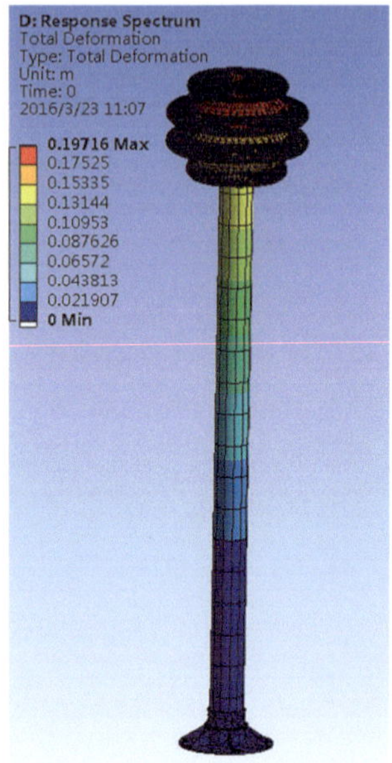

Fig. 97 Maximum displacement diagram of transformer under earthquake action

the required creepage distance according to the engineering voltage level and the creepage distance ratio of the project, and then design a suitable DC divider composite hollow insulator according to the creepage distance. When the voltage level is raised from ± 800 kV to ± 1100 kV, to meet the project creepage distance requirements, the dry arc distance of the composite hollow insulator needs to be increased by about 37.5% based on the ± 800 kV UHV DC transmission project.

Due to the increase in the length of the composite hollow insulator, the force changes. Mechanical simulation has been carried out on the elongated composite hollow insulator, and the simulation results are shown from Figs. 98, 99 and 100.

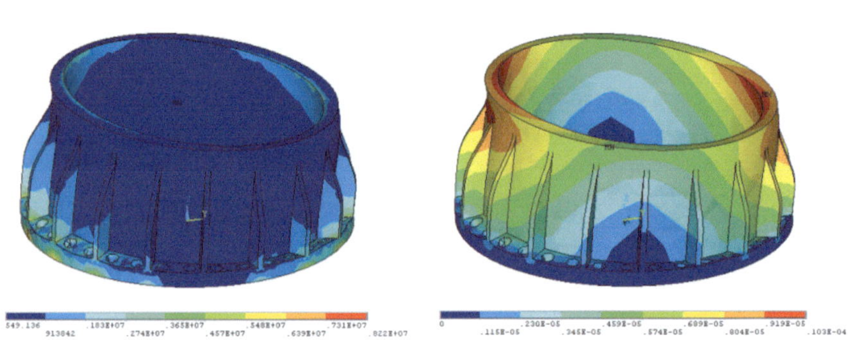

(a) stress nephogram (b) displacement nephogram

Fig. 98 Stress and displacement nephogram of ± 1100 kV lower flange

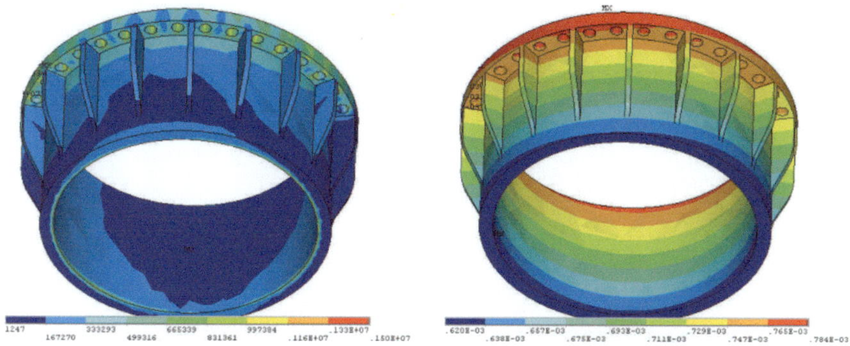

(a) stress nephogram (b) displacement nephogram

Fig. 99 Stress and displacement nephogram of ± 1100 kV upper flange

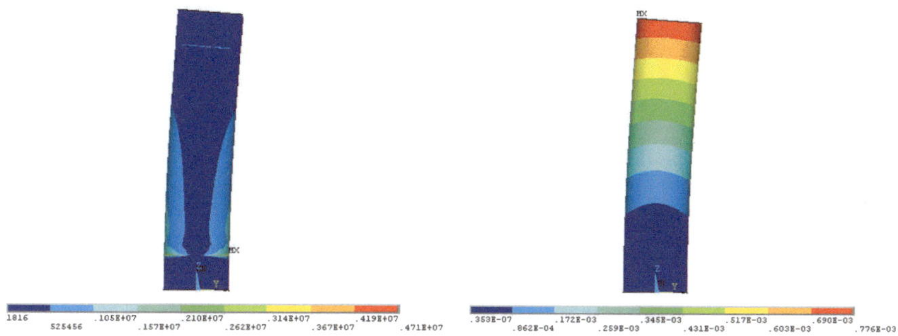

(a) stress nephogram (b) displacement nephogram

Fig. 100 Stress and displacement nephogram of ± 1100 kV epoxy pipe

Through simulation calculations, it is found that the maximum stress values of the upper/lower flanges and the epoxy tube have a large margin compared with the allowable strength values of component failure, and the mechanical performance meets the usage requirements.

As the voltage level increases, the electric field intensity of the DC divider increases. By reasonably designing the corona ring to reduce the equipment electric field intensity, it meets the requirements of the ± 1100 kV UHV DC transmission project. The size of the ± 800 kV corona ring is 2200 mm in diameter and 600 mm in tube diameter. The size of the ± 1100 kV corona ring is increased to 3000 mm in diameter and 800 mm in tube diameter. According to the actual size of the ± 1100 kV DC divider, the electric field simulation is carried out, and the simulation results are shown from Figs. 101, 102, and 103.

Through the simulation calculation of electric field of ± 1100 kV DC divider, the maximum field strength on the surface of the bus side corona ring is 768 V/mm, and the maximum field strength on the surface of the bracket side corona ring is 246 V/

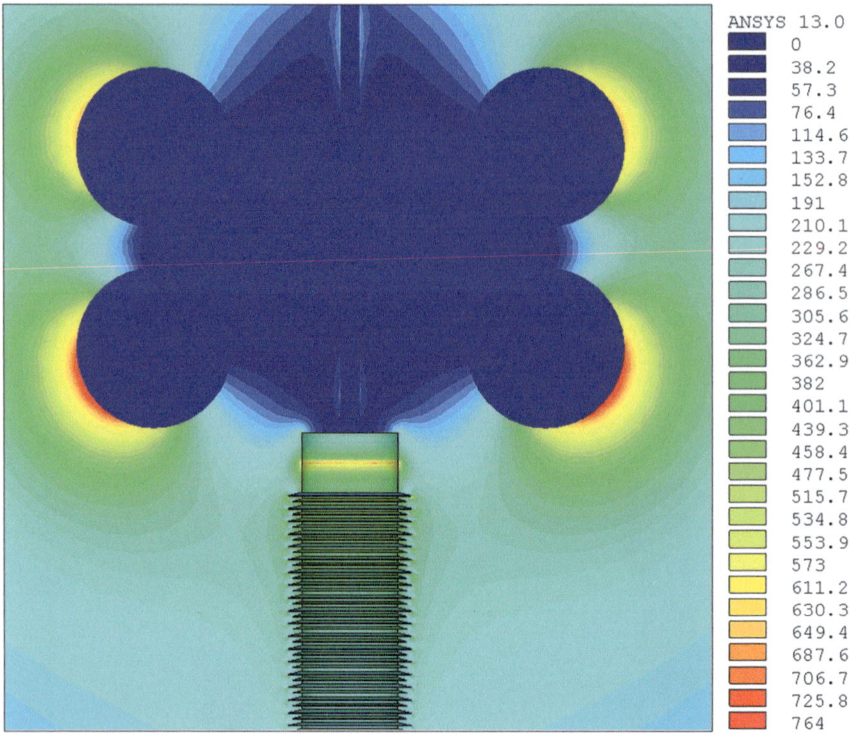

Fig. 101 Electric field distribution on ± 1100 kV busbar

mm, both of which are lower than the control field strength requirement of 1500 V/mm. The possibility of corona discharge is extremely small.

(II) Guquan Converter Station DC Divider

The DC divider used in the Guquan Converter Station mainly consists of an HV branch, a LV branch (low-voltage arm and resistance box), a remote module, a merging unit, and a multimode optical cable. The structure of the DC divider is installed in a composite silicone rubber bushing and uses an inflatable (SF_6) insulation structure. Each DC divider body is equipped with a pressure equalizing device at the top, flanges at the top and bottom, and 3 SF_6 gas pressure detection devices and one set of online monitoring devices at the bottom. The structure of the DC divider is shown in Fig. 104, and the principle is shown in Fig. 105.

As shown in Fig. 104, the parallel resistance R_{13} and capacitance C_1 are the HV branch, and the rest are the LV branch. R_{24} is the resistance of the LV arm, C_2 is the capacitance of the LV arm, the resistance R_3, and the capacitance value C_3 are the resistance and capacitance of the LV branch, of which is parallel connection of R_{24}, C_2 and the resistance and capacitance of the LV divider board, F_1 is the voltage limiting device.

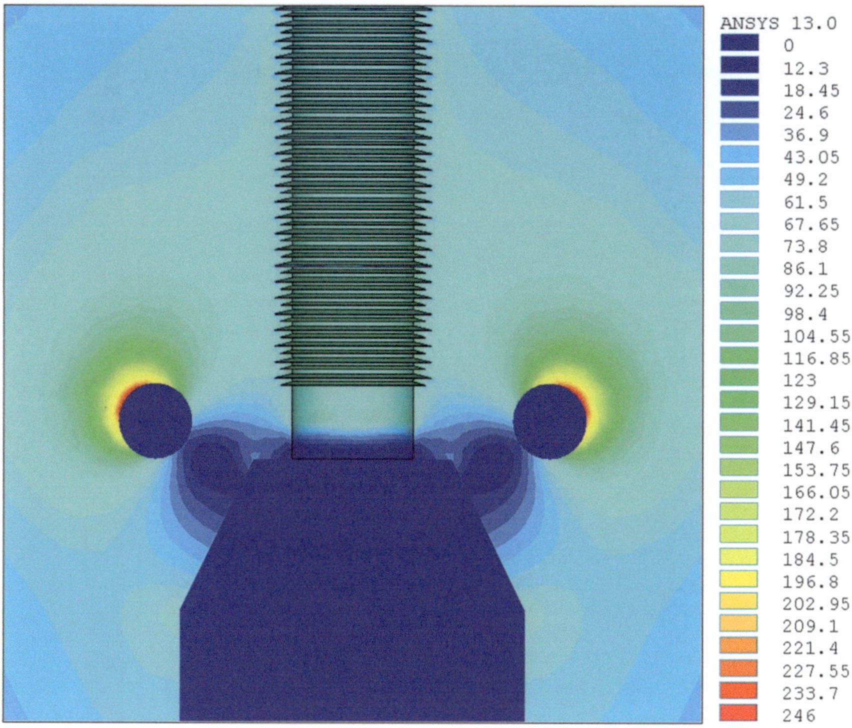

Fig. 102 Distribution of electric field on ± 1100 kV bracket side

The appearance of the pole bus DC divider is shown in Fig. 106.

7 1100 kV DC Filter

Compared with the DC filtering devices of 800 kV and below in traditional DC transmission projects, 1100 kV DC filtering equipment has problems in mechanical strength and anti-seismic design, electric field distribution in high-voltage capacitor tower, installation height limitation, component noise control, etc. Through optimizing the design scheme, the anti-seismic performance and stability are effectively improved, the surface electric field intensity and the overall installation height of the capacitor tower are reduced, and the equipment noise level is reduced.

I. Technical Requirements

(1) For environmental conditions, see the relevant requirements in the first section of this chapter.
(2) The parameters of the DC filter HP2/12 filter bank capacitor device are shown in Table 65.

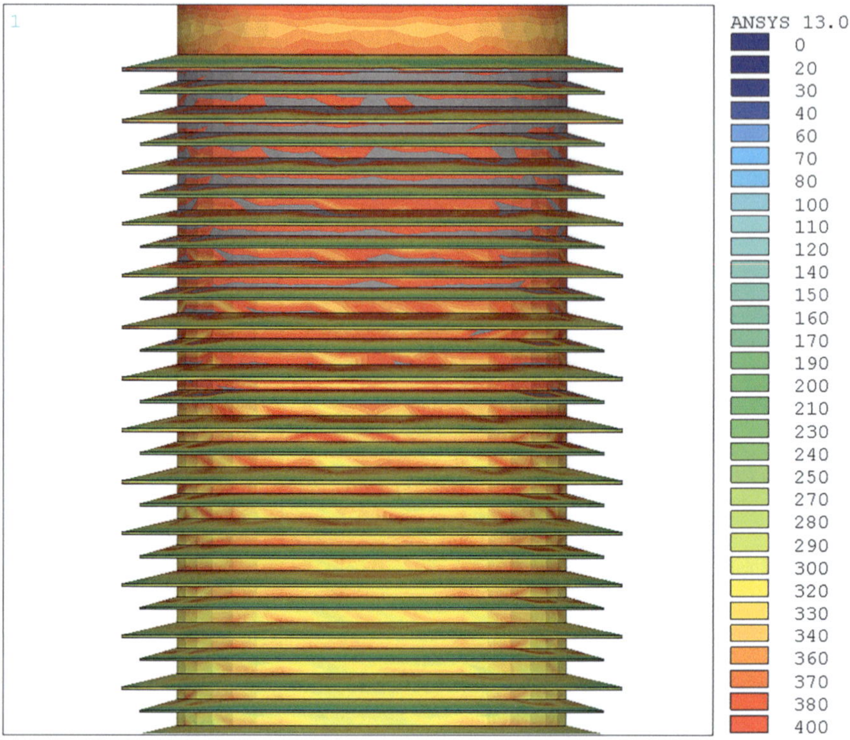

Fig. 103 Electric field distribution of ± 1100 kV umbrella skirt on busbar side

II. Key Technologies

(I) DC Filter Capacitor

Compared to the previous ± 800 kV DC filter capacitor devices, the ± 1100 kV DC filter capacitor device has a larger single group capacity, higher rated voltage, higher insulation level, taller capacitor tower, and is installed indoors. Therefore, the anti-seismic resistance of the capacitor device, surface electric field intensity, insulation coordination, tower height control, etc., are all prominent issues. Therefore, there are several difficulties in the design and manufacture of the ± 1100 kV DC filter capacitor:

(1) Mechanical strength and anti-seismic issues. The ± 1100 kV DC filter capacitor device of the Changji converter station is installed indoors, with each pole arranged in a single tower, the tower height reaches 25.95 m, and the weight reaches 65 t. The tower height and overall weight are not much different from the previous ± 800 kV DC filter capacitor devices. The previous ± 800 kV DC filter capacitor devices were arranged in a three-tower with parallel H-shape, with good anti-seismic performance and stability, but the anti-seismic requirement of the CG project is only 0.2g. Through mechanical force calculations on the

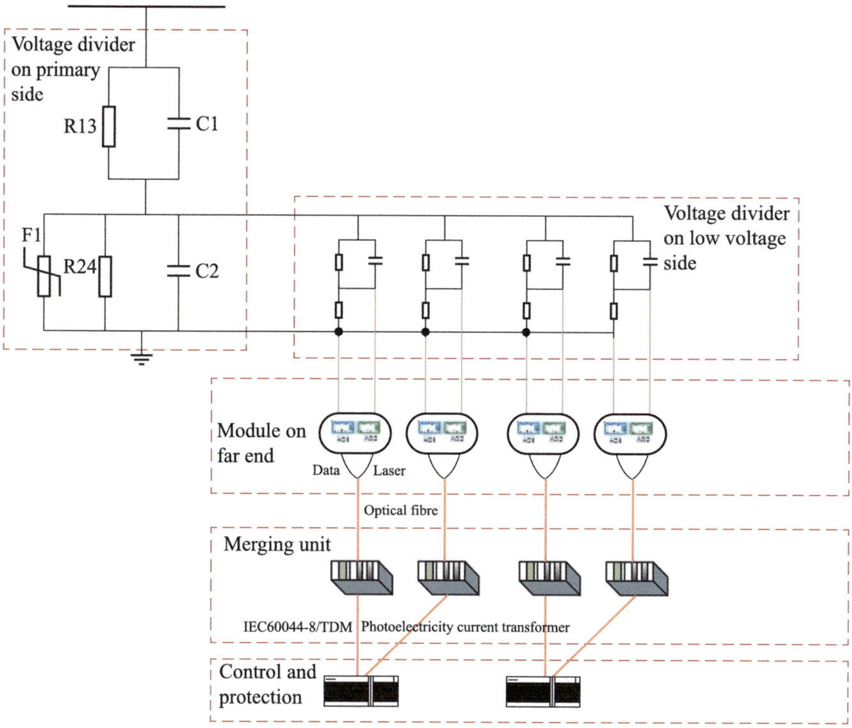

Fig. 104 Configuration and schematic diagram of DC voltage divider

capacitor device tower, including deflection and strength calculations, the design of the ± 800 kV DC filter capacitor device can be used to solve the problems of mechanical strength and anti-seismic resistance.

(2) Electric field intensity distribution problem. The insulation design of the indoor ± 1100 kV capacitor tower is quite tense, its spatial dimensions and complexity have greatly increased, and its electric field distribution is very sensitive to structural design. To prevent the device from discharging to the surrounding environment, the surface electric field intensity of the device is required to be less than 12 kV/cm. The surface electric field intensity of the previous ± 800 kV DC filter capacitor devices reached more than 20 kV/cm, and it is impossible to meet the requirements according to the previous design type of equalizing device. Therefore, it is necessary to optimize the design of the entire device's electric field distribution and redesign the grading and anti-corona device.

(3) Capacitor tower height limit problem. The previous ± 800 kV DC filter capacitor devices were installed outdoors, and the insulation level of the LV terminal

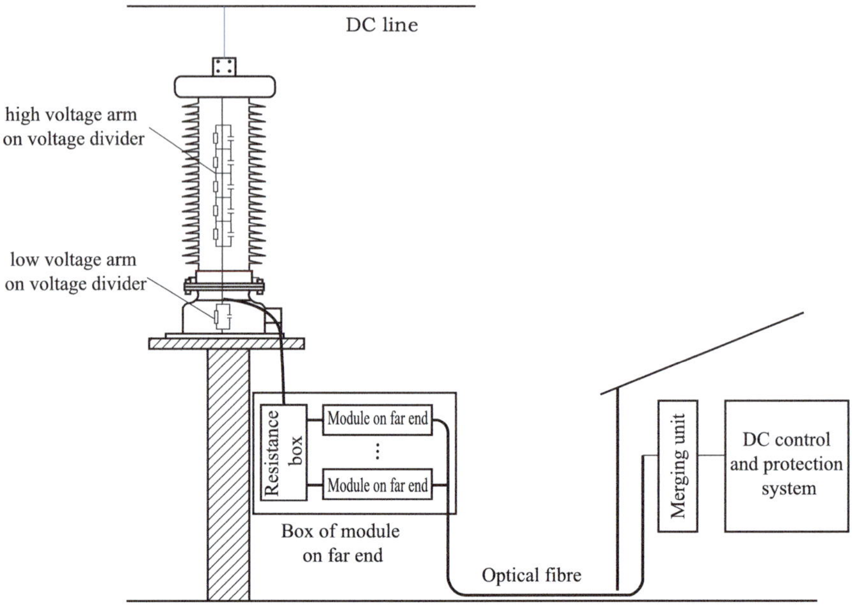

Fig. 105 Schematic and theoretical diagram of DC voltage divider

Fig. 106 Configuration of pole busbar DC voltage divider

Table 65 Capacitor bank parameters of DC HP2/12 filter

No.	Item	High Voltage Parameters	Low Voltage Parameters
1	Capacitance		
1.1	Rated capacitance (C_N) 25 °C (μF)	0.6	0.6
1.2	Manufacturing tolerance (%)	± 1.0	± 1.0
1.3	Range of capacitance parameter variation from lowest to highest ambient temperature (%)	− 2.185%/2.552%	− 2.185%/2.552%
2	Voltage		
2.1	Maximum sustained operating voltage (U_{Nb}) (kV, RMS)	1722	1722
2.2	Major harmonics (*n*/kV, RMS)	2/301.62	2/301.62
		12/23.22	12/23.22
		6/6.52	6/6.52
		3/3.73	3/3.73
		14/1.28	14/1.28
		10/1.27	10/1.27
2.3	Maximum temporary voltage, sustained 10 min (kV, RMS)	1725	1725
2.4	Maximum temporary voltage, sustained 30 s (kV, RMS)	1730	1730
2.5	Minimum sustained voltage, excluding harmonics (kV, RMS)	880	880
2.6	Voltage used for calculation of creepage distance		
2.6.1	Terminal-to-terminal (DC) (kV, RMS)	1163	1163
2.6.2	HV terminal to ground (DC) (kV, RMS)	1200	1200
2.6.3	LV terminal to ground (AC) (kV, RMS)	298.0	298.0
2.6.4	Voltage terminal to ground (DC) (kV, DC)	40	40
3	Rated resonant frequency (Hz)	100/600	100/600
4	Rated current		

(continued)

Table 65 (continued)

No.	Item	High Voltage Parameters	Low Voltage Parameters
4.1	Maximum current (A, RMS)	211.6	211.6
4.2	Major harmonics (*n*/A, RMS)	2/112.23	2/112.23
		12/54.96	12/54.96
		6/8.39	6/8.39
		24/6.59	24/6.59
		30/3.14	30/3.14
		14/3.09	14/3.09
5	Noise requirement		
5.1	Harmonics current used for calculation of noise (A, RMS)	137.8	137.8
5.2	Major harmonics (*n*/A, RMS)	2/61.01	2/61.01
		12/38.18	12/38.18
		6/9.60	6/9.60
		24/4.31	24/4.31
		30/3.78	30/3.78
		33/2.52	33/2.52
5.3	Maximum noise level (dB)	70	70
6	Loss angle (tanδ) (W/kvar)	0.3	0.3
7	Insulation level		
7.1	Switching impulse withstand level *SIWL*		
7.1.1	HV terminal (kV, RMS)	2100	2100
7.1.2	Terminal-to-terminal (kV, RMS)	3526	3526
7.1.3	LV terminal (kV, RMS)	1334	1334
7.2	Lightning impulse withstand level *LIWL*		
7.2.1	HV terminal (kV, RMS)	2700	2700
7.2.2	Terminal-to-terminal (kV, RMS)	4267	4267
7.2.3	LV terminal (kV, RMS)	1633	1633

of the device to the ground was relatively low, and the height was generally controlled at 25 m. The BIL/SIL of the LV terminal of the ± 1100 kV DC filter capacitor device is 1633 kV/1334 kV. According to the relevant standard query and calculation, the device needs a 750 kV voltage level insulator referred to the ground. And the device is installed indoors, the height of the capacitor tower is required to be controlled within 26 m, and the structure of the capacitor device needs to be optimized again.

1. Comparison of main technical parameters

The comparison of the main technical parameters of the ± 1100 kV and ± 800 kV DC filter capacitor devices is shown in Table 66.

2. Research on the electric field intensity distribution of the 1100 kV DC filter capacitor device

The rated voltage of the UHV DC filter capacitor banks of the CG project is currently the highest in the world's UHV DC transmission projects, and it is installed indoors, so the corona problem of the capacitor banks is quite prominent. To prevent the occurrence of corona, the maximum electric field intensity on the surface of the capacitor banks is required to be less than 12 kV/cm. Therefore, studying and analyzing the electric field distribution of the capacitor banks and designing a reasonable grading structure has become an important aspect in the design process of the capacitor banks.

(1) Preliminary design and simulation verification of grading structure. Based on the design experience of previous ± 800 kV UHV DC transmission projects, the grading structure of the capacitor group adopts the terminal spherical inlet tube busbar and corona ring, and measures to increase the diameter of the tube busbar and terminal sphere are taken. The maximum electric field strength calculated by the three-dimensional simulation model is 16.59 kV/cm, distributed on the terminal sphere, which is significantly different from the required value of 12 kV/cm.
(2) Design optimization and simulation verification of the grading structure.
    1) The terminal sphere is improved to an grading shield and its simulation verification. The structure of the terminal spherical inlet tube busbar and corona ring can no longer meet the requirements. To prevent the capacitor banks from generating corona, the grading structure must be fully optimized

Table 66 Comparison of main technical parameters of ± 1100 kV and ± 800 kV DC filter capacitor devices

Technical parameters	Zhalute HP12/30	Changji HP2/12
Capacitance (μF)	0.8	0.6
Series/parallel number	112 series/2 parallel	120 series/2 parallel
Height of capacitor tower (m)	22.0	25.95
Rated voltage (including harmonics) (kV)	1349	1732
Rated current (including harmonics) (A)	379.09	213.4
HV terminal to ground *BIL*/*SIL*	2075 kV/1703 kV	2700 kV/2100 kV
Terminal-to-terminal *BIL*/*SIL*	2927 kV/2651 kV	4267 kV/3526 kV
LV terminal to ground *BIL*/*SIL*	1086 kV/875 kV	1633 kV/1334 kV

and improved. Considering indoor installation, the terminal spherical inlet tube busbar is optimized to an equalizing shield. The maximum electric field strength calculated by the three-dimensional simulation model is 13.24 kV/cm, distributed at the lower edge of the grading shield chamfer, which does not meet the required value, but the electric field strength has significantly decreased compared to before the optimization.

2) Increase the chamfer radius of curvature of the grading shield and its simulation verification. The use of an grading shield can significantly reduce the surface electric field strength of the device. If you want to further reduce the electric field strength of the capacitor banks, you can consider increasing the diameter of the chamfer at the lower edge of the grading shield. After the diameter of the chamfer at the lower edge of the grading shield is increased by 100 mm, the maximum electric field strength calculated by the three-dimensional simulation model is 11.97 kV/cm, distributed at the lower edge of the grading shield chamfer, which meets the required value. The calculation results are detailed in Fig. 107.
3) The corona ring is improved to a trapezoidal structure and its simulation verification. If the five layers of corona rings under the grading shield are improved to a trapezoidal structure, and the outer edge of the corona ring does not exceed the outer edge of the grading shield, and the diameter of the three layers of corona rings below is kept unchanged. The diameter of the first and second layers of corona rings is doubled, and the outer edge of the

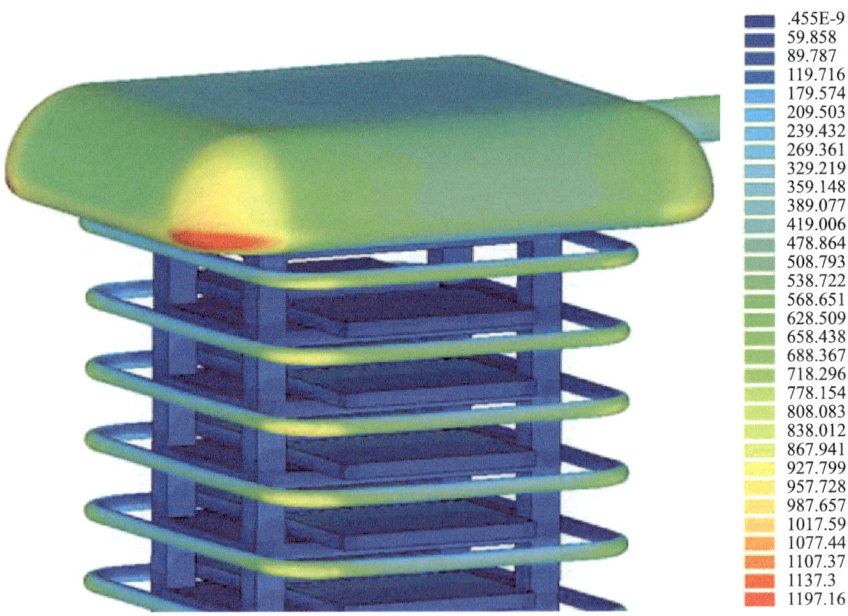

Fig. 107 Entire electric field distribution of capacitor bank (I) (kV/cm)

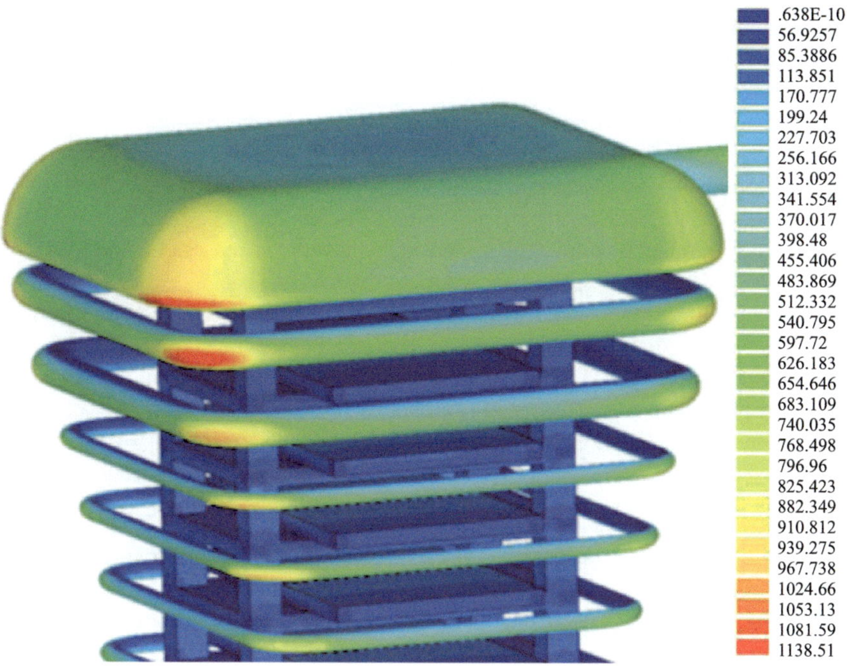

Fig. 108 Entire electric field distribution of capacitor bank (II) (kV/cm)

corona ring still does not exceed the outer edge of the grading shield. The maximum electric field strength calculated by the three-dimensional simulation model is 11.38 kV/cm, distributed at the lower edge of the grading shield chamfer and the first layer of the corona ring chamfer, which meets the required value. The calculation results are shown in Fig. 108.

In the research of device electric field design: By optimizing the grading structure to the coordination between grading shield and corona ring, the overall surface electric field distribution of the capacitor banks can be effectively improved; further, by increasing the radius of the grading shield and increasing the outer diameter of the corona ring, the overall surface electric field strength of the capacitor banks can be further reduced, and the maximum electric field strength is less than 12 kV/cm through the verification using finite element method three-dimensional model, which meets the required value.

3. Height control of 1100 kV DC filter capacitor tower

The ± 1100 kV DC filter capacitor device is installed indoors, with the highest DC voltage of 1122 kV, and the height requirement is controlled within 26 m. The height of the device is mainly composed of the base insulator, the capacitor between the HV terminal and the LV termimal, and the top grading shield.

(1) Design of the height of the base insulator. The device's LV terminal to ground BIL/SIL is 1633 kV/1334 kV, the creepage distance is 7775 mm, and it must meet BIL/SIL (1633 kV/1334 kV) and the corresponding clearance requirements, the base insulator height needs to be 5.6 m, equivalent to the insulator of the 750 kV voltage level plus adjustment plates, anchor bolts, base plates, spacer plates, etc., the total height of the base is about 5.9 m.
(2) Design of the interlayer insulator height. The device has a high rated voltage, a large number of unit strings, and many layers. In the case of many layers of the device, the design of the interlayer insulator plays a decisive role in the height of the device. For the insulator design of the DC filter capacitor device, the creepage distance and insulation level are mainly considered. The creepage distance of the interlayer insulator is: 1164 (creepage calculation voltage) $\times$ 25 (creepage ratio distance) $\div$ 29 (number of interlayer insulators) $=$ 1004 mm; the insulation level of the interlayer insulator BIL/SIL is 157 kV/130 kV. For this creepage distance and insulation level, an insulator of 500 mm height can meet the requirements.

The design of the interlayer insulator also needs to consider the safe clearance between the capacitor terminals of the upper and lower layers. Its insulation level is twice that of the interlayer insulator, so the BIL/SIL of the capacitor terminals between the upper and lower layers is 314 kV/260 kV. According to GB 311.1—2012 "Insulation Coordination Part 1: Definitions, Principles, and Rules", the insulation level corresponding to the lightning 325 kV is 66 kV. And the indoor safe clearance corresponding to the 66 kV voltage level in GB 50060—2008 "Design Specification for 3–110 kV High Voltage Distribution Devices" is 550 mm. To meet this clearance requirement, an insulator of at least 600 mm height is needed. Adding the steel support of the inclined tower part, the height of the interlayer platform flange, the capacitor platform between the high and low voltage terminals of the device, the total height is about 18.8 m.

(3) Design of grading shield on top of device and mounting bracket height. Under the condition of satisfying the surface electric field strength of 12 kV/cm, the height of the grading shield should be at least 1 m. Adding the safe clearance of 200 mm between the grading shield and the top layer capacitor platform, the total height of the grading shield and mounting bracket is 1.2 m. And a ϕ 400 mm inlet pipe busbar needs to be installed between the grading shield and the top layer capacitor platform, the height of the inlet pipe busbar plus the mounting hardware is 600 mm. At this time, the total height of the device is 26.5 m, which exceeds the height limit of 26 m. There is no room for optimization of the height of the device base and interlayer capacitor platform. In order to meet the height requirements of the device, the inlet method of the device can only be optimized and improved. An inlet hole is opened on the side of the grading shield, and the inlet is changed to pass through the grading shield, reducing the height of the device. The final total height of the device is 25.95 m, which meets the requirements.

(II) Noise Control of 1100 kV Filtering Device Components

The harmonic content of the 1100 kV DC filter equipment is relatively higher than that of previous projects, and because of its high capacitor tower height, the noise it generates is more difficult to control. The method of simply adding sound barriers in previous projects can no longer fully meet the needs of noise control, and special design and treatment of each noise component must be carried out.

1. Noise reduction measures for filter capacitors

Optimize the number of capacitors in series and parallel, and minimize the harmonic current of a single capacitor. At the same time, according to the actual harmonic current of the project, noise tests are conducted on a single capacitor, and different noise reduction measures are taken for capacitors of different frequencies. Figure 109 shows the results of the impact of different harmonic frequencies on noise. For different filter banks, the harmonic current flowing through is different, and the impact on the noise generated by the capacitor body is also different. Through a large number of tests, the noise impact rules caused by the capacitor unit due to different frequency harmonics are found out, guiding the design of capacitor noise reduction measures.

(1) Add a sound absorption device outside the unit. There are no charged components inside the shell wrapped in the bottom sound absorption cover, no hot spots, and the bottom temperature rise is minimized as much as possible.

As can be seen from Fig. 110, adding a capacitor cover on the box cover, 20 mm away from the box cover, within the design margin of the bushing creepage distance, ensuring the creepage distance while soundproofing, there are no hot spots at the

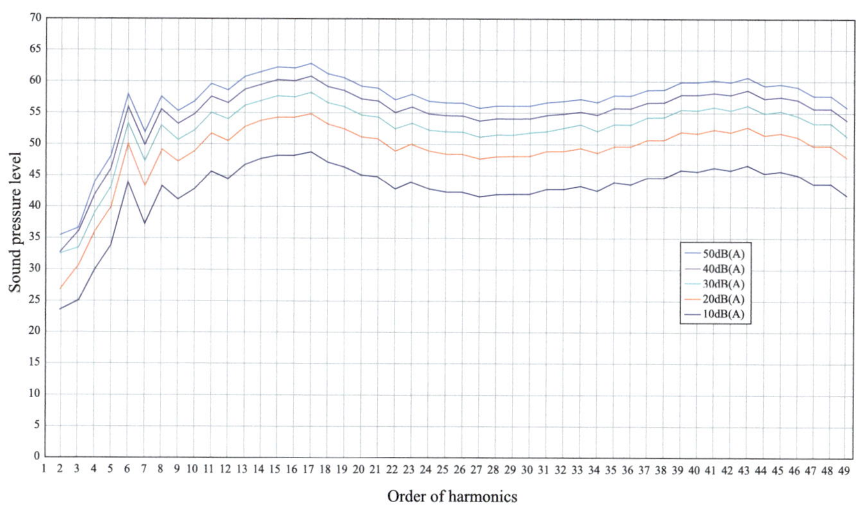

Fig. 109 Results of impact of different harmonic frequencies on noise

connection between the capacitor cover and the capacitor body, and there is enough heat dissipation distance. The sound-absorbing material has undergone flame retardant, acid resistant, water resistant, oil resistant and pollution tests, and all have passed the tests. Ensure that there is no risk in the use of materials. The top and bottom rubber rings are sealed with weather-resistant silicone rubber sealing rings, and are fastened with bolts, nuts and other connectors to achieve complete sealing and fastening effects, as shown in Fig. 111.

(2) Add noise reduction components inside the unit. The specific measures are as follows:

First, manufacture a metal cavity component made of stainless steel, as shown in Fig. 112. The overall thickness of the metal cavity is about 13 mm, the thickness of the metal plate used is about 3 mm, and the thickness of the cavity is about 7 mm.

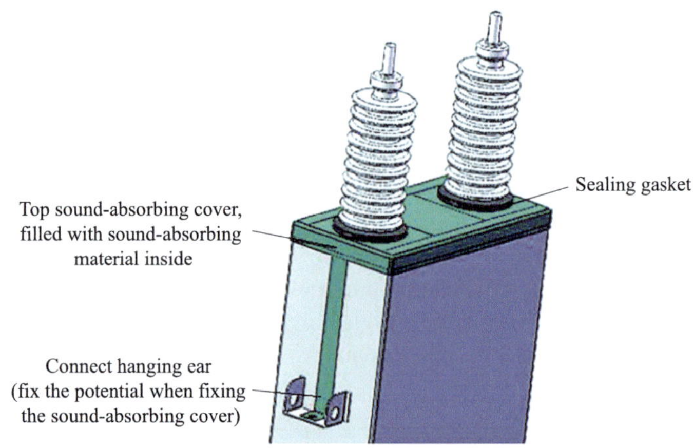

Fig. 110 Schematic diagram of installing sound absorbing device on top of capacitor unit

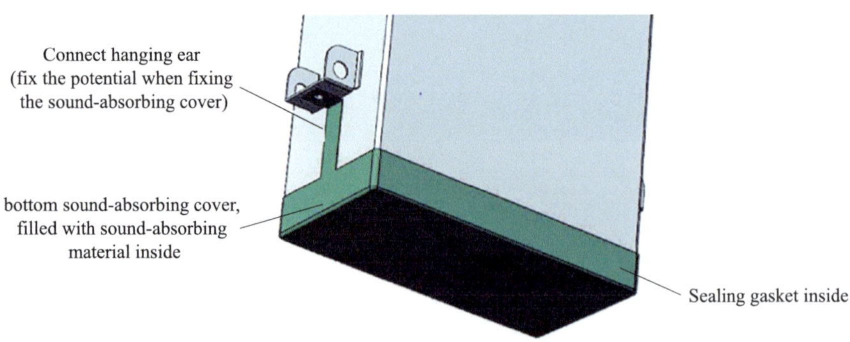

Fig. 111 Schematic diagram of installing sound absorbing device on bottom of capacitor unit

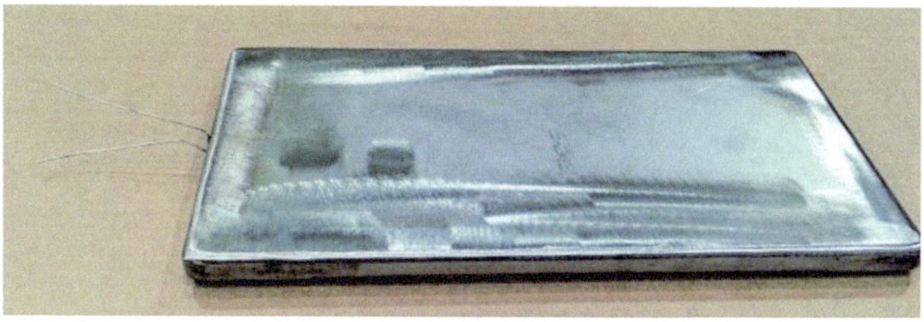

Fig. 112 Physical picture of metal cavity component

Secondly, insert the manufactured metal cavity component into the series section of the internal capacitive component of the capacitor, and connect it with other capacitor components to ensure the voltage stability of this metal cavity. The main positions where the metal cavity component is inserted are at the top and middle of the capacitor, and it is packaged together with other capacitor components, as shown in Fig. 113.

Again, using a stainless steel plate, first weld a cavity at the bottom of the capacitor, and then weld this bottom to the outer shell of the capacitor, as shown in Fig. 114.

Or directly add a sound insulation interlayer at the bottom of the capacitor unit shell, as shown in Fig. 115.

(3) Install a vibration damping pad between the capacitor and the frame (see Fig. 116) to absorb and reduce the amount of capacitor vibration energy transferred to the frame, thereby further reducing noise.

Currently, the corresponding capacitor noise reduction measures have been adopted in the already-operational Zhaqing and the upcoming Shangshan projects. The noise

Fig. 113 Encapsulation diagram of metal cavity components in capacitors

Fig. 114 Physical picture of a cavity welded at the bottom of a capacitor

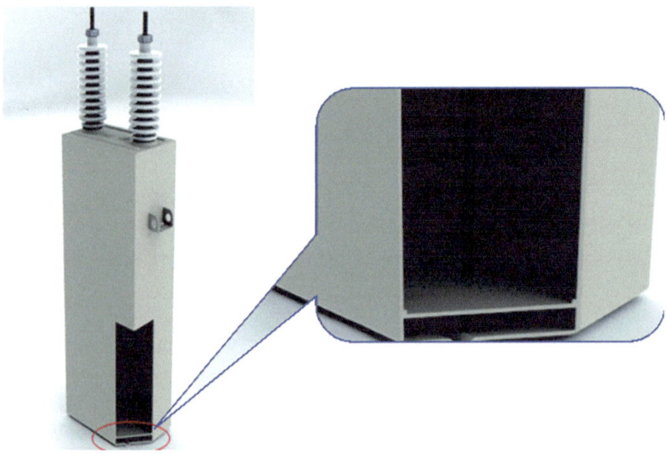

Fig. 115 Diagram of soundproof interlayer added to bottom of capacitor

Fig. 116 Physical picture of installing vibration damping pads between capacitors and frames

reduction effect can reach more than 5 dB, which well meets the project needs and saves the cost of the sound barrier.

2. Noise reduction measures for filter reactors

    (1) Noise control. For the noise problem of hollow reactors, due to the fundamental difference in structure between dry-type hollow reactors and iron-core reactors: there is no ferromagnetic circuit, so when the reactor is operating normally, a strong alternating magnetic field will be generated on the winding, and the wire in the magnetic field will vibrate due to the alternating magnetic field, thereby emitting sound, i.e., noise. In addition, by calculating the inherent vibration frequency of the reactor with a computer, this inherent frequency is adjusted to avoid the forced vibration frequency of each harmonic during the design process, to avoid mechanical resonance, thereby reducing noise.

In order to fundamentally control the level of noise, during the winding process, the entire reactor is ensured to be tightly sealed from a process perspective, leaving no gaps in the encapsulation. During the wire winding process, the wire is pre-treated with dipping glue, making each layer of winding very dense after curing. At the same time, non-woven fabric material is used on the outside of the wire, which can make the wire bond more tightly with the encapsulation epoxy layer. After the winding is cured at high temperature, the wire and the epoxy layer will form a whole, eliminating all kinds of looseness, also improving the integrity of the reactor, effectively controlling the vibration of the wire within a certain range, thereby reducing the noise level.

In addition to the above measures to improve the structural design of the reactor itself, a rainproof and soundproof cover device has also been installed on the outside of the reactor. The overall structure of this rainproof soundproof device adopts the form of "dressing and capping" the winding.

Design principle: As much as possible, keep it simple, applicable, and easy to install without affecting the ventilation and heat dissipation of the reactor. The soundproof cover outside the winding adopts the form of a pseudo-layer of the reactor, that is, an additional pure fiberglass cover without winding is added outside the main line of the reactor. This structure has good integrity, simpler process, stronger operability, and more importantly, it reduces the installation and connection steps, and reducing connection parts can better reduce the noise increase caused by loose connections. The advantages of this whole cover also include: no seams of the spliced cover, eliminating some safety hazards of electrical operation, and it can protect the winding from wind, rain, and ultraviolet rays.

This structure allows the sound wave to weaken gradually after multiple reflections in the cover, reducing noise to the maximum extent. These design measures have been widely used in the filter reactors of multiple converter station projects.

(2) Sound level simulation calculation. Based on the measured results of the completed filter reactor tests, the sound level of the CG project reactor was simulated and calculated as follows:

Table 67 Calculated results

Model and specifications	HP12/24 HV Side	HP12/24 LV Side	HP3
Test frequency (Hz)	50	50	50
Rated current (A)	305.66	387.04	256.03
Actual applied current (A)	916.5	916.5	916.5
Actual test sound pressure level [dB (A)]	44	45	43
Revised sound pressure level [dB (A)]	42	43	41
Convert current sound pressure level [dB (A)]	46	47	45

1) Sound level simulation calculation parameters: Laboratory surface area: 4894 m^2; Laboratory sound absorption coefficient a: 0.15; Environmental correction value K: 1.866; Sound absorption area A: 734.1 m^2.
2) The calculation method uses tested experimental results for conversion, and the calculation results are shown in Table 67.

Conclusion: The calculation results of the sound level meet the level$\leq$ 50dB (A) required by the tender.

8 DC Surge Arrester

The DC surge arrester is an important equipment of the ± 1100 kV UHV DC transmission project, which can effectively suppress lightning or switching overvoltage, and protect the equipment in the converter station from the impact of overvoltage caused by internal and external faults and other system disturbances. According to the purpose, it can be divided into two categories: DC converter station surge arresters and DC line surge arresters. The DC converter station surge arrester is detailed in Sect. 3 of Chap. 2, and this section focuses on the DC line surge arrester.

I. ± 1100 kV DC Line Surge Arrester with Gap

(I) Varistor Selection

The surge arresters made of high endurance varistor have been safely operated so far, and the situation is good. This time, the varistor used for the ± 1100 kV UHV DC line surge arrester is the internationally leading high-endurance DC system QE36 type varistor widely used in the ± 800, ± 660, ± 500 kV DC transmission system, and its main specifications are as follows:

(1) The size of varistor is 36 mm.
(2) The potential gradient is 220 V/mm.
(3) The absorption capacity of varistor is 41.35 kJ/piece.
(4) The 2 ms square wave withstand current capacity is 2000 A.
(5) The 4/10 μs high current impact capacity is 100 kA.

This high endurance varistor will be applied to the ± 1100 kV DC composite sheath line surge arrester.

(II) Electrical Parameter Design

1. Number of varistor

Each QE36 varistor has a DC 1 mA reference voltage center value of 7.80 kV. According to the minimum ratio of DC 2 mA to DC 1 mA reference voltage test data of 1.007, the DC 2 mA reference voltage target value of each high endurance varistor is 7.86 kV, then the surge arrester needs a total of $1250/7.86 \approx 159.03$ (pieces), rounded to 160 pieces.

The continuous operating voltage of each high endurance varistor is 7.02 kV, so the surge arrester needs a total of $1122/7.02 = 159.83$ pieces, rounded to 160 pieces.

2. Energy absorption capacity

According to the allowable absorption capacity of the varistor of 260 J/cm^3, the energy absorption capacity converted to the resistor chip is 41.35 kJ/piece, and the energy absorption capacity converted to the whole surge arrester is $160 \times 41.35 = 6616$ kJ ≈ 6.6 MJ, which meets the system's energy absorption requirements.

3. Impulse residual voltage calculation

As shown in Table 68, the given surge arrester has a nominal discharge current of 30 kA for lightning strikes. The pressure ratio of the QE36 type varistor under a lightning strike current of 40 kA is 1.842. Considering a grouping range of 1.04 times, 1247 kV/1.007 $\times$ 1.842 $\times$ 1.04 $\approx$ 2372 (kV) < 2500 kV. That is, even when calculated strictly according to 40 kA, the residual voltage of the entire surge arrester can meet the technical specification requirements.

The pressure ratio of the QE36 type varistor under a 40 kA steep wave impulse current is 1.974. Considering a grouping range of 1.04 times, the grouping range considered during grouping is 1.04, 1247 kV/1.007 $\times$ 1.974 $\times$ 1.04 $\approx$ 2542 (kV). That is, even when calculated strictly according to 40 kA, the steep wave residual voltage of the entire surge arrester can also meet the technical specification requirements.

4. Creepage distance

Due to the high pollution requirements of DC line insulators, a higher creepage ratio of 20 mm/kV is required, that is, the creepage distance required by the surge arrester is 1122 kV $\times$ 20 mm/kV = 22,440 (mm). The minimum creepage distance of the jacket designed according to the minimum height of the core body is 24,440 mm, which meets the creepage distance requirements.

5. Withstand voltage on surge arrester

The outer cover of the surge arrester body should be able to withstand a lightning impulse withstand voltage not less than 1.4 times the lightning impulse protection

Table 68 Type testing strategy of UHVDC line surge arrester

No.	Item		Testing sample	Quantity	Testing condition
1	Residual voltage test	Lightning	Varistor	12	5, 10, 15, 20, 40 kA
		Switching		12	500 A, 1 kA, 2 kA, 2.5 kA, 3 kA
		Steep wave		9	5, 10, 15, 20, 40 kA
2	Current impulse withstand test			3	2 ms square wave, 2000 A
				3	4/10 μs, 150 kA, 2 times
3	Action load test			3	6 set of energy, 75%, 30 min
4	Accelerated aging test			3	75% chargeability (DC 2 mA)
5	Appearance inspection of composite jacket		Main structure	1	Inspect surface defects of composite jacket
6	Main body creepage distance inspection			1	≥ 22 440 mm
7	DC reference voltage test			1	$U_{2mA.DC} \geq 1250$ kV
8	DC leakage current measurement			1	$0.75U_{2mA.DC} \leq 50$ μA
9	Partial discharge test				Power frequency test voltage is $1250 \times 0.75/\sqrt{2} \times 1.05$
10	Tensile load test			1	15 times self-weight
11	Moisture immersion test			1	Tensile load; cold and hot cycle; boil water for 42 h
12	Climate aging test			1	Consider not applying voltage after measuring the voltage divider ratio
13	Pressure release test			1	50 kA; 800 A
14	Sealing performance test			1	He mass spectrometer method; 6.65×10^{-5} (Pa L)/s
15	External insulation of main structure	DC withstand	Main structure and external Jacket	1	1683 kV, 1 min, humidity withstand
		Switching withstand		1	–
		Lightning withstand		1	1.3 times residual voltage, Positive and negative polarities with 15 times
16	Lightning impulse discharge voltage			1	3700 kV, $U_{50\%}$

(continued)

Table 68 (continued)

No.	Item	Testing sample	Quantity	Testing condition
17	Lightning impulse volt-second characteristics		1	Time is 1–10 μs, 4 points; actual installation
18	DC withstand voltage (wet)		1	Normal ± 1347 kV, fault ± 1122 kV
19	Switching impulse voltage (wet)		1	2000 kV, 15 times, Positive and negative polarities

level (corresponding to the nominal current of 30 kA), and a DC voltage not less than 1.5 times the continuous operating voltage.

The lightning impulse protection level of the surge arrester under 30 kA is 2500 kV, that is, the required lightning withstand voltage is 3500 kV. After calculation, the body can withstand a lightning impulse voltage of 3500 kV, which meets the requirements.

The continuous operating voltage of the surge arrester is 1122 kV (DC voltage), and the required DC withstand voltage is 1683 kV. After calculation, the withstand DC voltage of the body unit (1 section) is 705 kV, and the total DC withstand voltage of 4 sections is about 2820 kV, which is higher than the required value and meets the requirements.

6. Discharge voltage and withstand voltage of the entire surge arrester

The positive polarity lightning impulse 50% discharge voltage of the surge arrester should not exceed 3900 kV. The DC voltage withstand performance of the entire surge arrester considers two situations: body failure short circuit and non-failure short circuit:

(1) The DC withstand voltage under the condition of non-failure short circuit of the surge arrester body is not less than 1347 kV, and the DC withstand voltage under the condition of body failure short circuit is not less than 1122 kV.
(2) The positive polarity operating impulse withstand voltage of the entire surge arrester should ideally not be less than 2000 kV.

The above electrical performance will be met by adjusting the air gap distance.

(III) Mechanical Structure Design

The surge arrester is suspended and installed at the top of the pole tower. During the design, the tensile and bending strength of the body need to be considered. In addition, the height of the ± 1100 kV UHV DC transmission line pole tower is generally above 60 m, even exceeding 100 m. Due to the influence of wind, the top of the pole tower is in a state of micro-vibration for a long time, and the surge arrester fixed at the top of the pole tower will also vibrate with it. Therefore, the anti-vibration characteristics need to be considered when designing the surge arrester.

1. Tensile strength calculation

For the suspended composite sheath surge arrester, according to DL/T 815—2012 “Polymer-housed Metal Oxide Surge arresters for AC Transmission Lines”, the surge arrester should withstand at least 15 times the rated tensile load of the surge arrester’s own weight for 1 min without damage. The ± 1100 kV polymer-housed type line surge arrester uses a high-strength composite sheath. After reasonable optimization design, the total weight of the surge arrester body and the discharge gap ring is 675 kg. The tensile load applied to the surge arrester is 99.3 kN, and the tensile destruction load of the composite sheath used is 110 kN. The margin is 1.11 times, which can meet the tensile requirements.

2. Bending strength calculation

According to DL/T 815—2012, due to its own characteristics, the line polymer-housed surge arrester can be designed into two types of installation structures: suspension and horizontal. Its mechanical performance requirements vary with the structure type: when the surge arrester is installed in suspension, the mechanical performance is mainly assessed by the tensile load test; when the surge arrester is installed horizontally, it is mainly assessed by the bending load test; although the suspended surge arrester does not need to check the bending strength, considering that the UHV polymer-housed line surge arrester is installed on a pole tower of 60–100 m high, the wind pressure acting on the surge arrester at this time is very large, so it needs to be calculated according to the surge arrester standard requirement of 35 m/s. After calculation, the wind bending moment borne by the surge arrester is 6.0 kN·m, and the rated bending load of the polymer-housed sheath used MML is 6.8 kN·m, the margin is 1.13 times, when the sheath’s 1.5 MML allowable bending composite mechanical strength, the margin is 1.7 times, which can meet the bending strength requirements.

3. Calculation of surge arrester pressure release performance

The pressure release device of the UHV polymer-housed line surge arrester element introduces TOSHIBA’s unique design structure, which is widely used in AC 35–1000 kV porcelain sheath surge arrester products. Calculations show that when passing through a short-circuit current of 0.2 s, 50 kA, the actual maximum internal pressure of the polymer-housed sheath is 7.13 MPa, and the guaranteed value of the internal pressure of the polymer-housed sheath is 9.0 MPa. The safety margin is more than 1.26 times, which can ensure normal pressure release of the product and fully meet the product’s pressure release performance requirements. In addition, the AC 1000 kV UHV DC line surge arrester jointly researched by China Electric Power Research Institute and PTA uses the same sheath structure design, and the unit section length exceeds the unit section of the DC ± 1100 kV UHV DC line surge arrester developed this time. The related AC line surge arrester has passed the pressure release test.

(IV) Prototype Design Scheme

The prototype is a series gap structure (pure air gap), the body is a polymer-housed surge arrester, the body height is about 7.7 m, the body and electrode weight is about 700 kg, the gap distance is 2.8 m ± 0.2 m. The surge arrester installation schematic and structure size diagram are shown in Figs. 117 and 118.

(V) Experiment

1. Overall test plan

The overall test plan was formulated by integrating the relevant standards of China's external line surge arresters and combining the characteristics of the ± 1100 kV UHV DC line surge arresters. All tests were conducted on the prototype according to this test plan. The test results show that the developed prototype meets the performance parameter requirements of the ± 1100 kV UHV DC line surge arresters.

The type test items can be divided into body tests, body proportion unit tests, whole arrester electrical tests, whole arrester mechanical performance tests, pressure release tests, etc., depending on the test sample. The specific test plan for the ± 1100 kV UHV DC line surge arresters is shown in Table 68.

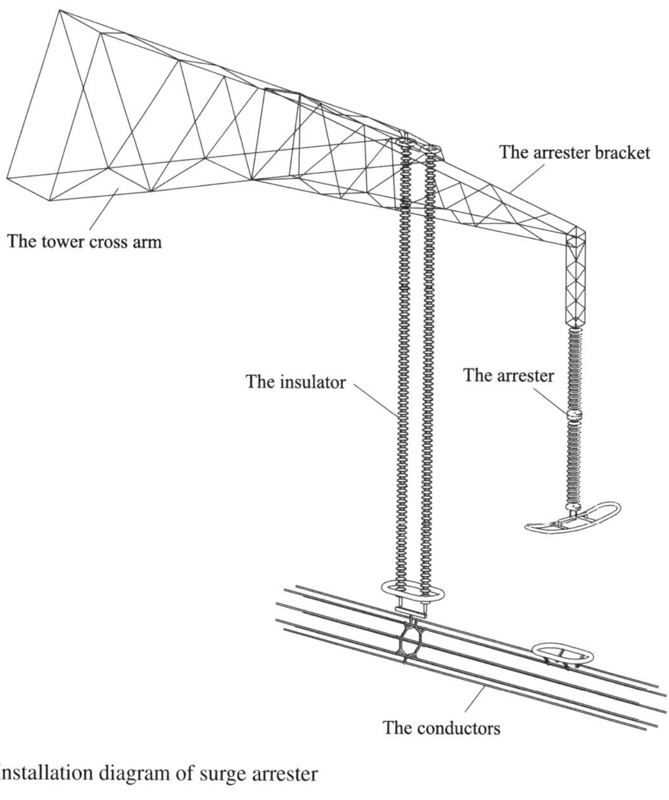

Fig. 117 Installation diagram of surge arrester

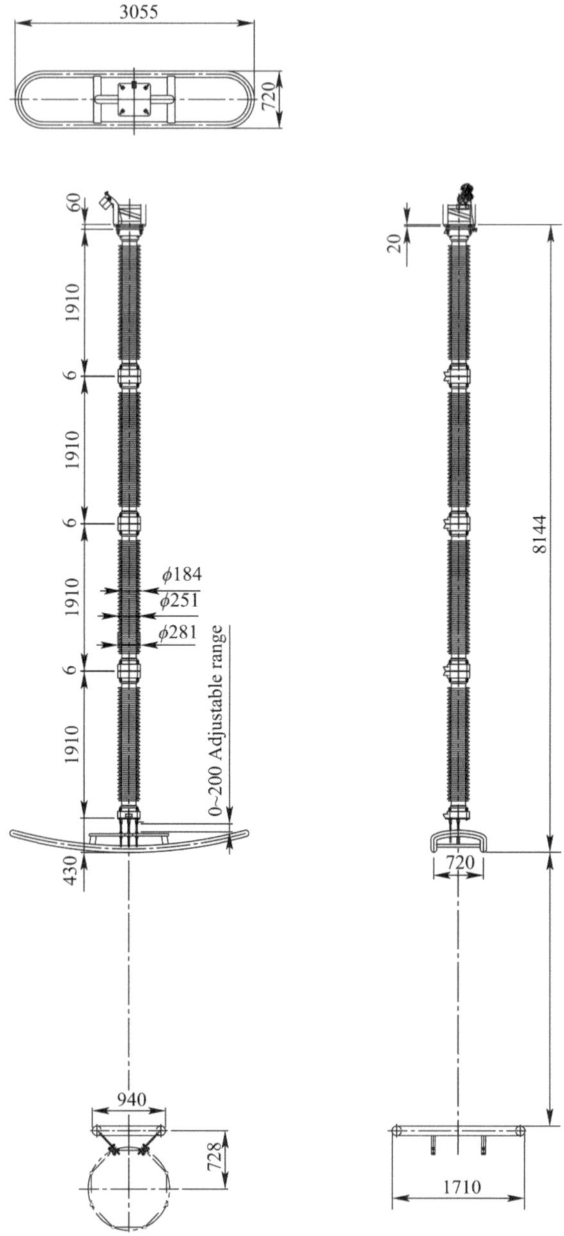

Fig. 118 Size and structure diagram of surge arrester

2. Residual voltage test

The specified lightning impulse residual voltage should not exceed 2500 kV (peak), and the steep-front impulse residual voltage should not exceed 2750 kV.

The test results conducted according to the regulations show that the lightning impulse residual voltage is 2356 kV (peak), and the steep wave impulse residual voltage is 2523 kV (peak), which meet the technical requirements. The data listed in Table 69, 70, and 71 are the test data.

3. DC reference voltage test

According to the characteristics of varistor of the developed ± 1100 kV line surge arrester, the DC reference current is selected as 2 mA, and the DC reference voltage is not less than 1247 kV.

Table 72 provides the DC reference voltage test data, and the results show that the whole DC reference voltage is 1267.7 kV, which meets the requirements.

4. Leakage current test under 0.75 times DC reference voltage

Table 73 provides the leakage current under 0.75 times the DC reference voltage, and the results show that the maximum leakage current in the four unit sections is 27 μA, which meets the requirements.

5. Internal partial discharge test

Table 74 provides the internal partial discharge test data, and the results show that the maximum partial discharge in the four unit sections is 5pC, which meets the requirements.

Table 69 Testing data of lightning impulse residual voltage

No.	Current value	Residual voltage of testing sample 1		Residual voltage of testing sample 2		Residual voltage of testing sample 3		Residual voltage of whole product (kV)
		Sample	Whole product	Sample	Whole product	Sample	Whole product	
1	1 mA (DC)	7.93	1267.7	7.9	1267.7	7.92	1267.7	1267.7
2	5 kA (8/20 μs)	11.75	1878.4	11.7	1877.5	11.7	1872.7	1878.4
3	10 kA (8/20 μs)	12.43	1987.1	12.43	1994.6	12.43	1989.6	1994.6
4	15 kA (8/20 μs)	13.01	2079.8	13.01	2087.7	13.01	2082.4	2087.7
5	20 kA (8/20 μs)	13.43	2146.9	13.37	2145.5	13.37	2140.0	2146.9
6	40 kA (8/20 μs)	14.68	2346.8	14.68	2355.7	14.68	2349.7	2355.7

Table 70 Testing data of switching impulse residual voltage

No.	Current value	Residual voltage of testing sample 1		Residual voltage of testing sample 2		Residual voltage of testing sample 3		Residual voltage of whole product (kV)
		Sample	Whole product	Sample	Whole product	Sample	Whole product	
1	1 mA (DC)	7.93	1267.7	7.9	1267.7	7.92	1267.7	1267.7
2	500 A (30/60 μs)	10.29	1645.0	10.29	1651.2	10.24	1639.0	1651.2
3	1000 (30/60 μs)	10.6	1694.5	10.55	1692.9	10.5	1680.7	1694.5
4	2000 (30/60 μs)	11.07	1769.7	11.07	1776.4	11.07	1771.9	1776.4
5	2500 (30/60 μs)	11.18	1787.2	11.18	1794.0	11.18	1789.5	1794.0
6	3000 (30/60 μs)	11.34	1812.8	11.34	1819.7	11.34	1815.1	1819.7

Table 71 Testing data of steep-front impulse residual voltage

No.	Current value	Residual voltage of testing sample 1		Residual voltage of testing sample 2		Residual voltage of testing sample 3		Residual voltage of whole product (kV)
		Sample	Whole product	Sample	Whole product	Sample	Whole product	
1	1 mA (DC)	7.93	1267.7	7.93	1267.7	7.93	1267.7	1267.7
2	5 kA (1/3 μs)	12.43	1987.1	12.43	1994.6	12.43	1989.6	1994.6
3	10 kA (1/3 μs)	13.27	2121.4	13.22	2121.4	13.22	2116.0	2121.4
4	15 kA (1/3 μs)	13.79	2204.5	13.79	2212.9	13.79	2207.3	2212.9
5	20 kA (1/3 μs)	14.31	2287.6	14.31	2296.3	14.26	2282.5	2296.3
6	40 kA (1/3 μs)	15.72	2513.0	15.72	2522.6	15.72	2516.2	2522.6

Table 72 Testing data of DC reference voltage

Section	DC 2 mA (kV)
1	316.3
2	316.7
3	317.3
4	317.4

Table 73 Testing data of 0.75 times DC reference voltage

Section	Leakage current at 0.75 times DC reference voltage (μA)
1	26
2	25
3	27
4	26

Table 74 Testing data of internal partial discharge

Section	Partial discharge (4pC)
1	5
2	5
3	5
4	5

6. Creepage distance check

GB/T 11032—2020 "Metal Oxide Surge Arresters without Gaps for AC Systems" and the technical specifications of the ± 1100 kV UHV DC line surge arresters stipulate: the creepage distance of the outer insulating part should not be less than 22,440 mm.

The measured creepage distance is 24,668 mm, which meets the standard requirements.

7. Polymer-housed jacket appearance inspection

As stipulated by GB/T 11032—2020 and ± 1100 kV UHV DC Line Surge Arrester Technical Specifications: Inspect the surface defects of the polymer-housed jacket. The area of a single defect on the jacket surface (such as missing glue, impurities, protrusions, etc.) should not exceed 25 mm^2, the depth should not be more than 1 mm, the protruding surface and the seam should be cleaned and leveled, the height of the protrusion should not exceed 0.8 mm, the height of the adhesive seam protrusion should not exceed 1.2 mm, and the total defect area should not exceed 0.2% of the total surface of the jacket.

The actual measurement shows that the appearance of the jacket meets the requirements.

Table 75 Testing data of sealing

Section	Leakage rate of sealing [(Pa L)/s]
1	1.8×10^{-5}
2	1.0×10^{-5}
3	2.2×10^{-5}
4	1.7×10^{-5}

8. Sealing test

The test is conducted using a helium mass spectrometer leakage detector. Table 75 provides the sealing test data, and the results show that the maximum partial discharge in the four unit sections is 5pC, which meets the requirements.

9. External housing dielectric test

According to GB/T 11032—2020 "Metal Oxide Surge Arresters without Gaps for AC Systems", during the lightning impulse withstand test, the surge arrester is subjected to 15 positive polarity lightning impulse voltage waves. The test voltage and waveform meet the requirements, that is, the peak (3062 kV) deviation does not exceed ± 3%, the wavefront time deviation does not exceed ± 30%, the half-peak time deviation does not exceed ± 20%, and there is no external insulation flashover or breakdown during the test. The test passed.

During the DC voltage withstand test, a DC voltage of 1690 kV is applied to the body of the surge arrester for 1 min; the test simulates rain conditions, and there is no external insulation flashover or breakdown during the test. The test passed.

10. High current impulse withstand test

The test results conducted according to GB/T 11032—2020 show: there was no breakdown, flashover, or damage during the test, and the maximum change rate of residual voltage under the nominal discharge current before and after the test was only 0.59%. The test passed. Table 76 provides the high current impulse withstand test data.

Table 76 Testing data of high current impulse withstand

No.		DZ-1	DZ-2	DZ-3
DC reference voltage		7.95	7.95	7.95
Residual voltage before test (kV, peak value)		13.59	13.57	13.56
High current impulse	First time (kA, peak value)	100.2	100.2	100.2
	Second time (kA, peak value)	100.2	100.2	100.2
Residual voltage after test (kV, peak value)		13.62	13.61	13.64
Percentage change in residual voltage (%)		+ 0.22	+ 0.29	+ 0.59

11. Rectangular surge discharge current test

The test results conducted according to GB/T 11032—2020 show: there was no breakdown, flashover, or damage during the test, and the maximum change rate of residual voltage under the nominal discharge current before and after the test was less than 5%. The test passed.

12. Action load test

Tables 77, 78, and 79 provide the aging test data, energy withstand test data, and action load test data respectively, and Fig. 119 provides the heat dissipation curve of the proportional unit and the entire surge arrester during the heat dissipation test.

13. Humidity immersion test

During the actual test, a tensile force of 31 kN was applied, and the tensile load remained unchanged until the end of the test. The electrical parameters before and after the test meet the standard requirements. The electrical parameters before and after the test are shown in Table 80.

14. Lightning impulse discharge voltage test

During the test, the surge arrester is arranged according to the actual installation situation, including the body, gap, simulated cross arm, simulated wire, polymer insulator or porcelain insulator.

The test shows that when the gap distance of the ± 1100 kV line surge arrester is 2.7 m, the positive polarity lightning impulse 50% discharge voltage $U_{50\%} =$ 3739 kV, when the gap distance is 2.6 m, the positive polarity lightning impulse 50% discharge voltage $U_{50\%} = 3335$ kV. According to linear conversion, when the gap distance is 2.74 m, the positive polarity lightning impulse 50% discharge voltage of the entire surge arrester should not exceed 3900 kV, which meets the requirements.

15. Lightning impulse volt-second characteristics test

The test is conducted according to the test method specified in IEC 60099–8:2017 "Metal-oxide surge arresters with external series gap (EGLA) for overhead transmission and distribution lines of a.c. systems above 1 kV". During the lightning impulse

Table 77 Testing data of aging

Sample No.	DC reference voltage U_{1mA} (kV, peak value)	Testing voltage U_{TA} (kV, RMS)	Chargeability (%)	Initial power P_{1TA} (W)	Final power P_{2TA} (W)	Temperature (°C)	Duration (h)
1	5.37	4.57	85	4.30	1.74	115	1008
2	5.37	4.57	85	2.79	1.02	115	1008
3	5.35	4.55	85	2.50	0.99	115	1008

Table 78 Testing data of energy withstand

Sample		Testing data: current impulse *I*, residual voltage *U*和 absorbing energy *W*								
1	*I* (A)	2575	2525	2539	2525	2462	2535	2520	2459	2426
	U (kV)	7.36	7.50	7.56	7.38	7.52	7.62	7.45	7.56	7.60
1	*W* (kJ)	45.58	45.55	46.12	44.95	44.64	46.55	45.22	44.82	44.48
	I (A)	2512	2454	2425	2507	2443	2415	2551	2521	2471
	U (kV)	7.41	7.56	7.62	7.39	7.51	7.58	7.32	7.46	7.51
	W (kJ)	44.91	44.73	44.50	44.69	44.26	44.11	45.06	45.39	44.71
2	*I* (A)	2513	2471	2442	2535	2484	2476	2514	2574	2533
	U (kV)	7.42	7.54	7.61	7.37	7.51	7.58	7.35	7.51	7.59
	W (kJ)	44.94	44.93	44.74	45.07	45.02	45.21	44.62	46.60	46.30
	I (A)	2514	2519	2493	2499	2425	2414	2527	2474	2451
	U (kV)	7.35	7.45	7.45	7.40	7.50	7.57	7.40	7.49	7.52
	W (kJ)	44.62	45.27	45.07	44.58	43.88	44.04	45.10	44.71	44.47
3	*I* (A)	2529	2464	2515	2522	2489	2457	2511	2451	2440
	U (kV)	7.40	7.54	7.63	7.42	7.52	7.59	7.42	7.54	7.59
	W (kJ)	45.17	44.83	46.23	45.14	45.10	44.94	44.89	44.57	44.64
	I (A)	2510	2511	2468	2487	2445	2414	2505	2472	2442
	U (kV)	7.36	7.42	7.53	7.39	7.52	7.58	7.41	7.51	7.54
	W (kJ)	44.57	44.89	44.83	44.32	44.34	44.18	44.73	44.75	44.41

volt-second characteristics test, the discharge path is all along the two electrodes of the surge arrester.

Table 81 provides the test data of the lightning impulse volt-second characteristics of the entire surge arrester.

16. Switching impulse voltage withstand test

The test results show that when the gap distance is 2.3 m and the body is not short-circuited, the switching overvoltage that the entire surge arrester can withstand is > 2000 kV, which meets the requirements.

17. DC voltage withstand test

The test results show that when the gap distance is 2 m and the body is not short-circuited, the positive or negative DC voltage that the entire surge arrester can withstand is not less than 1683 kV.

When the gap distance is 2.6 m and the body fails and short-circuits, the positive or negative DC voltage that the entire surge arrester can withstand is not less than 1122 kV. The test results meet the requirements.

Table 79 Testing data of action load

Sample No.		DZ-1	DZ-2	DZ-3
DC reference voltage U_{1mA} (kV, DC)		5.32	5.35	5.33
8/20 μs residual voltage (kV, peak value)		8.20	8.15	8.15
Continuous operating voltage of sample U_C (*CCOV*) (kV, DC)		4.52	4.55	4.53
Increased continuous operating voltage of sample U_{C*} (kV, RMS)		4.52	4.55	4.53
4/10 μs, 100 kA	Current value of first time (kA, peak value)	102	102	102
2 times impulse withstand	Current value of second time (kA, peak value)	101	102	102
Temperature of preheated test sample (°C)		61	61	61
2 times energy withstand	Energy withstand of first time (kJ)	44.91	44.48	44.58
	Energy withstand of second time (kJ)	44.74	44.47	44.70
After the second energy withstand, apply test substance for no more than 100 ms U_{C*}				
Continuous operation	Actually applied U_C (kV, DC)	4.52	4.55	4.53
Power loss under U_C (W)	Initialization	48.8	43.5	41.1
	10 min	13.9	12.0	16.1
	30 min	4.0	10.4	11.3
After test, residual voltage under 8/20 μs (kV, peak value)		8.31	8.31	8.36
Change rate of residual voltage before and after test (%)		+ 1.27	+ 1.92	+ 2.56
Situation of sample after test		No breakdown or flashover	No breakdown or flashover	No breakdown or flashover

18. Mechanical test

According to GB/T 11032—2020 requirements, the surge arrester should be able to withstand a tensile load of 15 times its own weight and a bending load of 2.5 times the wind pressure calculated under normal use conditions.

The body weight is about 675 kg, and the tensile load applied according to GB/T 11032—2020 should not be less than $675 \times 15 \times 9.8 = 99{,}225$ (kN). During the actual test, a tensile force of 93 kN was applied for a duration of 1 min.

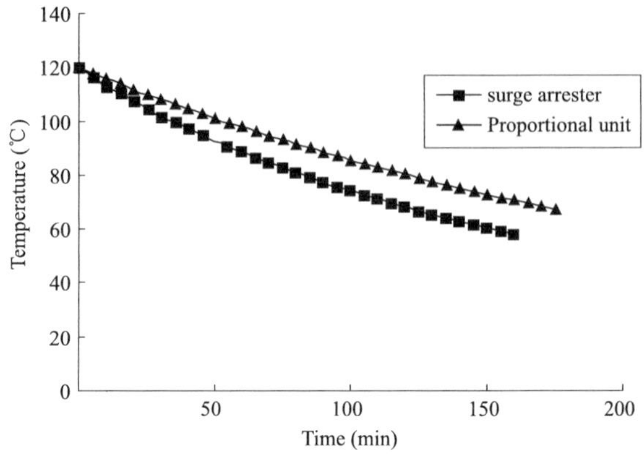

Fig. 119 Proportional unit and whole surge arrester heat dissipation curve in heat dissipation test

Table 80 Comparison of electrical parameters before and after test

Testing item	Before test	After test	Change quantity
U_{2mADC} (kV)	133.0	132.9	0.1
Leakage current under 0.75 times U_{2mADC} (μA)	7	9	2
Partial discharge (pC)	< 10	< 10	< 10

Table 81 Testing data of volt-second characteristic of entire surge arrester under lightning impulse (kV)

Positive/negative polarity	No.	1	2	3	4	5	平均
Positive polarity	Time 1	8.9	9.48	9.15	10.0	9.95	9.50
	Voltage 2	3726	3754	3755	3752	3745	3746
	Time 2	6.19	6.07	6.70	6.42	6.34	6.34
	Voltage 2	3965	3981	3990	3982	3971	3978
	Time 3	7.55	7.92	7.92	7.51	7.69	7.72
	Voltage 3	3861	3849	3805	3819	3831	3833
Negative polarity	Time 1	8.91	8.42	8.97	7.56	7.85	8.34
	Voltage 2	3836	3861	3859	3898	3885	3868
	Time 2	6.61	7.00	6.39	7.01	6.79	6.76
	Voltage 2	4050	4052	4039	4065	4063	4054
	Time 3	5.37	5.78	5.47	5.03	5.66	5.46
	Voltage 3	4276	4283	4227	4267	4227	4256

After the test, the partial discharge was less than 10pC, the DC reference voltage remained unchanged, meeting the requirements.

19. Pressure release test

(a) Pressure release test of small current (b) Pressure release test of large current

Fig. 120 Pressure release test of line surge arrester

The test setup was in accordance with the requirements of GB/T 11032—2020, for specific setup refer to Fig. 120.

The test showed that both the high current pressure release test (50 kA) and the low current pressure release test (800 A) did not cause the test sample to explode or rupture, meeting the requirements, and the test passed.

II. ± 1100 kV DC Line Surge Arrester without Gap

(I) Selection of Large Capacity Varistor

The Chinese power industry has developed surge arresters for AC systems with voltage levels of 1000 kV and below, and for DC systems with voltage levels of ± 800 kV and below. Based on the operation of Chinese surge arresters and the technical requirements of DC bus surge arresters, a large capacity varistor consistent with the ± 800 and ± 1100 kV DC bus surge arresters was selected. The specifications and characteristics of this resistor chip are as follows:

(1) Varistor: ϕ100 mm × 22 mm.
(2) Potential gradient: 230 V/mm.
(3) DC 1 mA reference voltage (center value): 5.1 kV.
(4) 2 ms square wave withstand current capacity: 2800 A.
(5) 4/10 μs high current impact capacity: 100 kA.

(II) Varistor Configuration

According to the requirement that the DC reference voltage of the surge arrester should not be less than 1308 kV, the surge arrester needs a total of 1308 × 1.015/5.18 ≈ 256 chips (1.015 times is the control value required by the routine test of the surge arrester, 5.18 is the average value of the DC reference voltage of a single varistor); the charge rate of the surge arrester $CCOV/U_{1mADC}$ = 1122/(256 × 5.18) = 84.6 < 85%, meeting the design requirements, so the minimum number of surge arrester valve disc in series is 256.

According to the requirement that the energy absorption of the surge arrester should not be less than 2000 kJ, the energy absorption capacity of a single PTA varistor is 44.5 kJ, and the energy absorption capacity of a single column is 44.5 × 256 = 11,392 (kJ) > 2000 kJ, a single column can meet the energy requirements.

Based on the structure of 256 varistor in series and 1 in parallel, the residual voltage of the surge arrester is detailed in Table 82.

The residual voltage of lightning (8/20 μs, 20 kA) should not exceed 2200 kV (peak), with a residual voltage margin of 1.016. The switching (30/60 μs, 1 kA) should not exceed 1826 kV (peak), with a residual voltage margin of 1.004. The steep wave (1/10 μs, 20 kA) should not exceed 2446 kV (peak), with a design margin of 1.069, which can meet the design requirements.

(III) Key Electrical Performance

1. High current impulse tolerance

   (1) Test requirements: The test sample consists of 3 proportional units (varistor), continuously subjected to 2 times of 100 kA high current impulse, with a time interval of 1 min between each two impulses.
   (2) Test judgment: The change in reference voltage before and after the test should not exceed 5%, and the change in residual voltage under 500 A switching impulse current should not exceed - 2% to + 5%. After the test, the test sample should be inspected, and the varistor should have no breakdown, flashover, or other obvious damage traces.
   (3) Test results: See Table 83.

The test results show that the varistor meets the requirements of high current 100 kA.

2. Rectangular wave impulse current tolerance

   (1) Test requirements: The test sample consists of 3 proportional units (varistor), continuously subjected to 18 rectangular waves of 2 ms, with an amplitude not less than 2000 A.
   (2) Test judgment: The change in reference voltage before and after the test should not exceed 5%, and the change in residual voltage under 500 A switching impulse current should not exceed − 2%– + 5%. After the test, the test sample should be inspected, and the varistor should have no breakdown, flashover, or other obvious damage traces.
   (3) Test results: The test was passed.

Table 82 Testing results of proportional unit in varistor

Sample No.	U_{1mADC}	Residual voltage of varistor			Ratio coefficient	Residual voltage of surge arrester		
		Lightning 8/20 μs (20 kA)	Switching 30/60 μs (0.5 kA)	Steep-front 1/10 μs (20 kA)		Lightning 8/20 μs (20 kA)	Switching 30/60 μs (0.5 kA)	Steep-front 1/10 μs (20 kA)
	(kV, DC)	(kV, peak value)	(kV, peak value)	(kV, peak value)	n	(kV, peak value)	(kV, peak value)	(kV, peak value)
7	5.21	8.44	6.56	8.92	254.8	2150.5	1671.5	2272.8
8	5.21	8.50	6.58	8.98	254.8	2165.8	1676.6	2288.1
9	5.18	8.38	6.5	8.9	256.3	2147.8	1666.0	2281.0

Table 83 High current impulse withstand test

Sample No.	18	19	20	Required value
U_{1mADC} (kV)	5.21	5.20	5.21	–
First impulse (kA)	100.0	100.4	100.2	100.0
Second impulse (kA)	100.2	100.0	100.2	100.0
Residual voltage of switching impulse before test (kV)	6.54	6.53	6.54	–
Residual voltage of 500 A Switching impulse after test (kV)	6.44	6.45	6.44	–
Change rate of residual voltage (%)	− 1.5	− 1.2	− 1.6	− 2 to + 5
External view after test	Qualified	Qualified	Qualified	Qualified

The test results show that the varistor meets the requirements of 2 ms rectangular wave 2000 A.

3. Operating duty test

The varistor is tested according to the operating duty test procedure in GB/T 22389—2008 "Guidelines of metal oxide surge arresters without gaps for HVDC converter stations", as follows.

(1) Operating duty test procedure:

1) Accelerated aging test, DC voltage charge rate 95%;
2) Thermal dissipation characteristics of the test sample;
3) Residual voltage test under 10 kA lightning impulse current;
4) 6 sets of energy tolerance, 3 times each set, cooled to ambient temperature between each set (each energy injection 45 kJ/2 ms);
5) Cooled to ambient temperature 25 ± 10 °C, assembled into 3 proportional units;
6) 2 times of 100 kA high current impulse (4/10 μs), with an interval of 50–60 s between the two impulses;
7) Assembled into a thermal proportional unit, preheated to 60 ± 3 °C;
8) 2 times of energy tolerance (each energy injection 45 kJ/2 ms), with an interval of 50–60 s between the two impulses;
9) Apply within 100 ms *CCOV* (DC Voltage, 85% charge rate), continuous for 30 min;
10) 10 kA lightning impulse residual voltage test, sample inspection.

(2) Test results: The valve parameters before and after the test meet the requirements, and the test is passed.

(IV) Introduction to the Basic Composition

The surge arrester is composed of 7 components in series, the inner diameter of the component bushing is 300 mm, the total height of the surge arrester is 19,660 mm, each component of the surge arrester adopts a single column structure, and the external dimensions of the DC line surge arrester are detailed in Fig. 121.

(V) External Insulation Design

1. External insulation creepage distance

The CG project requires a creepage distance of 63,910 mm, the creepage distance of the first component of the surge arrester is 9800 mm, the creepage distance of the 2nd to 7th sections of the surge arrester is 9280 mm, the design creepage distance of the surge arrester is 65,480 mm, which meets the creepage distance requirements.

2. External insulation withstand voltage

The requirements for the external insulation of the DC line surge arrester are as follows.

(1) Lightning impulse withstand voltage on external insulation. The lightning impulse withstand voltage of the surge arrester should not be less than 4500 kV, at altitude 2500 m, correction factor: 1.202, test requirement value: 5409 kV (peak). The lightning impulse withstand test is conducted in accordance with IEC 60099-4:2014 "Metal-oxide surge arresters without gaps for a.c. systems" 8.2.6 requirements if the sum of the dry arc distance or partial dry arc distance (in m) is greater than the test voltage (in kV) divided by 500 kV/m, then this test is not required.

The lightning impulse withstand voltage should not be less than 5409 kV (peak), then the dry arc distance d should be greater than 10.8 m (d = test voltage/500 = 5409/500 = 10.8); the measured dry arc distance is 18.8 m, so this external insulation can meet the requirements.

(2) External insulation switching impulse withstand voltage. The switching impulse withstand voltage of the surge arrester should not be less than 2200 kV, at altitude 2500 m, correction factor: 1.057, test requirement value: 2325 kV (peak).

The switching impulse withstand voltage test is conducted in accordance with IEC 60099—4:2014 requirements if the sum of the dry arc distance or partial dry arc distance (in m) is greater than the formula $d = 2.2 \times (e^{U/1069} - 1)$, then this test is not required.

If the switching impulse withstand voltage is not less than 2325 kV, then the dry arc distance d should be greater than 17.2 m[$d = 2.2 \times (e^{2325/1069} - 1)$].

The measured dry arc distance is 18.8 m, so the external insulation can meet the requirements.

(3) DC withstand voltage of external insulation. The DC (wet) withstand voltage of the arrester should not be less than 1150 kV, at altitude 2500m, correction factor: 1.202, test requirement value: 1382kV (peak value).

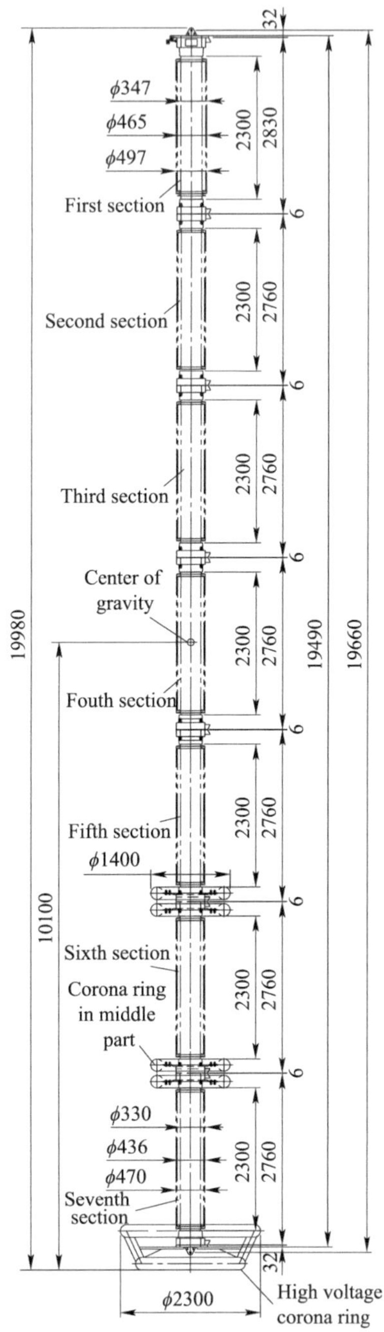

Fig. 121 External configuration of DC line surge arrester

(VI) Mechanical Structure Design

1. Pressure release performance

The pressure release current requires a large current of 40 kA and a small current of 600 A.

The arrester element used in the DC line is a polymer-housed element structure commonly used in UHV DC transmission projects. The technology is mature. This house element underwent a pressure release test at the Toshiba High Power Test Station in Japan in 2015 and passed the 65 kA, 25 kA, 12 kA, 600 A pressure release test. The situation before and after the test is shown in Fig. 122.

2. Tensile strength

According to the requirements of GB 11032—2020, the tensile load of the suspension type arrester should not be less than 15 times the weight of the arrester. The weight of the arrester is 2500 kg, that is, the tensile strength of the arrester should not be less than $15 \times 2500 \times 9.8 = 367\ 500$ (N) = 367.5 kN. The design value of the tensile strength of the arrester jacket is 400 kN, which meets the design requirements.

(VII) Installation Method of Arrester on Straight Line Tower

It is recommended to install the arrester on the outrigger bracket on both the straight line tower and the tension tower. The arrester on the straight line tower is suspended directly above the wire, and the arrester is connected to the special connecting plate through the hinge fitting; the arrester on the tension tower is connected to the cage-type hard jumper through the fitting and the clamp interval rod.

The grounding terminal of the arrester is connected to the outrigger arrester bracket through the hinge fitting, and the HV terminal is connected to the special connecting plate through the hinge fitting. Considering the complexity of the eight bundled wires system, the HV terminal of the arrester is only connected to four sub-wires with a special connecting plate to reduce the impact of the wire system on the arrester. Under normal conditions, the arrester itself needs to withstand wind load, and its HV terminal hinge fitting needs to have a margin to reduce the impact of the arrester wind load on the wire system and the impact of the load change of the wire system under various conditions on the arrester, as shown in Fig. 123.

When the surge arrester is suspended directly above the conductor, the projection length of the upper layer conductor V string is 12.4 m, which is less than the length between the single column surge arrester hanging points of 19.66 m. Therefore, the suspended surge arrester adopts the method of designing an overhanging bracket on the cross arm of the straight line tower conductor, and the hanging point of the surge arrester on the overhanging bracket needs to be raised by 8.0 m from the bottom surface of the cross arm (V string hanging point). The difference in the remaining

(a) Before 65kA test

(b) After 65kA test

(c) Before 25kA test

(d) After 25kA test

Fig. 122 Photos of pressure release test

(e) Before 12kA test

(f) After 12kA test

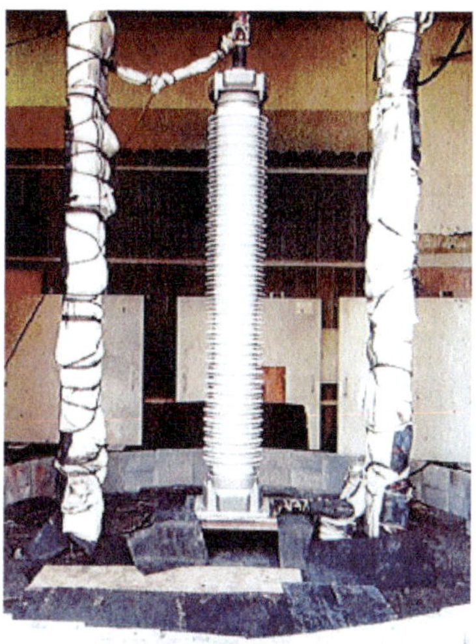

(g) Before 600A test

(h) After 600A test

Fig. 122 (continued)

length can be adjusted by adjusting the length of the hinge hardware, so that the distance from the overhanging bracket to the conductor is comparable to the length of the surge arrester.

Fig. 123 Installation method of straight line tower

The air gap values between the sectional surge arrester flange and the iron tower (grounding piece) are as follows: 1.3 m between the 7th and 6th sections, 2.6 m between the 6th and 5th sections, 3.9 m between the 5th and 4th sections, and 5.2 m between the 4th and 3rd sections. The surge arrester is divided into 7 sections, each section is 2.76 m long, and the sectional surge arrester flange has a pressure release port. During construction and installation, the pressure release port should be arranged to face away from the iron tower to increase the safety distance.

1. Surge arrester connection hardware

    (1) Design of drainage wire type and connection hardware. Under the switching overvoltage condition, the current flowing through the surge arrester is less than 300–500 A, so it is recommended to use a drainage wire to connect the line conductor and the surge arrester. The drainage wire uses LJ − 300 aluminum stranded wire, with a current carrying capacity of 550 A. To reduce the impact of corona, 2 drainage wires are used. The lower terminal of the drainage wire is crimped with a TL type wire clamp by hydraulic means, and the bolt section of the TL type wire clamp is installed on the upper sub-conductor between the surge arrester and the iron tower.

The upper terminal of the drainage wire is connected to the HV terminal (lower terminal) of the surge arrester with an SY hydraulic device wire clamp. The hardware manufacturer punches the SY drainage plate according to the surge arrester grounding terminal diagram before leaving the factory, and configures bolts and nuts. The SY drainage plate has a tilt angle of 30°, as shown in Fig. 124.

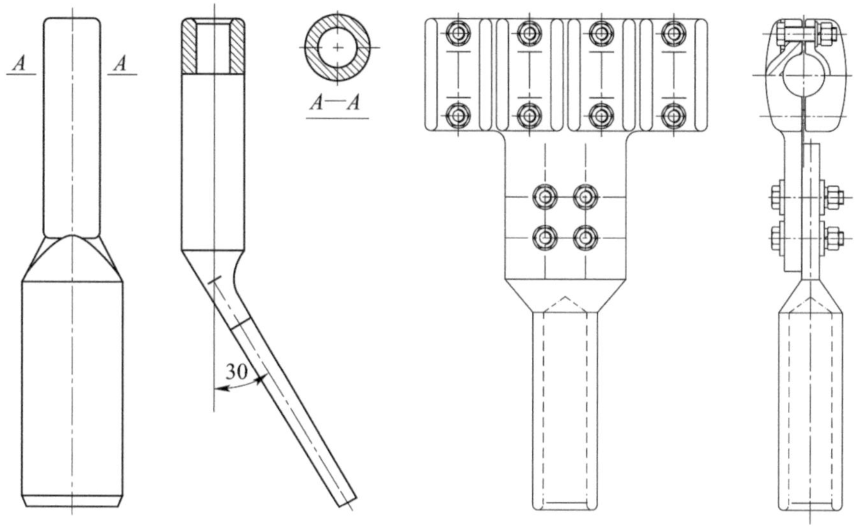

(a) SY hydraulic equipment clamp (b) TL type wire clamp

Fig. 124 Metallic connection of incoming feeder of HV terminal of surge arrester

The length of the drainage wire should be slightly longer than the hinge hardware, that is, while keeping the drainage wire unforced, the construction unit should compare and confirm on site before construction that the actual length of the hinge hardware installed at the HV terminal (lower terminal) of the surge arrester is less than the length of the drainage wire. For the two drainage wires and hinge hardware, care should be taken during installation to maintain a certain distance, they should not be entangled with each other, and both the drainage wire and the TL type wire clamp should be within the range covered by the surge arrester corona ring, and the remaining length of the drainage wire should be on the inside of the eight bundled conductor.

(2) Design of hinge connection hardware. When the surge arrester is suspended directly above the conductor, the grounding terminal (upper terminal) of the surge arrester is connected to the iron tower with an EB hanging plate, and the HV terminal (lower terminal) of the surge arrester is connected to the conductor with a four bundled special connecting plate, as shown in Fig. 125.

The special connecting plates A and B wire clamps are the same as the ordinary spacer bar FJGY-855/48D wire clamp, and come with a rubber pad; C and D wire clamps are the same as the pre-twisted spacer bar FJGY-855/48DY pre-twisted wire clamp; the single wire clamp grip strength is not less than 2.5 kN, the single wire clamp shaft (horizontal) and oblique strength is not less than 2.5 kN, the centripetal force and torque grip requirements are the same as the engineering spacer bar, and the connection size is the same as U-16 matches; When the dedicated busbar A and

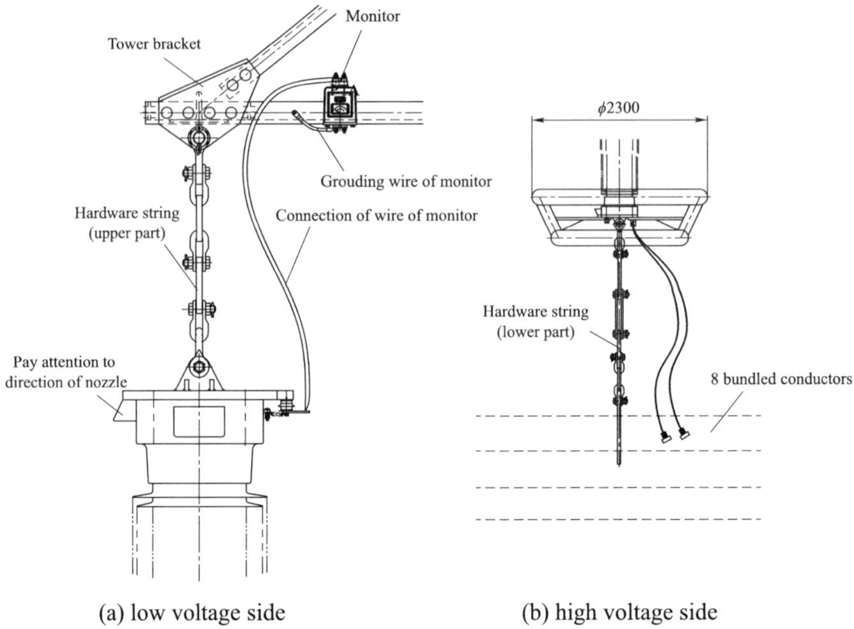

Fig. 125 Installation diagram on high and low voltage sides of surge arrester

B line clamps are tightened with the wire, the tightening bolt must face towards the inside of the surge arrester corona ring to reduce the effect of corona, as shown in Fig. 126.

2. Design of surge arrester counter installation

The surge arrester counter is installed on the horizontal member of the right triangle of the tower's external bracket using a clamp, and is connected to the grounding terminal (upper terminal) of the surge arrester through the surge arrester grounding wire, as shown in Fig. 127. During installation, the display screen of the surge arrester counter faces outward for easy observation, and the surge arrester grounding wire is kept loose. During construction and installation, the straight-line distance between the counter and the grounding terminal of the surge arrester is measured, the length of the surge arrester grounding wire is appropriately increased based on the straight-line distance, and it is ensured that the counter and the grounding terminal of the surge arrester are on the same side, and the hinge hardware and the grounding wire cannot cross. A grounding hole is reserved on the horizontal member of the right triangle, and the counter grounding wire is connected to the grounding hole with a bolt.

(VIII) Installation Method of Tension Tower Surge Arrester

The grounding terminal of the surge arrester is connected to the external surge arrester bracket through hinge hardware, and the HV terminal is connected to the squirrel

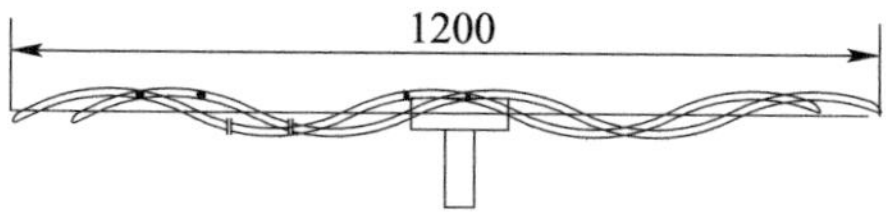

(a) Installation diagram of pre-twisted wire for specific connection plate C and D

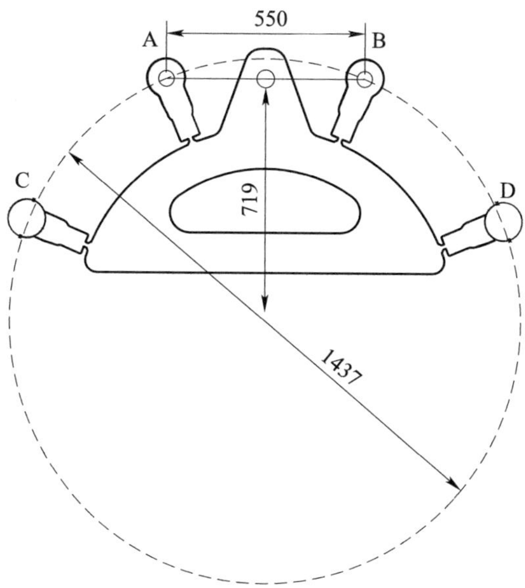

(b) Schematic diagram of dimensions for specific connection plates

Fig. 126 Specific connection plate of 4 bundled conductor (I)

cage hard jump special clamp through hinge hardware. Under normal conditions, the surge arrester itself needs to withstand wind load, and its HV terminal hinge hardware needs to have a margin to reduce the impact of the surge arrester wind load on the jumper system and the impact of the load change of the jumper system under various conditions on the surge arrester.

The surge arrester is connected to the wire system through a diversion wire. After the diversion wire connects the surge arrester and the wire, a certain corona will be generated. To reduce the impact of corona on the operation of the line, the length of the diversion wire is slightly longer than the hinge hardware, that is, under normal conditions, the diversion wire is not under force.

When the surge arrester is suspended directly above the squirrel cage rigid jumper, the length of the jumper V string projection is 13.7 m, which is less than the length between the single column surge arrester hanging points of 19.66 m. Therefore, the suspension of the surge arrester is designed on the crossbar of the tension tower wire, and the surge arrester hanging point on the external bracket needs to be raised 7.2 m

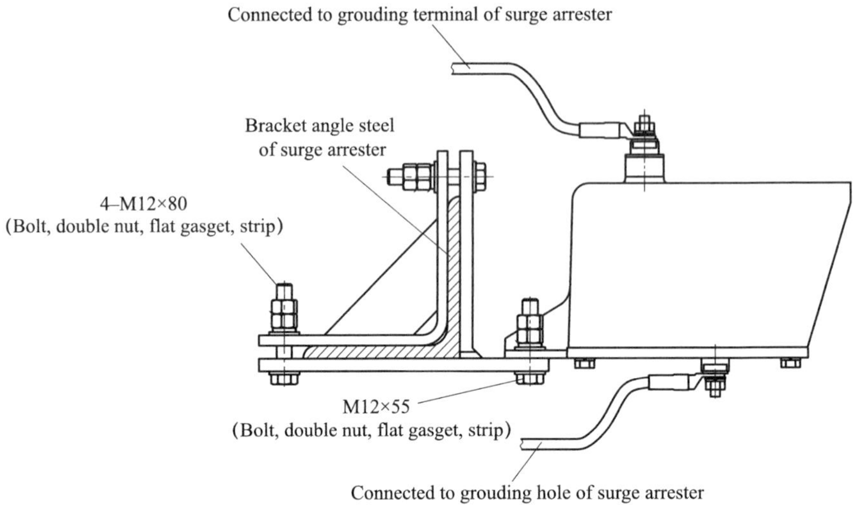

Fig. 127 Specific connection plate of 4 bundled conductor (II)

from the top surface of the crossbar. The remaining length difference can be adjusted by adjusting the length of the hinge hardware to make the distance from the external bracket to the wire equivalent to the length of the surge arrester. The installation gap requirements are the same as the straight line installation requirements. The design of the diversion wire type and connection hardware is the same as the straight line installation requirements.

When the surge arrester is suspended directly above the wire, the grounding terminal (upper terminal) of the surge arrester is connected to the iron tower with an EB hanging plate, and the HV terminal (lower terminal) of the surge arrester is connected to the squirrel cage jumper with a clamp-type spacer bar (dedicated busbar), as detailed in Fig. 125.

The clamp-type spacer bar (dedicated busbar) is modified based on the conventional clamp-type jumper spacer bar, adding an extension plate with a hole for connection with the U-shaped hanging ring (U-1695). The clamp-type spacer bar (dedicated busbar) and the ordinary spacer bar FJGY-The 855/48DZ line clamp is the same, and it comes with a rubber pad; the grip strength of a single line clamp is not less than 2.5 kN, the horizontal and oblique strength of a single line clamp is not less than 2.5 kN, the centripetal force and torque grip requirements are the same as the FJGY spacer bar of CG project-855/48DZ, the connection size matches with U-1695; when the A and B line clamps of the hoop-style spacer bar (special connecting plate) are fastened with the conductor, the direction of the fastening bolts must be towards the inside of the surge arrester corona ring to reduce the effect of corona, as shown in Fig. 128.

The installation design of the surge arrester counter is the same as the installation requirements of the straight-line tower.

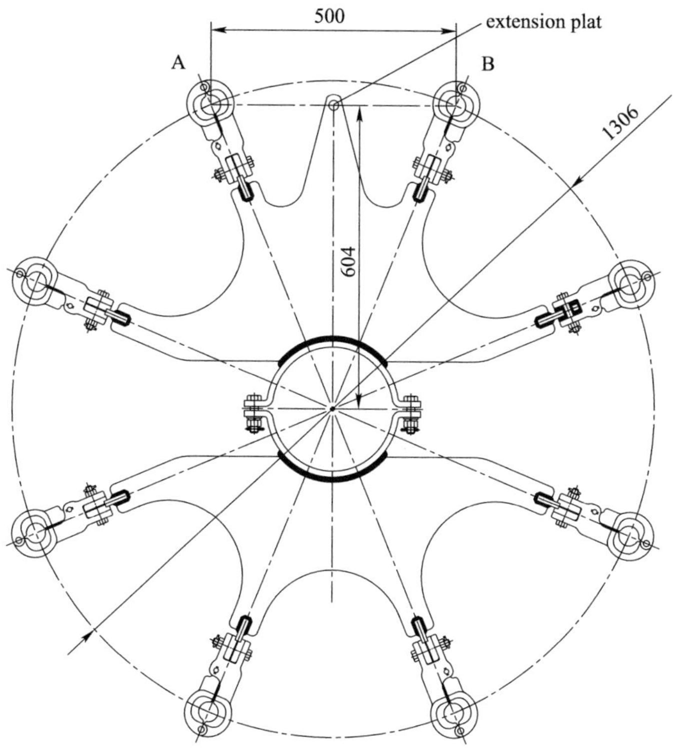

Fig. 128 Hold hoop spacer of surge arrester (specific connection plate)

Chapter 4
General Layout of ± 1100 kV Converter Station

Compared to ± 800 kV converter station, ± 1100 kV converter station has a larger transmission capacity and operates at higher voltage level. The AC yard, AC filter, converter yard, DC yard and other module areas of the station also need larger area, making it difficult to arrange reasonable layout scheme. Changji converter station, the sending station of the project, is co-constructed with the 750 kV substation in a same site. The HV and LV valve halls of each pole adopt a "face-to-face" layout, and the pole lines and DC filters adopt an internal layout scheme. Guquan converter station, the receiving station of the project, adopts a layered connection scheme for 500 kV and 1000 kV respectively. The valve hall adopts the single-line layout, and the pole lines and DC filters adopt an internal layout scheme. Also, two sets of synchronous condenser are configured inside. Layout schemes of these two stations basically contain all the possible factors of the layout schemes of China's UHV converter stations, which shows great reference significance for subsequent projects.

1 Layout of Converter Area

Valve hall and converter transformer area is the core area of the converter station, which are always placed in the central area of the converter station. The equipment height and air gap of the ± 1100 kV UHV converter station are much larger than those of ± 800 kV and below. This section analyzes and determines the layout scheme of the converter transformer area, the size of the valve hall, the longitudinal and lateral dimensions of the converter site, and the civil engineering and its ancillary design features of the valve hall.

I. Converter Transformer Area Layout

(I) "Face-to-face" Layout Scheme for HV and LV Side of Each Pole in the Converter Transformer Area

Z. Liu, ±*1100kV UHV DC Power Transmission Technology*,
https://doi.org/10.1007/978-981-96-3712-6_4

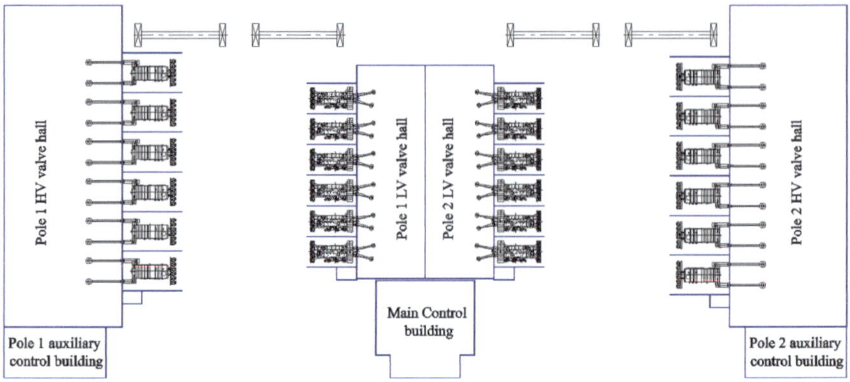

Fig. 1 The schematic diagram of the "face-to-face" layout scheme for the HV and LV valve halls of each pole

The schematic diagram of the "face-to-face" layout scheme for the HV and LV valve halls of each pole is shown in Fig. 1. The HV and LV valve halls of each pole adopt a "face-to-face" layout, where the HV valve halls of the two poles are arranged on the outside, and the LV valve halls are arranged "back-to-back" on the inside. All 8 sets (24 units) of converter transformers, among which 2 sets (6 units) in each hall, are arranged in a line close to the valve hall. The layout features are as follows:

(1) The "face-to-face" layout effectively blocks and absorbs the noise propagation from converter transformers, which is beneficial for noise control at the converter station's perimeter fence. However, the noise from the converter transformers spreads to both sides of the DC yard and AC yard, worsening the operational conditions of the DC yard.
(2) The lateral dimensions of the valve hall, converter transformer area, and DC yard are reduced, with the power flow direction of the AC yard, converter yard, and DC yard as the longitudinal dimensions.
(3) Auxiliary equipment is arranged by valve hall with clear functional zones, establishing a clear unit system.
(4) The DC wall bushing exits from the short axis direction of the valve hall, which is similar to the existing ± 800 kV UHV DC converter station.
(5) The choice of assembly site of the converter transformers is better placed at the HV and LV converter transformers of the same pole. The face-to-face installation and repair can be occupied at the same time, which also provides more flexibilities for operation and maintenance.
(6) The span of the converter transformer busbar is large, which results in high steel consumption of the busbar frame and poses greater challenges in fire protection."
(7) The spare converter transformers are distributed by pole, which results in long transportation distance when replacing the spare converter transformers of different poles. Multiple turns may also be required in transportation.

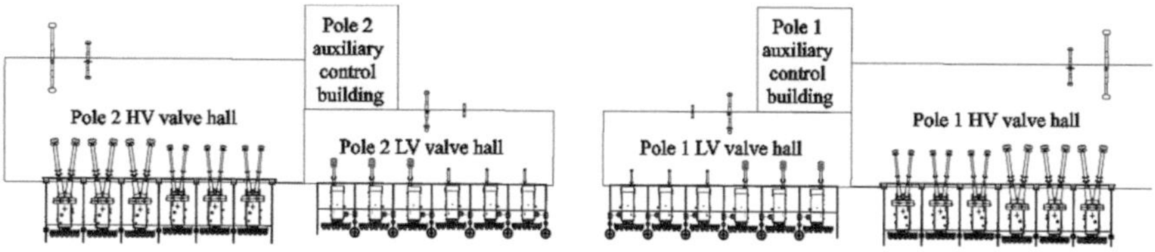

Fig. 2 The schematic diagram of single-line layout scheme of the valve hall

(8) When the LV converter transformer is connected to 1000 kV voltage level, the convergence of the converter transformer needs to be completed through the 1000 kV cross-line erected over the fire wall of the converter transformer. Due to the large phase-to-phase distance of the 1000 kV cross-line, the wiring of the converter transformer is relatively difficult and requires the installation of post insulators on the converter transformer courtyard.

(II) Single-line Layout Scheme for the Converter Transformer Area

The schematic diagram of linear layout scheme of the valve hall is shown in Fig. 2, with the HV and LV valve halls arranged side by side for each pole. The 8 groups (24 units) of converter transformers are arranged in a line on the same side of the valve hall, with the layout features as follows:

(1) The single-line layout of the valve hall effectively blocks the noise from converter transformers. The DC yard is essentially unaffected by the converter transformers. However, because 24 converter transformers are lined up facing the AC yard, their noise spreads to the AC yard and surroundings. The noise coverage is extensive, resulting in a relatively significant impact.
(2) The lateral dimensions of the DC yard and the valve hall are increased.
(3) The structure of incoming lines of the converter transformer is directly facing the valve hall, facilitating the wiring of the converter transformer.
(4) Auxiliary equipment is arranged by pole, establishing a clear unit system.
(5) The DC wall bushing exits from the long axis direction of the valve hall, similar to the existing ± 500 kV DC transmission project, making the wiring relatively simple.
(6) Each valve hall corresponds to 2 incoming line racks of the converter transformer, a total of 8 for the whole station. Since the converter area only considers the installation space of 1 converter transformer and the rear vehicle passage space, the span of the converter transformer incoming line is small.
(7) The assembly sites of Pole 1 and Pole 2 are interconnected. The transportation distance is shorter when replacing the spare converter transformer, and the layout direction of the HV and LV spare converter transformers is the same as that of the working transformer, which is quick and convenient without rotation when replacement.

Considering the system access, site geographical location and other conditions, Changji converter station at the sending end adopts the "face-to-face" layout, while Guquan converter station at the receiving end adopts the single-line layout.

II. Layout of HV and LV Valve Halls

The physical dimensions of the converter transformer, converter valve and other equipment used in the ± 1100 kV UHV DC transmission project are obviously different compared to the ± 800 kV projects. The control factor of the distance between the fire walls of the HV converter transformer is the required air gap for the valve group layout in the valve hall. When the AC side of the LV converter transformer at the receiving station is wiring to 1000 kV grid, the control factor of the distance between the fire walls of the converter transformer changes from the fan size of the converter transformer to the phase-to-phase air clearance of the AC 1000 kV lines. The electrical layout and dimensions of the valve hall of the ± 1100 kV UHV DC transmission project have changed significantly compared to the ± 800 kV projects.

(I) Electrical Layout Inside the Valve Hall

The principle of electrical layout inside the valve hall is similar to that of previous ± 800 kV projects. The electrical layouts inside the HV and LV valve halls are shown in Figs. 3 and 4.

(II) Optimization of Valve Hall Design

A patrol corridor is needed in the valve hall for the operation personnel to patrol. To ensure the safety of the operation personnel, the discharge standard deviation of is usually set as 5σ when calculating the safe air clearance from the converter valve to the patrol passage in the ± 800 kV UHV DC project. According to the principles of the ± 800 kV project, the distance between the converter valve and the patrol corridor of the ± 1100 kV UHV DC project needs to be at least 22 m, which will greatly increase the span of the valve hall, resulting in poor economy and increased difficulty for routine patrol. To address this issue, a double-layer shielded patrol corridor is proposed to protect the safety of patrolling personnel. With this optimized plan, the safe clearance between the converter valve and the patrol corridor only needs to be 15 m.

In addition, to meet fire protection requirements, a dedicated fire extinguishing access is added at a distance from the middle position of the converter valve. When a fire occurs in the converter valve, the operation personnel can enter the fire access to extinguish the fire after confirming that the converter valve has stopped operation (the fire access door is locked during normal operation and access prohibited to all personnel).

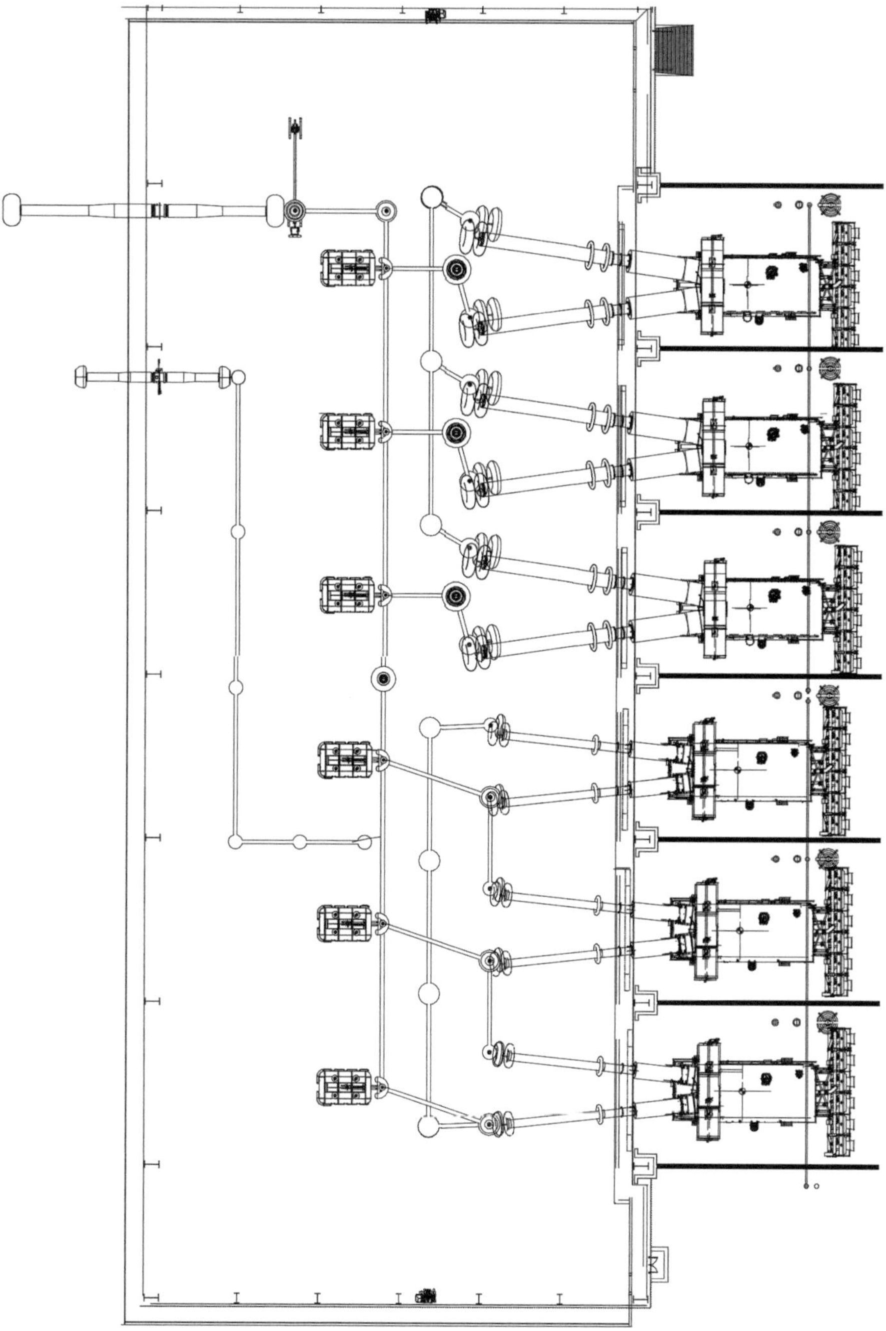

Fig. 3 The electrical layout inside the HV valve hall

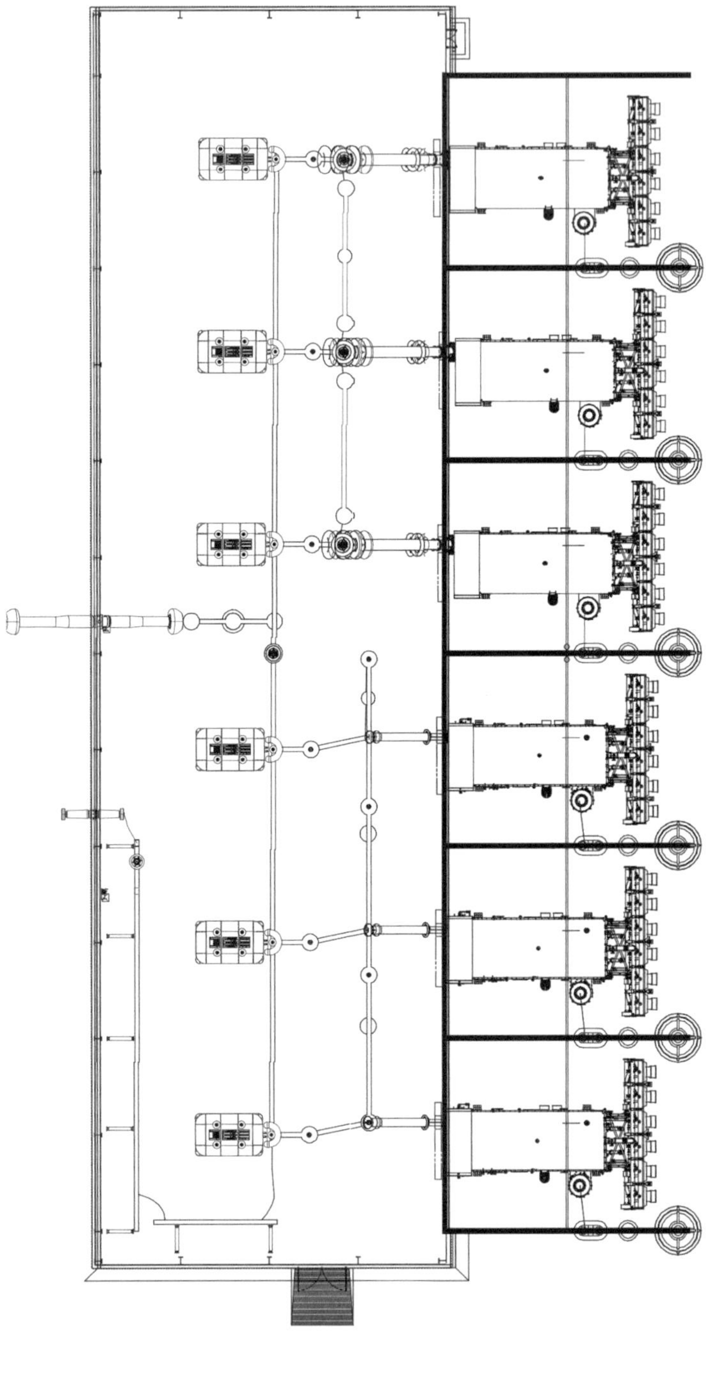

Fig. 4 The electrical layout inside the LV valve hall

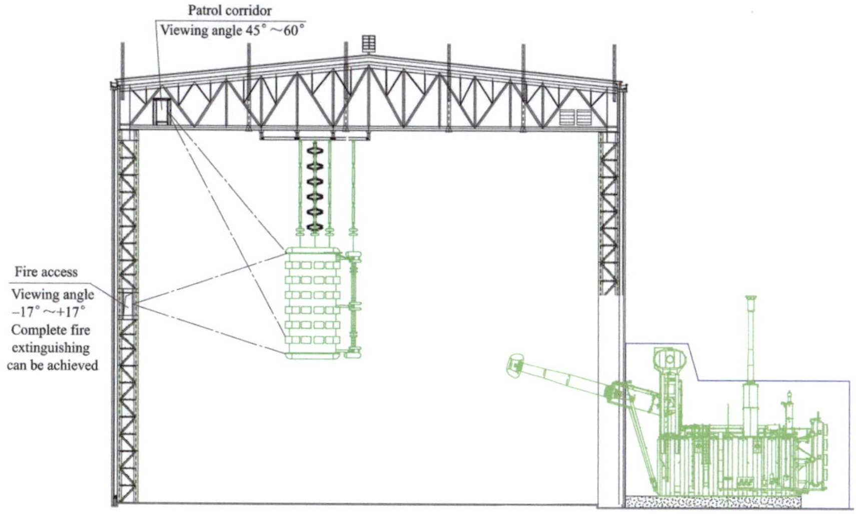

Fig. 5 The arrangement of the patrol corridor and fire access

The arrangement of the patrol corridor and fire access is shown in Fig. 5.

III. Dimensions of Converter Transformer Area

(I) Changji Converter Station

1. Dimensions of converter transformer

In Changji converter station, the voltage level of the converter transformer AC side is 750 kV, while that of the DC side is 1100 kV, which is different from the converter transformer with 750 kV on the grid side and 800 kV on the DC side. The dimensions and weight of the converter transformer have greatly increased compared to the ± 800 kV one, as shown in Fig. 6, Tables 1 and 2.

2. Dimensions of converter site

(1) Longitudinal dimension of the converter transformer operation replacement site (the distance from the outer wall of the valve hall to the center of the transport road). Longitudinal dimension of the converter transformer site is mainly controlled by the length of the valve hall, the length of the control building, and the distance between the fan and the control building when the spare converter transformer is transported on the transverse track. The schematic diagram of the longitudinal dimension of the converter transformer operation replacement site is shown in Fig. 7.

With the length of the HV valve hall increased to 118.5 m, the longitudinal dimension of the converter transformer site of this station is mainly controlled by the HV valve hall. The centerline of the side road in the converter transformer area is 13.1 m to the axis of the auxiliary control building. The auxiliary control building is 19.8 m wide, the HV valve hall is 118.5 m long, and the distance between the axis of the auxiliary

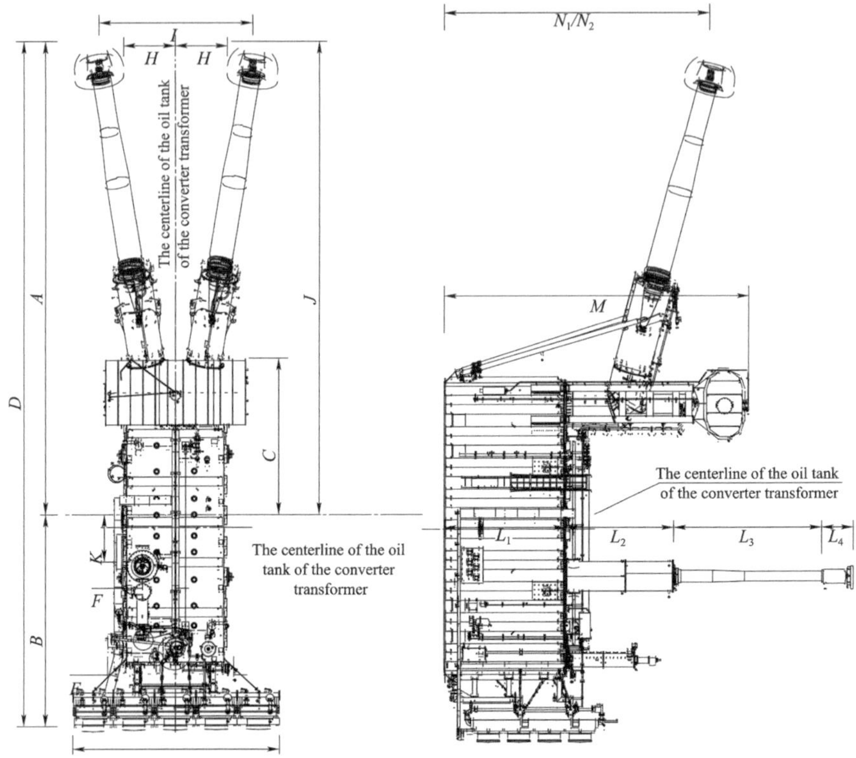

Fig. 6 The dimension diagram of the converter transformer

Table 1 Dimensions of the HV converter transformer in Changji converter station

Equipment	Dimensions (mm)							
	A	*B*	*C*	*D*	*E*	*F*	*G*	*H*
HV converter transformer	23,919	9488	9823	33,407	4316	2158	9521	3931

Equipment	Dimensions (mm)							
	I	*J*	*K*	$L_1 + L_2$	$L_3 + L_4$	*M*	N_1	N_2
HV converter transformer	7862	13,716	2109	10,486	8291	13,928	12,766	13,195

Table 2 Dimensions of the LV converter transformer in Changji converter station

Equipment	Dimensions (mm)							
	A	*B*	*C*	*D*	*E*	*F*	*G*	*H*
LV converter transformer	13,598	8818	6210	22,448	3502	1230	9590	2469

Equipment	Dimensions (mm)							
	I	*J*	*K*	$L_1 + L_2$	$L_3 + L_4$	*M*	N_1	N_2
LV converter transformer	5016	13,630	1946	9107	8182	11,006	6051	7087

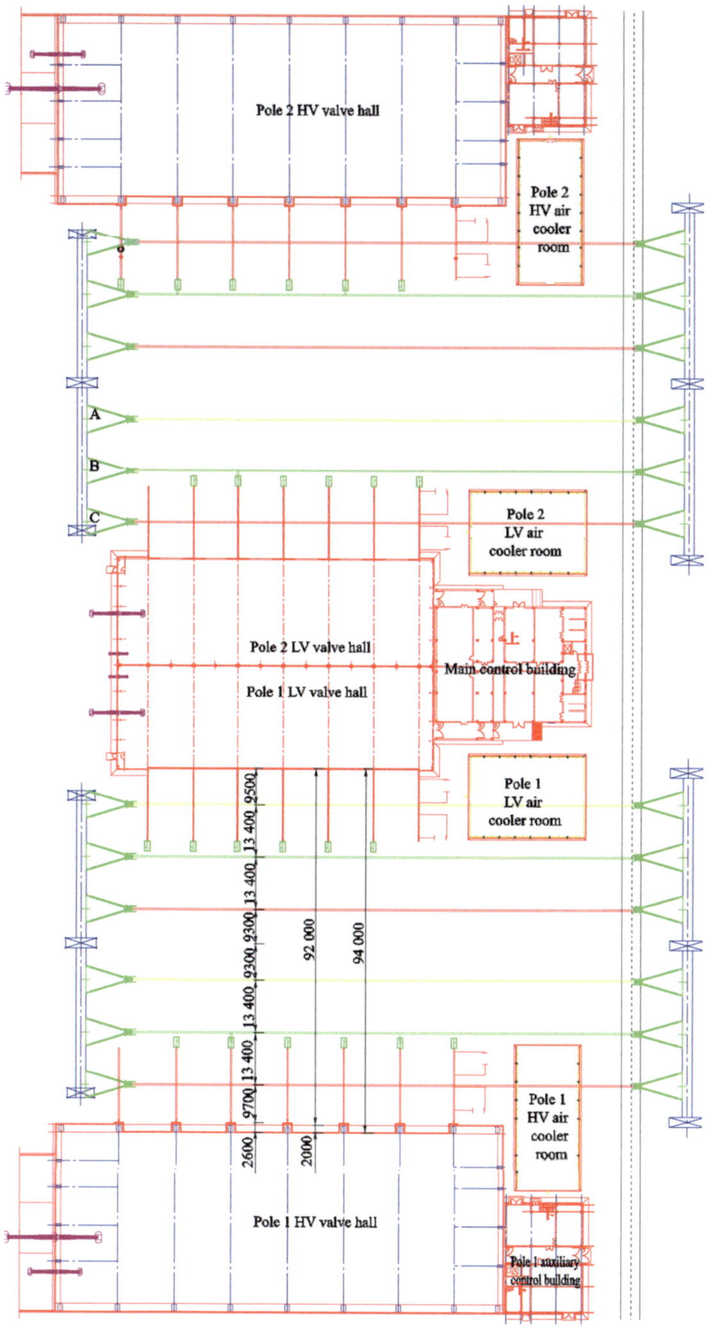

Fig. 7 The schematic diagram of the longitudinal dimension of the converter transformer operation replacement site

control building and the HV valve hall is 0.9 m. The longitudinal dimension to the converter area is calculated as follows: 13.1 + 19.8 + 0.9 + 118.5 = 152.3 (m). This value at the preliminary design stage was 157.5 m, while the length was saved 5.2 m at the construction drawing stage.

(2) Transverse dimension of the converter transformer operation replacement site (the distance between the outer wall surfaces of the HV and LV valve halls for each pole). The HV converter transformer uses the dimension of the converter transformer of which the valve-side bushing to lead out from the top of the oil tank, while the LV one uses the dimension of the converter transformer of which the valve-side bushing to lead out from the side of the oil tank. The physical dimensions of the converter transformer are listed in Tables 1 and 2.

According to "Standard for Design of Fire Protection for Fossil Fuel Power Plants and Substations (GB 50,229–2019)", the oil pit of the converter transformer should exceed 1 m from the top of fan, and top of the fire wall should exceed 1 m from the oil pit, that is, the length of the fire wall should be "B + C + 2". The HV fire wall is set to be 21.5 m long considering the margin, and the LV fire wall is set to be 17.5 m long. The schematic diagram of the specific determination of the transverse size of the converter transformer assembly site is shown in Fig. 8.

In Fig. 8, 21.5 m (length of the HV converter transformer fire wall) + 1 m (safe gap) + 24 m (A value of HV converter transformer) + 13.6 m (A value of LV converter transformer) + 1 m (safety gap) + 17.5 m (length of the LV converter transformer fire wall) = 78.6 (m).

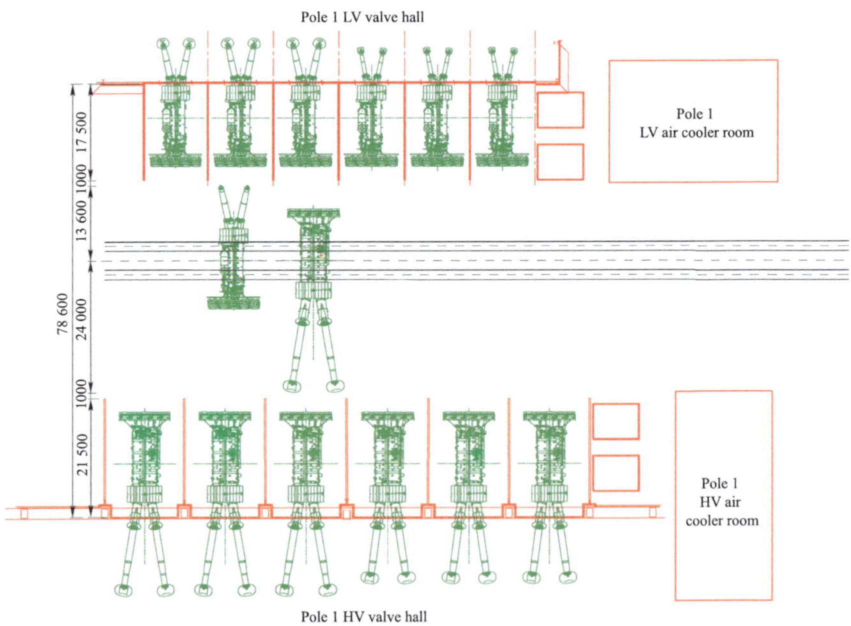

Fig. 8 The schematic diagram of the converter transformer assembly site

Therefore, during the transportation of the HV converter transformer, the distance from the top of fan of the converter transformer to the control cabinet is 4.5 m, which meets the set control safety distance. When the HV converter transformer is not equipped with fans, it also meets the requirement of transportation passage of 3 m.

(3) Courtyard dimensions determined by the converter transformer frame. At Changji converter station, the AC side of the converter transformer is connected to 750 kV. The 750 kV air clearance requirement is substantial, while the CG project features a significant span distance for the converter transformer's incoming lines. Also, strong wind in Changji leads to serious wind deviation and sway of the lines. As a result, courtyard dimensions of the converter transformer is mainly controlled by the phase-to-phase distance of 750 kV lines.

Considering the conductor tension, the sag of the converter transformer incoming line is taken as 6.5 m, and the calculation results of the wind deviation and sway of the converter transformer incoming line frame are shown in Table 3.

According to the phase-to-phase and phase-to-earth distance calculation results under above conditions, it is recommended that the phase-to-phase distance of the 750 kV converter transformer incoming line of the CG project is 13.4 m considering the characteristics of the 750 kV span conductor length (160 m), high voltage level, etc. with an appropriate margin. The phase-to-earth distance is 9.3 m, and the width of the incoming line interval frame is 45.4 m.

According to the design and operation experience of the already built converter station, the V-shaped insulator strings can effectively reduce the wind deviation and sway of the conductor. Therefore it is recommended to use the V-shaped insulator strings for the 750 kV converter transformer incoming line in the Changji converter station due to the large span of the converter transformer incoming line.

The phase-to-phase distance is determined by the wind deviation and sway. Therefore, the width of the converter transformer operation replacement site is determined to be 94 m. The schematic diagram of the transverse size of the converter transformer is shown in Figs. 9 and 10.

(II) Guquan Converter Station

1. Dimensions of converter transformer

In Guquan converter station, the voltage level of the converter transformer AC side is 500 kV/1000 kV, while that of the DC side is 1100 kV, which is different from the converter transformer with 500 kV on the grid side and 800 kV on the DC side. The dimensions and weight of the converter transformer have greatly increased compared to the ± 800 kV one, as shown in Figs. 11, 12, Tables 4 and 5.

2. Dimensions of converter site

The assembly site for the converter transformer is located in the area between the valve hall and the AC distribution device, with the converter transformer's incoming line located above the transformer, connected to the 1000 kV/500 kV AC distribution device via overhead lines.

Table 3 The phase-to-phase and phase-to-earth distances for the 750 kV converter transformer incoming line in Changji converter station (m)

Project	Atmospheric overvoltage (wind deviation, no short-circuit)	Switching overvoltage (wind deviation, no short-circuit)	Max. operating voltage (wind deviation, no short-circuit)	Max. operating voltage (wind deviation, short-circuit)	Requirement for 750 kV outgoing line	Requirement for corona radio interference	Verification of critical corona inception voltage	Requirement for electrical apparatus
Incoming lines of converter transformer	9.12	13.2	12.9	12.6	≥ 14.0	≥ 14.0	≥ 14.0	≥ 14.0

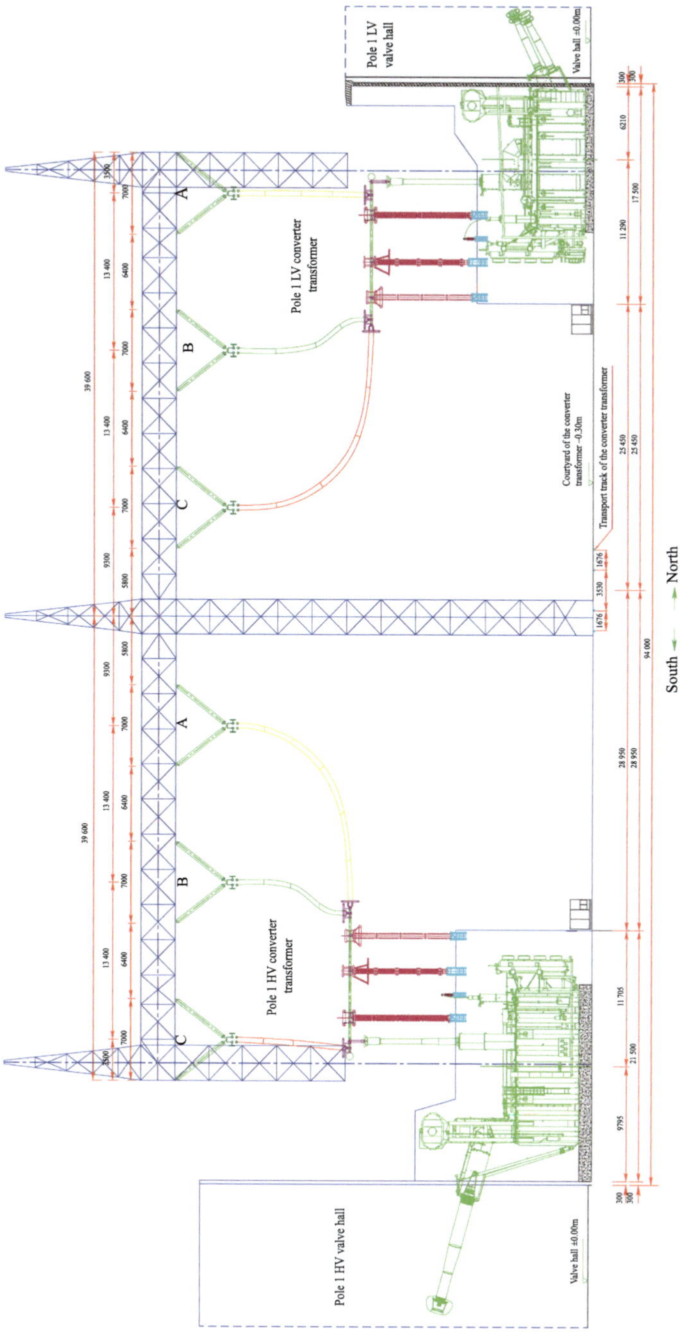

Fig. 9 The cross-section diagram of the converter transformer courtyard

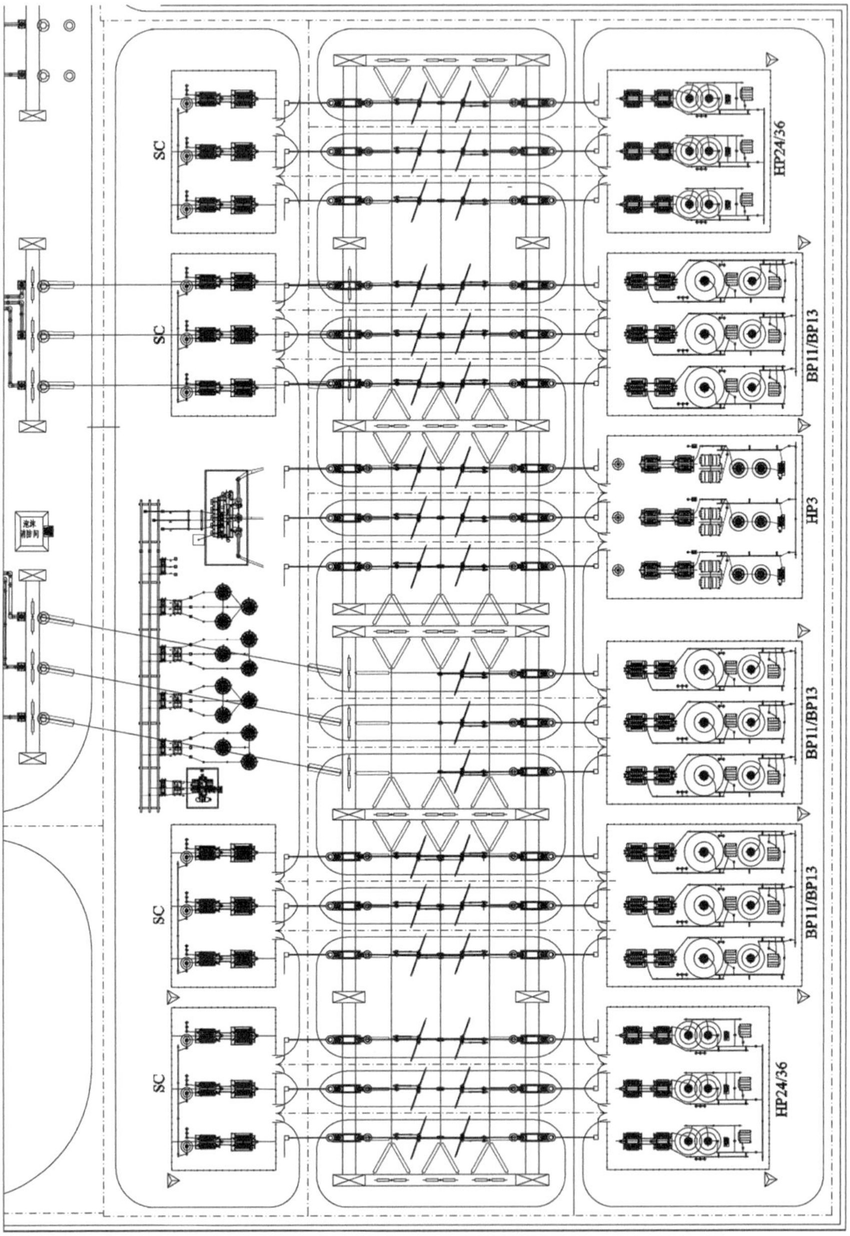

Fig. 10 The layout diagram of the converter transformer courtyard in the Changji converter station

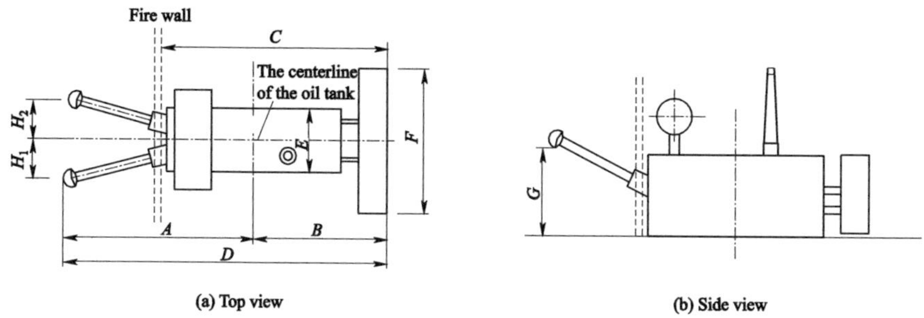

Fig. 11 The HV converter transformer in the Guquan converter station

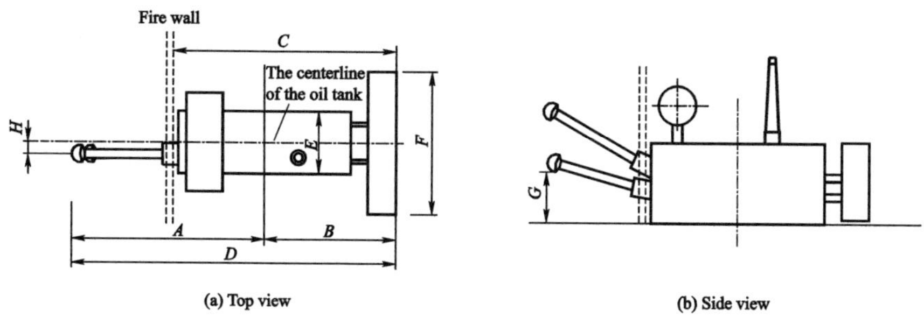

Fig. 12 The LV converter transformer in the Guquan converter station

Table 4 Dimensions of the HV converter transformer in Guquan converter station

DC transmission capacity (MW)	AC voltage (kV)	Dimensions (m)							
		A	*B*	*C*	*D*	*E*	*F*	*G*	H_1/H_2
12,000	500	25.47	10.62	22.12	37.42	5.42	12	11.09	4.18/4.18

Table 5 Dimensions of the LV converter transformer in Guquan converter station

DC transmission capacity (MW)	AC voltage (kV)	Dimensions (m)							
		A	*B*	*C*	*D*	*E*	*F*	*G*	*H*
12,000	1000	16.7	9.48	17.48	27.17	5	12	4.36	1.45

(1) Assembling principles for the converter transformer site. Based on the construction experience of previous ± 800 kV UHV DC projects, assembling principles for the converter transformer site for the CG project are as follows:

1) The biggest converter transformer can be delivered during operation.
2) The temporary placement of the converter transformer when switch off cannot affect the entry of the spare converter transformer into the assembly site.

3) When assembling one converter transformer, transportation channel of other converter transformers is no need to be considered.

(2) Selection of the converter transformer assembly site. Due to the single-line layout of equipment in the valve hall, the size of the assembly site is determined by the longest transformer assembly size and the safe clearance of HV transformer during transportation in the LV area.

The physical size of the HV Yy converter transformer (including bushing and cooler) is the longest about 37.5 m. The length of the HV converter transformer fire wall is determined to be 24.3 m based on this equipment size. However, since the voltage level of the grid side of the LV converter transformer reaches 1000 kV, it requires longer safe clearance. The size of the fire wall side is actually controlled by the LV converter transformer, and the length of the fire wall is determined to be 19.5 m based on the size of equipment, which meets the safe clearance requirement of not less than 7.85 m for the 1000 kV arresters during the transportation of HV converter transformers. According to the external dimensions of the equipment, the distance from the terminal board to the centerline of the oil tank is 25.5 m. According to the arrangement of the transformer's radiator on the road, the longest distance from the radiator to the centerline of the converter transformer oil tank is 11 m when the radiator does not exceed the right side of the road. Therefore, the assembly site (from the outer wall of the valve hall to the center of the transportation road) requires a distance of 59 m, which is designed to meet the requirements during assembly or transportation, and meet the site needs for the operation and replacement of the converter transformer. The site requirements for the installation of the converter transformer are shown in Fig. 13.

(3) Dimensions of the valve hall and converter transformer site

1) Longitudinal dimension. For the single-line layout of the valve hall, the longitudinal dimension is controlled by the layout size of the main control building, valve hall, converter transformer assembly site, and transportation channel. According to the determination of the converter transformer assembly site, the total longitudinal size of the valve hall and converter transformer assembly site is 59 m (width of the converter transformer assembly site) + 46 m (width of the HV valve hall) = 105 m.
2) Lateral dimension (perpendicular to the direction of the converter transformer incoming line, the centerline of the road on both sides of the valve hall and converter transformer site). Since the valve hall adopts the single-line layout, the total length of the 4 valve hall buildings exceeds 350 m. According to the "Code for Fire Protection Design of Buildings (2018 Edition) (GB 50,016—2014)", a fire access needs to be added in the valve hall. Therefore, the total lateral dimension is controlled by the layout size of the HV and LV valve halls, fire and transportation lanes. Considering the layout of the converter transformer transport track and the HV and LV valve halls, the total lateral dimension of the valve hall and converter transformer assembly site is: 2 × 119 m (length of the HV valve hall) + 2 × 98 m

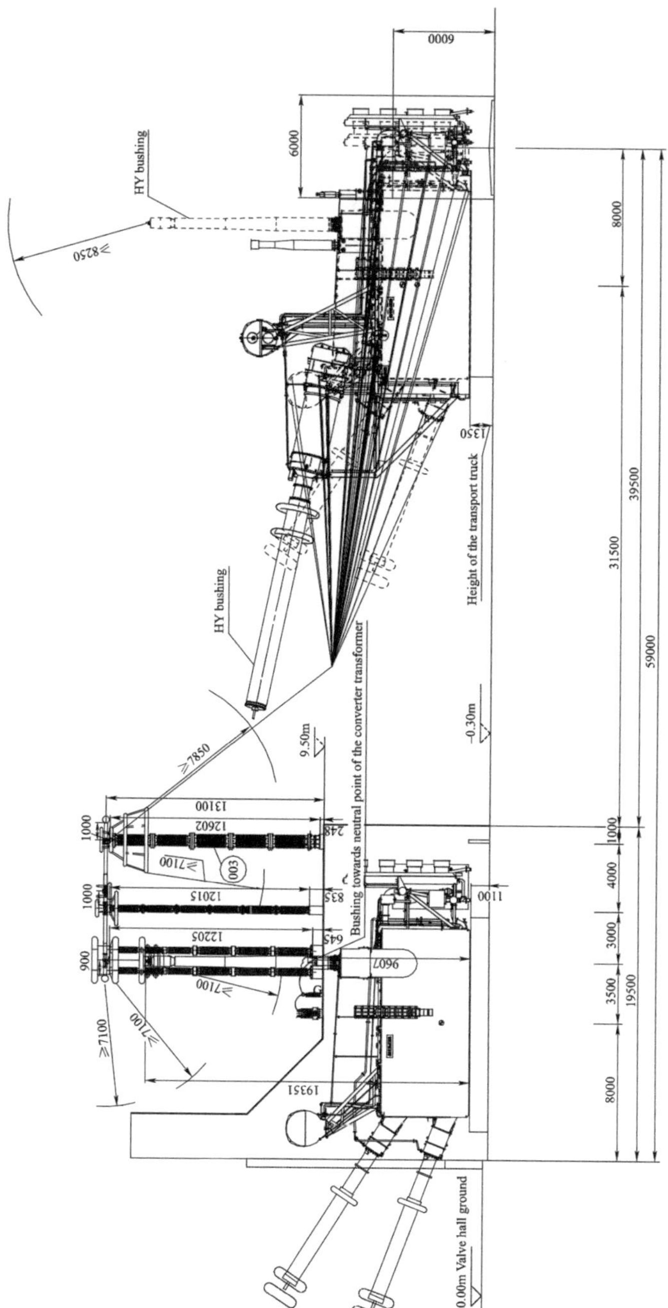

Fig. 13 The site requirements for the installation of the converter transformer in the Guquan converter station

(length of the LV valve hall) + 2 × 9 m (distance from the HV valve hall side wall to the road side ramp) + 2 × 11.5 m (distance from the LV valve hall side wall to the centerline of the road) + 5 m (width of the fire access) = 480 m.

In summary, the layout size of the valve hall and converter transformer site is 480 m × 105 m, see Fig. 14 for the layout diagram.

IV. Civil Engineering and Auxiliary Electrical Design of the Valve Hall

The purpose of the civil engineering and auxiliary electrical design of the valve hall is to ensure the normal and stable operation of the valve hall equipment based on the electrical layout requirements of the equipment/materials in the valve hall, supplemented by the building, structure, HVAC and auxiliary electrical design of the valve hall.

(I) Valve Hall Architectural Design

The building area and related dimensions of the valve hall in Changji converter station are shown in Table 6.

1. Principles of valve hall architectural design

    (1) Fire hazard classification of the production process: Class D.
    (2) Building fire resistance level: Level II.
    (3) Valve hall structure type: Steel structure.
    (4) Roof waterproofing level: Level I.
    (5) Building service life: 50 years.

2. Functions achieved by the valve hall architectural design

The architectural design should make the valve hall meet the requirements of electromagnetic compatibility, waterproofing and wind proofing, ensure the safety of the valve hall enclosure structure under severe weather conditions, and meet the requirements for airtightness and heat preservation of the valve hall under normal operating conditions. At the same time, the exterior facade of the valve hall should be aesthetically pleasing and simple.

(1) Windproof: Ensure the integrity and safety of the valve hall enclosure structure under severe wind conditions.
(2) Waterproof: Prevent leakage of the roof and side wall envelope of the valve hall from affecting the normal operation of the equipment under heavy rain conditions.
(3) Moisture-proof and heat preservation: Ensure the temperature and humidity inside the valve hall to meet all requirements under all weather conditions through the design of the valve hall enclosure structure.
(4) Maintain slight positive pressure: Ensure that the pressure inside the valve hall is 5 ~ 10 Pa higher than outside through the airtight design of the valve hall enclosure structure, maintaining a certain slight positive pressure.

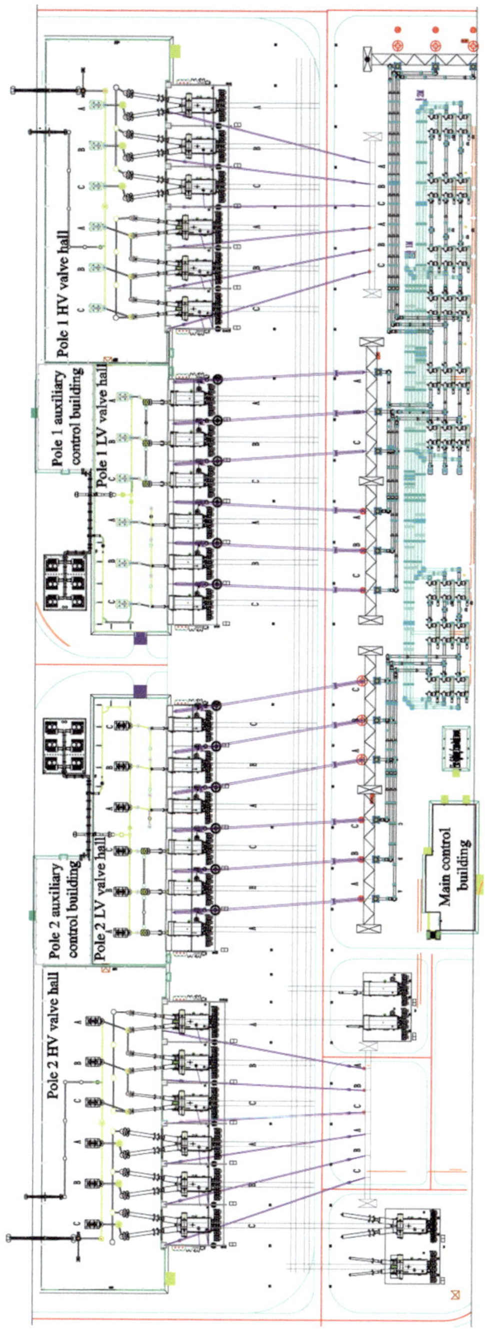

Fig. 14 The layout diagram of the valve hall and converter transformer site of the GQ station

Table 6 The building area and related dimensions of the valve halls in Changji converter station

No	Buildings	Building area (m^2)	Axis dimensions (m)
1	Pole 1 and 2 LV valve halls	4690	81.4 × (27.3 + 27.3)
2	Pole 1 HV valve hall	5660	118.5 × 48.1
3	Pole 2 HV valve hall	5660	118.5 × 48.1

(5) Electromagnetic shielding: Due to the high-frequency electromagnetic interference generated by the conduction and disconnection of the converter valve, it is necessary to set up good electromagnetic shielding in the enclosure structure, which ensures the normal operation of the equipment and the health of operators and effectively prevents the impact of electromagnetic radiation on the equipment outside the valve hall and the operators.

(6) Aesthetically pleasing: Through the combination and color matching of different tones of pressed steel plates, the facade of the valve hall is unified with the architectural style of the converter station.

(II) Valve Hall Structural Design

The structural design should ensure the design life of structure comparable to that of main equipment, reliable operation under harsh environmental conditions such as earthquakes, and maintenance-free throughout its life cycle.

The seismic precautionary intensity of Changji converter station is designed at 7 degrees, with a basic seismic acceleration value of 0.125 g, and a basic wind pressure of 0.73 kN/m^2 for a once-in-a-century event.

The structural design of the converter station valve hall follows the principles of advanced technology, safety, and economic rationality. During the structural design, factors such as the building's usage function, equipment process layout, load nature, material supply, fabrication and installation, construction conditions, and maintenance convenience are considered comprehensively, as follows.

1. HV valve hall

The HV valve hall adopts a full-steel structure, formed by welding steel column frames and planar trapezoidal steel roof frames, creating a large-span, high-level steel frame structure. The fire wall uses cast-in-place reinforced concrete wall panels. The steel structure of the valve hall is separated from the fire wall. The full-steel structure is conducive to the large span and high level of the HV valve hall. The 48.10 m span is reasonably light with a trapezoidal roof frame. The suspension valve tower and other suspended equipment transmit force to the roof frame through the steel beam suspension system, and then to the steel column frame. The full-steel structure is also conducive to the shielding and grounding requirements of the valve hall, and also conforms to the characteristics of the building exterior wall using pressed steel plates.

The fire wall made of reinforced concrete wall panels can effectively isolate the flammable converter transformer from the valve hall, and it is easy to install equipment embedded parts and post-installed steel equipment on the concrete wall panels.

(a) the schematic 1 (b) the schematic 2

Fig. 15 The schematic diagram of the HV valve hall structure

As one part of the valve hall exterior wall, the reinforced concrete wall panel can meet the waterproof and anti-seepage requirements of the valve hall. The schematic diagram of the HV valve hall structure is shown in Fig. 15.

2. LV valve hall

The LV valve hall consists of two back-to-back valve halls for Pole 1 and Pole 2. The structure of the LV valve hall is a hybrid structure composed of a steel roof frame and concrete wall columns, as shown in Fig. 16.

The roof frame uses a single-slope trapezoidal roof frame with a span of 27.30 m. The roof frames between the two back-to-back valve halls are designed to be separate, which is beneficial for fire protection in the valve hall. The suspension valve tower and other suspended equipment transmit force to the roof frame through the steel beam suspension system, and then to the concrete wall and columns. Like the HV valve hall, the outdoor fire wall of the LV valve hall also uses cast-in-place reinforced concrete wall panels.

The reinforced concrete frame and masonry infill walls serve as the fire wall in the middle of the two back-to-back valve halls, meeting the requirement of 3-h fire resistance.

3. Seismic design and measures for the valve hall structure

(1) Reasonably set up structural joints. The HV and LV valve halls are disconnected from the control tower with a 100 mm wide seismic joint according to the

Fig. 16 The schematic diagram of the LV valve hall structure

requirements of the valve hall structure's planar rules. The width of the seismic joint meets the requirements of structural deformation under major earthquakes.

(2) Reasonable selection and arrangement of structures. The structural system of the HV valve hall is a good seismic performance steel structure. Steel frame columns, trapezoidal steel roof trusses, roof support systems, wall beams, and diagonal braces between steel frame truss columns serve as the frontier of anti-seismic. The main structure is detached from the concrete fire wall to ensure the symmetry of the structural rigidity, avoiding the potential for structural damage due to severe asymmetry in structural rigidity.

The LV valve hall can adopt a combination structure scheme of concrete frame, shear wall, and steel roof truss due to the symmetrical fire walls on both sides of the structure. The structural anti-lateral force system mainly uses the strong rigidity of the concrete fire wall.

(3) Other seismic measures.

1) The structure has sufficient lateral rigidity. The displacement control value at the top of the column is taken as 1/400 under minor earthquake conditions.
2) Multiple seismic measures are set up to ensure the ductility of the structure. The displacement control value at the top of the column is taken as 1/50 under rare earthquake conditions.

3) A lightweight pressed steel plate wall roof system is used to reduce the input of earthquake action on the structure.

(III) HVAC Design of the Valve Hall

The HVAC system design should meet the operating environment requirements of the equipment under extreme operating conditions and extreme weather conditions, ensure the micro-positive pressure and cleanliness of the valve hall, maintain the necessary temperature and humidity, ensure no condensation in the valve hall, and ensure the correct linkage of the air conditioning system and the fire protection system under fire.

The basic principles and requirements of HVAC design for the valve hall are as follows:

(1) Air conditioning equipment is installed indoors. If there is no condition for outdoor arrangement, necessary protective measures should be taken to ensure the normal operation of the air conditioning equipment throughout its life cycle.
(2) The outdoor unit of the ventilation and air conditioning equipment should be arranged according to the specific positions of the valve hall and the control tower, minimizing the distance between the outdoor unit and the indoor equipment as much as possible.
(3) The linkage logic of air conditioning equipment and valve hall fire smoke exhaust equipment should be fully considered.
(4) Overall external conditions of air conditioning design.

   1) Long-term operating temperature of the valve hall: 10~50 °C.
   2) Relative humidity: 10~50%.
   3) Micro-positive pressure: 5~10 Pa.
   4) Valve hall exhaust volume: 0.5 times/day.
   5) Valve hall air circulation volume: 0.5 times/h.
   6) Valve hall ventilation and air conditioning filter level: F9.

(5) The direction of the air duct in the valve hall needs to cooperate with the process, ensuring the electrical distance between the air conditioning ventilation duct and other equipment in the valve hall.
(6) The ventilation and air conditioning system of the valve hall is designed with complete redundancy, ensuring 100% backup.
(7) An electric heating system is installed in the HV and LV valve halls, which is used for auxiliary heating in the valve hall under severe cold conditions such as winter.
(8) The temperature and humidity sensors of the air conditioning system should be reasonably positioned, ensuring relevant data being accurately collected and having enough electrical safety distance.
(9) The smoke exhaust outlet of the valve hall should be UV-resistant and opaque. The smoke exhaust outlet of the valve hall is equipped with a shielding steel mesh, and the reliability of these shielding meshes should be grounded.

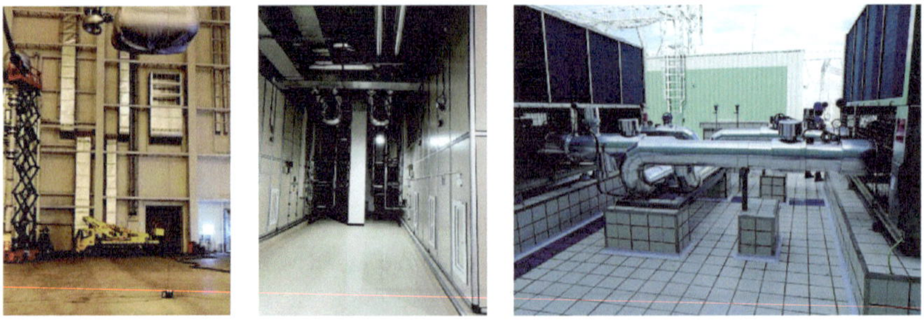

(a) Air ducts in valve hall (b) Air conditioning equipment room in valve hall (c) Air-conditioned outdoor units

Fig. 17 The composition of the valve hall ventilation and air conditioning system

The composition of the valve hall ventilation and air conditioning system is shown in Fig. 17.

(IV) Auxiliary Electrical Design of the Valve Hall

1. Valve hall shielding and lightning protection grounding

   (1) The valve hall should meet the requirements of hexahedral shielding. The surrounding and top shielding of the valve hall is achieved by the inner layer of the valve hall walls and the inner layer of the roof. The ground shielding of the valve hall is achieved by the metal mesh laid below the floor finish. In all aspects of the ground switch trench and cable trench of the valve hall, metal mesh are always needed covering and connected to the metal shielding mesh of the valve hall floor.
   (2) All metal components in the valve hall (steel trusses, steel columns, steel braces, steel purlins, bridges, steel wire troughs, steel ladders, patrol corridors, steel fences, checkered steel plates, air ducts, hangers, brackets, air conditioning unit shells, lamp shells, fire detector metal shells, camera metal shells, camera junction box metal shells, fire module box metal shells, lighting box shells, distribution box shells, socket box shells, etc.) must be reliably grounded with 35 mm^2 copper stranded wire.
   (3) The main steel structure grounding uses 150 mm^2 copper stranded wire, and the auxiliary metal component grounding uses 35 mm^2 copper stranded wire.
   (4) For the LV valve hall, a dedicated grounding busbar is set up at the filling wall of the middle frame to ensure that the purlins, embedded iron and other metal components behind the pressed steel plate are reliably grounded through this grounding busbar.
   (5) All metal frames at the wall penetration points of the valve hall (doors, inspection windows, smoke exhaust windows, air supply windows, air duct reserved holes, converter transformer valve-side bushing reserved holes, DC wall bushing reserved hole frames, etc.) must be reliably grounded at two points with 35 mm^2 copper stranded wire.

(6) The valve hall door and the interlocking device on the door are reliably grounded with 35 mm^2 copper stranded wire.
(7) The inspection windows, smoke exhaust windows, and air supply windows of the valve hall are equipped with 25 mm × 25 mm × 1.8 mm steel mesh to achieve shielding and grounding.

2. Valve hall lighting and power distribution

The lighting in the valve hall should meet an illuminance greater than 200 Lux, using energy-saving LED floodlights from the roof frame downwards. The lower edge of the floodlight should not be lower than the lower chord of the roof frame.

3. Valve hall fire detection system

The fire detection system in the valve hall uses a multiplexing setup, including an early smoke detection system, UV detection system, and sound and light alarm system, etc.

4. Valve hall control system

(1) The power supply and control system of the valve hall ventilation and air conditioning system are designed in a fully redundant manner. The signals from the valve hall ventilation and air conditioning system to the station SCADA system are divided into A and B types of alarms. A alarm is an emergency accident alarm, and B alarm is a general event alarm. The electric heating equipment in the HV and LV valve halls is also controlled through the valve hall ventilation and air conditioning system.
(2) Each valve hall is equipped with a set of redundant power supply smoke exhaust fans for post-disaster, which can be remotely or locally started to exhaust smoke after the fire is confirmed to be extinguished.

5. Valve hall image monitoring system

Each HV valve hall is equipped with 7 track-type infrared temperature measuring cameras, and each LV valve hall is equipped with 7 track-type infrared temperature measuring cameras, and 10 network high-speed dome cameras are installed in each valve hall to monitor the operation of the equipment in each valve hall.

6. Fireproof sealing module with electromagnetic shielding

All cables, optical cables, refrigerant pipes, etc., that pass through the valve hall/ control tower wall above 0 m in the valve hall must be shielded and grounded and sealed through a fireproof sealing module with electromagnetic shielding. Each fireproof sealing module frame with electromagnetic shielding embedded in the wall of the valve hall is integrated with other metal components in the valve hall or grounding copper wire through two 120 mm^2 copper twisted wires.

2 Layout Scheme of DC Yard

The UHV DC yard has characteristics such as high voltage level of equipment installed, large air gap requirements, diverse equipment types, and complex wiring. For the $\pm$ 1100 kV UHV converter station, due to requirement for longer air gap and higher equipment, the influence of environmental conditions on the DC yard pole line equipment has increased. Therefore, a detailed study needs to be conducted on the layout scheme of the UHV DC yard in the early stage of the project.

I. Research on DC Yard Layout Scheme

(1) Technically, both 1100 kV indoor/outdoor DC yards are feasible. The insulator technical indicators of indoor DC yard equipment are nearly the same with that of the $\pm$ 800 kV outdoor DC yard equipment. The equipment is relatively mature, and most manufacturers have developed prototypes. The insulator technical indicators of $\pm$ 1100 kV outdoor DC yard equipment need to be significantly improved compared to that of the $\pm$ 800 kV outdoor DC yard. Therefore, many equipment need to be redeveloped, and there are certain R&D risks for equipment such as DC voltage dividers.

(2) Adopting the indoor DC yard scheme, the equipment is not restricted by external environmental conditions, which can reduce the probability of insulator pollution flashover, reduce the overall height of the equipment, better ensure the stability of the equipment in terms of mechanical performance, have more space to adapt to harsher environments, and obtain a larger design margin. The design risk is also smaller to reach higher reliability of the project.

(3) The indoor DC yard scheme has a higher initial investment cost, which is about 12% more than the outdoor scheme, mainly coming from the investment in the building construction. On the contrary, considering factors such as equipment operating environment, project reliability, and operation and maintenance, the indoor scheme is more advantageous.

(4) The main economic differences between indoor and outdoor scheme are in two aspects. One is the cost difference of pole-side equipment. The other is the difference in civil construction costs of two schemes, mainly coming from the investment in the building of the indoor scheme.

(5) If the HVAC scheme only considers ventilation but not air conditioning, the indoor scheme can save about 10 million RMB. The cost of using a single air conditioner is about 20 million yuan, and the cost of using single ventilation is 8 million to 10 million RMB.

II. Air Clearance of the DC Yard

The DC yard of the CG project uses the indoor DC yard scheme, placing the DC pole line equipment (excluding the DC wall bushing and its outdoor side six-column parallel insulators), by-pass equipment in the HV valve hall, and the high voltage capacitor tower of the DC filter all indoors. According to the insulation coordination analyses, the air clearance of the DC yard is determined according to the long wavefront characteristics of the ± 1100 kV DC system when overvoltage, with the adjustment for indoor temperature and humidity. (see Table 7).

III. Specific Layout Scheme of the DC Yard

The layout of the DC yard is adapted to the layout of the valve hall of the converter station. The indoor DC yard is closely arranged with the valve hall, located at the two poles of the DC yard. The neutral line equipment is arranged outdoors in the middle of the DC yard. The pole line and high-voltage capacitor tower with some 550 kV tubular busbars are placed in the DC distribution room. The DC distribution room is arranged on both sides of the DC yard. The indoor high-voltage capacitor tower C_1 is connected to the low-voltage side of the filter through the wall bushing. The 550 kV busbar of the high voltage by-pass circuit of the valve group is connected to the LV valve hall 550 kV by-pass circuit after passing through the wall bushing of the DC distribution room. The pole line is led out through the outgoing tower after passing through the wall bushing. The design features of ± 1100 kV indoor DC yard are detailed as follows.

1. ± 1100 kV disconnector

The ± 1100 kV disconnector is designed as a double-column horizontal telescopic type. Considering space saving, disconnectors on the pole line-bypass circuit and pole line-metal circuit adopt a common fixed contact combination layout. The external dimensions of the ± 1100 kV disconnector are shown in Fig. 18.

2. Patrol corridor

The indoor DC yard is designed to operate normally without human intervention. The indoor DC yard sets up a shielding cage only to the position of the disconnector operating box, which is convenient for on-site operation. Also, observation of key equipment can be easily achieved at the same time.

The indoor patrol corridor is set up according to a fully enclosed Faraday cage. The Faraday cage must maintain a minimum air clearance of 15 m from the energized parts of the equipment. Because the height of the pole bus conductor is about 18 m, the total height of the Faraday cage is 2.1 m.

The shielding cage is designed as a double-layer metal mesh cage, which can protect operating personnel and meets the operation needs such as opening the terminal box, as shown in Fig. 19.

For routine opening the terminal box in daily maintenance, it is needed to allow inspection personnel to approach the terminal box. When the operation and inspection personnel open the terminal box, their hands should not be exposed outside the shielding cage. Therefore, when the ground main patrol corridor passes under the equipment bracket with a terminal box (operation box), a small T-shaped branch needs to be made to cover the equipment terminal box and protect the operation and inspection personnel, as shown in Fig. 20.

Table 7 The air clearance of the DC yard

No	Typical Gap	*SIWL*(kV)	Gap coefficient	Min. clearance (m)
I. Indoor gap (45 °C)				
1	Between ± 550 kV equipment and ground	1180	1.15	5.0
2	Between ± 1100 kV and ± 550 kV equipment	1290	1.2	5.2
3	Between ± 1100 kV powered corona ring and non-powered corona ring	2100	1.3	12.0
4	Between powered parts of ± 1100 kV equipment and support structure (including adjacent supports)	2100	1.25	13.0
5	Between normal ± 1100 kV equipment, tubular busbar and fences and enclosure cage	2100	1.2	15.0
6	Between ± 1100 kV flexible conductor and ground	2100	1	18.0
7	Between ± 1100 kV flexible conductor and steel structure	2100	1	18.0
8	Between ± 1100 kV line and support structure (including adjacent supports)	2100	1	18.0
9	Between ± 1100 kV line and fences and enclosure cage	2100	1	18.0
10	Between powered parts to personnel when partial voltage maintenance (5σ)	1265	1	10.5
11	Between corona ring on the ± 1100 kV bushing and ± 550 kV tubular busbar	1290	1.2	5.2

(continued)

Table 7 (continued)

No	Typical Gap	*SIWL*(kV)	Gap coefficient	Min. clearance (m)
12	Between corona ring the ± 1100 kV pole lines and equipment on the low voltage side of the DC filter	2100	1.28	10.8
13	Between low voltage side of the ± 1100 kV HV capacitor C1 and ground	1334	1.2	5.5 5.4
14	Between suspension ball on the incoming line of the HV capacitor and rooftop	2100	1.25	13
15	Between suspension ball on the incoming line of the HV capacitor and unbalanced TA/capacitor tower	1049	1	5
16	Between unbalanced TA of the HV capacitor and wall	1675	1.15	8.5
17	Clearance of ± 1100 kV by-pass circuit breaker	2100		11.9
18	Between the indoor neutral line and ground	550 (*LIWV* = 600 kV)	1.1	1.2
II. Outdoor gap				
19	Main gap between outdoor 1100 kV parallel multi-column (Critical wavefront)	2100	1	17.6
20	Between outdoor 1100 kV equipment and ground (Critical wavefront)	2100	1.05	16.0
21	Between outdoor 1100 kV equipment and personnel	2100	1.05	29.0
22	Between outdoor 1100 kV and 550 kV equipment	1290	1.2	5.2
23	Between outdoor ± 550 kV equipment and ground	1180	1.15	5
24	Between metallic return line / neutral line and ground	550(*LIWV* = 600 kV)	1.05	1.2

(continued)

Table 7 (continued)

No	Typical Gap	*SIWL*(kV)	Gap coefficient	Min. clearance (m)
25	Between HV-end of DC filter L_1/R_1 and ground	1334	1.2	4.5
26	Between HV-end of DC filter L_1/R_1 and fence	1334	1.23	4.0
27	Between LV-end of DC filter L_1 and ground	1099	1.2	3.5
28	Between LV-end of DC filter L_2 and ground	550($LIWV = 600$ kV)	1.15	1.2
29	Between LV-end of DC filter L_1 and fence	1099	1.2	3.5
30	Between LV-end of DC filter L_2 and fence	550($LIWV = 600$ kV)	1.05	1.2
31	Between HV- and LV-end of DC filter L_1	948	1.15	3.5
32	Between HV- and LV-end of DC filter L_2	998	1.15	3.5
33	Between metallic return line and HV-end of DC filter L_1	1516	1.3	5.5
34	Between metallic return line and HV-end of DC filter L_2	1559	1.3	5.6

3. Maintenance space

The indoor DC yard is not designed for personnel entry during normal operation. Only a shielding cage is provided from the isolator to the operating mechanism box to facilitate the on-site operation needs of the operating personnel. Due to the large air clearance requirement for full voltage maintenance of the high voltage capacitor tower of the filter, the indoor building dimension is difficult to meet the needs. Considering the half-voltage maintenance condition of the high voltage capacitor tower (when the high voltage capacitor is out of maintenance, the corresponding pole is only allowed to run at half voltage), a maintenance fence is set on the indoor ground. Maintenance is only allowed to work within the fence. The maintenance diagram of the indoor DC yard of the Guquan converter station is shown in Fig. 21.

3 Layout of AC Yard

Based on the layout scheme of the valve hall and converter transformer site, as well as the scale and direction of the AC outgoing lines, the AC yard of the Changji converter station adopts a single-line layout scheme for the 750 kV outdoor GIS; while that

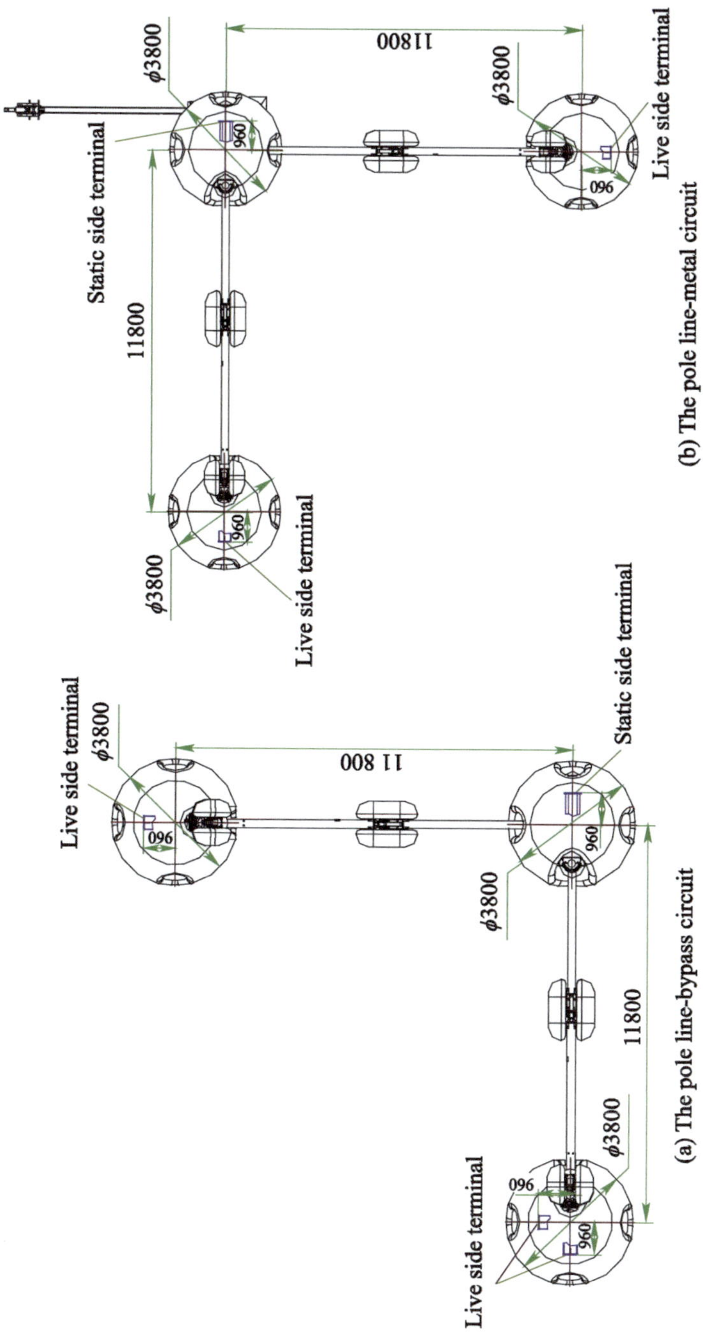

Fig. 18 The external dimensions of the ± 1100 kV disconnector

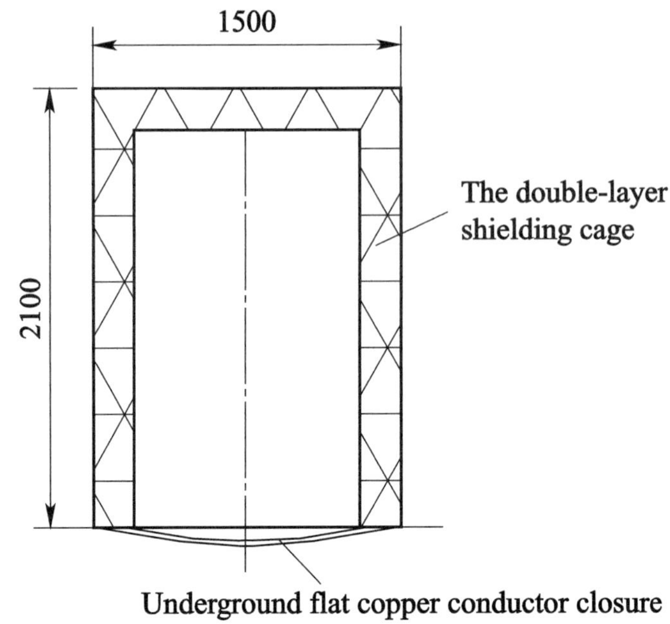

Fig. 19 The double-layer metal mesh shielding cage on the ground

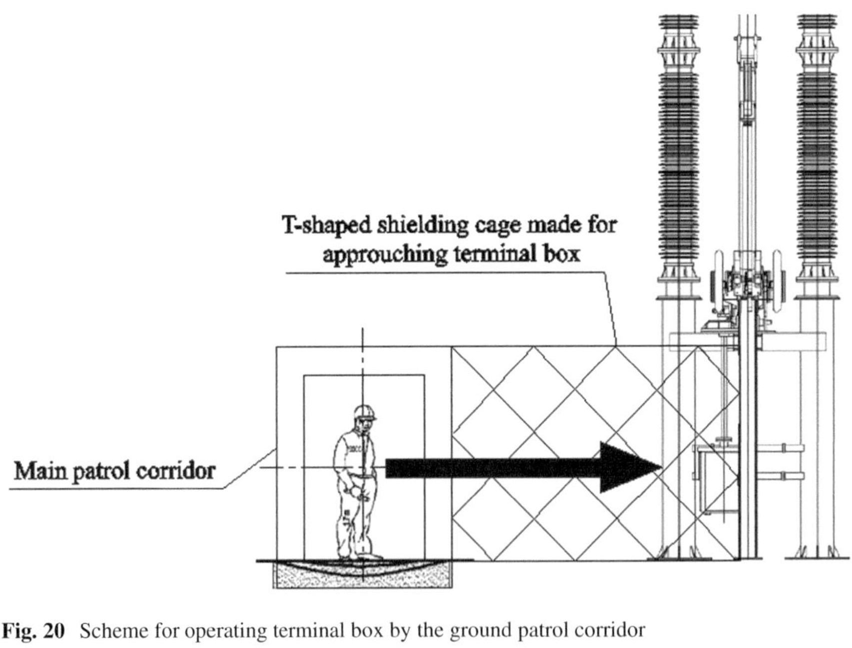

Fig. 20 Scheme for operating terminal box by the ground patrol corridor

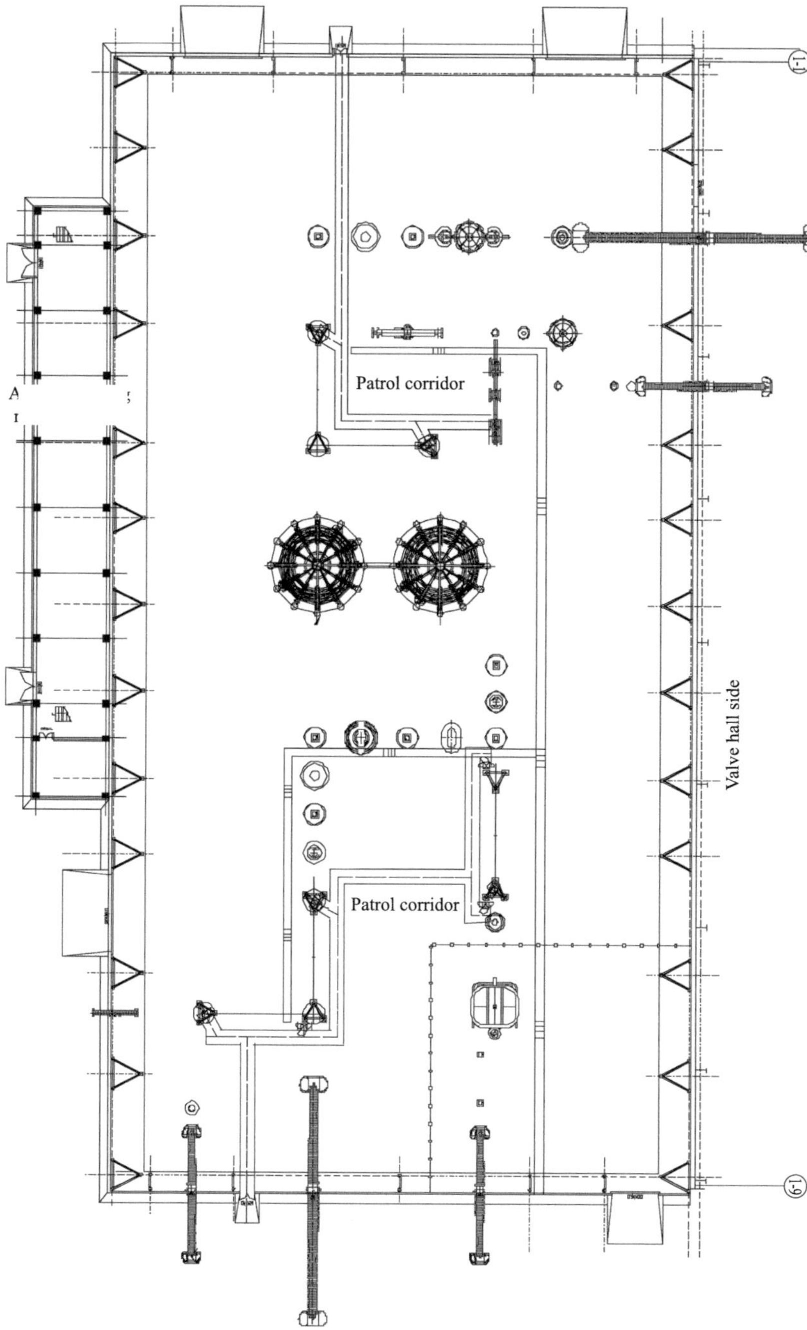

Fig. 21 The maintenance diagram of the indoor DC yard of the Guquan converter station

of the Guquan converter station adopts a linear layout scheme for both the 1000 kV outdoor GIS and the 500 kV indoor GIS.

I. Changji Converter Station

The 750 kV AC switchyard of Changji converter station adopts a 3/2 circuit breaker connection. There are 13 outgoing lines (3 to the joint substation, 10 to the power station), 4 incoming lines for the converter transformer, 4 incoming lines for the AC filter, a total of 21 components, forming 10 complete strings and 1 incomplete string. The 800 kV GIS adopts a single-lin layout scheme with a centralized busbar. The circuit breaker unit is close to the side of the liaison transformer transportation road, and the lifting and transportation of the equipment can use the liaison transformer transportation road, which is convenient for operation inspection.

II. Guquan Converter Station

The 1000 kV AC switchyard of Guquan converter station adopts a 3/2 circuit breaker connection, with 2 outgoing lines (without expansion), 2 incoming lines for the LV converter transformer, 2 banks of AC filters, a total of 6 components, forming 3 complete strings. There are 8 outgoing lines for 500 kV, 2 incoming lines for the HV converter transformer, 3 banks of AC filters, 2 groups of 500 kV/10 kV station transformers (directly connected to the busbar), 2 groups connected to the phase shifter, a total of 17 incoming and outgoing line components (2 of which are connected to the busbar), forming 7 complete strings and 1 incomplete string. The 500 kV AC distribution device of Guquan converter station adopts an indoor GIS layout, and the 1000 kV AC distribution device adopts an outdoor GIS layout. The size of the 500 kV GIS room is 197 m × 16.6 m (length × width), located on the south side of the station area. 500 kV AC outgoing lines are directed to the south both in current stage and future expansion. The 1000 kV GIS is located on the north side of the station area, arranged in a north–south direction. The 1000 kV AC outgoing lines are directed to the north.

The electrical layout of the 500 and 1000 kV AC distribution devices at the Guquan converter station are shown in Figs. 22 and 23 respectively.

4 Layout of AC Filter Yard

The total capacity of 750 kV capacitive reactive power compensation in Changji converter station is 6700Mvar, which is divided into 4 banks and 20 sub-banks. The total capacity of 1000 kV capacitive reactive power compensation in Guquan converter station is 4080Mvar, which is divided into 2 banks and 12 sub-banks. The total capacity of 500 kV capacitive reactive power compensation in Guquan converter station is 3990Mvar, which is divided into 3 banks and 14 sub-banks. Both stations use a newly improved layout for the AC filter yard.

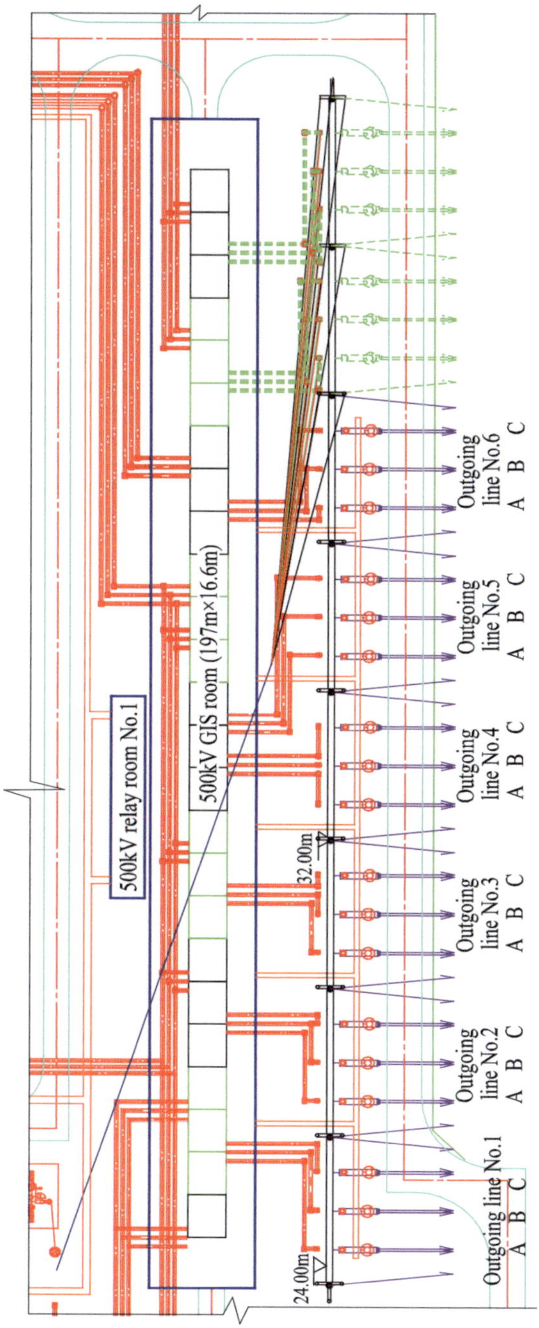

Fig. 22 The electrical layout of the 500 kV AC distribution devices at the Guquan converter station

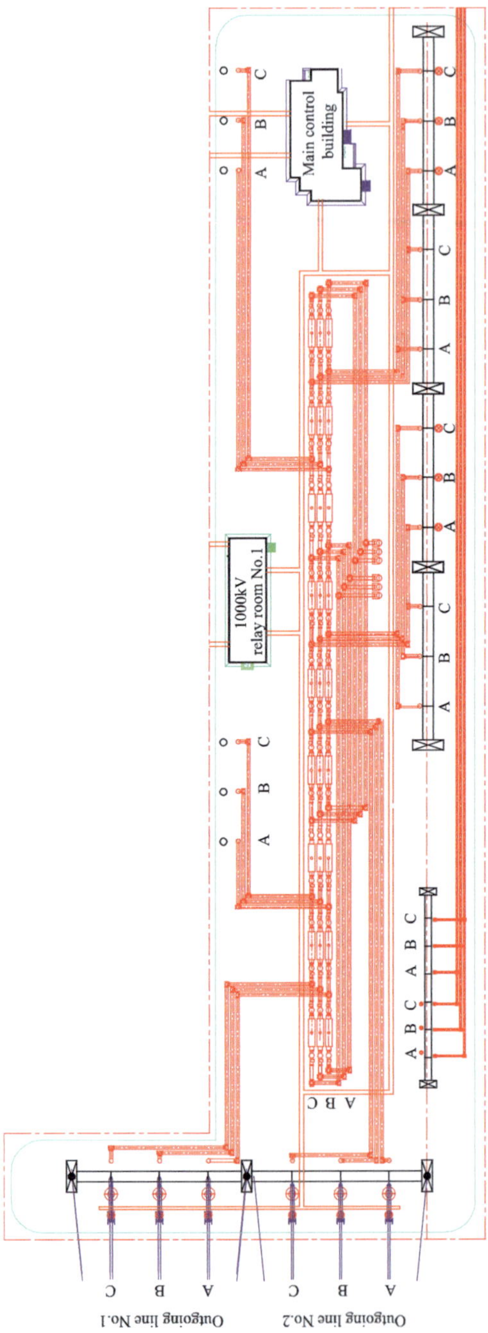

Fig. 23 The electrical layout of the 1000 kV AC distribution devices at the Guquan converter station

Table 8 Dimensions of the filter fence at the Changji converter station (m × m)

Filter	HP24/36	BP11/BP13	HP3	SC
Fence dimensions ($L \times W$)	37.5 × 36.5	45 × 36.5	45 × 36.5	24 × 36.5 (with damping)

I. Changji Converter Station

The total capacity of capacitive reactive power compensation in Changji converter station is 6700Mvar. All AC filters and parallel capacitors are divided into 4 banks and 20 sub-banks. 12 sub-banks are filters, each with a capacity of 305Mvar, while 8 sub-banks are parallel capacitor, each with a capacity of 380Mvar. Each bank is connected as an electrical component into the AC 3/2 circuit breaker wiring string. Single busbar wiring is used within the AC filter bank, and the sub-bank and bank busbar equipment are directly connected to the busbar.

Devices within the AC filter yard fence include capacitor towers, reactors, current transformers, arresters, etc. The dimensions within the fence mainly consider the installation, maintenance, and inspection of each device, and meet the air gap requirements of "Code for design of high voltage electrical switchgear (DL/T 5352—2018)". The dimensions of the filter fence at the Changji converter station are shown in Table 8.

1. Layout of the AC filter bank

The layout of the AC filter yard at the current converter station is arranged by several banks, which can be classified into three types: single-line layout, conventional crisscross layout,[1] and improved grid layout. Based on construction experience, the filter yard in the Changji converter station uses a newly improved layout.

The newly improved layout is based on the conventional crisscross grid layout, changing the filter inlet from crossing over the busbar to crossing over from the SC filter bank inlet. This makes the position of the filter bank inlet more flexible so that it can be placed above any sub-bank and is beneficial for saving GIL pipelines. On the other hand, the number of frames is greatly reduced compared to the previous improved grid layout, saving investment and space.

The typical scheme of the newly improved layout of the AC filter yard is shown in Fig. 24, and the sectional view shown in Fig. 25.

2. Important dimensions for AC filter layout

The 750 kV filter bank busbar of the Changji converter station uses flexible conductors, with a busbar frame width of 42 m, a frame height of 41.5 m, a phase-to-phase distance of 11.25 m, and a phase-to-earth distance of 9.75 m. The filter inlet frame width is 41 m, the frame height is 30 m, the phase-to-phase distance is 11.25 m,

[1] The scheme arranges the four sub-banks of each filter bank back-to-back on both sides of the busbar, to save space. More details can be found in the article "± 800 kV UHV DC Converter Station AC Filter Yard Layout Optimization, Electric Power Construction, 2010".

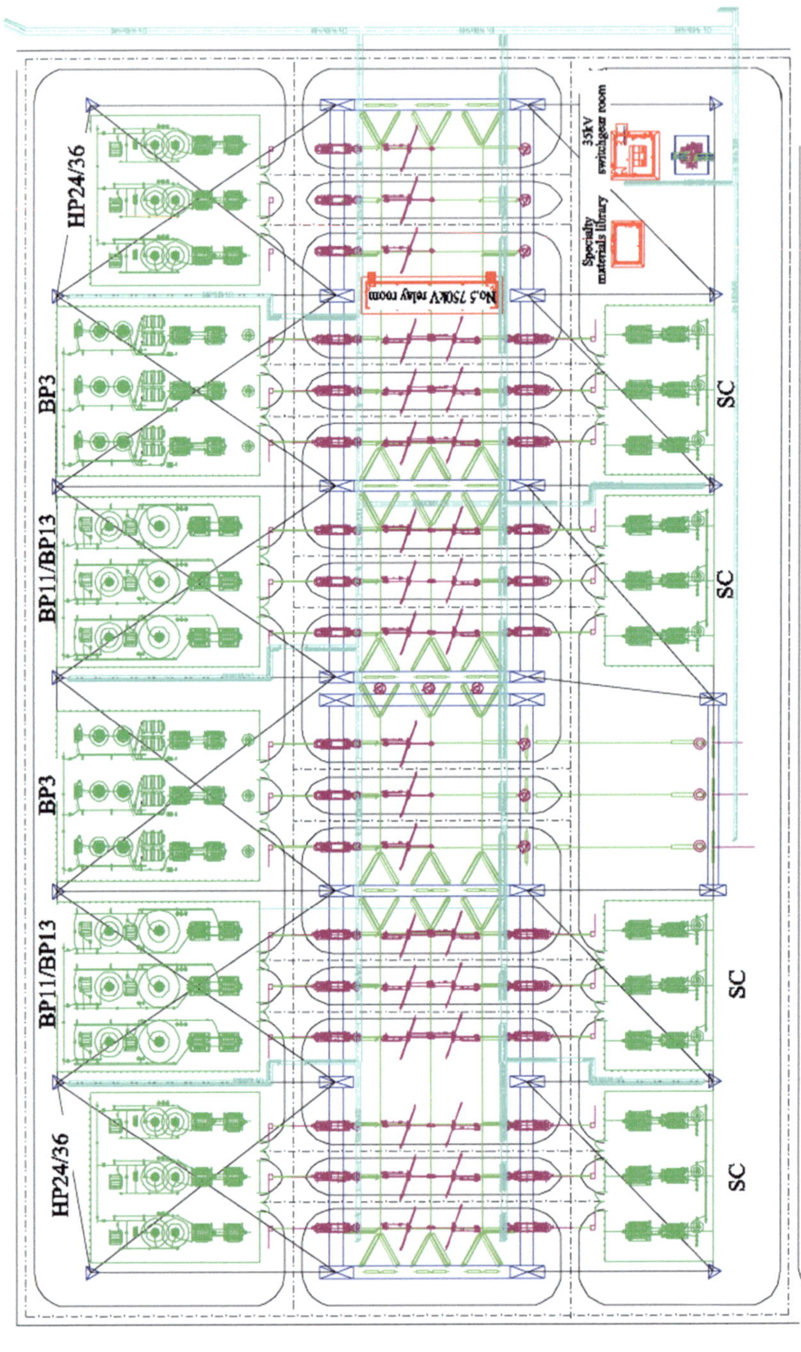

Fig. 24 The diagram of the newly improved layout of the AC filter yard in the Changji converter station

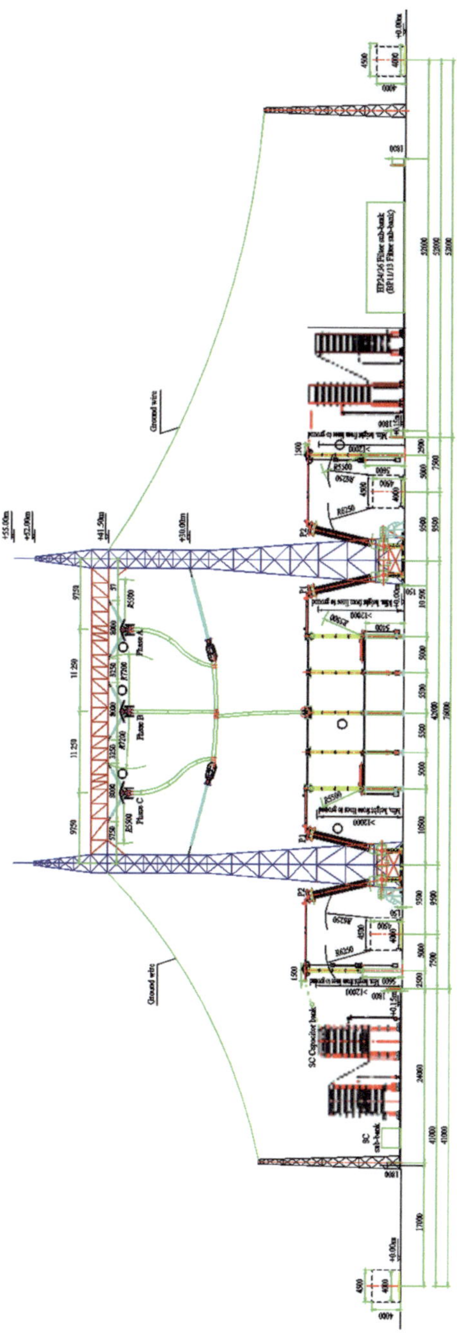

Fig. 25 The typical sectional view of the newly improved layout of the AC filter yard in the Changji converter station

and the phase-to-earth distance is 9.25 m. A combined five-column horizontal telescopic disconnector is used through optimization. Therefore, the distance between the fences of the two groups of filters facing each other is compressed to 76 m.

II. Guquan Converter Station

The total capacity of 1000 kV reactive power compensation is 4080Mvar. All devices are divided into 2 banks and 12 sub-banks. Each bank uses a single busbar connection and is connected as one component in the AC string. The total capacity of 500 kV reactive power compensation is 3990Mvar. All devices are divided into 3 banks and 14 sub-banks. Each bank uses a single busbar connection and is connected as one component in the AC string.

Devices inside the AC filter yard fence include capacitor towers, reactors, current transformers, arresters, etc. The dimensions within the fence mainly consider the installation, maintenance, and inspection of each device, and meet the air gap requirements of "Code for design of high voltage electrical switchgear (DL/T 5352—2018)". The dimensions of the filter fence at the Guquan converter station are shown in Tables 9 and 10.

1. Arrangement of AC filter bank

The typical scheme and sectional views of the newly improved layout of the 1000 kV and 500 kV AC filter yards of the Guquan converter station are shown in Figs. 26, 27, 28 and 29.

2. Important dimensions of AC filter arrangement

In the Guquan converter station, the height of hanging point of the 1000 kV filter bank inlet lines is 59 m. The height of hanging point of lower layer crossing line is 42 m. The width of the bank busbar frame is 51 m. The distance between the upper layer crossing line phases is 14.2 m, while between phase and ground is 11.3 m. The interval width of the AC filter sub-bank is 53 m, of which the phase-to-phase distance is 14.2 m and phase-to-earth distance is 12.3 m. The disconnector is chosen as a combined five-column horizontal telescopic one through optimization.

In the Guquan converter station, the centerline height of the 500 kV filter bank tubular busbar is 16.8 m, of which hanging point height of the interconnection tubular busbar in the sub-bank is 21 m, while that of the upper layer crossing line is 28 m.

Table 9 Dimensions of the 1000 kV filter fence at the Guquan converter station (m × m)

Filter name	HP12/24	HP3
Fence dimensions ($L \times W$)	45 × 49	48 × 49

Table 10 Dimensions of the 500 kV filter fence at the Guquan converter station (m × m)

Filter name	HP12/24	HP3	SC
Fence dimensions ($L \times W$)	36 × 28	46 × 28	25 × 28

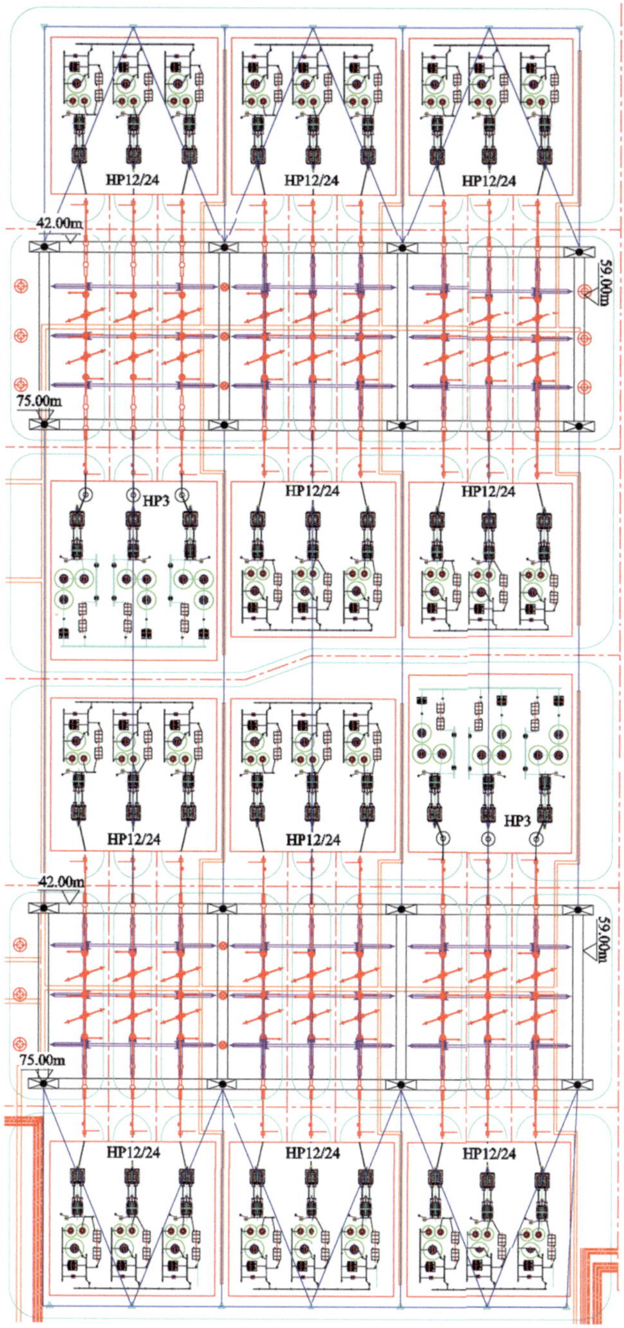

Fig. 26 The diagram of the newly improved layout of the 1000 kV AC filter yard in the Guquan converter station

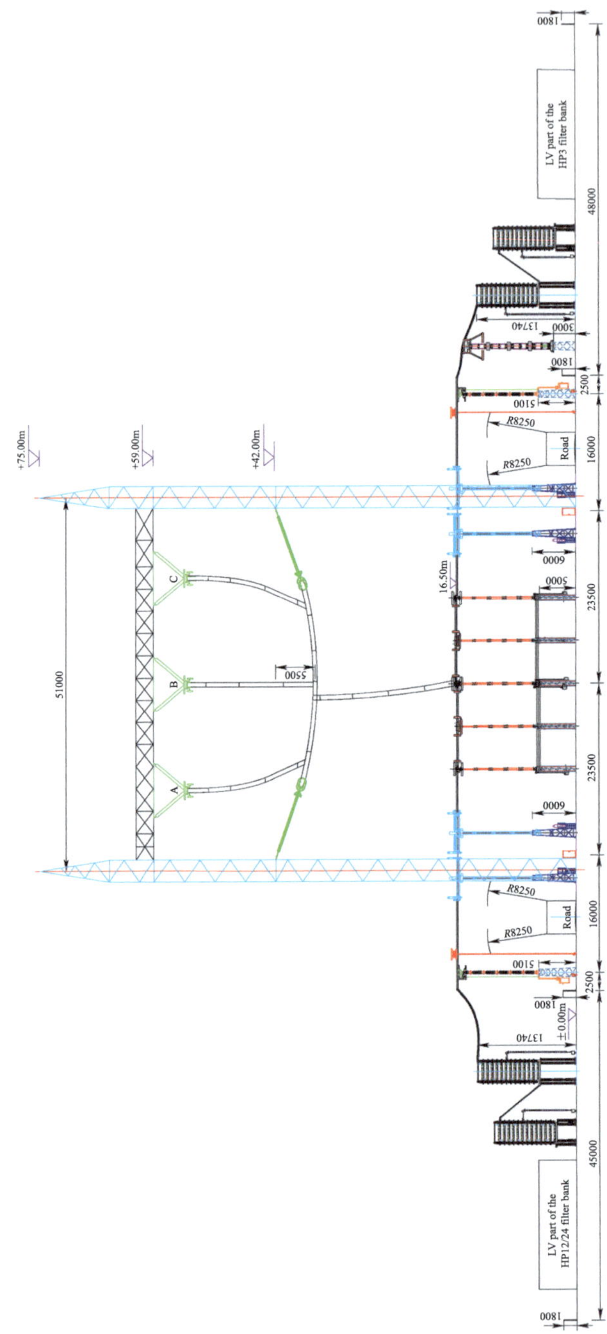

Fig. 27 The typical sectional view of the newly improved layout of the 1000 kV AC filter yard in the Guquan converter station

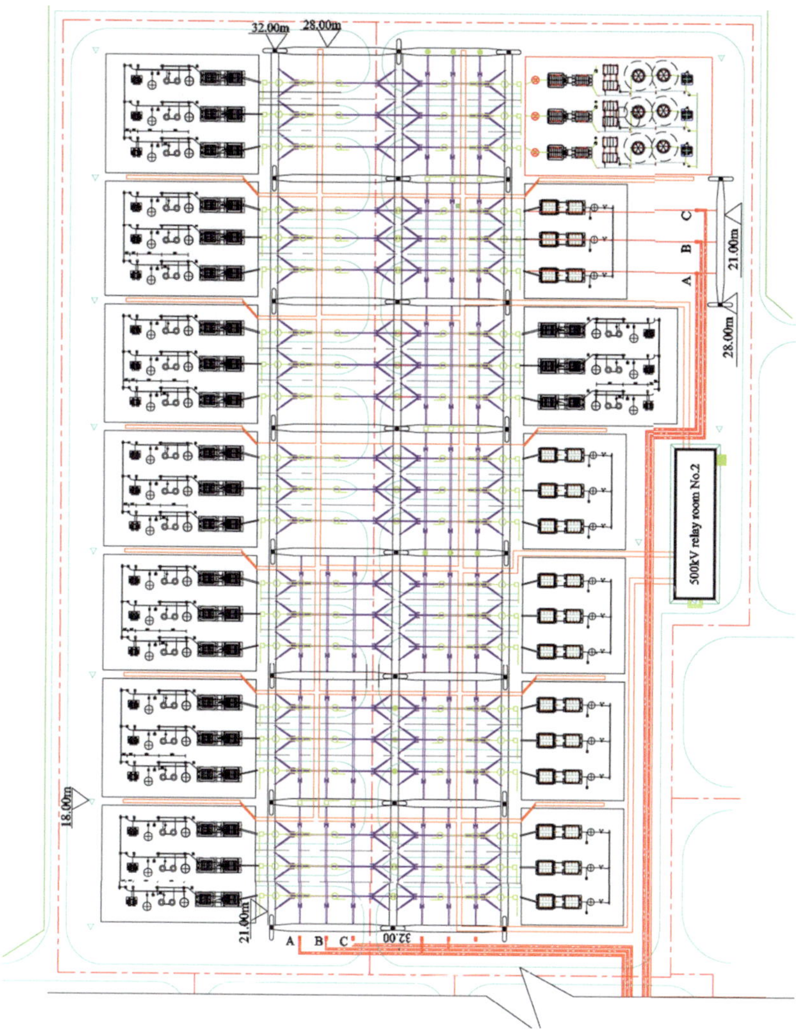

Fig. 28 The diagram of the newly improved layout of the 500 kV AC filter yard in the Guquan converter station

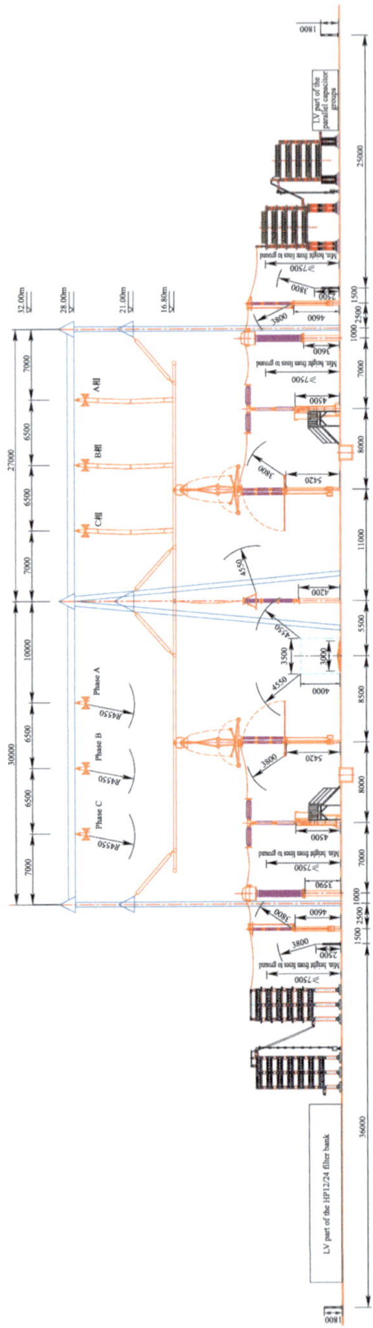

Fig. 29 The typical sectional view of the newly improved layout of the 500 kV AC filter yard in the Guquan converter station

Considering the need of the maintenance lane under the busbar frame of one circuit line in the bank, the widths of two circuits bank busbar frames are 27 m and 30 m respectively. The phase-to-phase distance of the upper layer crossing line is 6.5 m. The phase-to-phase distance of the suspended tubular busbar is 7.5 m, while phase-to-earth distance is 7.0 m. The interval width of the AC filter sub-bank is 30 m, of which the phase-to-phase distance is 7.5 m and phase-to-earth distance is 7.0 m.

5 General Layout

The general electrical layout is determined considering the factors including the main electrical wiring, direction of the AC and DC outgoing line, the type of converter transformer, valve hall and distribution device, the entrance road, the terrain of the converter station and the layout of the buildings.

The principles of the general layout: On the determined site, the general electrical layout should be consistent with the requirements of system planning, reasonably plan the outgoing lines, ensure the safe operation of electrical equipment, meet the needs of operation and maintenance, etc. The general electrical layout is made according to the following principles:

(1) The direction of the outgoing line should adapt to the requirements of the AC and DC line corridors, minimize the outgoing corner towers and avoid line crossings.
(2) The AC distribution device adopts the AC GIS layout scheme.
(3) The valve hall adopts a double valve tower, and the HV valve hall, LV valve hall, and converter transformer adopt a face-to-face/single-line layout scheme.
(4) The DC yard adopts an indoor DC yard.

I. Changji Converter Station

(I) Joint-Construction of the Converter Station and the Substation

The Changji converter station is built together with the Wucaiwan 750 kV substation with the principles of joint-construction as follows:

(1) The AC side of the converter station is connected to the 750 kV open distribution device of the substation by an overhead tubular busbar. Space for installing a 750 kV current limiting reactor is reserved on the 3-loop contact line of the converter station.
(2) The converter station is located on the south side of the 750 kV substation. The original south wall of the substation is demolished. The centerline of the wall is used as the entrance road, while the west wall of the converter station is next to the west wall of the substation.
(3) Considering that the Wucaiwan 750 kV substation had been basically constructed, the original design of the main control building in the substation cannot be used as the administration building of the converter station. Also, the area in front of the substation is relatively compact. It is difficult to expand the

administration building adjacent to the main control building. As a result, there still needs to build administration building in the converter station.

(4) The converter station assembly and test hall are planned to be built within the wall of the converter station, with the engineering quantity included in the main body of the converter station.

(II) Key Issues in the General Layout

In the Changji converter station, the voltage level of the grid side of the converter transformer is 750 kV. And there are 13 outgoing lines, 4 AC filter banks, and 4 incoming lines connected to the grid side of the converter transformer. How to reasonably optimize the land occupation of the converter station is one of the key issues in the overall electrical layout design. Since the 750 kV pipeline is expensive, reducing the length of the pipeline busbar becomes a key issue in layout optimization.

Based on the characteristics of the project, the general electrical layout focuses on the following issues:

(1) Reasonably optimize the layout and coordinate with the Wucaiwan 750 kV substation.
(2) Optimize the incoming line of the filter bank to reduce the length of the GIL branch pipeline.
(3) Reasonably arrange the position of the AC filter to reduce the length of the GIL branch pipeline.
(4) The equipment layout and selection are convenient for operation and maintenance.
(5) The design and equipment selection of the converter station should fully consider wind and sand prevention.
(6) Reduce the volume of the indoor DC yard to reduce investment.

1. Reasonably optimize the layout to save land

The significance of saving land in the converter station is not only to save land resources, but more importantly, it will result in a more compact general layout of the converter station, greatly reducing auxiliary materials used in grounding and cables, thereby reducing project costs. The following work has been done in the land-saving scheme:

(1) Use the newly improved layout to optimize the filter bank arrangement. This scheme optimizes the wiring of the filter bank to avoid the crossing problem of different bank buses in the past improved grid layout scheme. Instead, bank buses are crossing from the upper part of the SC filter. Also, this layout scheme saves 0.76 hm^2 of land compared to the traditional grid layout scheme.
(2) Optimize the layout of the converter site to save land to the maximum extent. The design of the converter site is based on the principle of being able to adapt to various types of converter transformers. By analyzing the assembly site required for the transportation and handling of the converter transformer (78.6 m, between the HV and LV valve hall axes), and the phase-to-phase air clearance size required for the 750 kV frame incoming line, the size between

the HV and LV valve hall axes is controlled to 94 m by using V-shaped tension insulator strings and other methods to limit wire wind deviation and swing, effectively reducing the land occupation of the converter site.

2. Optimize the layout of the 750 kV GIS circuit breaker to reduce pipelines

The 750 kV GIS circuit breaker adopts a side-phase circuit breaker folding layout to save land.

3. Optimize the size of the sub-bank filter fence to reduce land occupation

By comparing the characteristics of the 750 kV AC filter bank fence size controlled by the phase-to-phase distance, the fence size of the 750 kV AC filter bank has been greatly reduced compared to the NZ project. The sizes of the AC filter fences are as follows:

(1) HP3 AC filter sub-bank: 36.5 m × 45 m.
(2) BP11/BP13 AC filter sub-bank: 36.50 m × 45 m.
(3) HP24/36 AC filter sub-bank: 36.5 m: × 37.5 m.
(4) SC parallel capacitor sub-bank: 36.5 m: × 24 m.

4. Optimize the design of the DC yard to reduce the land occupation

Considering the design of the indoor DC yard of the project to make full use of the lateral size, an L-shaped layout and an similar irregular layout are used to make full use of the area of the neutral line area to compress the longitudinal size of the DC yard.

(III) General Layout

The general layout of the Changji converter station is arranged from north to south according to the process flow of "Wucaiwan 750 kV substation—750 kV AC filter bank—750 kV AC switchyard (east)—valve hall and converter transformer courtyard (middle), DC switchyard (west), AC filter bank (east)". The administration building, comprehensive water pump room, garage, etc. are arranged on the southwest side of the station. The entrance road enters the station from the west side.

The whole station layout is compact with clear zoning, which has the characteristics of small land occupation, short GIL pipeline, less earthwork, smooth entrance road, good lighting of the administration building, good coordination with the 750 kV substation, short transportation distance of the converter transformer group, etc.. The land occupation within the wall is about 26.51 hm^2.

II. Guquan Converter Station

(I) Principles of Electrical General Layout

The electrical general layout is conducted according to the following principles:

(1) The valve hall and converter transformer site adopt the single-line layout scheme with the sequence of double valve tower, HV valve hall, and LV valve hall.
(2) The 1000 kV AC distribution device adopts a new outdoor AC GIS layout scheme, and the 500 kV AC distribution device adopts an indoor AC GIS layout scheme.

(3) The 1000 kV AC filter adopts the newly improved layout scheme, and the 500 kV AC filter adopts the newly improved layout scheme, reducing the land occupation.
(4) The general layout should try to compress the land size and reduce the occupation of basic farmland.
(II) Electrical General Layout Scheme

The distribution device in the DC yard is arranged in a pole symmetry layout, with the DC neutral point equipment and grounding pole outgoing line equipment arranged in the center of the device, and the DC high-voltage pole line equipment arranged on both sides of the device. Each pole has 1 set of DC filter banks arranged between the DC neutral point equipment and the DC high voltage pole line equipment, each filter bank is a double-tuned wiring. The smoothing reactor adopts a dry-type reactor, and 2 dry-type smoothing reactors are connected in series on the pole bus and neutral bus, arranged in a single-line manner, and installed close to the valve group.

The valve hall adopts a single-line layout scheme, with the arrangement sequence of "High Pole 1—Low Pole 1—Low Pole 2—High Pole 2". At the same time, the HV and LV valve halls are arranged adjacently. Rooms such as valve cooling, valve control, valve hall air conditioning, and valve group distribution equipment are centrally arranged between the valve hall and the DC yard, and are arranged close to the LV valve hall. The 8 sets (24 units) of converter transformers in the whole station are arranged in a line on the same side of the valve hall. The converter transformer assembly site is located in the open space between the valve hall and the AC distribution device, and the converter transformer incoming line is located above the converter transformer, and is led to the AC distribution device by overhead lines.

The 1000 kV AC distribution device adopts a single-line layout scheme with the sequence of outdoor GIS and the circuit breaker. The connection between the Yy and Yd converter transformers of the same group is achieved through the 1000 kV GIS bus. The 500 kV AC distribution device adopts an indoor GIS, and the circuit breaker adopts a single-column layout scheme.

The 1000 kV AC filter adopts the newly improved layout, and the AC filters of the same banks are arranged on both sides of the bank bus. The bank bus frame and the sub-bank incoming line frame jointly adopt a double-layer frame form, with the bank bus using flexible conductors arranged on the upper layer, and the sub-bank incoming lines using flexible conductors arranged on the lower layer. The sub-bank incoming line circuit disconnector adopts a five-column horizontal isolator. The disconnector is connected to the lower layer conductor through a down lead.

The 500 kV AC filter bank also adopts the newly improved layout, and the AC filter sub-banks and parallel capacitor groups of the same bank are arranged on both sides of the bank bus. The bank bus frame and the sub-bank incoming line frame jointly adopt a double-layer frame form, with the bank bus using flexible conductors arranged on the upper layer, and the sub-bank incoming line using tubular bus arranged on the lower layer. The sub-bank incoming line circuit disconnector adopts a single-column vertical telescopic form.

The DC yard is located on the west side of the station. $\pm$ 1100 kV DC line goes out to the west. The 1000 kV AC yard is located in the north-central part of the station. 500 kV AC yard is located on the south side of the station, and the 500 kV AC line goes out to the south. The 1000 kV AC filter yard and the 500 kV AC filter yard are located on the east side of the station. The valve hall and converter transformer site are located between the AC and DC yards. The DC yard, valve hall, converter transformer, and AC yard are arranged in three columns from west to east. The main entrance to the station is located on the south side. The entrance road is connected from the south side, directly connecting to the transportation road of the converter transformer, which is conducive to the transportation and maintenance of the converter transformer. The overall layout of the entire station is compact with clear functional zoning and continuous connections between distribution devices. The total area within the station's wall is 27.43 hm^2.

Chapter 5
Key Construction Techniques

This chapter summarizes representative key construction techniques in the construction process of the CG project, including civil engineering, electrical engineering, and smart construction. It details the construction considerations, difficulties, control measures, and control effects of each key technique from seven aspects: large-span steel structure hoisting, valve-side outlet device and bushing installation of the HV converter transformer, converter transformer installation tools and work platform, dispose of converter transformer oil, BIM technology, and central control platform of the converter station.

1 Large-Span Steel Structure Hoisting

The CG project not only represents a voltage upgrade but also reflects large equipment volume and large electrical clearance, which in turn affects the building space volume far exceeding conventional DC projects. The steel structure tonnage and span of the two station buildings are both large. Especially, the indoor DC yard is a single-layer, large-span, full steel structure, open industrial building with no internal partition. The maximum span of the building exceeds 80 m, and the height exceeds 46 m. This building is joint-designed with the valve hall, which makes it the largest span substation steel structure building. Construction of the valve hall and indoor DC yard needs numerous auxiliary steel structures and accessories with high installation precision. The indoor DC yard is all high-altitude welding, with stringent installation conditions, extremely difficult construction conditions, and high construction risk. This section uses the Changji converter station valve hall and indoor DC yard as examples to introduce steel structure hoisting construction technology.

Z. Liu, *±1100kV UHV DC Power Transmission Technology*,
https://doi.org/10.1007/978-981-96-3712-6_5

I. Overview and Engineering Quantity of the CG Project

The indoor DC yard of Changji converter station is located on the west side of the HV valve hall, using a spatial steel pipe truss structure to form an L shape, with a lateral span of 80 m in the north and 62.5 m in the south. The north–south distance reaches 99 m, and the roof coverage area is 7075 m^2. The indoor ± 0.00 m is equivalent to an absolute elevation of 513.05 m. The eaves height of the roof is 43.00 m, the ridge height is 45.50 m. A three-dimensional truss is used, with the width of the truss is 3 m, the height of the truss column is 3 m, and the height of the truss beam gradually changes from 4.5 m to 3 m to 4.5 m. The lowest eaves height inside is 38.50 m, and the axis distance is 11 m, 12 m, and 13 m. The truss column is a 3 m three-dimensional triangular tube truss, and the wind-resistant column is a 1.5 m plane tube truss structure. The total weight of the steel structure is 1310 t.

Using a portal rigid frame lightweight building steel structure, the HV valve hall is located on the west side of the auxiliary control building with a lateral span of 48.1 m and a longitudinal span of 118.5 m. The roof coverage area is 5672.5 m^2. The indoor ± 0.00 m is equivalent to an absolute elevation of 513.05 m. The eaves height of the roof is 39.4 m, and the ridge height is 41.50 m. GZ-1, GZ-2 columns are 1200 mm × 400 mm × 25 mm × 30 mm H-shaped steel double-component columns, with a distance of 1600 mm. The double-column lacing bars are mainly angle steel welded to the column. The wind-resistant column is a grid column structure. The double columns are all angle steel and steel plates welded into "[" shaped column structure, symmetrically arranged with the axis distance of 1400 mm. The double-column tie bars are angle steel welded to the column. The total weight of the steel structure is about 2176 t.

II. Key Considerations and Difficulties in Construction of the CG Project

The large span of the steel structure of the indoor DC yard leads to limited construction site. Also, there are many technical difficulties during the construction, including:

(1) The maximum span of the indoor DC yard truss beam is 80 m, and the maximum lifting height is 47.5 m, which are all biggest among all UHV projects under construction. Compared with the indoor DC yard of the Guquan converter station (maximum span 61.6 m), the span is 18.4 m larger. Requirements for the hoisting process are higher than the international advanced level.
(2) The steel columns and roof truss beams of the indoor DC yard are all triangular space truss structures. When welding in the air, adjustments and corrections are needed (each column and roof truss beam has 3 main connection points and 2 diagonal brace key connection points, a total of 5 key connection points). The control of the deformation amount caused by the structural processing size and the hoisting process is quite strict.
(3) One end of the Z2-2 axis roof truss beam in the Pole 2 of the indoor DC yard needs to be fixed at the contact truss. Therefore, the contact truss needs to be immediately installed after the steel columns at the Z2-1 axis and Z2-3 axis are hoisted, which increases the difficulty of aerial docking welding.

(4) The span of the roof truss beam is large and some roof trusses are asymmetrical. During the double-machine lifting process, the coordination requirements of the two crawler cranes are much higher. Also, the deformation amount during the roof truss hoisting process needs to be controlled by calculation, which is difficult to calculate.
(5) Due to the project is located in the eastern part of the Junggar Basin in Xinjiang, the temperature difference between day and night during construction is large (the lowest temperature at night is 18 °C, the highest temperature during the day is 65 °C, and the temperature difference between day and night reaches about 50 °C), causing large temperature fluctuations in the surface of the steel structure. Therefore it is difficult to control material deformation caused by ambient temperature.
(6) The single-column steel column is high and heavy (the height of a single steel column is 44.9 m with 30 t in weight), resulting in difficult control of the column verticality during the construction.

The construction period of the steel structure installation of the HV valve hall is short with cross-operation situation with the fire wall, leading to many technical difficulties during the construction, including:

(1) The steel columns and wind-resistant columns of the HV valve hall are all double-component column structures. The span of the HV valve hall truss beam is 48.1 m, and the maximum lifting height of the steel column is 44.55 m. The control of the deformation amount caused by the structural processing size and the hoisting process is quite strict.
(2) Due to the requirement of the first-level gridding plan, the installation period of the HV valve hall steel structure is short, and it overlaps with the construction of the Pole 2 HV fire wall for some time. When hoisting the A-axis steel column, wind-resistant column and column support, there is a cross-operation situation with the fire wall structure construction. It is necessary to plan the installation of the steel structure and the construction process of the fire wall structure in advance to avoid adverse effects on construction safety and construction period.
(3) Due to the project is located in the eastern part of the Junggar Basin in Xinjiang, the temperature difference between day and night during construction is large (the lowest temperature at night is 18 °C, the highest temperature during the day is 65 °C, and the temperature difference between day and night reaches about 50 °C), causing large temperature fluctuations in the surface of the steel structure. Therefore it is difficult to control material deformation caused by ambient temperature.
(4) The single-column steel column is high and heavy (the height of a single steel column is 42.55 m with 51 t in maximum weight), resulting in difficult control of the column verticality during the construction.
(5) Construction call for higher cooperation requirements for the steel structure manufacturer. The sequence and continuity of the steel structure components arriving on site must meet the site installation progress. The assembly speed of

the steel structure columns and roof trusses must meet the site lifting requirements. The quality of the steel structure assembly must be qualified at one time.

III. Working Procedure

(I) Situation in Guquan Converter Station

1. Roof truss assembly
   (1) The assembly of the roof truss is better to be completed in the corresponding vacant space under the installation axis.
   (2) The assembled roof truss must be leveled with sleepers to prevent deformation.
   (3) Check the quality of the assembled roof truss, including the geometric dimensions of the components and permanent deformation. Anti-deformation measures during the assembly process includes:
      1) Place sleepers before assembly, level with a level to prevent lateral deformation.
      2) Measure the overall length and height of the component before tightening the bolts, to prevent difficulties in positioning due to size deviations.
      3) During the assembly process, hang a horizontal line at both ends of the component to ensure the straightness of the component.
      4) Set the pre-arch of the roof truss according to the design requirements.
2. Setting of lifting points

The number of lifting points is determined according to the span and quality of the roof truss. It is better to set special lifting lugs or lifting holes on the components. Before lifting, the force at each lifting point should be calculated to ensure the overall stability of the roof steel truss during the lifting process.

Before the roof truss is lifted, all column supports and connecting beams in this span must have been lifted, and the steel columns and roof trusses have passed the welding flaw detection. It is strictly forbidden to lift the truss when the frame structure has not formed a stable structure.

(II) Situation in Changji Converter Station

1. Steel structure lifting construction scheme and process flow of the indoor dc yard

A spatial steel pipe truss structure is used in construction of the indoor DC yard. Through BIM visual simulation, assembly using on-site vertical assembly platform, and the use of independent high-altitude welding platforms, construction is much safer and more convenient. The assembly accuracy and the quality of the force-bearing welds will be improved. And the construction costs and period can both be reduced. The three-dimensional diagram of the indoor DC yard steel structure is shown in Fig. 1.

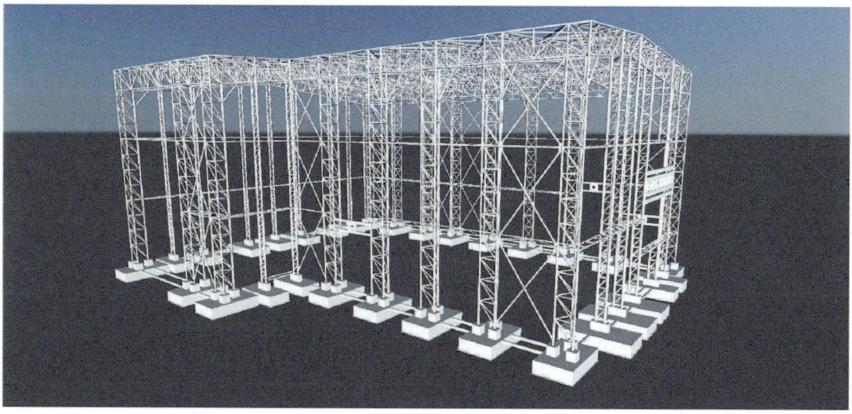

Fig. 1 The three-dimensional diagram of the indoor DC yard steel structure

(1) Arrangement of lifting sequence. The lifting starts from the steel column then the roof truss, from the inside to the outside.
(2) Lifting of spatial steel pipe truss column.
    1) A single-column steel column is hoisted using a 150 t crawler crane tied below the top of the column at 13.32 m for the main hoisting, and a 50 t truck crane tied above the bottom of the column at 13.32 m for auxiliary hoisting. The main crane's hook is 14 m away from the top of the column. During the hoisting process, the main crane performs the vertical hoisting of the steel column, and the auxiliary crane serves to balance both ends of the steel column to prevent deformation. The hoisting stops when the steel column is lifted 100 ~ 200 mm off the ground. After waiting for 10 min, it is necessary for professional crane operators to observe the stress state of the steel wire rope, whether there is any sliding at the hoisting point, etc. (trial hoisting). The hoisting will continue when all operation confirmed to be correct. In the subsequent process of the steel column standing vertically, the auxiliary crane only serves to prevent deformation of the steel column and no longer participates in the vertical hoisting of the steel column. After the steel column stands vertically, the auxiliary crane unloads the hook, and the subsequent docking work of the steel column and the foundation is carried out by the main crane. After the steel column and the foundation are docked, the steel column is temporarily fixed with balance ropes and ground anchors, and the initial tightening of the anchor bolts is carried out. The main crane unloads the hook thereafter.
    2) Correction for the steel column installation.
        a. The correction of the column includes planar position, elevation, and verticality.
        b. The column correction is carried out simultaneously with measuring tools (theodolite, level). The theodolite is used to correct the verticality

of the column, and attention should also be paid to the effects of wind and temperature. The level is used to measure the flatness of the steel column and the anchor bolts.

c. After the steel column is hoisted and the column base is inserted into the foundation bolts, the column correction is carried out mainly for adjusting the elevation and correcting the verticality. Correction can use theodolite, balance rope, jack, crowbar and other tools to apply vertical and lateral forces to the steel column, and adjust the nut at the bottom of the column. After adjusting and correcting between the bottom plate of the column and the foundation, it is fixed with bolts and using double nuts to prevent loosening. The crane can loosen the steel wire rope after the balance rope is fixed. The schematic diagram of the steel column hoisting process is shown in Fig. 2, and the hoisting correction schematic diagram is shown in Fig. 3.

(3) Hoisting of spatial steel pipe truss roof frame.

1) There are nine frames in the indoor DC yard steel roof frame, including four 80 m span steel roof frames and five 62.5 m steel roof frames.
2) The truss is transported to the site in 2 sections from the processing factory and is vertically assembled on a pre-assembled on-site vertical assembly platform. First, use crane to position the truss in the corresponding position, fix the positioning block, adjust the adjustment plate, and ensure the relative positions of the steel components all correct. A welding shrinkage allowance must be reserved before determining the relative position of the main pipe. A centerline of each node of the main pipe is drawn on the assembly platform. Before the overall welding, the branch pipe is assembled and position welded. Position welding the branch pipe needs no less than 4 points. After positioning, full-position welding is performed on the main truss. To ensure the quality of welding, welds of the 3 main pipes must be alternately welded and one main pipe cannot be fully welded before another in all process of overhead position welding, vertical position welding, and welding. The welding of the upper chord support cross should be carried out first. It is much better to conduct position welding with the main pipe after shrink. Since the toe part is a full penetration weld, the root part is a fillet weld, and the side transitions from a penetration weld to a fillet weld, the weld sequence is arranged as toe part first, then the root part, and finally the side considering the welding deformation. Before the truss is hoisted, the weld is subjected to non-destructive flaw detection and is qualified. The scaffolding, scaffolding board, horizontal lock fixing, etc. are checked and qualified. The assembly of the steel truss is shown in Fig. 4.
3) The 80 m truss hoisting uses two 150 t crawler cranes for the main hoisting. Before the hoisting starts, 8 hoisting points (8 steel wire ropes are used to evenly distribute the force, and the length of the steel wire rope during the truss hoisting process is calculated quantitatively to ensure the safety of the hoisting) are tied with steel wire ropes. The crane slowly lifts the hook to tighten the hoisting

Fig. 2 The schematic diagram of the steel column hoisting process

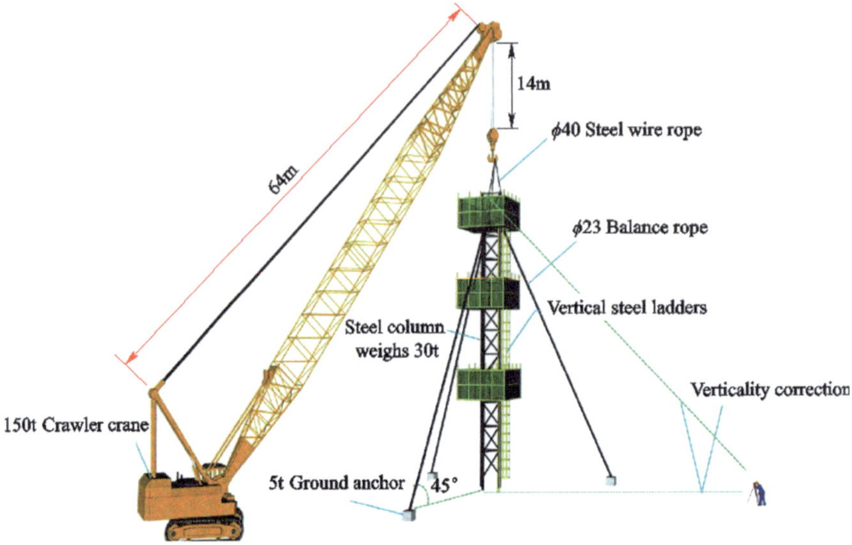

Fig. 3 The schematic diagram of the hoisting correction

Fig. 4 The assembly of the steel truss

Fig. 5 The 62.5 m span truss hoisting diagram

steel wire rope and put it under stress. A 20 t chain hoist is required for the truss leveling work. In this process, a reverse chain is used to pull the steel wire rope on both sides to find the correct position for the truss hoisting. Also, it needs to check the deviation of various data of the steel truss. The 62.5 m span truss hoisting diagram is shown in Fig. 5, and the 80 m truss double machine lifting construction is shown in Fig. 6.

4) When the truss is lifted 100 ~ 200 mm off the ground, the hoisting worker should conduct a comprehensive check on the force of the steel wire rope, then continue hoisting after check. If tilting occurs, the steel wire rope should be immediately pulled to find the correct position to prevent safety accidents.
5) After the truss is lifted, first slowly let it approach one end to the steel column connecting plate. When approaching, the installation personnel hold the truss beam, cooperate with the hoisting worker to position one end of the truss beam. The other end is docked by dragging the balance rope. Then, the nodes at the beam-column joint are welded. To eliminate the stress generated during the welding process, the 6 welds of the beam-column joint in the air must be welded simultaneously. When the nodes between the beam and column stably connected by welding, the main crane can loosen the hook. The deviation of various data of the truss should meet the site control deviation standard.

2. Construction plan and process flow of HV valve hall steel structure hoisting

The steel structure hoisting plan adopts the piece-by-piece installation method: First, a 450 t crawler crane is used to install all the columns. The columns are corrected

Fig. 6 The 80 m truss double machine lifting construction

and finally fixed. The connecting beams and column supports are installed using a 50 t truck crane. After the columns and connecting beams are hoisted, the 450 t crawler crane and truck crane are used to install the roof trusses and roof supports section by section.

A 450 t crawler crane is used for the hoisting of the HV valve hall steel columns and wind-resistant columns. After a single-column steel column is erected and adjusted according to the foundation cup axis position, it is temporarily fixed with balance ropes and ground anchors, and then the main crane is unhooked.

A 450 t crawler crane is also used for the hoisting of the HV valve hall steel roof trusses. The crane is unhooked only when all the high-strength bolts at the connection point between the truss and the steel column are installed and preliminarily tightened. When installing the first truss, since a stable structure has not yet been formed, it is necessary to set up ϕ 17 mm balance ropes at both ends of the truss to prevent sudden strong winds. Check the weather forecast the day before hoisting, and continuously monitor the wind speed 1 h before hoisting. If a sudden strong wind occurs during the hoisting of the truss, it should be landed in time if possible. When find it cannot be landed, it should be quickly approached to the fixed components and temporarily secured with steel wire ropes. When the hoisted to the position, it can be temporarily fixed and set up balance ropes in time. During the hoisting process, it is necessary to add 20 t chain hoists at both ends to prevent deformation adjustment, and rigid internal supports are added at the ends.

The operation platform at the top of the steel column is built on the horizontal lacing bar at 34.5 m of the steel column. The platform is covered with scaffolding tubes, and the scaffolding boards extend 200 mm from each end of the steel column. Then, a 1.2 m high, 1.5 m spaced vertical pipe, and 600 mm spaced horizontal pipe are built on the scaffolding board platform to form a circular passage. The outer edge of the scaffolding pipe is covered with a dense mesh. When working at the connection point of the steel column and the beam, the personnel work on the built safety platform. The platform fixes the mobile power box and places the welding machine and fire extinguisher (the secondary line length of the welding machine should not exceed 30 m).

Use the welded steel tube on the upper chord of the roof truss as anchor points for the horizontal rope. The steel columns are equipped with vertical pull ropes, steel ladders, and safety self-lockers. The setting of the steel beam safety rope is shown in Fig. 7.

(1) The hoisting method and requirements of the steel column. The hoisting of the HV valve hall steel column starts from the J-9 axis and ends at the J-8 axis counterclockwise. The J-9 axis corner column is hoisted first, followed by the 9-axis wind-resistant column, and then the A-axis steel column. After the hoisting is completed, it is temporarily fixed with ground anchors and balance ropes, and the crane is unhooked after the temporary fixation is completed.

The maximum weight of the HV valve hall steel column is about 51 t. The maximum hoisting weight is $51 \times 1.2 = 61.2$ (t) (1.2 is the sum of the power factor and safety factor). The length of the main arm of 450 t crawler crane reaches 61 m, with the

Fig. 7 The setting of the steel beam safety rope

working radius of 16 m and the rated lifting load of 88.2 t. The load rate is 69.4%, which meets the hoisting requirements. The ϕ of the steel wire rope is selected 52 mm, 10 times safety factor. When hoisting the steel column, each hoisting point is bound with 4 steel wire ropes. And the 4 steel wire ropes are stressed at the same time. When hoisting the wind-resistant column, the steel wire rope is tied below the top of the grid column L/3 m (about 13 m, L is the length of the steel column), the main steel column hoisting uses two types of hoisting points: one is the hoisting point tied below the column top L/3 m (about 13 m), the other is to use the high-strength bolt hole on the side of the end steel column to install the lifting lug. To prevent the steel wire rope from damaging the paint of the steel structure due to friction, anti-slip rubber needs to be added at the contact point between the steel wire rope and the steel column. Before hoisting, check the foundation surface elevation and report to the project supervision unit for acceptance. The hoisting height is greater than the limit height of "in-place height + steel wire rope height + hook". After verification, the 450 t crawler crane with a boom end height of 61.8 m meets the hoisting requirements. The schematic diagram of steel column hoisting point 1 is shown in Fig. 8, the schematic diagram of steel column hoisting point 2 is shown in Fig. 9, and the schematic diagram of the ear plate is shown in Fig. 10.

After the steel column is assembled on the ground, the project supervision unit organizes the acceptance of the steel column. After passing the acceptance inspection, the hoisting is carried out. Before the hook is lifted, the balance ropes and vertical steel ladder are tied near the top of the steel column, and the vertical lock is tied on the ladder. At the same time, the installation center lines of the upper and lower ends of the steel column and the + 1 m elevation line of the column lower end are drawn.

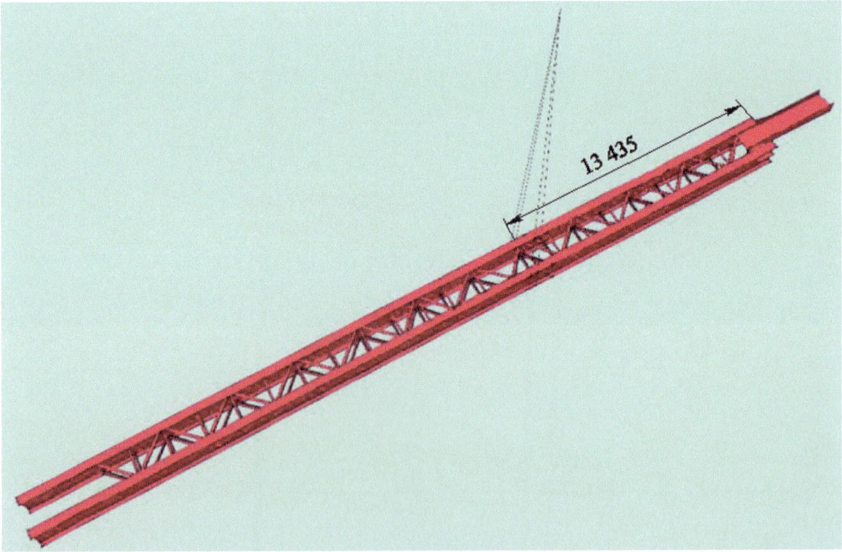

Fig. 8 The schematic diagram of steel column hoisting point 1

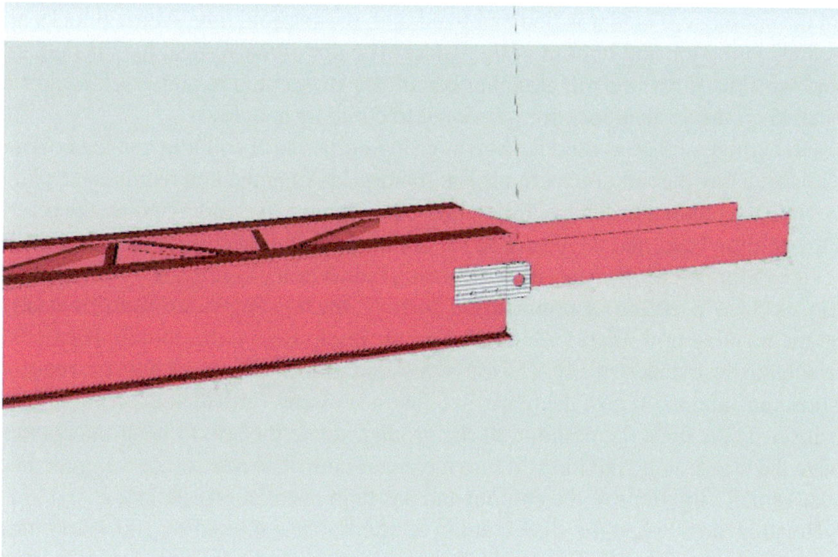

Fig. 9 The schematic diagram of steel column hoisting point 2

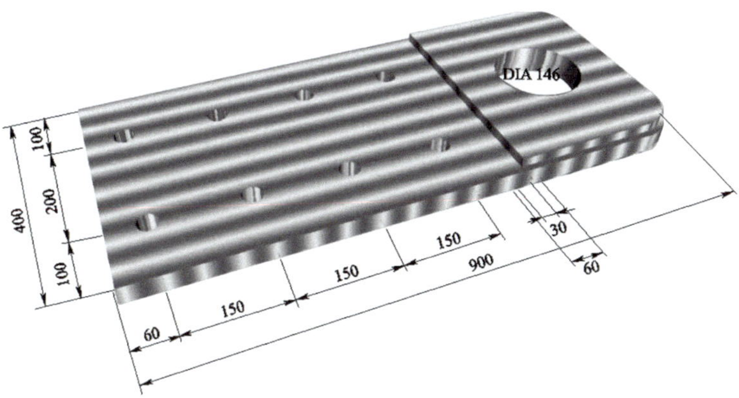

Fig. 10 The schematic diagram of the ear plate

Dusts, oil stains, and dirt on the surface of the components need to be cleaned up. Check the ink line of the column frame on the top surface of the foundation, and prepare for the hook after all above work completed.

The vertical steel ladder is prefabricated on site, with 40 mm × 4 mm galvanized flat iron on both sides of the ladder, spaced 400 mm apart. ϕ 16 mm round steel is used for personnel climbing up and down, with a spacing of 300 mm between adjacent round steels. The round steel is welded to the flat iron through holes. The ladder is tied on the lacing bars between the columns, the ladder is hung on the scaffold tube, and the subsequent ladder is tied and fixed with the scaffold tube. Vertical locks are installed at the top and bottom of the ladder. The top of the vertical lock is tied and fixed with the most horizontal lacing bar of the steel column, and a self-locker is installed on the vertical lock for personnel to climb up and down.

450 t crawler crane is used for hoisting single truss steel column, and the column is hoisted using the rotation method. The rotation hoisting method requires the plane layout of the column to achieve following requirements: the binding point, the center of the column base, and the center of the column foundation cup are all on the same arc (the radius of the arc is set as the hoisting radius of the column R). The "column base" is close to the foundation. During hoisting, the hoisting radius remains steady, and the hoisting arm rotates while hoisting the hook. Before the column is erected, the column base stands steady. The top of the column gradually rises with the rotation of the crane and the rise of the hook, making the column vertical at the column base position. Then, hoist the column off the ground, rotate the arm to hoist the column above the cup mouth, and insert it into the cup mouth. The rotation hoisting method causes small vibrations to the column and has high installation efficiency.

Hoisting stops when the steel column is lifted off the ground by 100 ~ 200 mm. After 10 min waiting, specialists of hoisting will observe the force status of the steel wire rope and whether there is any sliding at the lifting point (trial lifting). Hoisting will continue when all risks are checked and confirmed. After the steel column is docked with the foundation, use balance ropes and ground anchors to temporarily fix the steel column. With all above work completed, the main crane unloads the hook. The sequence of hoisting a single steel column is shown in Fig. 11.

To avoid the free swing of the hoisted steel column, it should be tied with a hemp rope 2 m away from the bottom of the column as a restraining rope to adjust the direction. When the steel column is hoisted to the corresponding installation foundation, the crane is commanded to slowly descend. When the bottom of the column is 100 ~ 200 mm away from the foundation position, adjust the column and the two reference lines of the foundation to the accurate position, and command the crane to descend to the cup mouth.

After the steel column is hoisted into the cup mouth foundation and accurately positioned, use a total station or theodolite to adjust the verticality of the steel column (two directions form an angle greater than 90° and less than 180° for simultaneous monitoring and adjustment). When the column is straightened, weld a ϕ 28 mm rebar on all four sides of the steel column, so that the rebar is topped on the inner wall of the cup mouth, and then tighten the balance ropes to temporarily fix the steel column, as shown in Fig. 12.

Two schemes are prepared and used in temporary reinforcement. The first one uses completed structures such as fire walls and auxiliary control building frames to make a temporary reinforcement point with rebar or section steel with another balance rope set on the ground anchor. The other scheme uses steel columns far from the building. According to the hoisting sequence, immediately install the support beam between the previously hoisted steel columns, with another balance rope set on the ground anchor, as shown in Fig. 3.

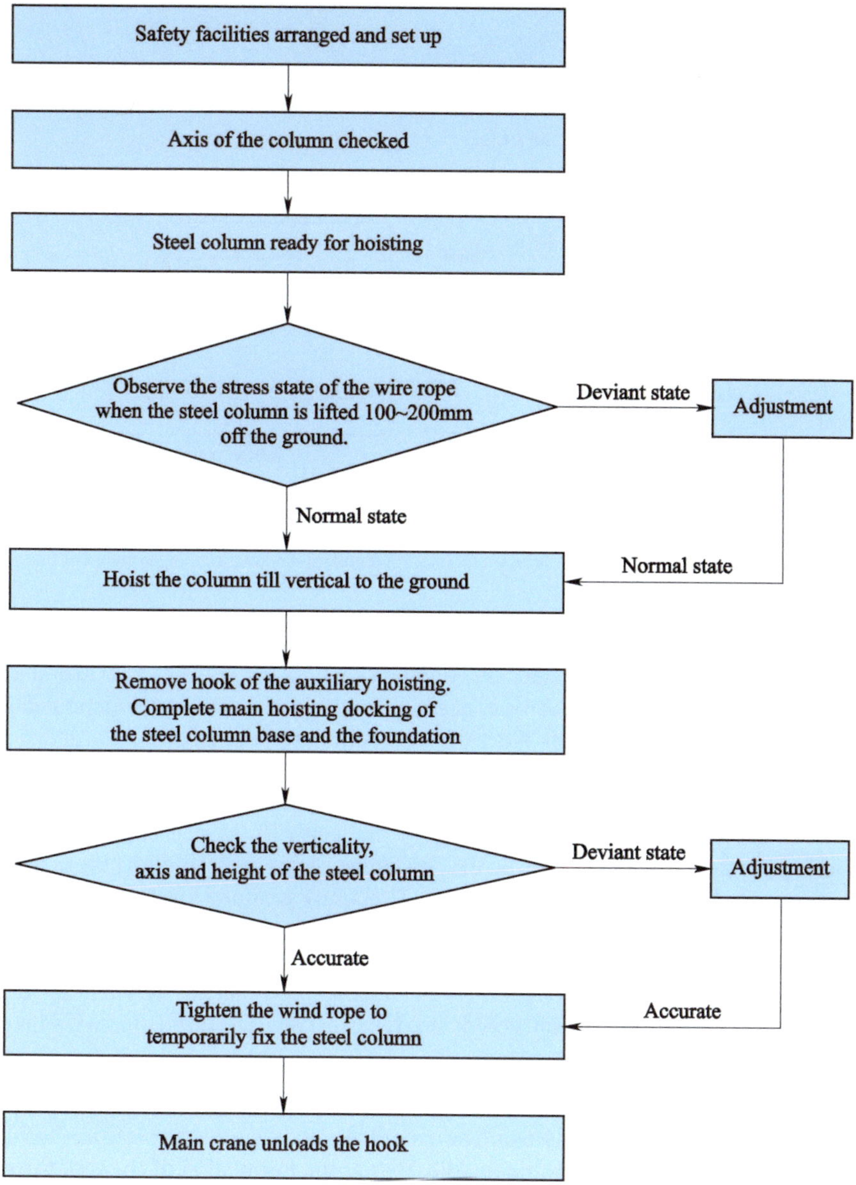

Fig. 11 The sequence of hoisting a single steel column

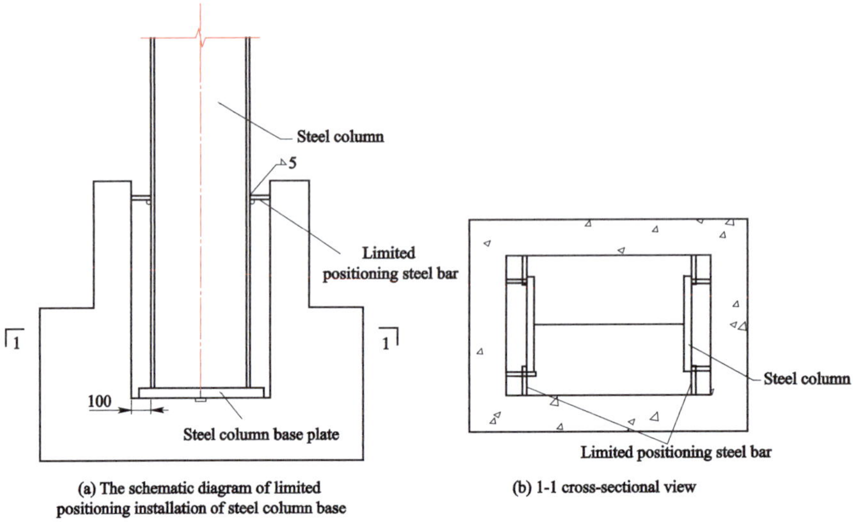

Fig. 12 The schematic diagram of temporary fixation measure for hoisting single steel column

Using the above method, the adjacent column is hoisted. After the second column is hoisted and temporarily fixed, the supports or tie rods between the two steel columns are installed immediately, connecting two columns to form a structural unit to enhance stability. And so on. After the steel columns of each axis are hoisted and connected into a whole structure with tie rods or supports, the overall acceptance of the axis is carried out. After passing the acceptance, the secondary fine stone concrete grouting of the column base is carried out to permanently fix the steel column. Before pouring concrete, clean and moisten the cup mouth. After the strength of the poured concrete reaches 70%, the balance rope or temporary reinforcement measures can be removed.

According to the hoisting sequence, the second axis steel column is hoisted. After the overall hoisting is completed with all the supports and tie rods between the columns are installed, the overall acceptance is carried out. The next column hoisting is proceeded after passing acceptance.

(2) Installation of the steel roof truss. The HV valve hall steel roof truss consists of nine sections, spanning 46.6 m. The overall installation sequence is from the 9th axis and 1st axis towards the middle. Before the installation of the roof truss, it is necessary to communicate with the manufacturer about the arrival of the equipment, the on-site assembly, and the site issues.

The truss is transported to the site in sections by the manufacturer and assembled at the construction site.

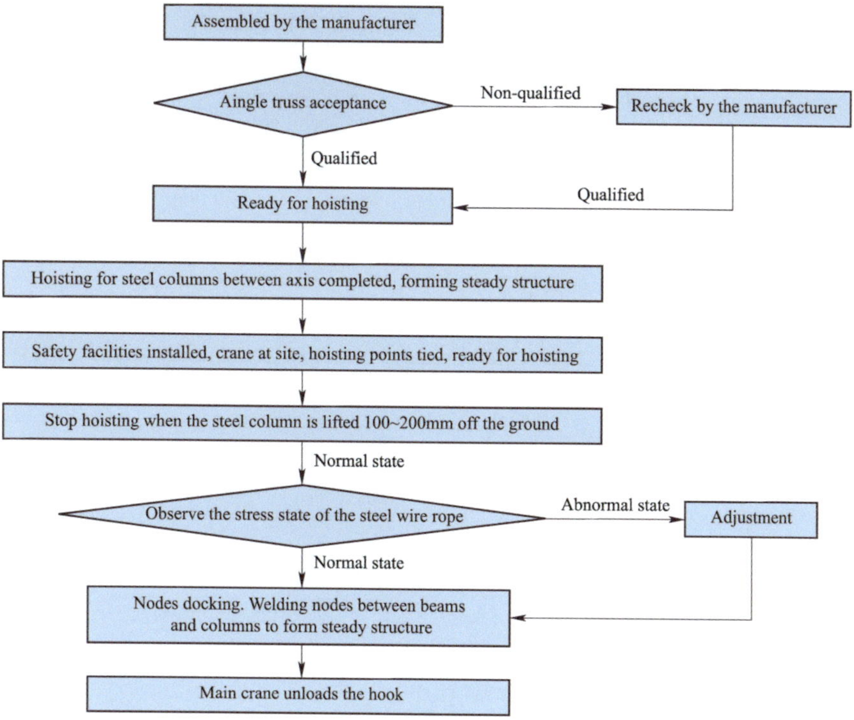

Fig. 13 The installation sequence of a single truss

The installation of the steel roof truss begins after the steel column is installed. Before the installation of the steel roof truss, it needs to pass the acceptance organized by the project supervision unit. The scaffolding, footboards, and horizontal locks needed for the installation of the truss are fixed in place. The installation sequence of a single truss is shown in Fig. 13.

The installation of the truss uses a 450 t crawler crane for the main hoist. Before the start of the hoisting, the four hoisting points are tied with steel wire ropes, and the crane slowly lifts the hook to tighten the hoisting steel wire ropes and put them under stress. During this process, it is necessary to use a chain to pull the steel wire ropes on both sides for the truss hoisting and alignment work, check the deviation of various data of the steel roof truss, and form document data for transfer. The steel wire rope binding node diagram is shown in Fig. 14, and the schematic diagram of the 48.1 m span truss hoisting point is shown in Fig. 15.

When the truss is hoisted 100 ~ 200 mm off the ground, the hoisting worker should conduct a comprehensive check on the force of the steel wire rope. Hoisting can be continued after all checked. If tilting occurs, it is necessary to immediately use a chain to pull the steel wire rope for alignment to prevent safety accidents.

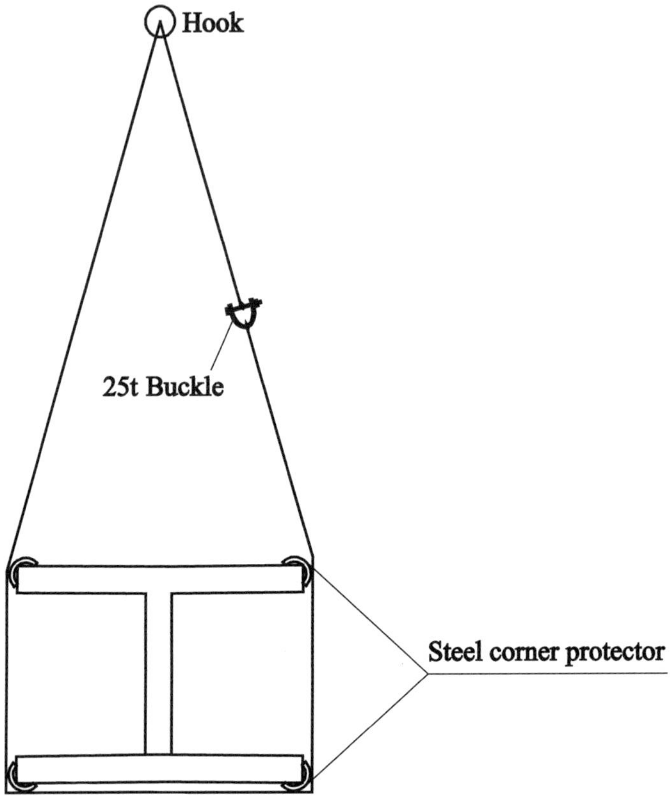

Fig. 14 The steel wire rope tied node diagram

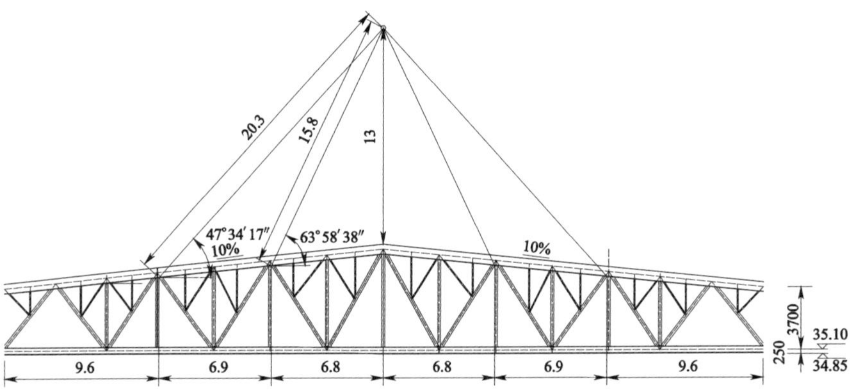

Fig. 15 The schematic diagram of the 48.1 m span truss hoisting point

When the truss is hoisted, first slowly approach one end to the steel column connecting plate. When approaching, the installation personnel hold the truss beam, cooperate with the hoisting worker to position one end of the truss beam. The other end is docked by dragging the balance rope to make the beam-column connection node dock, then install high-strength bolts. Only when all high-strength bolts are installed and initially tightened as well as the node between the beam and column is stably connected, the main crane can loosen the hook. The deviation of various data of the truss should meet the site control deviation standard. The connection node diagram of the truss and steel column is shown in Fig. 16.

After the first truss is installed, the balance ropes should be tightened and ground anchors should be set up for temporary fixation. When the first truss on the 1st and 9th axes is installed, in addition to setting up the balance ropes in time, 20 mm diameter round steels are used to temporarily weld connect with the indoor DC yard steel column and auxiliary control building structure. After the horizontal support between adjacent trusses is installed to stabilize the connection between the trusses, the truss balance ropes can be removed.

After the trusses on the adjacent two axes are all installed, 50 t truck crane is used to hoist the horizontal support and diagonal brace between the trusses.

(3) Installation of column supports, inter-column trusses, horizontal supports between trusses, and diagonal braces. After the adjacent steel structure columns are installed, the installation of column supports and inter-column trusses begins. The installation of column supports uses 50 t truck crane. During the installation process, attention should be paid to the installation direction of the column supports.

(III) Welding

The connections between roof truss and steel columns, roof truss and horizontal supports, and adjacent columns are all connected by welding. High height of the indoor DC yard results in unstable high-altitude airflow and air pressure. High-altitude welding is carried out using manual arc welding. During the welding of the chord rods between the roof trusses, a factory-processed high-altitude welding operation platform is suspended on the roof truss tube. The welders work inside the high-altitude welding operation platform. After each weld is completed, it is inspected by a professional inspection unit for non-destructive testing. Construction proceeds to the next process and after passing.

(IV) Fireproof Coating and Topcoat Spraying

After the components are welded and pass the non-destructive testing, they are painted with anti-rust primer. The fireproof coating is applied next. The spray gun used is a gravity-type paint spray gun with a nozzle diameter of 6 ~ 10 mm (using an adjustable nozzle) and air pressure controlled at 0.4 ~ 0.6 MPa. The nozzle is

Fig. 16 The connection node diagram of the truss and steel column

25 ~ 30 cm away from the spray surface, and is basically perpendicular to the base surface. The direction of the spray gun movement is parallel to the surface of the base material. During construction, the thickness of the coating is checked with a thickness gauge after each spray until it meets the thickness specified in the design. The coating after spraying should be properly repaired. Noticeable nodules should be removed with tools such as plastering knives to ensure the uniformity of the coating surface.

IV. Key Technologies and Innovations

(I) Key Technologies

1. Hoisting of spatial steel tubular truss column

A 150 t crawler crane is used for the hoisting single truss steel column, with the main hoisting point set 13.32 m below the top of the column. During the hoisting process, the main crane performs vertical lifting of the steel column. After the steel column is docked and positioned with the foundation, the steel column is temporarily fixed with balance ropes and ground anchors, and the anchor bolts are initially tightened. The main crane can unload the hook when all above work completed.

The maximum weight of the main steel column is 30 t. When hoisting the steel column, each hoisting point is tied with 2 steel wire ropes, that is, the main steel column is hoisted with 6 steel wire ropes, and the wind-resistant column is lifted with 4 steel wire ropes. The hoisting point is located at 0.3 L of the total length of the column below the top of the column, that is, the hoisting point is 13.32 m away from the top of the column. The hook is 14 m vertical distance from the top of the column, that is, the distance H from the hook to the hoisting plane is 27.32 m. Therefore, it can be known that the length L of a single steel wire rope from the hook to the hoisting point during the hoisting of the wind-resistant column is 27.36 m. Two cranes work simultaneously. The auxiliary crane is a 50 t truck crane. The maximum force on the crane is half of the maximum steel column (15 t), while the force on the auxiliary crane decreases from 15 to 0 t. According to the working condition table of the 50 t truck crane, the maximum hoisting capacity at a radius of 6 m is 28 t, which meets the hoisting requirements.

2. Hoisting of spatial steel pipe truss roof beam

(1) The maximum weight of roof beam of the indoor DC yard is 44 t. The roof truss is hoisted by two 150 t crawler cranes, each crane hoisting 50% of the roof truss weight. The crane model selection meets the hoisting needs and specification requirements.

The roof beam hoisting all adopts double-crane hoisting. Each crane connects 4 lifting points, and the 4 lifting points are stressed at the same time. Therefore the steel wire rope that can be used in the 80 m span roof truss hoisting can also meet the requirement of 62.5 m span roof truss hoisting.

(2) The roof truss beam is transported to the site in 2 sections (40 m of each section) from the processing factory and assembled vertically on the pre-assembled on-site assembly platform. First, use crane to position thc truss in the corresponding position, fix the positioning block, adjust the adjustment plate, and ensure the relative positions of the steel components all correct. A welding shrinkage allowance must be reserved before determining the relative position of the main pipe. A centerline of each node of the main pipe is drawn on the assembly platform. Before the overall welding, the branch pipe is assembled and position welded. Position welding the branch pipe needs no less than 4 points. After positioning, full-position welding is performed on the main truss. To ensure the quality of welding, welds of the 3 main pipes must be alternately welded

and one main pipe cannot be fully welded before another in all process of overhead position welding, vertical position welding, and welding. The welding of the upper chord support cross should be carried out first. It is much better to conduct position welding with the main pipe after shrink. Since the toe part is a penetration weld, the root part is a fillet weld, and the side transitions from a penetration weld to a fillet weld, the weld sequence is arranged as toe part first, then the root part, and finally the side considering the welding deformation. Before the truss is hoisted, the weld is subjected to non-destructive flaw detection and is qualified. The scaffolding, scaffolding board, horizontal lock fixing, etc. are checked and qualified. The schematic diagram of the on-site vertical assembly platform is shown in Fig. 17, and the schematic diagram of the on-site assembly of the roof truss is shown in Fig. 18.

(II) Innovations

(1) BIM technology is used to perform visual simulation analysis of important links in the construction of spatial steel pipe truss structures. Installation plan is simulated and optimized according to the construction progress. The entry and construction plan of steel components are also scheduled.
(2) Factory-based modular processing of steel components is utilized with special assembly molds designed. The roof truss steel structure is assembled on the on-site vertical assembly platform. Direct vertical hoisting is used to avoid deformation when hoisted, reducing the amount of high-altitude work and safety risks, and improving work efficiency at the same time.
(3) Factory-processed independent high-altitude welding platforms are established, which can reduce the safety risks brought by high-altitude erecting and dismantling of scaffolding safety protection platforms, and reduce the labor costs and construction period of erecting and dismantle scaffolding platforms.

V. Quality Control Standards and Measures

(I) Quality Inspection of Components before Installation

The purpose of acceptance on steel components and materials is to count the quantity of components and handle defective components on the ground. Therefore these defective components do not enter the installation process. After the steel components arrive at the site, check the quantity and numbering of the components according to the freight bill. Any problems found should be recorded on the return receipt in time and feedback to the manufacturing factory, so that the factory can replace and supplement the components. According to the design drawings, specifications, and the quality inspection report of the manufacturing factory, the quality of the components is checked and accepted with the inspection records kept. Components that exceed the specification error in production or are deformed or damaged during transportation should be returned to factory for repair.

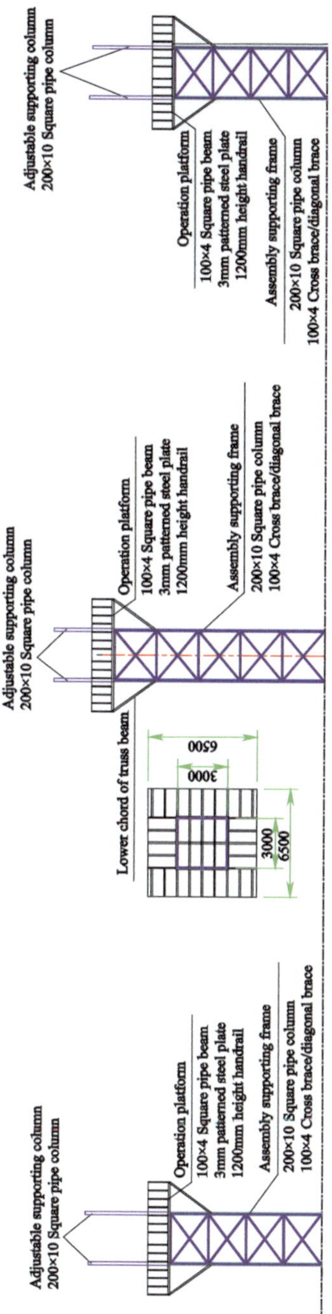

Fig. 17 The schematic diagram of the on-site vertical assembly platform

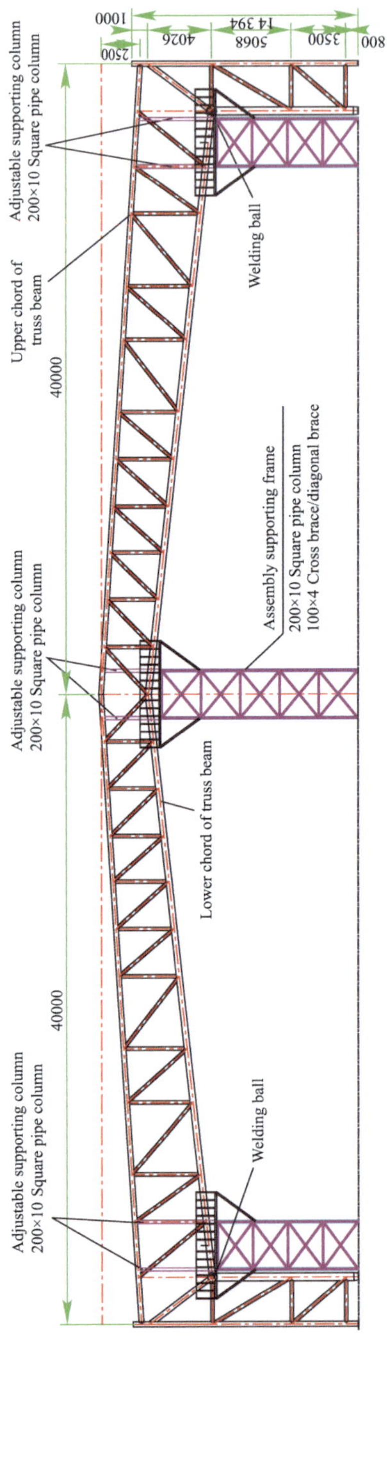

Fig. 18 The schematic diagram of the on-site assembly of the roof truss

(II) Installation Quality Control Measures

(1) The installation process should strictly follow the scheme. Installation procedure must ensure structural stability and not cause permanent deformation.
(2) The component storage site should be flat and solid without standing water. Components should be stored by type, model, and installation order. The bottom layer cushion should have enough large support surface. When stacking the same model, the support points of each layer should be on the same vertical line, and prevent steel components from being crushed and deformed.
(3) The nodes and contact surfaces that are tightly designed in the structure should be in close contact with 70% of the surface. Check with a 0.3 mm thick feeler gauge, the sum of the areas that can be inserted should not exceed 30% of the total area of the contact surface. The maximum edge spacing should not exceed 0.8 mm. The detection of steel structure installation deviation should be carried out after the structure forms a space stiffness unit and is connected and fixed.

(IV) Welding Quality Control Measures

(1) Strengthen the management of welding quality. Before welding begins, quality assurance is required to ensure the pass rate.
(2) Preparation before welding.
   1) The welding platform should be well set up. The height and width of the platform should be conducive to the comfort and convenience of the welder's operation. Windproof measures are needed.
   2) The welding handle wire should be well insulated. Any damages should be wrapped with insulating cloth to avoid sparking with the parent material when dragging the welding handle wire. The welding equipment should be correctly wired and debugged. Trial welding should be carried out before formal welding, adjusting the voltage and current to the appropriate range.
   3) Check the quality of groove assembly. Oxides, moisture, oil stains and other impurities that affect the quality of the weld should be removed from the groove area. If the groove has been cut with oxygen-acetylene, it should also be polished with a grinder until the metal luster exposed. When the groove gap exceeds the allowable deviation, it should be made to meet the requirements by building up weld on one or both sides of the groove.
(3) Position welding. Depending on the diameter of the steel tube, position welding is generally done at 2–4 places. Before position welding, check whether the pipe end is completely aligned with the spherical or pipe surface. The position welding must be fully penetrated, and the length is generally about 300 mm. Depending on the pipe diameter, the position weld should not be too thick.

(4) Formal Welding

1) The intersecting weld seam is an arc-shaped weld seam with the centerline of the pipe, and the ring-shaped weld seam is divided into two joints in the order of overhead, vertical, and flat position welding. This method can ensure good separation of molten iron and slag, and the depth of fusion is also relatively easy to control.
2) When the first pass of the root weld is completed, the slag must be carefully cleaned. Any missed welding defects should be checked. Only when there are no errors found can the second pass of weld begin.
3) The welding rods should be baked before welding. The baking temperature is generally 70–150 °C, and drying is within 2 h. The remaining welding rods must be collected after work and baked at a constant temperature.
4) When the wall thickness of the rod is greater than 4 mm, the welding thickness should be 1.5 times the wall thickness. When the wall thickness of the rod is less than 4 mm, the welding thickness should be 1.2 times the wall thickness. When the wall thickness of the rod is less than 4 mm, the entire weld is formed in 2 passes. When the wall thickness of the rod is 4–8 mm, the entire weld is formed in 3 passes. When the wall thickness of the rod is 8–12 mm, the entire weld is formed in 4 passes. The first pass of weld (the root weld) must be fully penetrated. No slag inclusion, porosity, cracks, etc. are allowed. After careful cleaning and inspection, subsequent welding is carried out. The last pass is the shaping weld. The surface of the weld after shaping is required to be flat, with uniform width and weld ripples. No defect such as slag inclusion, porosity, overlaps, undercut, etc. is found, as well as slag and metal spatter around the weld.
5) Use appropriate welding groove to reduce welding filler. No force assembly is allowed during component installation, causing initial assembly stress. Use reasonable welding sequence, symmetric welding, and segment welding. The joint with larger shrinkage should be welded first, followed by the joint with smaller shrinkage. Welding should be done under as little restraint as possible.

(5) Welding quality inspection and flaw detection.

1) After welding, carefully inspect the appearance of the weld, the surface of the weld should not have defects such as cracks and overlaps. First and second level welds should not have surface porosity, slag inclusion, crater cracks, arc scratches, etc., and first level welds should not have undercut, incomplete penetration, root shrinkage, etc.
2) Non-destructive testing is carried out on the basis of satisfactory visual inspection. After the weld is completed and cooled for 24 h, 100% inspection is carried out on the full penetration first level weld

as per design requirements. The method and evaluation standard for ultrasonic flaw detection are in accordance with "Method for Manual Ultrasonic Testing and Classification of Testing Results for Ferritic Steel Welds (GB/T 11345—2013)".

(6) Measures for dealing with welding non-conformities. For internal defects of the weld, locate the defects using the ultrasonic testing (UT) method, and remove them with a carbon planer. For cracks, the carbon planing area should extend 50 mm outward at both ends of the weld metal. When re-welding, preheating and post-heating treatment must be carried out according to the original process for thick plates. The preheating temperature should be increased by 20 °C based on the previous process. The re-welding of the same part should not exceed 2 times. If it exceeds 2 times, a special re-welding process should be developed and approved by the supervision engineer.

(IV) Quality Control Measures for High-Strength Bolts

(1) Strictly implement the construction technical regulations for high-strength bolt connections, and conduct special technical briefings for the team.
(2) Assign a specialized operator to be responsible for the calibration and re-inspection of the torque wrench, and check the construction quality and initial tightening and final tightening marks of high-strength bolts.
(3) Implement the on-site management of high-strength bolts, with a specialized operator for material picking, checking, issuing and recycling. Keep good records to ensure the correct model and position of high-strength bolts. Keep the warehouse ventilated and dry, and store by category and model code.
(4) The relationship between the tightening of high-strength bolts and welding: initial tightening—final tightening—welding.
(5) All high-strength bolt connections must be re-inspected in batches before construction, and non-conforming products must not be used.
(6) The final tightening of high-strength bolts must ensure the construction torque, and the high-strength bolts installed on the same day must be fully tightened. At the same time, the construction torque must be rechecked within 24 h.
(7) The insertion of high-strength bolts should be carried out after adjusting the structural center. The direction of insertion should be convenient for construction, striving for consistency. Pay attention to the front and back of the washer during installation. The side of the nut with a conical surface should face the chamfered side of the washer.
(8) Strictly control the length of high-strength bolts during installation to avoid insufficient strength and bolt confusion caused by substituting long for short or short for long. After the final tightening, ensure that 2–3 threads are exposed on the outer ring of the nut.

(9) The tightening axial force of the connecting parts and the anti-slip coefficient of the friction surface are tested by the manufacturer in the factory. At the same time, the manufacturer provides test pieces according to the specifications, and the installation unit conducts the anti-slip coefficient test of the friction surface on site. The bolts used for re-inspection of the connecting parts should be randomly selected from the batch of bolts waiting to be installed on the construction site.

(V) Measures to Prevent Falling Objects

(1) When transporting objects from high altitude to the ground, they should be tied with ropes and lowered. During hoisting, do not pile or hang scattered objects on the components. Scattered materials and objects must be hoisted and passed after being firmly tied with a lifting cage or steel wire rope, safety rope, and materials and tools should not be thrown at will to prevent slipping and injuring people or accidents.
(2) Components must be securely tied, the hoisting point should pass through the center of gravity of the component. The hoisting should be smooth to avoid vibration or swinging.
(3) The speed of hoisting components should not be too fast. Hoisted objects should not stay in the air for too long. Sudden hoisting and dropping are strictly prohibited to prevent the components from falling off.

(VI) Measures to Prevent Crane Overturning

(1) The walking area of the crawler crane must be flat, firm, and reliable. Specialized operators should check the ground conditions before starting work each day.
(2) Cranes should not be parked on slopes for work. It is prohibited for the two tracks or legs of the crane to be at different heights or the hard and soft soil. The crane equipment must be fixed, and the operators must hold a certificate to work.
(3) When hoisting components, the sling should be kept vertical and should not be dragged diagonally beyond the crane's turning radius to avoid overload and the steel wire rope slipping or breaking the rope, causing the crane to become unstable. When hoisting heavy components, a pulling rope should be set.
(4) During crane operation, the boom lifting, lowering, and rotation should be steady, which should not sway in the air. At the same time, try to avoid sudden braking or impact vibration. Without reliable technical measures and approval from the relevant technical department, the crane is strictly prohibited from overloading to avoid accelerating the wear of mechanical parts and causing the crane to overturn.
(5) Cranes should avoid full load travel as much as possible. When fully loaded or close to full load, it is strictly forbidden to perform two actions of hoisting and rotating (hoisting and horizontal rotation or hoisting and

traveling) at the same time to avoid crane overload due to uneven roads or inertia, causing a rollover accident.

(6) During hoisting, there should be specialized operators in charge of unified command. The commander should be located where the operator can clearly view the entire hoisting process. Crane drivers must be familiar with signals and operate according to various signals from the commander. The command signals should be predefined, and the signals issued should be clear and accurate.

(7) When the crane stops working, the rotation and walking mechanisms should be braked, and the driver's cabin door should be locked. No components should be suspended on the hook, and it should be raised to a high place to prevent swinging injuries and crane instability.

(VII) Measures to Prevent Hoisting Structure Instability

(1) Component hoisting should be carried out according to the prescribed hoisting process and procedures. Without calculation, reliable technical measures, and approvals, the installation structure components should not be arbitrarily changed or reversed.

(2) After hoisted into place, components can only be unhooked after preliminary calibration and reliable connection. The temporary fixing tools can only be removed after final fixation.

(3) After fixed, components should not be arbitrarily pried or moved. If recalibration is needed, it must be re-hooked.

(VIII) Measures to Prevent Lightning Strikes

(1) There should be specialized operators in charge of installing, maintaining, and managing electrical lines and equipment in the hoisting area. Insulation wrapping measures should be taken on all types of power lines in high places to prevent accidental leakage of electricity to the steel structure.

(2) Crawler cranes should have lightning and electric shock prevention measures. All kinds of electrical machinery must have good grounding or zero connection, the grounding resistance should not be greater than 4 Ω, and the grounding electrode resistance should be regularly tested.

2 Installation of Valve-Side Outlet Devices and Bushings of HV Converter Transformer

The 825 and 1100 kV converter transformers are developed and applied for the first time, which are of the highest voltage level in the world currently. The equipment volume, accessory quality, and installation difficulties and precision requirements all exceed previous DC projects. Among them, bushings and outlet devices on the valve side of the converter transformer are the most representative, with large volume and different structure from the past. This section introduces the installation technology of the valve-side outlet devices and bushings of 825 kV and 1100 kV HV converter transformers.

I. Construction Overview

The valve-side outlet devices of HV converter transformers in the CG project totals (24 + 4) sets [Changji and Guquan station each has (12 + 2) sets]. Due to the different internal structure of the HV converter transformer equipment at Changji converter station, the installation methods of the valve-side outlet devices and bushings of 1100 kV HV converter transformer are different from that of the 825 kV HV converter transformer. The valve-side outlet devices of 1100 kV converter transformer are transported as a whole, "L" shaped assembled, with specialized valve-side bushings and installation platform. The outlet devices of 825 kV converter transformer are transported in 3 sections. Outlet devices and bushings do not need to be docked on the installation platform. The 1100 kV converter transformer outlet devices weigh 11.2 t, while the 825 kV ones weigh 13.3 t with the heaviest single section (the 2nd section) of 5 t. The size of the outlet devices of high 1100 kV HV converter transformer at Guquan converter station is 8179 mm × 4481 mm × 2851 mm, weighing 17.3 t, which will be transported as a whole and "L" shaped assembled. Specialized valve-side bushings and installation platform will also been equipped.

There are (24 + 4) pieces of valve side bushings on the HV converter transformer in the CG project. Valve side bushings of the 1100 kV converter transformer at Changji converter station totals (12 + 2) pieces. Single bushing weighs up to 9.87 t with 15.3 m in length. Valve-side bushings on the 1100 kV HV converter transformer at the Guquan converter station is 20.19 m long, and weighing 15.5 t. Hoisting auxiliaries from the manufacturer weigh 5.5 t with a maximum hoisting weight of 21 t.

II. Key Points and Difficulties of the Project

The ± 1100 kV converter transformer is used for the first time. Currently, it is the converter transformer with the highest DC voltage level in the world. The transformer has large volume with high installation angle of the valve side bushing and the elevated seat. The whole station has 7 HY converter transformers and 7 HD converter transformers. Single HY HV converter transformer has a total oil weight of 280 t, while single converter transformer has a total weight of about 909 t. Single HD HV converter transformer has a total oil weight of 247 t, while single converter transformer has a total weight of about 845 t. The oil service workload is huge. The weight of the accessories, installation difficulty and precision requirements all exceed those of the previous DC transmission projects.

III. Brief Description of Construction

Construction preparations should be carried out first, including technical preparation, site preparation, construction machinery, tools, materials and safety equipment preparation. Then, carry out tests and installations on valve-side outlet devices and bushings. When the converter transformer is in place to the foundation, carry out bushing vacuum extraction (leakage test). SF_6 gas filling will be conducted finally after the vacuum extraction.

IV. Construction Planning
(I) Technical Preparation

(1) Before installation, check whether the bushing installation drawings, technical documents, product technical agreements, relevant acceptance specifications and records for installation and debugging are prepared.
(2) Before installation, the technical director should read the product's installation manual, assembly general drawing, accessory list, as well as the technical instructions and product technical agreement of each accessory, understand the structure, performance, parameters and installation technical provisions and requirements of the product and its accessories, and give detailed technical and safety briefings to the construction personnel, while keeping a record of the briefings.
(3) The construction personnel shall install according to the technical measures and technical briefing requirements, understand the installation procedures, methods and technical requirements, and be familiar with the manufacturer's data, installation drawings, technical measures and relevant regulations and standards. Focus should be paid on the following operation points:
   1) Before installation, site inspection should be carried out first. The impact recorder should record the impact situation during the transportation and loading/unloading, as well as the gas pressure in the converter transformer. The impact in the longitudinal, transverse, and vertical directions should not exceed 3 g, while the gas pressure inside the oil tank should be maintained at 0.02 ~ 0.03 MPa. The withstand voltage of the oil sample inside the transformer should not be less than 50 kV, and the water content should not exceed 15 mg/L.
   2) The transformer inspection has been completed with no abnormal results. Tests on the valve side outlet devices and bushings (insulation resistance, dielectric loss, capacitance, etc.) have been completed with qualified results.
   3) Installation should not be carried out in rain, snow, wind (above level 4) and relative humidity above 75% (the working environment must also meet the requirements of the manufacturer's technical documents).
   4) During the installation process, dry air with a dew point of − 55 °C should be continuously added to the transformer. The rate of dry air replenishment should comply with the product technical documentation. The air pressure inside the transformer should be maintained at a slight positive pressure.
   5) The time the transformer is exposed to air (blowing dry air into the oil tank) should be minimized as much as possible. The maximum allowed exposure time to air (from the start of opening the cover plate

and breaking the product seal to re-evacuation) should comply with the specifications and product technical requirements.

6) If the assembly work continues and the exposure time meets the above requirements, during the non-assembly time at night, dry air (with a dew point not exceeding − 55 °C) should be filled into the transformer and preserved at 20 ~ 30 kPa. If the assembly work continues and the exposure time meets the requirements, it should be immediately sealed and evacuated to 100 Pa and maintained for a duration exceeding the specified time, then filled with dry air with a dew point not exceeding − 55 °C and preserved at a slight positive pressure of 20 kPa.

(4) Communicate with manufacturer personnel with testing and installation matters.

(II) Site Preparation

The foundation of the converter transformer equipment, the courtyard pavement and the track system have all been completed, meeting the construction requirements of the valve-side outlet devices and bushings of the converter transformer. At the same time, the converter transformer courtyard is formed by one-time casting by the civil construction, to ensure the quality of the square concrete pavement process, prevent pollution and damage to the square ground surface during the lifting and storage of accessories. The installation area of the outlet devices and bushing is fenced off. The ground of oil service area is covered with plastic cloth or oil-proof cloth for protection, and is isolated with fences. Fire-fighting equipment is reasonably placed.

For the installation of the outlet devices and bushing, reasonable placement of bushing, testing equipment, materials, tools, and auxiliary materials is carried out. After reserving the position of lifting machinery next to the converter transformer, the site layout is carried out reasonably according to the construction operation time of each process, to ensure the reasonable use of hoisting space during the installation process of the bushing.

(III) Preparation of Construction Power Supply

The construction power supply box uses three self-provided terminal distribution boxes, utilizing "one machine, one switch, one protection". According to the progress of the site construction, the construction power supply can also use the converter transformer courtyard maintenance box.

(IV) Construction Organization

According to the installation characteristics, workload, and process flow of the outlet devices and bushings, there are 1 general manager, 1 site work leader, 1 technical leader, 1 safety guardian, 2 crane commanders, 2 crane drivers, 2 testers, 4 ground workers, 3 high-altitude workers, etc.

(V) Preparation of Machinery, Tools, Materials, and Safety Equipment

Before installation, the installation machinery, tools, safety equipment, traction equipment and consumable materials of the converter transformer should be fully prepared, and the main construction machinery should be tested and qualified, and the quantity and specifications meet the construction requirements as shown in Table 1.

(VI) Equipment Arrival Inspection and Storage

(1) Upon arrival of the converter transformer's valve-side outlet device and bushing at the site, conduct a verification of the product nameplate, accessories, and manufacturer's documentation in accordance with the purchase agreement.
(2) Inspect and document the impact events recorded by the shock recorder during the transportation and loading/unloading.
(3) Ensure that the oil level in the oil-filled bushing is within the normal range, with no signs of oil leakage and that the porcelain body shows no damage. For the gas-filled bushing, confirm that the gas pressure is in compliance with the manufacturer's specified standards.
(4) Ensure that all accessories of valve-side outlet device and bushing are complete, free from rust and mechanical damage, well-sealed, and with no peeling of the outer surface paint. Monitor the gas pressure and temperature, comparing these readings with the initial factory settings according to the temperature curve, and keep the gas pressure at 0.02 ~ 0.03 MPa. Conduct at least two inspections daily during storage and maintain thorough records. If the pressure indicated by the pressure gauge drops quickly, the cause must be identified, properly handled, and the pressure should be replenished to the specified position in time.
(5) Upon the arrival of the valve-side outlet devices and bushings arrive at the site, prepare for installation as soon as possible to minimize storage time.
(6) Store lifting components carefully to prevent saturation from rainwater and avoid rust and contamination.
(7) The valve-side outlet devices, bushings, and other external parts can be stored outdoors, and should be supported on wooden blocks.
(8) Erect temporary barriers around the equipment to shield the bushings from potential external damage, thereby preventing gas and oil leaks.
(9) The arrangement of construction machinery and materials on-site should be orderly, adhering to the standards of fixed-position management.

V. Optimization of Construction Process

(I) Changji Converter Station: Installation of the Valve-side Outlet Devices of 825 kV Converter Transformer on the oil tank

The valve-side outlet device of 825 kV converter transformer is divided into three sections, see Fig. 19. The total weight of the outlet device is 13.3 t, of which the second section is the heaviest of 5 t. The position of the crane refers to the unchanged position of the 1100 kV bushing hoisting. The selected Shackle is 15 t. Four nylon slings of 16 t × 6 m, and two 15 t chain hoists are used for hoisting, see Fig. 20.

Table 1 The list of machinery, tools, materials, and equipment for hoisting

No.	Name	Specification and model	Quantity	Remarks
1	Vacuum pump	5000 m^3/h	2	
2	Lift truck	18 m in height	1	
3	Dry air generator	Dew point − 55 °C	2	
4	Crane	80 t	1	
		25 t	2	One with operation platform
5	Shackle	30 t	2	
		20 t	4	
6	Steel wire rope head	ϕ 22.0 mm × 4 m	8	
7	Chain hoist	10 t/5 t	Each 2	
8	Wooden blocks	100 mm × 150 mm × 1000 mm	15	
9	Slings	30 t/10 t/3 t/2 t	Each 2	
10	Steel wire rope	ϕ 21.5 mm, 8 m, 2 m	Each 2	
11	Pry bar		2	
12	Electronic vacuum gauge		1	Need to be checked for quilified
13	Temperature/humidity meter		2	
14	Pressure gauge	0 ~ 1 MPa	2	
15	Adjustable wrench	6 ~ 32 mm	30	
16	Fixed wrench	6 ~ 32 mm	30	
17	Torque wrench	400 N m	1	
		310 N m	1	
18	Multimeter		1	
19	Flashlight		2	
20	Manufacturer's special hoisting tools		1	
21	Power box		2	
22	Dry powder fire extinguisher	20 kg	2	
23	Mobile bench		10	
24	Protective fence		200 (m)	
25	Hammer		2	
26	Wrench	Big, small	2	
27	Screwdriver		4	
28	File		2	
29	White coir rope	ϕ 6 mm	4	

(continued)

Table 1 (continued)

No.	Name	Specification and model	Quantity	Remarks
30	Spirit level		1	
31	Angle gauge		2	
32	Tape measure	10 m, 5 m	Each 2	
33	Plumb line		2	
34	Emergency light		3	
35	Anhydrous alcohol	99.99%	5	
36	Vacuum tube	Metal wire tube	100 (m)	
37	Dry air tube	Metal wire tube	100 (m)	
38	White gauze tape, crepe paper	Insulating material	Proper amount	Offered by manufacturer
39	Plastic sheet		50 (m^2)	
40	White cloth		20 (m^2)	
41	Cotton yarn		20 (kg)	
42	Omnidirectional seat belt		5	
43	Grounding wires		50 (m)	
44	Fire protection facilities		2	

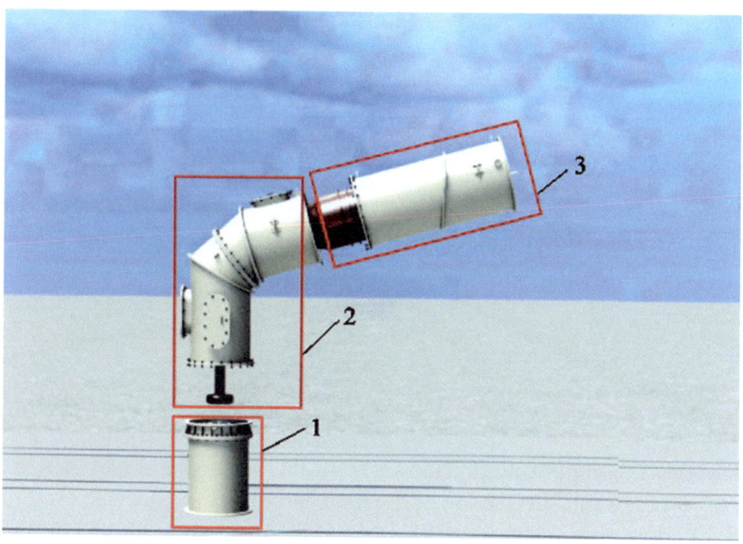

Fig. 19 Three parts of the valve-side outlet device

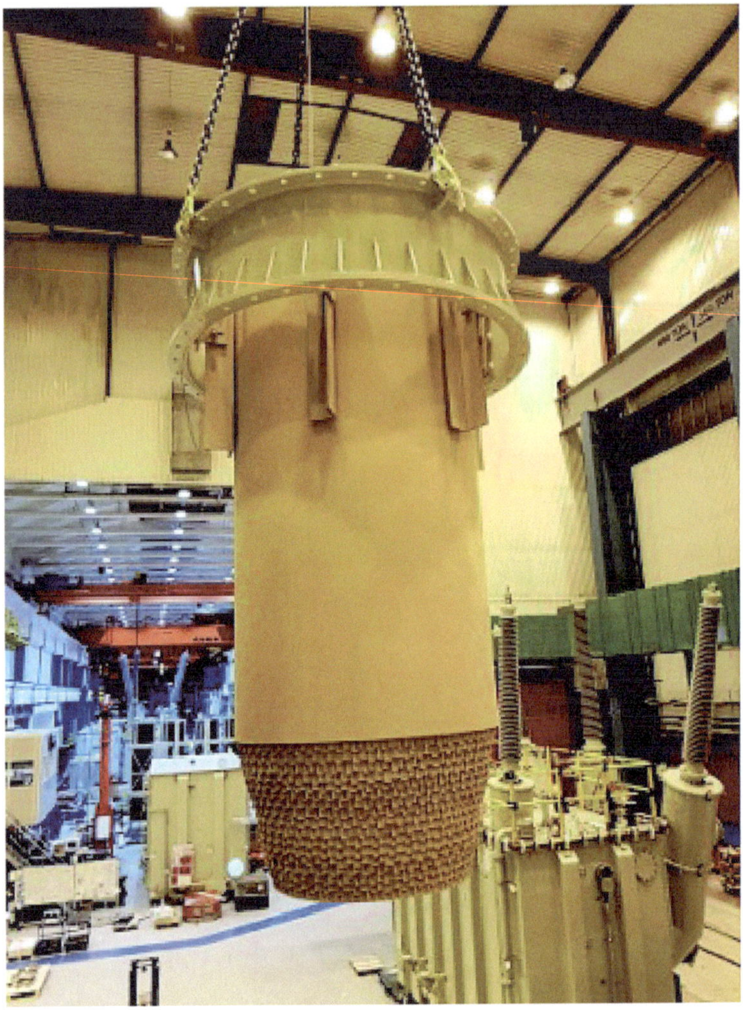

Fig. 20 Hoisted outlet devices

(1) Remove the upper transport cover from the valve-side outlet device 1 and carefully hoist the device out of the transport barrel.
(2) Before lowering valve-side outlet device 1, extract six leads and one equipotential line from the valve outlet inner tube using a tag line (or white cloth strap), as depicted in Fig. 21. Ensure that the equipotential line is bundled with a conductor wire and secured with a white cloth strap, leaving ample length at the bolt connection point to prevent damage to the equipotential line during extraction.

(a) 6 leads and 1 equipotential line inside the converter transformer (b) Extract 6 leads and 1 equipotential line

Fig. 21 Tag line

(3) Once the leads are extracted, lower valve-side outlet device 1 into place and securely fasten the connecting bolts. Position the leads within the space between the valve outlet insulation and the valve-side outlet device barrel.
(4) Install valve-side outlet device 1 onto the oil tank, ensuring proper alignment with the tank cover during installation, as verified by the manufacturer.
(5) Before installing valve-side outlet device 2, remove the transport covers from both ends and install the conductive rod pulley device and the grading ball, taking care to cover with plastic sheeting during the process. Thread the pull rope through the pulley device into the shielding tube and draw it out, ensuring the rope end at the hook connection is wrapped with soft leather or other protective material to prevent scratching the inner wall of the shielding tube. Connect the other end of the pull rope to a counterweight. See Fig. 22 for the installation of the conductive rod and Fig. 23 for the installation of the grading ball.
(6) Install the support rod for the valve-side outlet device, but temporarily do not fix it to the device itself. See Fig. 24.
(7) Installation of valve-side outlet device 2. Slowly lift the device horizontally. Position it beside the flange of the valve outlet on the tank cover. Securely attach one end of the pull rope to the lead terminal, wrapped with soft leather to prevent damage to the shielding tube's interior during lead extraction. As the device descends, usc the pull rope to draw the leads out of the shielding tube. After the device is fully lowered, tighten the flange bolts. See Fig. 25 for the hoisting diagram. Subsequently, remove the pulley device from the top of valve-side outlet device 2, lead the terminals out of the shielding tube, and remove the pull rope and counterweight. Install the conductive rod and grading ball according to the manufacturer's schematic. See Fig. 26.

Fig. 22 Installation of the conductive rod

(8) Installation of valve-side outlet device 3. Remove the device from the transport barrel and install it onto device 2. Before installation, fit a plastic sheath onto the conductive rod. Verify the installation position of device against the drawings during the process. See Fig. 27.

(9) Secure the support rod of the valve-side outlet device.

(II) Changji Converter Station: Bushing Testing and Installation

(1) A joint inspection is conducted by the owner, supervisor, material management, construction team, and manufacturer to verify the contents of the package (factory documentation, product serial numbers, equipment appearance, shock records, etc.). Before bushing installation, ensure that the three-dimensional impact values recorded during transportation are all below 2 g. All products need to be cross-verified with the manufacturer's factory test reports and the stamped nameplate. Once confirmed, the valve bushings should be carefully hoisted from the crate following the bushing installation and operation manual, and then visually inspected and cleaned. Before installation, the transformer oil from the transport barrel and the valve bushing oil chamber should be drained (ensuring that oil cloths fully cover the oil equipment area during this process).

Fig. 23 Installation of the grading ball

(2) Remove the transport barrel from the rear of the valve bushings, hoist the bushings out of the barrel, and use the manual chain hoist on the hoisting rope to maintain a horizontal position of the bushing, as illustrated in Fig. 28.
(3) Refer to the factory test report and standard test specifications to perform test on the single valve bushing. It can only be hoisted after passing the test, as shown in Fig. 29.
(4) Install the valve bushing hoisting auxiliaries and the conductive rod traction special tooling. Attach the traction tooling to the conductive rod and push the rod towards the bushing's end.
(5) Remove the flange blind in the center of the transport cover on the valve-side outlet device, loosen the bolts that secure the conductive rod internally, and install the special hoisting auxiliaries onto the transport cover. Throughout the process of removing the cover, continuously blow dry air into the oil tank to minimize moisture absorption by the insulator.
(6) Install the special hoisting auxiliaries on the transport cover, loosen the bolts connecting the transport cover to the valve side outlet device flange, and lift the cover, support tube, and tooling using the lifting lugs at the top of the cover. Carefully move the assembly out horizontally to prevent

Fig. 24 Installation of the support rod for the valve-side outlet device

damage to the insulation within the valve-side outlet device. After the transport cover is lifted out, dismantle the special hoisting auxiliaries, reattach the cover, and wrap the cover and support tube in plastic cloth and stored properly.

(7) Inspect the positioning and integrity of the internal outlet insulation of the valve-side outlet device. Install the special hoisting auxiliaries at the lifting lug on the upper part of the TA flange to ensure vertical lifting of the TA. When installing the TA onto the valve-side outlet device, check the seal for any deformities or damage and replace it if necessary. After installation, tighten the flange bolts and then cover the flange surface with plastic, as shown in Fig. 30.

(8) Remove the transport barrel from the end of the valve bushing, hoist the bushing out of the barrel, and use the manual chain hoist on the hoisting rope to keep the bushing level. Coordinate the use of a 75 t and a 25 t crane to lift the 9.87 t valve bushing, securing it with 4-point fixation using two ropes to protect against wind, with each point weighted with 30 kg counterweights.

(9) During installation, the valve bushing remains stationary while the valve-side outlet device is placed on a crab. Move the crab towards the valve bushing to complete the mating and installation of the bushing and outlet device, controlled by a chain hoist for forward and backward movement.

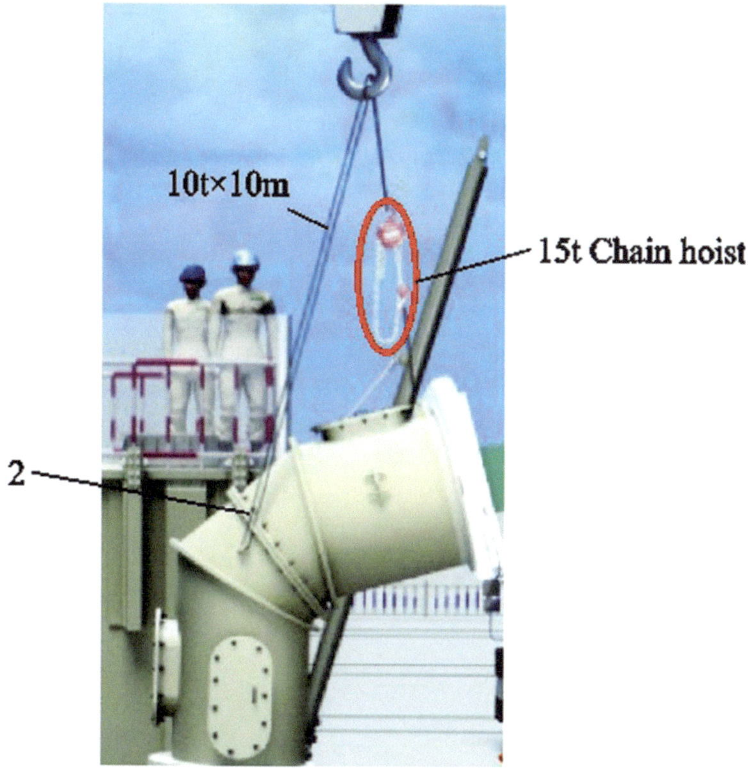

Fig. 25 The hoisting diagram of valve-side outlet device 2

Given the approximately 3.2 m length of the bushing's end, measure the distance from the valve side outlet device flange face to the track top before mating to ensure it meets the installation requirements. A dustproof shelter must be used during the docking process to ensure cleanliness of the mating surfaces, as shown in Figs. 31 and 32.

(10) Slowly dock the valve-side outlet device and the valve bushing horizontally. When the bushing's end approaches the flange face of the outlet device, connect the conductive rod. Then carefully insert the valve bushing into the outlet insulation within the valve-side outlet device. Throughout the installation, check carefully to avoid damaging the outlet insulation. As the bushing's end is inserted, gradually retract the conductive rod. Once the bushing's flange aligns and connects with the valve side outlet device's flange, tighten the flange bolts. Carefully release the hoisting ropes, inspect the support tooling of the outlet device. After ensuring that there are no errors, remove the ropes. Avoid any swaying of the valve bushing and ensure horizontal alignment of the flanges before mating, as shown in Fig. 33.

Fig. 26 Installation of the conductive rod and the grading ball

Fig. 27 Docking of the No.3 and No.2 valve-side outlet devices

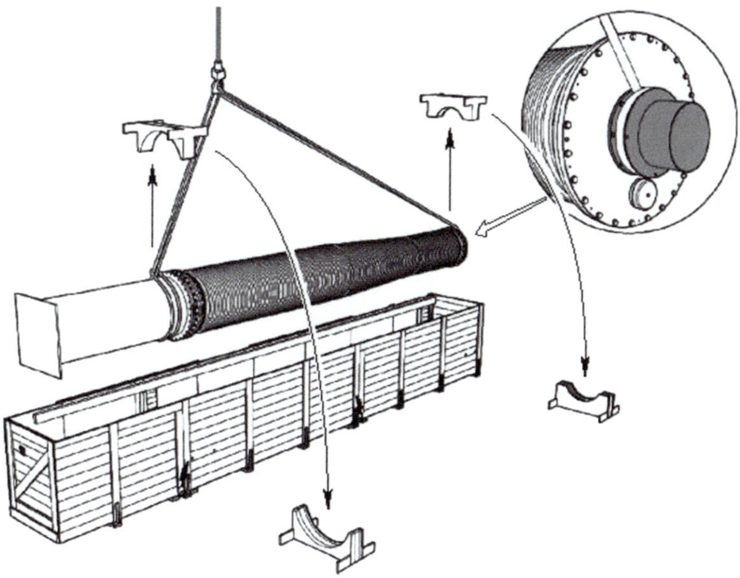

Fig. 28 Hoisting the bushing from the packing box

Fig. 29 Test on the single valve bushing

(11) After the valve bushing installation is complete, install a sealing cap at the bushing's head and fill the valve-side outlet device with dry air, which has a dew point of no higher than − 55 °C. Stop inflation once the pressure inside the outlet device reaches a range of 20–22 kPa, and then maintain the inflation.

Fig. 30 Hoisting TA with specific hoisting auxiliaries

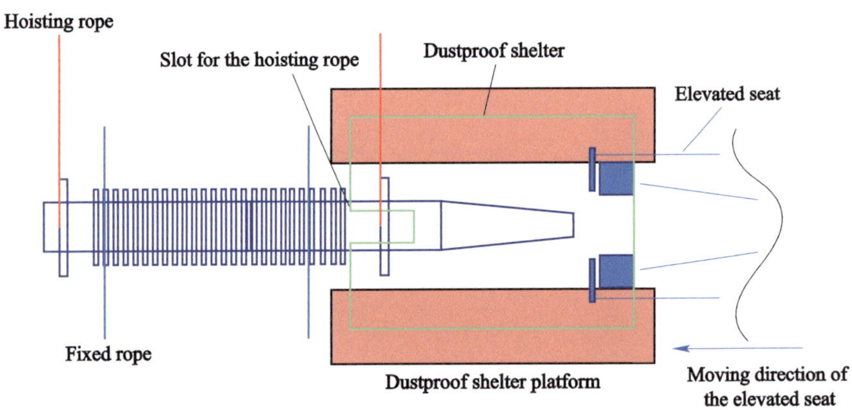

Fig. 31 Dustproof shelter used in docking

(III) Changji Converter Station: Installation of ± 1100 kV Valve-side Outlet Device and Bushing as a Whole on the Oil Tank

(1) Before the installation of the valve-side outlet device and bushing, the oil within the product should be drained in advance. During the oil draining process, dry air with a dew point not exceeding – 55 °C should be continuously blown into the oil tank.
(2) Install the valve-side outlet device's positioning columns at the valve outlet position on the oil tank cover (two positioning columns should be installed initially according to the sequence of the valve-side outlet device installation).

Fig. 32 Photogragh of the dustproof shelter

Fig. 33 Real-time measurement during docking

(3) Install 4 positioning frames on the valve-side outlet device to align with the positioning columns on the oil tank, aiding in the positioning during the installation process of the valve-side outlet device.
(4) Install two manual chain hoists at the lug place locations on the short-axis side wall of the valve-side of the oil tank to temporarily fix the support rods of the valve-side outlet device, as shown in Fig. 34.
(5) Connect the support rod of the valve-side outlet device with the temporary support rod using screws. After the connection, install both support

Fig. 34 Installation of manual chain hoists

rods as a single unit onto the two supports of the oil tank, and use the manual chain hoists for temporary traction and fixation.

(6) Install two manual chain hoists at the lug place locations for temporary fixation, and connecting the support rods as described, then connecting to the two supports, referring to Figs. 35 and 36.

(7) Loosen the bolts on one of the transport cover plates of the valve outlet flange on the oil tank cover, lift and remove the transport cover, install the safety pedal tooling onto the valve outlet flange, and then cover it with plastic cloth, as shown in Fig. 37.

(8) Before lifting, install the bushing hoisting auxiliaries and the conductive rod tensioning tool at the head of the valve bushing. Connect a steel wire rope from the guide pulley of the conductive rod tensioning tool to the ground, attach the other end of the steel wire rope to a manual chain hoist, and then place a 200 kg counterweight beneath the manual chain hoist.

(9) Utilize a 75 t crane (with a 12 t counterweight) for the installation of the valve-side outlet device and bushing.

(10) The crane tautens the lifting ropes on the valve-side outlet device and bushing, and after confirmation, loosens and removes the four fixed tension bars attached to the valve-side outlet device.

(11) For the hoisting of the $\pm$ 1100 kV converter transformer valve-side bushing and outlet device, which weighs 20 t and is assembled on the ground installation platform, use four 16 t $\times$ 6 m nylon slings, two 15 t

Fig. 35 Temporary support rods

Fig. 36 Temporary fixation support rods for manual chain hoists

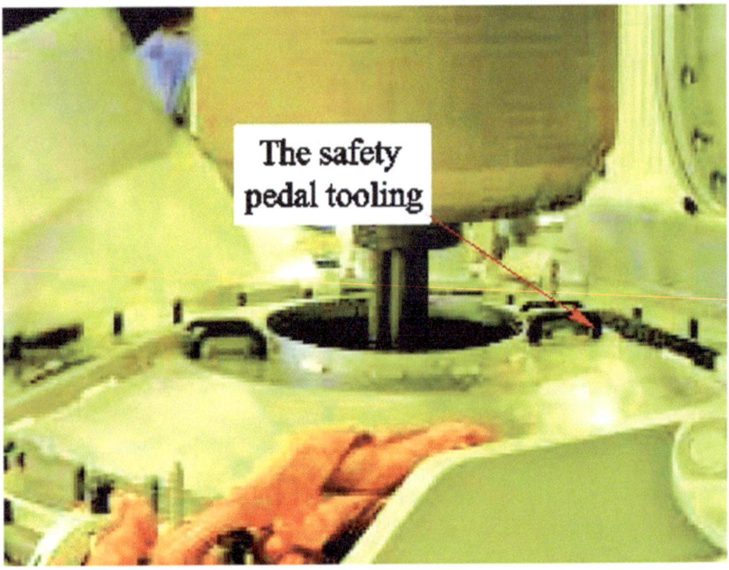

Fig. 37 The safety pedal tooling

manual chain hoists, and three 15 t shackles, with a total weight of the hoisting auxiliaries amounting to 1 t.

Lifting Step One: Adjust the distance from the two lifting points on the valve-side outlet device to the lifting hook to 7 m in length. Only commence lifting when the hook is aligned vertically with the center of gravity of the valve side. The crane operator follows the lifting commands, slowly lifting the assembly. Once the valve-side outlet device is under tension, release the fixed tension bar of the installation platform. According to the 1:1 CAD drawing, the angle between the nylon sling at the top of the bushing and the horizontal is 30°, and at the valve-side outlet device is 81°, as shown in Fig. 38.

Lifting Step Two: After lifting approximately 1 m, adjust the crane at the top of the bushing to raise it to an angle of 15° with the horizontal ground. Remove the lower transport cover plate of the valve-side outlet device and then wrap the outlet seal with clean plastic cloth. According to the 1:1 CAD drawing, the angle between the nylon sling at the top of the bushing and the horizontal is 17°, and at the valve-side outlet device is 81°, as shown in Fig. 39.

In summary, the angle adjustments during the hoisting process ensure that the overall center of gravity of the valve side remains within the lifting points on both sides of the valve-side outlet device and the bushing, eliminating the possibility of overall overturning due to significant re-offset. Moreover, based on the hoisting load calculations, all hoisting auxiliaries meet the required conditions. To prevent the sling from unhooking during the hoisting process, all crane hooks must be equipped

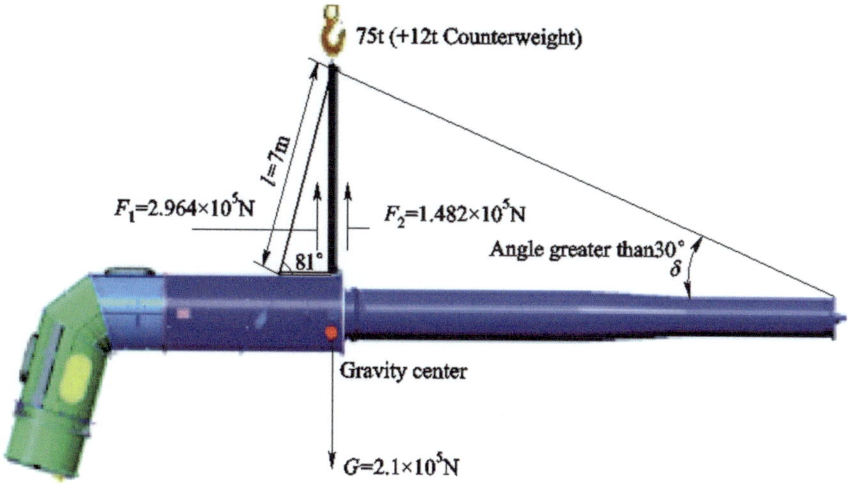

Fig. 38 Step one for hoisting the valve-side bushing and outlet device

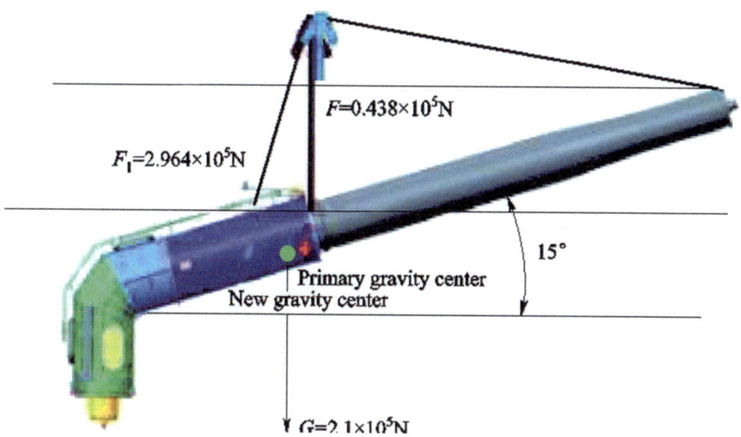

Fig. 39 Step two for hoisting the valve-side bushing and outlet device

with anti-unhooking devices, which should be inspected for normal function before each lift, as shown in Fig. 40.

(12) After lifting above the tank cover, move the valve-side outlet device to the corresponding position above the cover, and slowly lower it into the two positioning columns on the tank cover. Once the device is in place, insert and secure the four positioning pins. Concurrently, attach the counterweight to the conductive rod tensioning steel wire rope, and gradually pull out the conductive rod using a 200 kg counterweight as the valve-side outlet device descends.

Fig. 40 The diagram of the hook

(13) Subsequently, install the two support rods of the valve-side outlet device to the lifting rings on both sides of the outlet device and secure them.
(14) Verify the positioning of the valve outlet and the position of the terminal post against the drawing requirements. When confirmed, adjust the terminal post to the suitable position by guiding the conductive rod. Protect the connection area to prevent bolts from accidentally falling into the oil tank during the connection process. Connect the 6 wires together with bolts, and perform a pre-insertion positioning check on the valve-side outlet device insulation. Remove all auxiliary materials such as plastic cloth from the flange area.
(15) Lift the 200 kg counterweight to a position more than 2 m above the ground, remove the four positioning pins, and dismantle the two support rods of the valve-side outlet device from their supports as a single unit. Then, remove the lifting ropes tied to the lower part of the valve-side outlet device. With the crane, continue to slowly lower the valve-side outlet device and bushing until the flange of the valve-side outlet device connects with the flange of the oil tank cover, and secure the flange bolts.

(16) Finally, remove the temporary support rods from the main support rods, install the formal lower section support rods of the product, and then mount the assembly onto the supports and tighten all bolts. After the product has been filled with oil, apply a hydraulic pressure of 25 t at the middle connection screw of the support rods (this step is carried out after oil filling), and subsequently tighten the screws.

(IV) Installation of Valve-side Outlet Device and Bushing at Guquan Converter Station

1. Installation of valve-side outlet device

Overall lifting lugs: The valve-side outlet device is equipped with numerous small, segmented lifting lugs. However, these are not robust enough to support lifting the entire assembly. Only the reinforced lifting lugs located at the top of the valve-side outlet device (indicated in red in Fig. 41), should be utilized for lifting the entire unit.

Special shackles: Due to the substantial thickness of the overall lifting lug, standard shackles are inadequate. Special shackles must be employed to ensure a secure connection. These special shackles should have a pin diameter not exceeding 50 mm, an opening length of 90 mm, and a minimum load-bearing capacity of 20 t each, with a total of four required.

Employ two truck cranes, with lifting capacities of 80 t and 25 t respectively, to lift and flip the valve-side outlet device. Remove the transport tube and brackets at the device's end, using a chain fall to adjust the installation angle and prevent any collisions. Post-flip, inspect the position of the internal outlet device to confirm it aligns with the drawing specifications before proceeding with installation. Continuously blow dry air with a qualified dew point into the interior of the product. Sequentially open the covers of the valve-side outlet device and commence the installation process.

Note: During the installation process, it is essential to prevent the insulating components from absorbing moisture: 1. When opening the main body to install the bushing and connect the internal leads, continuously blow dry air into the transformer. 2. During the lifting and adjustment of the valve-side outlet device, the transformer's

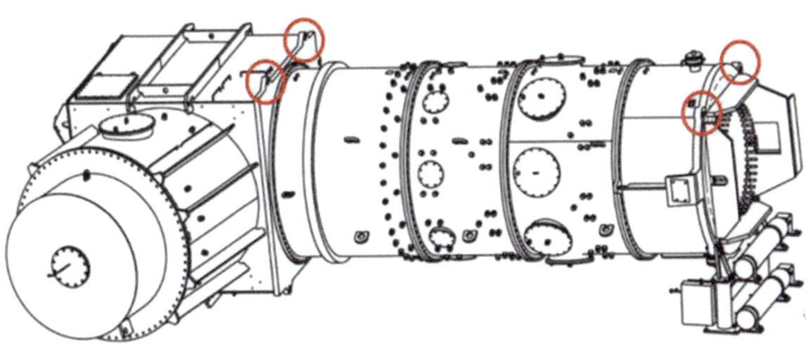

Fig. 41 The diagram of the reinforced lifting lugs

cover must not be opened prematurely. It can only be opened just before the mating process to ensure that the exposure time is minimized. 3. After opening the cover of the converter transformer's valve-side outlet device, continue to blow dry air continuously into the transformer.

2. Installation of valve-side bushing

Utilize two truck cranes, one with an 80 t and the other with a 25 t lifting capacity, in tandem to hoist the valve-side bushing crane box. Once lifted, avoid placing it directly on the ground. Instead, remove the internal bracing from within the bushing box or elevate the flange surface using sleepers, and take protective measures to prevent the bushing from making contact with the ground. Subsequently, proceed to dismantle the tail protection tube.

When removing the tail protection tube, it is crucial to prevent dust and moisture ingress, as the process of adjusting the angle of the bushing and the dimensions between the bushing and the hoisting auxiliaries is time-consuming. Shield the tail end of the bushing with plastic sheeting or plastic wrap to protect it.

Adjustment between the bushing and the hoisting auxiliaries: After the bushing has been lifted to a height of 2 m, conduct a size verification and make necessary adjustments. This includes aligning the angle of the bushing and the distance between the bushing and the hoisting auxiliaries. Before installation, ensure that the high voltage capacitive bushing is thoroughly cleaned before lifting, with particular attention to the flange and the lower porcelain sleeve, which should be wiped with a clean cloth until spotless. The oil level indicator of an oil-filled bushing should be positioned to face the direction of operational inspection.

Note: During installation, measures should be taken to prevent insulation parts from absorbing moisture: when opening the transformer to install the bushing and connect the internal leads, dry air should be continuously blown.

During the bushing lifting and adjustment process, the transformer cover must not be opened prematurely. It can only be opened immediately prior to mating to minimize the exposure time.

(5) Bushing Filled with SF_6 Gas

Only after the converter transformer's hot oil has circulated for 48 h should the bushing be filled with SF_6 gas. Attach the vacuum unit to evacuate the bushing to a pressure of 100 Pa, then fill it with SF_6 to the required pressure specified by requirement, which is approximately 106.9 kg. During the inflation of the bushing, open the bushing inflation valve and connect the SF_6 inflation hose to the gas cylinder for the filling process. The SF_6 gas must have a moisture content of no more than 25 μg/g at (− 36 °C) and a purity of no less than 99.999%. After the inflation, regularly employ an SF_6 detection device to monitor for any potential leaks and prevent SF_6 gas leakage. When the SF_6 gas pressure fall below 300 kPa, it will be necessary to replenish the gas.

Note: After filling the bushing with SF_6 gas, it needs to be left still for at least 6 h before field testing is allowed.

VI. Selection of Construction Machinery

(I) Calculation for the Selection of the Crane for the Valve-side Bushing

Referring to Table 2, an 80 t crane with a main arm length of 27.2 m can lift a payload of 25 t at a operating radius of 8 m. When calculated at 90% of the rated capacity, the results in a lifting capacity of 22.5 t, with a maximum lifting height of 27 m.

According to the design schematic parameters, the HY high voltage valve-side bushing is the most substantial in both weight and height, with the tallest uppermost

Table 2 80 t Crane performance parameter

Operating range (m)	Main arm						
	Cylinder I extended to 100%, fully extended legs, side and rear operation, 3 t fixed weight + 8.5 t movable weight (kg)						
	12.1 m	16.4 m	20.7 m	27.2 m	33.6 m	40.1 m	46.5 m
3.0	80,000	65,000					
3.5	75,000	63,000	44,000				
4.0	68,000	61,000	44,000				
4.5	65,000	60,000	44,000	32,000			
5.0	58,000	54,000	42,000	32,000			
5.5	54,000	50,000	40,000	31,000	25,000		
6.0	50,000	46,000	38,000	30,000	24,000		
7.0	41,500	40,000	33,700	28,000	22,000		
8.0	35,000	34,500	30,200	25,000	20,000	17,000	
9.0	28,000	27,300	27,300	23,200	18,800	15,800	
10.0		22,600	22,000	21,700	17,500	14,800	12,500
11.0		19,000	18,500	19,900	16,200	13,600	12,000
12.0		16,000	15,600	17,000	15,100	12,800	11,000
14.0			11,400	12,800	13,200	11,400	10,000
16.0			8400	9800	10,700	10,000	9000
18.0				7600	8400	9000	8000
20.0				5900	6700	7300	7200
22.0				4700	5400	6000	6400
24.0					4400	4900	5200
26.0					3500	4000	4400
28.0					2700	3200	3600
30.0						2600	3000
32.0						2100	2400
34.0						1600	2000
36.0							1500

point reaching approximately 13.5 m. The heaviest bushing weighs around 15.5 t. The hoisting auxiliaries provided by the manufacturer, weighs about 5.5 tons based on their data. Consequently, the combined weight of the bushing and the hoisting auxiliaries is 21 t, which is less than the 22.5 t capacity. The bushing itself measures approximately 20.19 m in length. The calculated lifting height required for the crane, which includes the installation height plus the sling height, is roughly 20.8 m, well within the 27 m capability. Based on the conditions at the site, it has been determined that the crane's working radius at the installation location for the converter transformer bushing is less than 8 m. This confirms that the crane's performance is adequate to handle the lifting requirements for the heaviest bushing and is therefore deemed suitable for use.

(II) Selection of Slings

For the lifting of the HV converter transformer valve-side bushing, the converter transformer manufacturer provides matching hoisting auxiliaries and slings. It is recommended to carry out simulated lifting, see Table 3.

VII. Other Security Measures
(I) Heat Prevention
(1) During installation, confirmation of torque-fastened connections, which are previously tightened in the factory, should be made according to the torque values specified by the manufacturer. This practice prevents local heating during operation that may arise from connections not meeting the required torque values.

Prevention measures: Strictly adhere to marking vertical lines on areas where torque bolts have been applied.

(2) When connecting equipment and fittings, ample anti-loosening measures should be integrated to prevent connection points from becoming loose over extended periods of operation. The connection for the converter transformer terminal board follows the "1 bolt + 2 flat washers + 1 spring washer + 2 nuts" configuration.

Prevention measures: Procure bolts in accordance with above specifications.

(3) After drilling holes in fittings, the drilled places should be properly deburred.

Table 3 Parameters of hoisting auxiliaries and slings

Applicable product category	Crane capacity	Leading chain	Sling	Orchid bolt	Hoop
HSPGSETF 2079/945-2906	Hoisting weight $\geq$ 75 t, height $\geq$ 24 m	30 t electric-driven	25 t sling, 9 m, 5 m each	5 t	1100 kV

Prevention measures: Fittings should be smoothed using an aluminum file after drilling. A quality inspector must conduct a check before installation, with installation only proceeding upon passing inspection.

(4) Discrepancies between the equipment terminal board and the design drawings.

Prevention measures: When unboxing the equipment, verify the terminal board with the design. If inconsistencies are found, promptly report to the supervision engineer, request clarification from the manufacturer, and await approval from the design institute before proceeding with installation.

(5) The cross-sectional area of the fittings is smaller than that required for the equipment terminal board.

Prevention measures: Construction personnel should verify the fittings with the equipment terminal boards. Replace fittings if they do not match the terminal connectors.

(6) Electrical joint compound does not meet requirements and be uniformly distributed.

Prevention measures: Establish a protocol for applying the compound, which includes using 400-grit sandpaper to remove surface oxidation and imperfections, cleaning the sanded area with alcohol, wiping the fittings clean with a lint-free cloth, applying a layer of electrical joint compound approximately 0.2 mm thick, smoothing it with a stainless steel straightedge, tightening in the prescribed sequence, and removing any excess paste.

(7) DC resistance does not meet requirements when tested.

Prevention measures: Quality inspectors should oversee the testing unit to ensure all equipment within the contracted scope undergoes DC resistance measurement, also ensure that the testing unit accurately completes the "Interface Part Installation Process Quality Control Form."

(II) Quality Control

1. Inspection and storage before the installation of the valve-side outlet devices and bushings

   (1) During the handover process, the impacts recorded during the transportation and loading/unloading of the valve-side outlet device should adhere to the product's technical specifications. In the absence of specific requirements, the impacts in the longitudinal, transverse, and vertical directions should not exceed 3 g, with air transport pressure maintained between 0.02 and 0.03 MPa.
   (2) The surfaces of valve-side outlet device and bushing should be free of cracks or scratches.
   (3) The inner walls of the bushing, the flange neck, and the grading ball should be thoroughly cleaned.

(4) The oil-filled bushing should show no signs of oil leakage, and the oil level indicator should function correctly. The gas-filled bushing should maintain normal gas pressure.
(5) The valve-side outlet device and bushing must pass the required tests.

2. Installation requirements for the valve-side outlet device

(1) Before installation, the current transformer should be tested. The transformer's ratio, polarity, and arrangement should comply with the design specifications. The insulation from the outlet terminal to the bushing should be adequate. The washers under the connection bolts and fixtures should be securely tightened. The terminal board should be well sealed, with no oil leakage.
(2) When installing the valve-side outlet device, the position for the air release plug should be at the highest point of the device.
(3) The centerline of the current transformer should be aligned with that of the valve-side outlet device.
(4) The insulating tube should be firmly and securely installed.
(5) During the installation of the valve-side outlet device, the angle should be properly adjusted before connecting to transformer.
(6) The installation of the valve-side outlet device must conform to the product's technical specifications.

3. Installation requirements for the bushing

(1) When lifting the bushing, ensure that the lifting points and tools comply with the technical specifications.
(2) When the bushing is lifted, it should be connected to the valve-side outlet device once the angles are aligned. The sealing gasket at the top structure of the bushing must be correctly installed with a good seal. The lead wire connections should be reliable, bolts should be tightened to the specified torque values, and the insertion dimension of the conductive rod at the bushing's end should adhere to the technical specifications.
(3) For inflatable bushings, test the gas for micro water content and leakage rate to meet the requirements. During gas filling, verify that all pressure connection points operate correctly. After installation, inspect the release valve in the bushing's oil–gas separation chamber for any oil seepage or gas leaks. The bushing's terminal screen should have a good ground connection.
(4) The oil level indicator of an oil-filled bushing should be positioned to face outward, and the terminal screen should be well grounded.

4. Common quality problems

(1) The leakage in oil (gas)-filled equipment is primary at the flange connections. Before installation, thoroughly inspect the sealing ring material and the flange surface flatness. The bolt tightening torque should meet the manufacturer's manual specifications. After the installation of the

main transformer's fire extinguishing device connection pipes, a pressure test must be conducted (this can be done separately for this section of the pipeline after sealing the connection point, or in conjunction with the main transformer. Reference test method: After the main transformer is filled with oil, open the nitrogen fire extinguishing device pipeline valve, apply a pressure of 0.03 MPa from the oil storage tank, and ensure no leakage occurs over 24 h).

(2) In equipment installation, the protruding lengths of the bolts on both sides of the through-bolts are inconsistent.

(3) When connecting an oil-filled equipment bushing with a hard conductor, the bushing terminal should be under proper stress.

(III) Safety Control

(1) The vacuum pump should be properly lubricated, with ample cooling water flow, and maintained by a specialized personel for maintenance.

(2) When connecting the valve-side bushing and converter transformer side, ensure good ventilation and safety lighting, with a designated supervisor present. Workers must wear work uniforms without buttons or pockets, and specialized protective equipment such as oil-resistant, non-slip boots to prevent tools and debris from falling onto the equipment.

(3) The construction site should be equipped with adequate and reliable fire-fighting equipment. Clear fire safety responsibility system should be established. The site should be kept level and clean, with no open flames or flammable/explosive materials within a 10 m radius.

(4) Prior to powering the converter transformer, the main body, bushings and accessories should be reliably grounded. Spare secondary terminals of the current transformers should be short-circuited and grounded, and all electrical tests must be passed.

(IV) Requirements for Workers

(1) Construction personnel must wear identification badges when entering the construction site, be neatly dressed, and correctly wear personal safety protective equipment.

(2) Special operations and specialized trades should only be performed by trained and certified personnel. Unauthorized individuals are prohibited from engaging in such tasks.

(3) Before construction, safety measures outlined in the construction plan must be verified and implemented. Work should only proceed if safety measures are in place, strictly following the construction plan and adhering to safety and civil construction discipline without any violations.

(4) Protect all safety and civil construction facilities at the construction site, following usage regulations. The disassembly, movement, or misuse of safety and civil construction facilities is strictly forbidden without the approval of the site's safety management personnel. For facilities that

require temporary removed, appropriate temporary measures should be taken, and they should be promptly restored afterward.

(5) Construction personnel should be conscious of protecting finished and semi-finished products, maintaining the integrity of construction outcomes and protective facilities, and strictly refraining from unauthorized disassembly, removal, defacement, or misuse.

(6) In accordance with site management requirements, organize training sessions for construction personnel before the installation of each converter transformer.

(V) Construction Operation Requirements

(1) During lifting operations, construction personnel is prohibited to enter under the crane arm and the vertical area below the lifting parts, or the inner angle side of the stressed steel wire rope. Proper hoisting auxiliaries should be used. "Use small ones instead of large ones" is prohibited. When working near construction machinery, construction personnel is prohibited to linger, walk, or work within the operating radius of the machinery.

(2) For vehicle transportation operations, the vehicle condition should be checked to be normal. Driving without a license or drunk drinking is strictly prohibited, as well as speeding or overloading. The loads should be securely tied, and mixing people with goods or carrying people in dump trucks is strictly prohibited.

(3) Workers working at heights must not lose protection during the entire process and must have measures to prevent tools and materials from falling.

(4) Construction safety protective equipment should be used exclusively by designated personnel, properly stored when not in use, and regularly inspected. The inspection cycle should be determined based on the frequency of individual use, but should not be less than once a month.

(VI) Requirements for Height Working

(1) Use ladders or elevated work platforms for tasks performed at height.

(2) Equip construction personnel of height working with fall-arrest safety harnesses.

(3) Establish a signaling or communication system linked to the ground, with specialized operators.

(4) Stop outdoor height working during severe weather conditions such as wind scale of strong breeze or higher, heavy rain, thunderstorms, hail, heavy snow, dense fog, or sandstorms.

Install a horizontal safety rope above the converter transformer and erect a temporary rigid barrier at the perimeter (refer to Fig. 42). Workers at height must wear safety harnesses with safety ropes attached to secure and reliable anchor points above. Work attire should be agile, with sleeves and pant legs secured, and soft-soled shoes

Fig. 42 Temporary rigid barrier at the perimeter

should be worn. Workers must continuously check the security of their harnesses and maintain protection when relocating. A safety supervisor should be appointed.

(5) Workers at height should utilize tool bags, secure larger tools with safety ropes, and use designated ropes for passing items to prevent dropping.
(6) During operations, all work pieces and offcuts should be placed in secure locations with precautions taken to prevent falls.
(7) Delineate the construction area with fences and clearly mark it with safety signs indicating no entry. Personnel are prohibited from lingering or passing through hazardous zones.

(VII) Prevention for Backflow of Vacuum Pump

Ensure the vacuum pump is in proper working order. A high vacuum ball valve and check valve should be installed at the pump's outlet to prevent backflow of oil and gas in the event of a power outage. The power supply must be reliable, controlled separately, and managed by a designated individual. Unauthorized personnel are not permitted to operate the control switches. When evacuating the system, initiate the cooling water flow before starting the vacuum pump. Once the pump is operating smoothly, gradually open the gate and butterfly valves. The shutdown procedure should be conducted in reverse order. The vacuum gauge should not be positioned atop the oil tank, and there should be a high vacuum ball valve in front of the gauge. A designated operator should read the vacuum level, slowly opening the ball valve. The vacuum gauge should not be set too high to avoid mercury ingress into the vacuum line. When adding oil using vacuum methods, strictly regulate the vacuum level to prevent over-evacuation, and ensure the capsule is connected to the oil reservoir. Monitor the deformation of the tank walls during vacuuming.

(VIII) Safety Measures for Low Transportation and High Construction

(1) When working in the vicinity of powered equipment, steel tape measures, cloth measuring tapes, or any line measures (those containing metal wires) are forbidden. Utilize insulated measuring tools or instruments instead.
(2) Operating near powered parts needs to take work permit. The entire construction process is viewed by a specialized supervisor.
(3) In cases where construction equipment may generate induced voltage due to its parallel or close proximity to powered equipment, personal safety grounding wires must be installed. These personal safety grounding wires should be documented on the work permit and are installed and removed by construction workers.
(4) When working in close proximity to powered components, the normal activity area of the construction personnel and the safe distance from the powered equipment should comply with the specifications outlined in Table 4.
(5) For construction machinery such as cranes and aerial work platforms operating near powered parts, the safe distance from the powered equipment during normal activities should not exceed the limits specified in Table 5.
(6) Regardless of whether high voltage equipment is powered, personnel is prohibited to move or bypass barriers to work alone. When it is necessary to move a barrier, a supervisor must be present. The safe distance as stipulated in Table 6 must be maintained.
(7) In outdoor areas with high-voltage distribution equipment rated at 10 kV or higher, driving lanes should have vehicle safety height limit indicators installed in accordance with Table 7.

Table 4 Normal activity area of the construction personnel and the safe distance from the powered equipment

Voltage level (kV)	Safe distance (m)	Voltage level (kV)	Safe distance (m)
10 and below	0.70	1000	9.50
20, 35	1.00	± 50 and below	1.50
60, 110	1.50	± 400	6.70
220	3.00	± 500	6.80
330	4.00	± 660	9.00
500	5.00	± 800	10.10
750	8.00		

Table 5 The safe distance from the powered equipment during normal activities

Voltage level (kV)	Safe distance (m)	Voltage level (kV)	Safe distance (m)
10 and below	3.00	1000	13.00
20, 35	4.00	± 50 and below	4.50
60, 110	4.50	± 400	9.70
220	6.00	± 500	10.00
330	7.00	± 660	12.00
500	8.00	± 800	13.10
750	11.00		

Table 6 The safe distance when equipment is energized

Voltage level (kV)	Safe distance (m)	Voltage level (kV)	Safe distance (m)
≤ 10	0.70	1000	8.70
20 ~ 40	1.00	± 50 and below	1.50
60 ~ 110	1.50	± 400	5.90
220	3.00	± 500	6.00
330	4.00	± 660	8.40
500	5.00	± 800	9.30
750	7.20		

Table 7 The safe distance between trucks (including loads) and uncovered powered parts

Voltage level (kV)	Safe distance (m)	Voltage level (kV)	Safe distance (m)
10	0.95	750	6.70
20	1.05	1000	8.25
35	1.15	± 50 and below	1.65
66	1.40	± 400	5.45
110	1.65 (1.75)	± 500	5.60
220	2.55	± 660	8.00
330	3.25	± 800	9.00
500	4.55		

(8) The established fences should be conspicuous and sturdy. Grounding wires, safety signs, and other safety protection facilities must not be arbitrarily moved or dismantled.

(9) Safety signs, fences, and other protective measures should be correctly and promptly set up, and should be removed in a timely manner after work is completed.

(10) Do not perform work in powered areas during adverse weather conditions such as rain, fog, or high winds.

(11) Install safety distance warning tapes in the installation area of HV converter transformers to alert of the necessary safety distances.
(12) The operational load of lifting equipment, slings, and other hoisting auxiliaries must not surpass the limits specified on their nameplates.
(13) The lifting and transportation of all significant objects should be managed by experienced personnel. Prior to the operation, provide safety and technical briefings to all involved workers to ensure familiarity with the construction plan and safety measures. During lifting operations, only one person should be in command, with intermediate support staff assigned to relay signals as needed. Lifting command signals should be clear, unified, and unambiguous, with a clear division of labor.
(14) Ensure that items to be lifted are securely fastened, and that the hook is positioned directly above the item's center of gravity.
(15) Prohibit outdoor lifting work when winds exceed level 6. For objects with large windage areas, avoid lifting when winds are at level 5 or higher.
(16) Refrain from conducting lifting operations in conditions of heavy fog, poor lighting, or when the command personnel cannot clearly see all work areas, or when the lifting operators have not received effective commands.
(17) Standing on lifted objects is prohibited, and workers should not use hooks for ascending or descending.
(18) Unauthorized personnel are forbidden from entering or lingering within the lifting work area.
(19) Do not allow lifted heavy objects to be left suspended in the air for extended periods. When a heavy object is under control, operators are not allowed to leave their cabs or other working areas.
(20) When using lifting machinery within the station, install a grounding device. The grounding wire should be made from multi-strand soft copper, with a cross-sectional area sufficient to meet the grounding short-circuit capacity requirements, but not less than 16 mm^2.
(21) All construction workers must also adhere to the relevant requirements set by the operating entity.
(IX) Safety Measures for Low Transportation and High Construction cranes

Establish a hazard warning zone with caution tape. The minimum safe distance between the crane and the converter transformer busbar is 13 m. A safety forbidden line is placed at a horizontal distance of 13 m from the LV busbar's position on the converter transformer courtyard. A specialized guardian is assigned to monitor the area. In the event the crane's boom approaches the forbidden line, a "whistle" alarm is used to warn the crane operator and the crane commander, ensuring that they remain alert throughout the installation process.

A warning sign indicating the boom extension length is posted inside the crane's operator cabin to remind the crane driver that the busbar is powered and to maintain a safe distance of 13 m from the LV converter transformer busbar.

VIII. Environmental Protection Measures, Safe and Civilized Construction, and Finished Product Protection Measures

(I) Environmental Protection

(1) Adhere strictly to national and local government laws, regulations and rules on environmental protection during construction. Enhance the control and management of waste, including construction material packaging, and accept supervision and inspection by relevant authorities.
(2) All solid waste at the construction site must be promptly cleaned up and sorted into designated bins within the work area to facilitate centralized recycling and disposal.
(3) Paints and coatings should be centrally collected for recycling, with a specialized person responsible for the collection of used paints and coatings.
(4) Keep construction activities within the scope permitted by the project. Arrange and standardize the enclosures rationally. Ensure all signs are complete, uniform, and that various markings are accurate and clear. Maintain a clean and civilized construction site.
(5) Strictly monitor construction noise to ensure there is no disruption on the living and working order of residents in the vicinity of the substation.
(6) The construction should conform to the "Green Construction Guideline (CQ [2007] No. 223)" and the planning requirements for creating a green construction demonstration project at the Guquan Converter Station. Strictly follow national and local government laws, regulations, and rules on environmental protection.
(7) Construction machinery requires regular maintenance and must be inspected and approved by the responsible management department. Vehicles that fail to meet emission standards are prohibited from use.
(8) During construction, minimize the adverse impact on the surrounding environment. After completing construction tasks, enhance measures for the protection of finished products and ensure that waste and debris within the construction area are promptly removed.

(II) Safe and Civilized Construction

(1) Construction personnel must wear ID badges, work shoes, and uniforms, maintain neat attire, and correctly wear personal safety protective equipment when entering the construction site.
(2) Individuals engaged in special operations or trades must be trained and certified before they can work in their respective positions. Unqualified personnel are prohibited from performing related tasks.
(3) Before commencing construction work, safety measures outlined in the construction plan must be inspected and implemented. Work should not

start if these measures are not in place. Adherence to the construction plan and compliance with safe and civilized construction protocols. No violations are allowed.

(4) Protect all safety and civilized construction facilities at the site. Follow usage regulations, and do not dismantle, move, or misuse these facilities without approval from the site's safety management personnel. For facilities that require temporary removal, implement corresponding temporary measures and restore them promptly afterward.

(5) Construction workers should be conscious of protecting finished and semi-finished products. They must maintain the integrity of finished and semi-finished construction outcomes and protective measures, strictly prohibiting unauthorized disassembly, removal, painting, or defacement.

(6) The construction site should be secured with enclosures, isolation, and barriers, implementing regional management. The site should be divided into designated areas such as construction work zones and equipment/material storage areas, based on the type of work being performed.

(7) Implement fixed-position management at the work site by planning and creating layout diagrams to standardize the placement of equipment, materials, and tools.

(8) Select the necessary signs for the site based on the four basic types: prohibition, warning, instruction, and informational signs. When multiple signs are displayed together, they should be arranged in the order of warning, prohibition, instruction, and informational signs, from left to right and top to bottom, avoiding any contradictions or repetitions. Use multiple signs as needed based on actual conditions.

(9) The construction site should be equipped with first aid kits and firefighting equipment. Drinking stations and designated smoking areas should be established in appropriate locations.

(10) Use steel tube and shackle assembly safety fences for enclosing and isolating the construction area. Promptness barriers should be used to segregate the lifting operation areas, and set up risk control information boards for the construction site.

(III) Protection Measures on Finished Products

(1) During installation, pay attention to the protection of finished products such as fire walls, foundations, roads, tracks, and equipment. Control the equipment hoisting process to prevent damage to the foundations and equipment. Prohibit direct contact between tools like pry bars and jacks and the ground. Instead, use steel or wooden padding.

(2) Use slings for lifting accessories to prevent surface paint damage. Implement special measures for bushing hoisting to prevent damage from hoists.

(3) All bolts must be tightened according to the torque values specified by the manufacturer.

(4) All cable installations on the converter transformer should utilize troughs for transition, ensuring no exposed cables and that all seals are complete and aesthetically pleasing.
(5) Conduct education on finished product protection for construction workers to strengthen their awareness and commitment to preserving the integrity of completed work.
(6) For oil-related tasks, lay down oil-resistant cloths in the work area to ensure that insulating oil does not contaminate the foundation surfaces.
(7) Handle installation tools with care. Use designated tool bags for passing tools up and down to prevent accidental drops that could damage the foundation surfaces and transformers.
(8) The converter transformer's equipment foundation, courtyard pavement, and track system have been completed, meeting the construction requirements. To ensure the quality of the courtyard's concrete surface and to prevent contamination and damage during the lifting, storage, and placement of accessories, a hard fence has been erected in the transformer area for isolation. After the crane enters the hoisting position, an oil-resistant cloth is laid on the ground for protection. A cushioning pad is then placed on top of the cloth for each individual converter transformer. For accessories made of iron, a plastic cloth should be laid on the ground.
(9) Before a crane enters the converter transformer courtyard, check all conditions being OK. If there is any oil leakage, entry is prohibited. Once a crane is positioned within the courtyard, lay down an oil-resistant cloth underneath to prevent oil leakage from contaminating the ground. Cover the courtyard ground under the crane's path with colored cloth or plastic to prevent tire marks from causing pollution.
(10) Adhere to the converter transformer courtyard management system established for the Guquan converter station during the installation and construction period.
(11) Environmental monitoring system. Install devices for detecting temperature, humidity, and dust levels, along with display screens. This includes temperature probes, humidity sensors, dust detectors, data collection devices, and display screens. Data is collected by the probes and transmitted to the data collection devices, which then show the information on the screens, providing a clear picture of the on-site installation environment. Assign specialized operators to maintain the monitoring and environmental monitoring equipment.

IX. Emergency Response Plan

(I) Encounter Adverse Weather Such as Rain during the Installation of Bushing

Install dust and rain covers at the bushing installation joint and the inspection port of the main body to prevent the intrusion of rainwater and dust. In the event of adverse weather conditions such as rain, it is crucial to promptly and reliably seal any removed covers. Subsequently, blow dry air with a dew point below – 55 °C into the system to achieve a pressure of approximately 0.03 MPa for positive pressure storage. In the event of continuous rainy weather, monitor and record the readings of the main body's pressure gauge. Address any sealing issues that are detected promptly. Should rain occur during vacuuming and poor sealing results in an inadequate vacuum, break the vacuum with dry air. Implement rainproof measures to rectify the poorly sealed areas as quickly as possible, then continue to fill with dry air to the specified positive pressure for storage. Once the weather clears, resume vacuum treatment.

(II) Handling of Sudden Power Outage or Equipment Damage During Installation

During the vacuuming process, if a sudden power outage occurs or the vacuum unit and dry air generator suffers damage, thereby impairing the normal operation of the mechanical equipment, the following control measures should be implemented:

(1) The site should be equipped with an adequate supply of emergency flashlights. In the event of a sudden power outage during nighttime vacuuming duty, these emergency lights should be deployed for immediate response, the machinery should be halted, and the relevant valves should be securely closed.
(2) In the event of a power outage or equipment malfunction during the vacuuming process, it is imperative to promptly shut down the vacuum valves and cease the operation of the vacuum pump.

3 Installation Tools and Work Platforms for Converter Transformers

Converter transformers used in the CG project have large volume, (the oil of single HV converter transformer weights over 240 t), large oil pollution treatment volume, high installation precision requirement, and stringent installation conditions. This section introduces the construction techniques of three safety and quality measures: the special shed for internal inspection of the converter transformer, the vacuum oil filter with capacity of 20,000 L/h, and the traction force monitoring alarm system.

I. Special Shed for Internal Inspection of Converter Transformer

To ensure that the internal inspection of the converter transformer can be proceeded successfully in humid (exceeding standard operating environment humidity) or light rain conditions, while ensuring process quality and meeting construction progress requirements, a special tool for internal inspection of the converter transformer installation is developed. Based on this, a method for using the special tool is researched and developed, solving the problem of installing converter transformers in the rainy and humid conditions (like in the Southern China).

The internal inspection of the converter transformer is conducted within a specialized air purification system. This system includes air conditioning, a dust shower room, a dust-proof shed, and other modules. An air purifier is installed inside the dust-proof shed, and the system has dust showering, dehumidification, and air purification functions. The on-site processed dust-proof shed measures 3000 mm × 4000

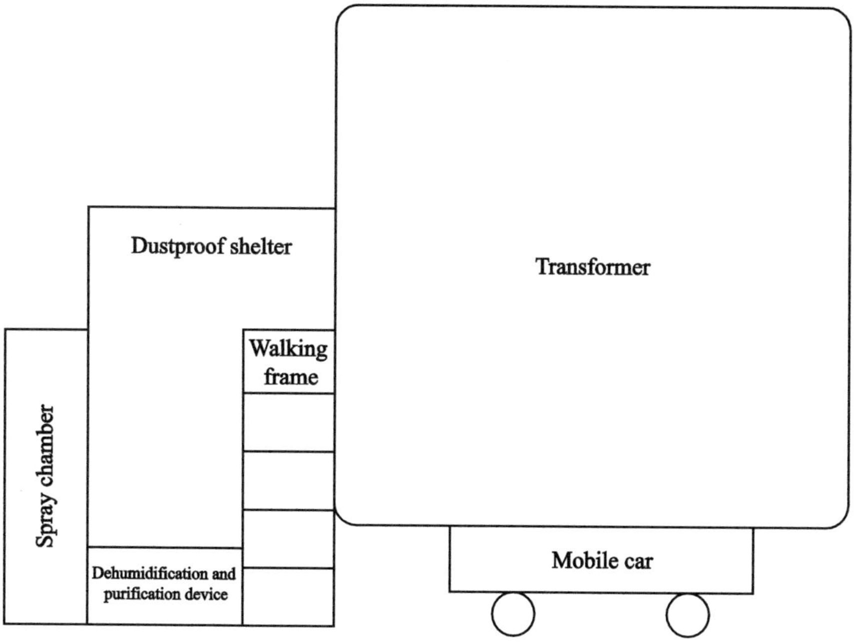

Fig. 43 Outside view of the special shed for internal inspection

mm, is covered with transparent plastic cloth, and is processed into a sealed dust-proof space. After the dust-proof shed is connected to the dust shower room, it is sealed with tape all around, achieving the requirement for a controllable internal inspection environment for the converter transformer. See Figs. 43, 44 and 45 for details.

II. 20,000 L/h Vacuum Oil Filter

To improve oil service efficiency and ensure the quality of the converter transformer's insulating oil, insulation strength, oil filtering efficiency, and oil filtering precision are focused on the construction site. The construction entity is required to optimize and upgrade construction equipment and purchase a 20,000 L/h vacuum oil filter.

The capacity of the vacuum oil filter has been increased from 12,000 to 20,000 L/h, and oil service treatment has been upgraded from two-stage filtering to four-stage filtering, improving installation efficiency by 20%. A high-flow oil filter is used to precisely control the heater temperature of the oil filter, reliably increasing the outlet oil temperature of the oil filter, which reduces the hot oil circulation time from the previous 8–6 days. In terms of insulation grade, the electrical strength of the insulating oil is improved, from 70 kV/2.5 mm to 75 kV/2.5 mm, achieving a dual improvement in insulation and precision.

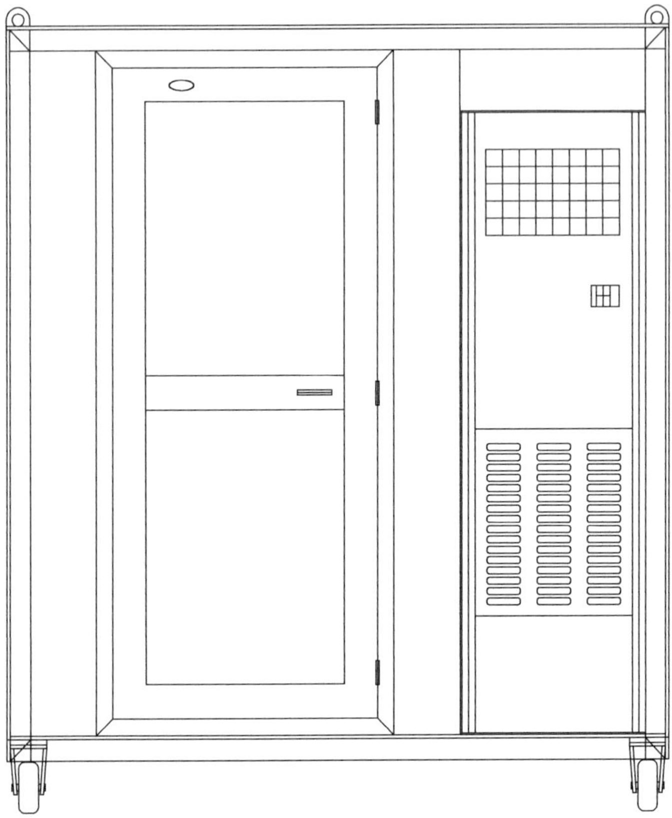

Fig. 44 Outside view of the dust shower room

III. Alarm System for Traction Force Monitoring

During the traction process of the converter transformer, there may be a problem of deviation due to uneven force on the tag lines on both sides. A traction force monitoring alarm system is developed and deployed to prevent this problem, ensuring the safety and quality of the converter transformer's positioning to the foundation construction.

When the converter transformer is positioned to the foundation construction, a set of traction force monitoring alarm devices is installed on both sides. The monitoring alarm device displays real-time traction force data, allowing for a direct and accurate judgment of whether the traction force on both sides is balanced through data, preventing the converter transformer from deviating due to uneven force on both sides, and thus avoiding corresponding safety and quality hazards.

Fig. 45 The photo of the special shed for internal inspection

4 Dispose of Converter Transformer Oil

In the installation of converter transformers, the insulating oil can be treated to the construction acceptance standard outside the station according to the contract, effectively improving the subsequent process flow of the converter transformer. At the same time, The capacity of the vacuum oil filter has been increased from 12,000 to 20,000 L/h, and oil service treatment has been upgraded from two-stage filtering to four-stage filtering, improving installation efficiency by 20%. A high-flow oil filter is used to precisely control the heater temperature of the oil filter, reliably increasing the outlet oil temperature of the oil filter, which reduces the hot oil circulation time from the previous 8–6 days.

I. Construction Preparation

(I) Technical Preparation

(1) Before the treatment of insulating oil, check whether the factory technical documents, product technical agreements, relevant acceptance specifications, and insulating oil dispose record forms of the converter transformer insulating oil are complete.

(2) The technical director needs to read the product manual, construction contract, etc. in detail, understands the product performance, main parameters, technical provisions and requirements, and gives a detailed technical and safety briefing to the construction personnel, while keeping a record of the briefing.

(3) The construction personnel work according to the technical measures and technical briefing requirements, have a clear understanding of the insulating oil dispose procedure, method and technical requirements, and are familiar with the manufacturer's information, technical measures and relevant regulations and standards.
(4) Communicate and familiarize with the manufacturer.

(II) Site Preparation

The foundation of the converter transformer equipment, the courtyard pavement and the track system have all been completed, meeting the construction requirements of the valve side outlet device of the converter transformer. At the same time, the courtyard of the converter transformer is cast in one go by the civil construction, to ensure the quality of the courtyard concrete pavement process, and to prevent the storage of insulating oil from polluting and damaging the courtyard ground surface. The oil service area ground is covered with plastic cloth or oil-proof cloth for protection, and is isolated by a fence, with reasonable arrangement of fire-fighting equipment.

(III) Construction Power Supply Preparation

The construction power supply box uses three self-provided terminal distribution boxes, utilizing "one machine, one switch, one protection". According to the progress of the site construction, the construction power supply can also use the converter transformer courtyard maintenance box.

(IV) Preparation of Machinery, Tools, Safety Equipment

Before installation, the installation machinery, tools, safety equipment, traction equipment and consumable materials of the converter transformer should be fully prepared, and the main construction machinery should be tested and qualified, and the quantity and specifications meet the construction requirements as shown in Table 8.

II. Construction Process and Key Points of Operation
(I) Changji Converter Station
1. Three-stage oil filling system
(1) The schematic diagram of the three-stage oil filling system is shown in Fig. 46.
(2) Connect the pipeline according to Fig. 46: The oil flow direction of the No.2 and No.3 oil tanks is from top to bottom, where the oil inlet is inserted into the oil tank for 1.5 m, and a humidifier needs to be connected at the breathing port. Except for the position of the breathing port, the oil tank is not sealed elsewhere.
(3) Filling oil into Tank No.2: Connect the crude oil tank to the oil inlet pipeline of the second-stage oil filter, open the oil drain valve of the crude oil tank, start the first-stage oil filter for self-circulation, and when the oil temperature reaches 65 °C, fill oil into Tank No.2 at atmospheric pressure. The heater temperature of the first-stage oil filter is set at (70 ± 5) °C (Before filling oil into the crude oil tank, in addition to meeting the performance requirements, heating treatment must be carried out. When starting to fill oil into the converter transformer, the oil temperature in the crude oil tank should not be less than 30 °C).

Table 8 List of machinery, tools, safety equipment

No	Shackle	Specification and model	Quantity
1	Oil filter	20,000 m^3/h	2
2	Vacuum pump	5000 m^3/h	2
3	Crane	25 t	1
4	Buckle	5 t	4
5	Slings	5 t	2
6	Steel wire rope	ϕ 18 mm, 6 m	4
7	Pressure gauge	0 ~ 1 MPa	2
8	Multimeter		1
9	Flashlight		2
10	Power box		2
11	Protective fence		200 (m)
12	Emergency light		3
13	Filter oil tube	Metal wire tube	100 (m)
14	Vacuum tube	Metal wire tube	100 (m)
15	Grounding wires		50 (m)
16	Fire protection facilities		2

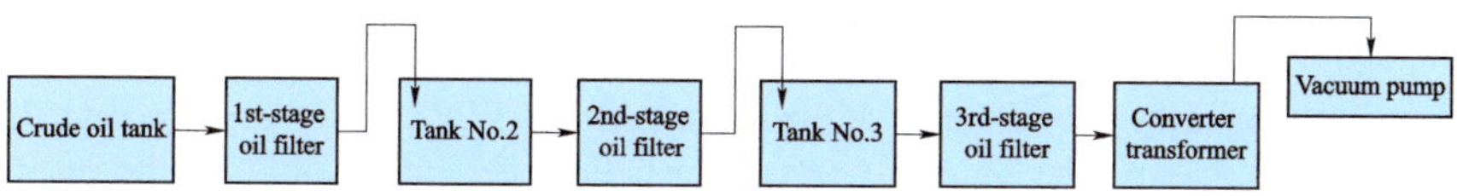

Fig. 46 The schematic diagram of the three-stage oil filling system

(4) Filling oil into Tank No.3: Start the second-stage oil filter for self-circulation, when the oil level of Tank No.2 is about 1/2 ~ 3/4 height, fill oil from Tank No.2 to Tank No.3 at atmospheric pressure. The heater temperature of the second-level oil filter is set at not less than 75 °C, and the outlet oil temperature is not less than 70 °C. While filling oil into Tank No.3, Tank No.2 maintains the oil filled state.

(5) Filling oil into the transformer body: When Tank No.3 is filled with oil to 1/2 ~ 3/4, start the third-stage oil filter and use self-circulation + oil filling method (about 2/3 flow of the oil filter drain pump is used for self-circulation, 1/3 flow is used for oil filling) to fill oil into the converter transformer (the heater of the third-stage oil filter can be turned off to avoid concentrated heating of the oil to form oil vapor and increase the vacuum degree of the degassing cylinder). When the inlet oil temperature of the third-stage oil filter is 60 °C, the actual vacuum degree in the degassing cylinder should not be more than 35 Pa. When the inlet oil temperature is 60–65 °C, the actual vacuum degree in the degassing cylinder should not be more than 45 Pa. When the inlet oil temperature is 65–70 °C,

the actual vacuum degree in the degassing cylinder should not be more than 50 Pa. The oil filling speed is 5–8 m^3/h, and during the oil filling process, the oil level in Tank No.3 should be kept above 1/2. While filling oil into the converter transformer, the crude oil tank and Tank No.2 continue to maintain the oil filling state.

The indicators of oil filling into the converter transformer must meet Table 9.

(6) Oil filling should use V-shaped joint tooling, and the gas in the oil filling pipeline should be discharged through the exhaust valve on the oil filling V-shaped joint tooling. When there are no bubbles in the oil in the oil filling pipeline, close the exhaust valve of the oil filling V-shaped joint tooling. Open the oil filling valve of the converter transformer, adjust the opening degree of the oil filling valve, so that the pressure gauge on the V-shaped joint tooling always maintains a slight positive pressure to fill oil into the transformer.
(7) Vacuum monitoring during oil filling: During the oil filling process of the converter transformer, the vacuum level in the oil tank should always be no more than 40 Pa.
(8) Oil level monitoring during oil filling: When the oil level of the converter transformer is about to reach the neutral point valve side outlet device, grid side outlet device, upper valve of the valve side outlet device, arrange a specialized operator to control the valve. After the oil surface exposed by each valve side outlet device valve is about 100 mm above the valve outlet, close each valve side outlet device valve, and arrange a specialized operator to observe the oil outflow situation of the vacuum tube of the valve bushing. When the oil surface appears on the vacuum tube, carefully observe whether there are bubbles in the oil. If the oil surface rises steadily and there are no bubbles in the oil, it proves that the oil filling quality of the bushing is good, wait until the oil surface exceeds the observation window, then close the valve.

After the bushing is filled with oil, continue to fill the oil tank. When the pressure of the tank cover reaches 55 kPa, open the valve on the main manifold leading to the oil storage cabinet. The oil storage cabinet capsule is pre-inflated to 10 kPa. The vent valves at both ends of the top of the oil storage cabinet must be open, close after oil

Table 9 The indicators of oil filling into the converter transformer

Dielectric loss factor (90 °C)	Electrical strength (kV)	Water content (μL/L)	Gas content (%)	Chromatographic analysis (No acetylene) (μL/L)				Particle size (≥ 5 μm)
				H_2	CO	CO_2	Total hydrocarbons	
≤ 0.25%	≥ 80	≤ 5	≤ 0.5	< 1	< 20	< 100	< 1	≤ 1000/1000 mL

outflow, fill the oil to the standard oil level of the oil storage cabinet, stop filling, and close the valve.

(9) The switch is evacuated and filled with oil along with the converter transformer.

2. Hot oil circulation

After vacuum oil filling, hot oil circulation work can be carried out. The oil is required to enter from the top and exit from the bottom. To ensure the oil filter temperature of the converter transformer, the HV converter transformer uses two 20,000 L/h oil filter machines for hot oil circulation work. During the hot oil circulation process of the converter transformer body, note:

(1) Before hot oil circulation, vacuum the oil pipe, use the oil filter machine to perform long-axis diagonal hot oil circulation on the converter transformer, and the oil temperature at the outlet of the oil filter machine should be set to (65 ± 5) °C.
(2) The oil temperature in the oil tank is maintained at 50 ~ 60 °C. After reaching this temperature, the circulation time needs to reach 96 h and the total circulation oil volume reaches 3 ~ 5 times the total oil volume of the product.
(3) In addition to the hot oil circulation time must comply with the regulations, the sign of the complete is that the samples taken at the end of the circulation meet the following requirements: the electrical strength is not less than 75 kV/2.5 mm; the water content is no more than 8 mg/L; the dielectric loss factor $\tan \delta$ (90 °C) is no more than 0.1%, the gas content is no more than 0.5%; the impurity particles are no more than 1000 per 100 mL (5 ~ 100 μm particles, no particles above 100 μm). Otherwise, the hot oil circulation should continue until the above standards are met.
(4) Connect the pipeline of the hot oil circulation system, sequentially open the valves between the radiator, switch oil chamber and the converter transformer, and then let the hot oil enter the oil tank from the butterfly valve on the top cover of the oil tank, and flow back to the treatment device from the oil filter valve of the lower section of the oil tank.
(5) After the hot oil circulation is completed, take oil samples from the lower end for comprehensive inspection tests. The oil that has undergone hot oil circulation should meet the relevant procedural requirements.

(II) Guquan Converter Station

1. Before insulating oil filling

(1) Before the converter transformer oil is filled into the converter transformer oil tank, it should be dehydrated and purified. Only after the treatment is qualified can it be filled into the converter transformer. After each batch of insulating oil arrives at the site, the quantity should match the contract and there should be a factory test report. A simplified analysis should be taken for sampling. A full analysis should be done if necessary.

The insulating oil filled into the converter transformer must comply with "Fluids for electrotechnical applications-Unused mineral insulating oils for transformers and switchgear (GB 2536—2011)" and the following conditions:

1) The electrical strength should not be less than 75 kV/2.5 mm.
2) The water content should not be more than 8 mg/L.
3) tan δ should not be more than 0.5% (90 °C).
4) The particle size should not be more than 1000 per 100 mL (5 ~ 100 μm particles, no particles above 100 μm).
5) No acetylene tested in chromatographic analysis.

(3) When the new oil sample taken on site does not meet the above standards, connect the oil filter and the upper and lower flanges of the 15 t oil tank, check for no oil leakage, and start filtering oil by opening the flange respirator. The oil filtering process is as follows:

1) Turn on the main switch in the power operation box. The power indicator is normal.
2) Fully open the oil inlet valve of the coarse filter tank of the oil filter, fully open the primary valve of the vacuum oil filter, slightly open the secondary valve, and check that there is oil flow in the oil pipe.
3) Press the start button of the oil temperature heater, check that the heating indicator displays normally, and the automatic control heating system operates normally.
4) Thoroughly check that the oil filter is operating normally.

When the oil temperature at the outlet of the oil filter body reaches (70 ± 5) °C, the tank oil temperature is controlled at 55 °C, the speed does not exceed 10,000 L/h, and the circulation is greater than or equal to 5 times the oil volume, oil can be taken for testing after meeting the requirements.

(4) Insulating oil test. Open the respirator, then open the sampling valve to take oil. The oil sampling is carried out under the supervision of the project supervisor and sent for inspection. If it does not meet the standard, filter the oil again until it passes test. Then seal the oil tank.

After the oil sample test and hot oil circulation, the insulating oil should meet:

1) The electrical strength should not be less than 75 kV/2.5 mm.
2) The water content should not be more than 8 mg/L.
3) tan δ should not be more than 0.5% (90 °C).
4) The particle size should not be more than 1000 particles/100 mL (5 ~ 100 μm particles, no particles above 100 μm). In the gas content test, hydrogen should not be more than 30 μL/L, acetylene is 0 μL/L, and total hydrocarbons should not be more than 20 μL/L. Other status all meet the specifications and standard requirements. The oil can only be filled when the converter transformer installation meets requirements.

2. Filling insulating oil into the converter transformer

The transformer should maintain in vacuum during the entire oil filling process. The oil filling pipeline should always maintain positive pressure, with filling speed not exceed 6000 L/h. During the oil filling, the vacuum level in the oil tank should be maintained at less than 200 Pa. The oil temperature at the outlet of the oil filter is controlled within the range of (65 ± 5) °C, and the oil temperature at the inlet is not less than 55 °C. If necessary, insulation measures should be taken for the oil tank and oil filling pipeline. The insulating oil filling into the converter transformer must comply with GB 2536 and the requirements of clause 3.

3. Post-treatment after oil filling into the converter transformer

(1) During the hot oil circulation process, the oil temperature at the outlet of the oil filter is controlled within the range of (70 ± 5) °C. When the oil temperature at the outlet of the converter transformer is not less than 55 °C, start timing. The hot oil circulation volume is not less than 5 times the total oil volume of the converter transformer, and the circulation time is not less than 96 h (refer to the manufacturer's operation manual). When the oil temperature at the outlet of the converter transformer keeping above 55 °C for 24 h, open the cooling oil pump circulation in two batches lasting for 6 h. Continue to circulate the entire converter transformer. When the circulation time is up, oil samples should be taken for inspection test. If the oil sample indicators meet the testing requirements, the hot oil circulation can be stopped.

(2) After the hot oil circulation, the relevant indicators of the insulating oil should meet following requirements: ① The electrical strength should not be less than 75 kV/2.5 mm; ② The water content should not be more than 8 mg/L; ③ tan δ should not be more than 0.5% (90 °C); ④ The particle size should not exceed 1000 particles/100 mL (5 ~ 100 μm particles, no particles above 100 μm), and the gas content should not exceed 1%. After the hot oil circulation, the transformer is left to stand and degassed as required. The standing time for the LV converter transformer is 7 days, while for the HV converter transformer is 10 days (refer to the manufacturer's operation manual). After the standing period, field tests are conducted according to the acceptance standards, and the transformer can only be put into operation after passing the tests.

(3) During the process of insulating oil filling and hot oil circulation, a specialized operator is required to monitor the operation of the unit. At the same time, the outlet oil temperature of the oil filter and the outlet oil temperature of the converter transformer are recorded in paper form every 1 h.

III. Quality Control

The key points of quality control are as follows:

(1) Transformer oil plays the vital role in insulation and cooling. Its performance indicators need to meet the specification requirements. Before filled into the converter transformer, a simplified test must be conducted. A full analysis of the oil must be carried out after the hot oil circulation.

(2) Vacuuming is required before and during oil filling. The purpose of vacuuming is to dry the body of the converter transformer.

(3) Before mixing different grades of insulating oil or adding new oil of the same grade, a test on the mixed oil must be conducted.

(4) The filter element of the oil filter machine needs to be replaced frequently. When the pressure value of the filter element reaches 0.75, the filter element must be replaced.

(5) When adding supplementary oil, it should be filled through the dedicated oil filling valve on the oil storage tank, and should be filled through the oil purifier. During oil filling, the air in the body and accessories should be discharged.
(6) After the hot oil circulation, multiple degassing should be carried out from the top of the bushing, the top of the valve side outlet device, the top of the oil storage tank, the top of the cooling device, the connecting pipe, the pressure relief device, and other related parts of the converter transformer during the standing period.

5 BIM Technology in Converter Station Construction

During the construction of the CG project, the construction team used BIM technology to introduce three-dimensional animation and VR technology in the construction of templates, decoration, equipment installation, and steel structure construction. This technology was used for planning, briefing, and other tasks. Use 3D simulation of the building in advance to better guide the on-site construction. This section introduces the application of BIM technology in the CG project.

I. Application of BIM Technology in Base Design for Engineering

The base design for engineering is a multi-process, multi-coordination, highly complex sub-project that runs through the entire process of structural engineering construction. Using BIM technology, the entire process of base design for engineering construction is simulated. Through simulation analysis, the logical relationship between each specialty is derived, thereby determining a reasonable construction plan to guide construction.

In BIM three-dimensional design of template engineering, the properties of components can be defined, such as the diameter of steel tubes, wall thickness, elastic modulus of templates, etc. Property definition can provide data support for the safety check of components, used in component safety checks, support calculation books, etc. When the three-dimensional design of the template support system is completed, the usage of each material in the entire model can be counted, such as the area of the template, the length of the steel tube, the volume of the wood, the number of fasteners, etc. The statistics of material usage are beneficial for the site to manage materials and resources uniformly, achieving refined management of base design for engineering. During the construction of base design for engineering, carpenters need to count the component sizes on the two-dimensional drawings on site, and cut the templates centrally, which not only reduces the construction efficiency but also causes waste of templates. Input the template size used on site to the established BIM model, automate the template matching can be achieved, which is beneficial to improve construction efficiency and avoid waste of templates, see Fig. 47.

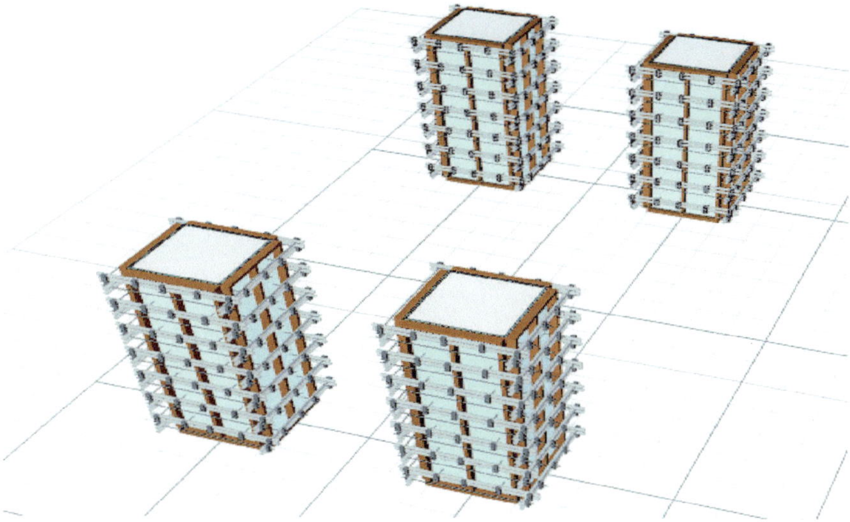

Fig. 47 The schematic diagram of BIM in base design for engineering

II. Application of BIM Technology in Foundation Pit Construction

By creating a BIM model of the foundation pit, the barriers between foundation pit design, construction, and monitoring are broken down, providing an intuitive representation of the entire project, enabling unobstructed information sharing among multiple parties, and allowing different teams to work in the same environment. Through three-dimensional visualization communication, a comprehensive assessment of the foundation pit project is made, leading to more scientific management, improving work efficiency, and reducing investment, see Fig. 48.

III. Application of BIM Technology in Scaffolding Engineering

BIM technology can be used in the creation of a building ground double-row scaffolding model and providing the construction plan. Detailed statistical management of scaffolding materials is carried out. Using simulation in the scaffolding safety system, construction risks can be minimized, also achieving to reduce construction costs and improve construction quality. In addition, BIM technology is used for three-dimensional safety technical disclosure of scaffolding.

IV. Application of BIM Technology in the Selection of Interior Decoration and Renovation Schemes

During the scheme selection, BIM is used to assess the layout, viewing, lighting, coloring, etc. of the decoration and renovation schemes of the main control building, administration building, auxiliary control building, etc. BIM can scrutinize the details of a part of the building and quickly analyze the problems that may need to be addressed in design and construction. The scheme selection uses BIM to provide

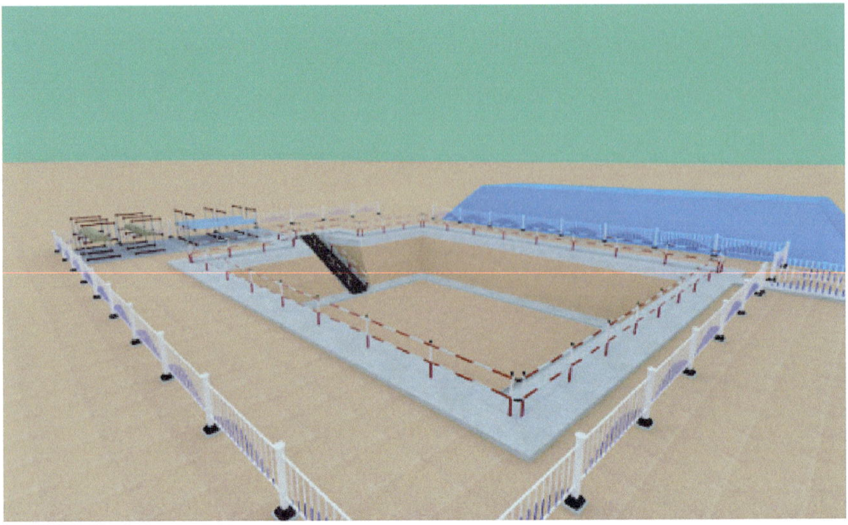

Fig. 48 The schematic diagram of BIM in foundation pit construction

convenient, low-cost different solutions for the construction unit selection, find the pros and cons of different solutions through data comparison and simulation analysis, and help the project construction unit quickly evaluate the cost and time of the building investment scheme. On the BIM platform, the focus issues of all parties in the project are easier to be intuitively discussed and quickly reach a consensus, which helps achieve much quicker decision-making than before, see Fig. 49.

V. Application of BIM Technology in Steel Structure Engineering

Applying BIM technology to the steel structures installation such as indoor DC yards and valve halls, establishing a three-dimensional model to deepen the analysis of each node, problems that may be encountered during the installation process can be discovered and solved in advance. By simulating the installation and factory pre-assembly stages to set the installation order and formulate the installation process, each component will be set in a fixed installation order, ensuring that each component is accurately installed in place, see Fig. 50.

VI. Application of BIM Technology in the Equipment Installation of Indoor DC Yard

The building BIM model is a complete information model that can integrate engineering information, processes, and resources at different construction stages of the project into one model. Through three-dimensional digital technology, it simulates the real information possessed by the building, providing a coordinated and internally consistent information model for engineering design and construction, which makes the model achieve the integration of design and construction, thereby reducing the production cost of the project and ensuring the project is completed on time and with

Fig. 49 The schematic diagram of BIM in interior decoration and renovation

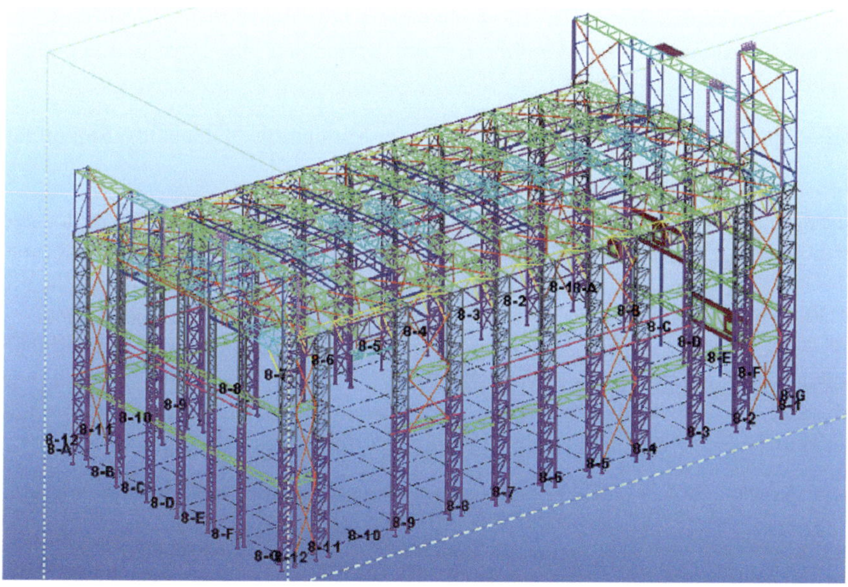

Fig. 50 The schematic diagram of BIM in steel structure engineering

Fig. 51 The schematic diagram of BIM in the equipment installation of indoor DC yard

quality. There are many equipment installed in indoor DC yard. The hoisting environment is complex. Using the BIM technology to simulate hoisting, the installation error of the equipment can be within ± 1 mm, which meets the construction process standards, see Fig. 51.

VII. Application of BIM Technology in the Simulation of Hoisting Smoothing Reactor in Pole Line

The smoothing reactors in pole line in the ± 1100 kV converter station are located in the indoor DC yard. A 500 t crane is used for hoisting indoors for the first time. The hoisting weight of the smoothing reactor is 116 t with an installation height of 20.45 m. Factors of heavy equipment, high requirement for installation accuracy and hoisting anti-collision and positioning, limited hoisting site result in difficult hoisting. It is necessary to carry out three-dimensional BIM simulation hoisting in advance. Through simulation, key hoisting data are obtained, and the final crane position, lifting height, and boom angle are determined, as shown in Fig. 52.

VIII. Application of BIM Technology in DC Wall Bushing Hoisting Simulation

In Guquan converter station, the DC wall bushing is for the first time installed from the valve hall to the indoor DC yard side, as well as the "anti-center of gravity" side hoisting. Due to the large number of equipment in the valve hall, it is necessary to prevent collision with the converter valve and other equipment during hoisting, and ensure that the bushing can be successfully hoisted.

Fig. 52 The schematic diagram of BIM in the simulation of hoisting smoothing reactor

Through three-dimensional BIM simulation, parameters such as crane position and radius of boom swing direction are obtained. The bushing removal process is also simulated, providing important information for emergency rescue and equipment maintenance.

IX. Using BIM Technology to Complete the 3D and VR Video Display of the Valve Hall and Indoor DC Yard

Through 3D technology, spatial collisions during equipment installation can be effectively checked. Through 3D interactive interface, the construction front line is better briefed. By using animations and graphics to clarify the issues that need to be paid attention to during construction, the construction plans and techniques are more effectively to be implemented.

X. Application of BIM Technology in Converter Transformer Installation Engineering

Using 3D modeling, the installation process of the converter transformer is displayed in 3D video, which can help to arrange the crane position and on-site equipment placement position in advance, clear show the process and technical difficulties in the converter transformer installation, save installation time and provide technical support for construction personnel, as shown in Figs. 53 and 54.

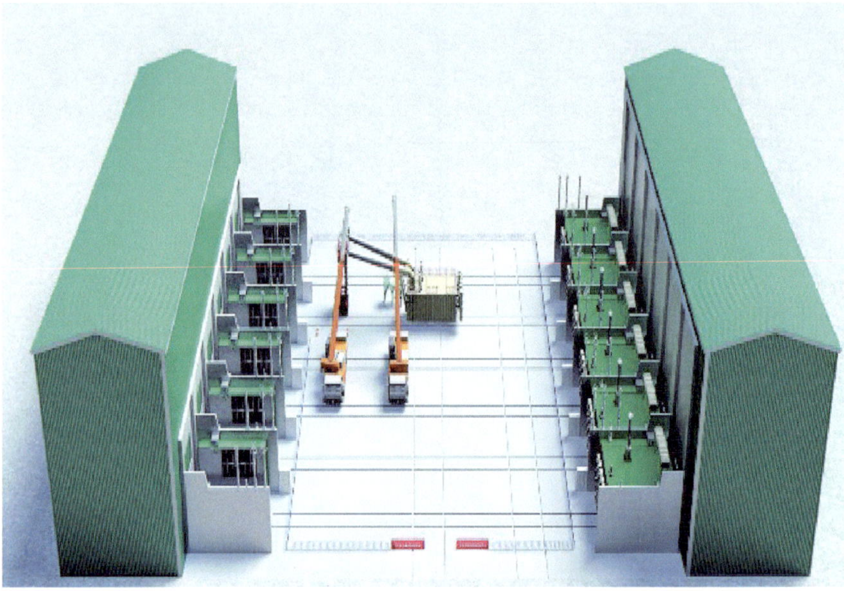

Fig. 53 The 3D display of using BIM to simulate transformer installation (I)

Fig. 54 The 3D display of using BIM to simulate transformer installation (II)

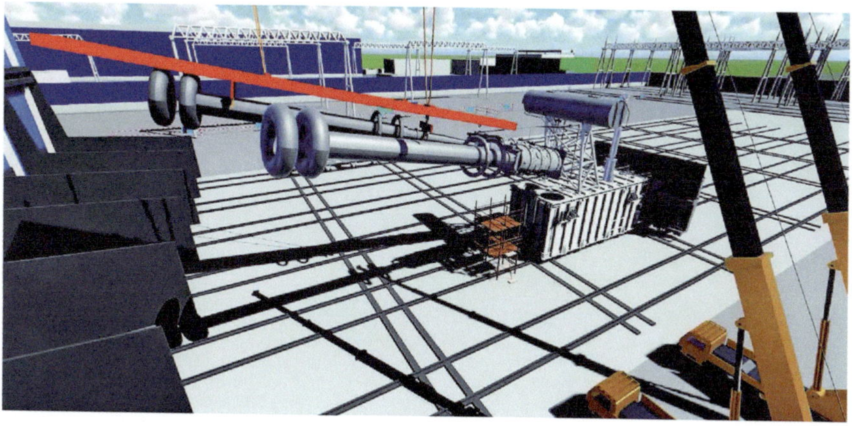

Fig. 55 The display of BIM simulated converter transformer installation

Before the installation of HV converter transformers, BIM simulation technology is used to demonstrate the overall installation of the converter transformer in animation. The hoisting position and hoisting method of the crane are simulated in advance, which is applied to guide the installation of the converter transformer on site, ensuring the successful implementation of the plan, as shown in Fig. 55.

XI. Application of BIM Technology in the Erection of 1000 kV Framework

Before the construction of the 1000 kV framework of the Guquan converter station, BIM 3D video technology is used to demonstrate the animation, decomposing the 1:1 nodes according to the process, showing the dynamic construction process, simulating the pre-arch of the framework beam and the selection of the hoisting point position in advance instead of the traditional text and tables. BIM is applied to guide the on-site assembly of the framework, ensuring the successful implementation of the plan, as shown in Fig. 56.

6 Converter Station Central Control Platform

The Smart Construction Command and Service (SCCS) platform utilizes internet and big data technologies to analyze the progress, safety, quality, and risk characteristics and patterns of the construction site. It deeply integrates the virtual business system with the real-world scenarios, time, personnel, and material information, achieving coordinated construction management and control services.

The SCCS platform has implemented real-name registration for all construction site personnel and precise control of entry and exit, full-element data monitoring of machinery, equipment, materials, and environment in construction site, comprehensive integration of collected data, entry and exit data, and construction business data,

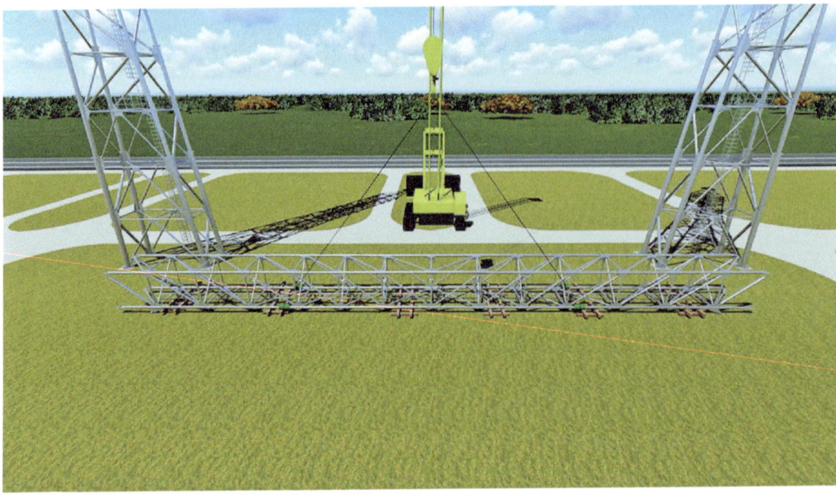

Fig. 56 BIM simulated erection of framework

real-time visualization control of the construction site. This integrated monitoring command platform can realize big data analysis display of the entire construction process, and form data assets of the smart construction site.

I. System Construction Goals

(I) Implement Real-name Registration for All Personnel in Construction Site and Precise Control of Entry and Exit

Based on mobile IT and IoT technology, the SCCS platform will collect information of on-site construction personnel's entry and exit, safety tool wearing and usage, construction work surface location and personnel change, safety inspection, violation photos, supervision inspection information, etc. By analyzing the collected data, the system dynamically monitors the trend of construction personnel input, main safety hazards, high-risk areas, problems, patterns, etc., and can perceive the inspection trajectory of project supervision personnel, the near-field perception during the inspection process, work efficiency evaluation, provide management basis for all management levels, to enhance on-site safety and comprehensive construction control level, and achieve precise control of on-site labor real-name system and entry and exit.

(II) Implement Full-element Data Monitoring of Machinery, Equipment, Materials, and Environment in Construction Site

Using technologies such as LoRa, NB – IoT, WiFi, Bluetooth, GPS, video collection, etc., the SCCS platform achieves dynamic perception of on-site construction, forming video, image, index parameter data, real-time, precise full-element control of machinery, equipment, materials, external environment, internal environment, personnel data during the construction process, improving the management level of construction projects.

(III) Implement Comprehensive Integration of Collected Data, Entry and Exit Data, and Construction Business Data

Using the dynamic perception data collected on-site such as video, environmental, personnel, foundation pit, material equipment, etc. and traditional data of construction engineering, personnel, equipment, material, environmental and other business data for integration, according to the various aspects of construction to achieve closed-loop management, the SCCS platform achieves dynamic monitoring and visualization control of all elements of construction.

(IV) Implement Real-Time Visualization Control of the Construction Site in all Aspects, and Build an Integrated Monitoring Command Platform

By integrating data such as personnel status, environment and micrometeorology, real-time dynamic impact, behavior of large machinery and equipment, risk dynamic division, on-site inspection and construction progress, an integrated command platform will be built to achieve dynamic monitoring and visualization control of all elements of construction.

(V) Implement Big Data Analysis Display of the Entire Construction Process, Forming Data Assets of the Smart Construction Site

By integrating collected data from various equipment and construction business, using big data technology to build big data analysis models, big data analysis of construction site quality, safety, risk, progress, etc. will be conducted. Therefore, data assets of the smart construction site can be formed.

II. System Architecture

The system architecture of the SCCS platform is divided into four layers: the perception layer, the data layer, the application layer, and the access layer. The construction of the system is based on industry standards and the company's system to ensure the quality and safety of the system, as shown in Fig. 57.

Command and Service Platform.

(1) Perception layer: Electronic devices such as cameras, turnstiles, environmental monitors, and GPS locators are used to monitor and collect data, and the monitoring results are transmitted to the database via internet.
(2) Data layer: By interfacing with existing video surveillance systems, meteorological monitoring systems, access control systems, outdoor electronic screen display systems, converter station entry and exit control platforms, tower crane safety monitoring systems, and government press release platforms, integrated monitoring data is obtained. This data, along with drone panoramic aerial photography data, three-dimensional model data, and construction progress plan data, is integrated either automatically or manually and stored uniformly in the database. The database is divided into three major categories, namely, device collected data, core database, and core business data.

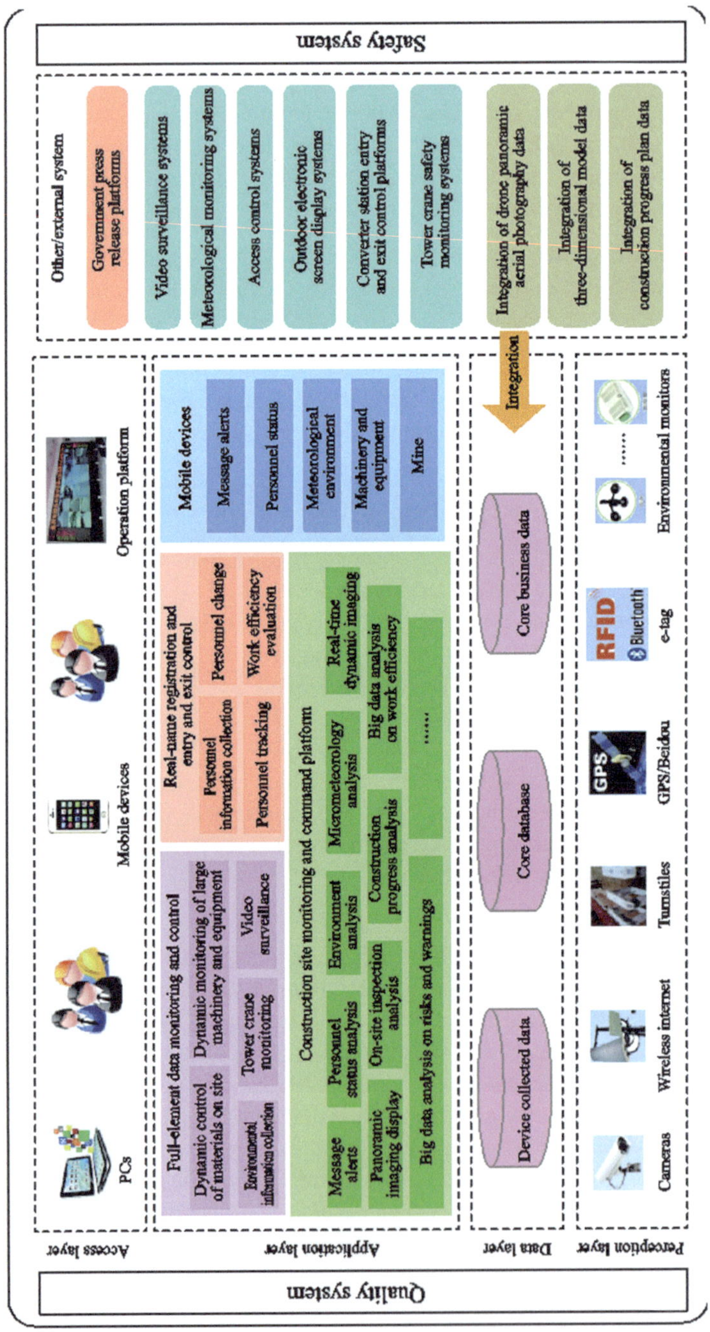

Fig. 57 The system architecture of the Smart Construction

(3) Application layer: The Platform includes both a Web and mobile version. The Web application includes three parts: data monitoring and control, labor real-name system and entry and exit control, and construction site monitoring and command service platform. The mobile version includes five parts: homepage, personnel status, meteorological environment, machinery and equipment, and personal center.
(4) Access layer: The access layer is the interface where users directly interact. Users can operate the Platform directly through PCs, mobile devices, and large display screens.

III. Comprehensive Data Monitoring and Control

(I) Dynamic Control of Materials on Site

Once the materials have arrived on site, the list of materials and physical photos are collected through the Smart Construction Site (SCS) App. The arrival time, acceptance person, and location information are automatically obtained through the mobile App.

(II) Dynamic Monitoring of Large Machinery and Equipment

There are many large machinery and equipment on the construction site, which are managed in the following three ways.

1. On-site lifting machinery and construction machinery

The display function of the layout diagram in SCS App supports querying according to machinery classification, showing the distribution of existing tools and equipment on the site and the dynamic changes in historical distribution. Clicking on a device, the device's entry time, inspection status, responsible person, etc. can be easily viewed.

2. Positioning management of owner-supplied equipment

For owner-supplied equipment, the storage location information of the owner-supplied equipment materials is recorded by smart beacons or QR codes according to the site conditions.

3. Equipment registration management, installation of GPS smart beacons

Install GPS smart beacons when very important equipment enters the site, to record real-time location information and trajectory information.

4. QR code scanning

On-site inspectors can use the scanning method to view the basic information of existing tools, such as the person in charge, the responsible unit, the entry time, the inspection time, etc.

(III) Environmental Information Collection

(1) Social environment (public opinion, major events, public events). Recent social issues, major events, public events, and public opinion information are collected through local government websites.

(2) Regional environment (within the work area). Using embedded technology of micro weather stations, various meteorological data such as wind speed, wind direction, temperature, humidity, atmospheric pressure, total radiation, rainfall, evaporation, soil temperature, and soil moisture are measured. The system adopts a modular design and can flexibly add or subtract corresponding modules and sensors according to user needs (measured meteorological elements), conveniently and quickly meeting the needs of various users. The system comes with display, automatic saving, real-time clock, data communication and other functions. This automatic weather station has the advantages of advanced technology, high measurement accuracy, large data capacity, long telemetry distance, friendly human–machine interface, and high reliability.

(IV) Tower Crane Monitoring

Tower crane monitoring is provided for construction safety supervision and management departments, construction enterprises, project departments, project supervision units, and tower crane property owners, etc., to achieve remote supervision and management of tower cranes by multiple parties. The tower crane installation equipment diagram is shown in Fig. 58.

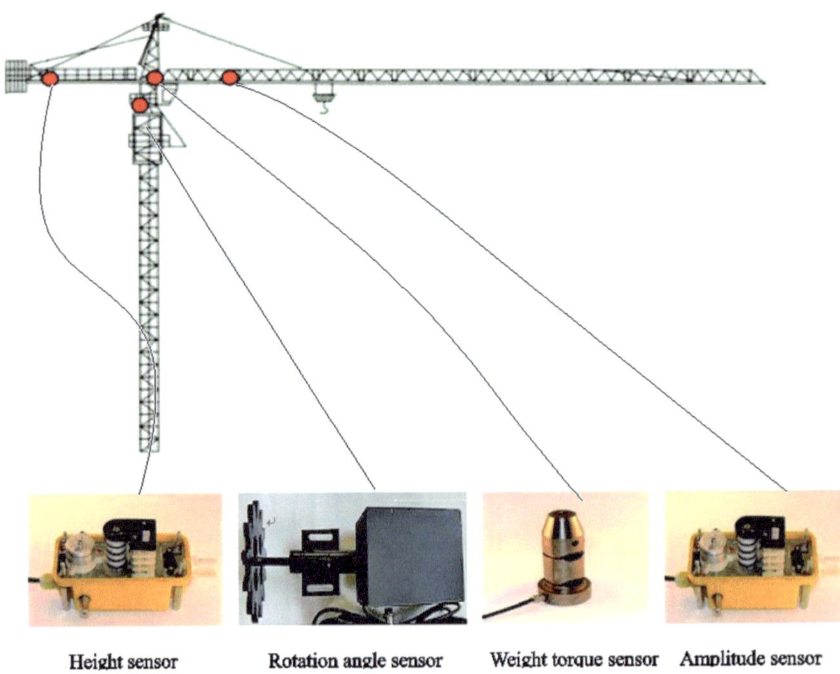

Fig. 58 The tower crane installation equipment diagram

(1) Intelligent display of tower crane location map. One-click access to the actual installation geographical location of the tower crane, project information, working status, and other related information, intuitive and clear.
(2) Real-time display of tower crane operation parameters. Obtain various performance parameters of tower crane operation directly through network transmission, as if the management personnel were on the scene, the system integrates a mobile phone SMS alarm module to promptly inform the on-site alarm information.

(V) Video Surveillance

Video surveillance information is shown in Fig. 59, which can track the on-site operation dynamics of the construction unit in real time, focusing on monitoring major safety risks such as deep foundation pit excavation, scaffolding erection, large template support and electrical equipment hoisting, installation; strictly control the construction quality of key nodes, retain concealed project construction video information, can query historical records, assist in tracing the cause of the accident. Its features are as follows:

(1) 24 h round-the-clock real-time monitoring.
(2) Specifically increase attention to key protection areas.
(3) Management department staff can monitor centrally in the control room, if problems are found, they can communicate directly with on-site personnel, on-site safety personnel record evidence in real time and issue rectification instructions. Remote monitoring and on-site inspections complement each other, making safety management more rigorous and efficient.

(VI) Homepage

The elements displayed on the homepage interface of the SCCS platform are as follows:

(1) User login information: The system displays the logged-in user's avatar in the upper right corner. The user's detailed information can be viewed by clicking.
(2) Message alerts, warning notifications: The latest messages or warning information are displayed in the upper right corner of the page. Different statuses are used to alert when there are messages or warnings. The details and access more information can be viewed by clicking.
(3) Project overview: The homepage displays the project name, project manager, proportion of area, construction period, current stage, and total project investment information.
(4) Total number of project personnel and active personnel (total number of people entering and leaving the site): The homepage displays the current number of people on site, the number of large machinery, the number of risk points, and the number of safe production days in large font.
(5) Micrometeorology: The homepage displays the micrometeorology information collected, including the day's weather information at the construction

Fig. 59 Video surveillance information

site, indoor temperature, humidity, visibility, PM2.5 index information, and tomorrow's weather forecast. The [Micrometeorology] tab can be accessed by clicking on the Details.

(6) Social environment: The social environment information of the construction project collected from the government public information collection platform is dynamically displayed in the upper right corner of the homepage. The [Off-site Environment] display tab can be accessed by clicking on 'view more'.

(7) Video surveillance: The construction site is displayed in multiple windows showing video information of several key surveillance locations. This video is regularly updated dynamically.

(8) Chart display area: The number of people entering and leaving the site, the number of active large machinery, dynamic risk point analysis, and weather forecast information are displayed in chart form.

    1) Outbound personnel information: The number of personnel entering and leaving the site each month during the construction period is displayed, shown in a line graph.
    2) Active large machinery quantity information: The number of active large machinery each month during the construction period is displayed, shown in a bar graph.
    3) Risk point dynamic analysis: The number of risk points in each area within the construction zone is displayed, represented in a pie chart.
    4) Weather forecast information: The meteorological forecast information for the next 15 days in the construction area is displayed, including weather, temperature, and wind direction information.

(9) Safety quality information: The homepage displays the number of safety inspections this month, the number of quality inspections this month, the number of issues, and the number of issues awaiting rectification. The [Patrol Management] tab can be accessed by clicking for more details.

(VII) Project Overview

Access the project overview page through the project introduction link on the homepage, the project overview, aerial video information of the project, picture information, etc. will be shown, serving as a snapshot view of the project.

(VIII) Environmental Monitoring and Micrometeorology

1. Social environment

The construction site monitoring platform will display recent related news collected, categorized into social issues, major events, public events, and public sentiment information.

It supports searching by time and title keywords, and viewing historical social environment information.

The homepage's banner is also one of the entrances to the [Social Environment] information display.

2. Micrometeorology

Real-time environmental information collected by environmental monitoring equipment, such as temperature, humidity, wind direction, precipitation, noise, dust, etc., in the meteorological environment is displayed in a list format. Data collected from different collection points can be viewed separately.

(IX) Real-time Dynamic Imaging

Dynamic imaging display includes four parts, namely panoramic imaging, multi-dimensional electronic photo wall, video surveillance, and three-dimensional model display.

1. Panoramic imaging

Panoramic maps, also known as 360° panoramic maps or panoramic surround view maps, refer to maps that simulate three-dimensional images into the three-dimensional effects of real objects. Viewers can drag the map to view the effects of real objects from different angles, also view historical panoramic image information by week through the timeline, assisting in project management.

2. Multi-dimensional electronic photo wall

The multi-dimensional electronic photo wall displays photo information according to units, specialties/trades, and timelines. It integrates the inspection results of various units such as owners, supervisors, designers, constructors, manufacturers, and others during construction, displaying problems and highlights in the construction inspection results. Use diversified methods to praise excellence and urge problem improvement.

3. Video surveillance

The video surveillance function displays videos of major safety risk monitoring sites such as deep foundation pit excavation, scaffold erection, large template support, and electrical equipment hoisting and installation in a multi-window format. The video is dynamic and can be changed in real time, strictly controlling the construction quality of key nodes and retaining concealed project construction video information.

The video surveillance function can provide the historical records of monitoring points to assist in tracing the cause of accidents.

Video surveillance can remotely monitor equipment.

4. Three-dimensional model display

The three-dimensional model display shows the digital three-dimensional design and digital construction of the substation, realizing digital and visual management of the design process. The display content includes four types of digital models: large-scale transportation simulation, large- span steel structure hoisting simulation, large-scale equipment installation, and maintenance simulation.

The three-dimensional model display can be viewed and managed separately according to the design unit. Figures 60 and 61 show the three-dimensional model display effect diagrams of two design institutes respectively.

(X) Personnel Status

1. Personnel status analysis

Fig. 60 3D model display (I)

Fig. 61 3D model display (II)

Personnel status analysis manages the number of personnel in progress, including the total number of project personnel, the number of people entering the station on the date, and displays the number of people entering the construction site every day, displayed in a line char.

2. Personnel distribution

Personnel distribution management is achieved through intelligent beacons or GPS locators, displaying the distribution information of personnel on the construction site in real time on the main interface map of the construction. Personnel from different units are displayed in different colors, making it easy for managers to view the status of construction personnel.

(XI) Analysis of Large Machinery Equipment Behavior

1. Positioning management analysis of owner-supplied materials

For owner-supplied equipment, intelligent beacons or QR codes are installed on or around the equipment to record the storage location of the owner-supplied materials. The storage location of the owner-supplied materials is displayed on the site layout map with bright spots, and different materials are displayed in different colors or shapes.

If a specific material is selected, it is easy to view the photo information of its location.

2. Distribution of large equipment

Through the locator installed on large machinery equipment, the location information of large equipment is recorded in real time. Based on the collected data, the distribution of large equipment is analyzed. The information of a single mechanical device (such as a tower crane) can be viewed in real time. When selecting a device, the photos taken on site of the device, as well as view the archive information/activity track, maintenance record/detection data can all be viewed.

3. Active status of large equipment

The system calculates the usage frequency of key large machinery equipment through the use data of large machinery equipment, analyzes the active status of each large equipment and displays it in a bar chart. The historical active status of large equipment can also be checked, see Fig. 62.

(XII) On-Site Inspection Display Analysis

On-site inspection display analysis includes problem distribution and highlight distribution.

1. Problem distribution

Problems are divided into inspection problems, three-level risk problems, and tool problems.

The problem distribution function marks the problems with type at various locations on the engineering plan based on the inspection problem information. Different

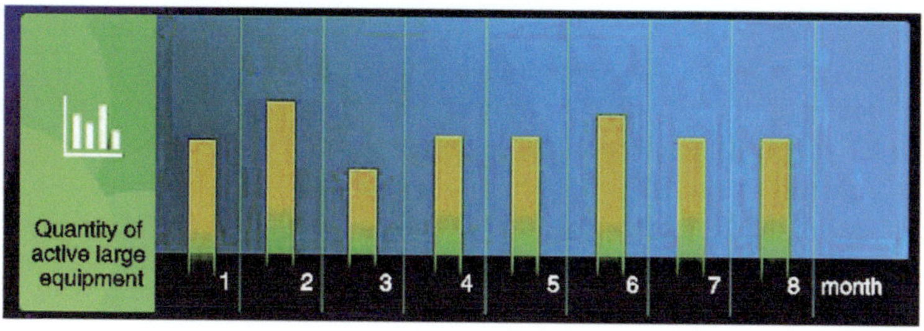

Fig. 62 The diagram of active status analysis of large equipment

categories of problems are marked with different icons. The detailed information of any problem is shown when clicking, including the responsible unit, inspection unit, problem title, rectification requirements, on-site photo information, and complete the rectification content and post-rectification photo information after the rectification is completed.

2. Highlight distribution

The highlights of the on-site inspection results are displayed in original shapes, which are divided into three levels of highlights. The higher the highlight level is, the larger the pattern is. The detailed information of each highlights can be viewed, including the highlight title, the progress of the highlight, the responsible unit, the excellent level, and the highlight photos.

(XIII) Construction Progress Analysis

The construction progress analysis shows the construction information and analyzes the factors that may affect the construction, predicting whether the construction will be ahead of schedule or delayed.

If the project is delayed, it will analyze the reasons that affect the progress and provide feasible suggestions for catching up with the schedule.

(XIV) Dynamic Analysis of Risks

According to the risk list, the specific locations where risks occur are marked on the construction distribution map, and the distribution of risks is automatically marked on the map. Different risk levels are identified by different colors and icons, and the number of risk points checked and missed are counted.

Any risk point can be clicked to view details, including the time of occurrence of the risk, the area, detailed coordinate information, the progress of the construction, and the risk unit.

(XV) User Information

User information has user avatar, contact information, affiliated unit, personal tasks, bonus points, and other functions. Among them, the red packet points function is a way to encourage users to actively use the system. Every time a user scans a QR code, completes a patrol, uploads equipment and material information, construction images, models, and other data, the system will automatically reward the user with points. Every week, a points ranking is conducted, and the top 10 workers will receive bonus of different amounts.

The points data is reset every week, cleared on the weekend, and scoring begins anew on Monday.

Chapter 6
Commissioning

The commissioning of DC power transmission projects is divided into FAT test, equipment commissioning, subsystem commissioning, station auxiliary system commissioning, and system commissioning. The purpose of combination test is to comprehensively inspect the functionality, performance, and interface functions of equipment in the secondary system from the point of whole power system. The purpose of equipment commissioning is to confirm the equipment integrity during transportation, to check the status and installation quality of the equipment, to test whether the equipment can safely be charged, loaded and start, and whether the performance of the equipment can meet the requirements of the contract and technical specifications. Subsystem commissioning is the energizing or startup test of all independent subsystems in the converter station, the purpose of which is to prove that several components or devices can be combined into a subsystem to operate safely, and to check whether their functions and performance can meet the requirements of the contract and technical specifications. The purpose of station auxiliary system commissioning is to check the function of a single converter station according to the requirements of the contract and technical specifications, and also to prepare for end-to-end system commissioning. The purpose of system commissioning is to comprehensively assess the performance of all equipments and the secondary control and protection system of the UHV DC project, to test whether the performance indicators of the DC power transmission system can meet those stipulated in the contract and technical specifications, and to ensure the security and reliability of the equipments and system after the project is put into operation. With system commissioning, the system performance of the DC power transmission project can be comprehensively and correctly understood and assessed.

Among the five stages, FAT and equipment commissioning are the basis for subsystem commissioning, subsystem commissioning is the basis for station auxiliary system commissioning, station auxiliary system commissioning is the basis for system commissioning, and system commissioning is the overall inspection on the system and equipment performance of the DC power transmission project, and is the

Z. Liu, ±*1100kV UHV DC Power Transmission Technology*,
https://doi.org/10.1007/978-981-96-3712-6_6

final check before the project is put into operation. Therefore, all aspects of engineering commissioning are interlinked and verificated from each other. The purpose and role of engineering commissioning is to carry out a comprehensive inspection on the equipment and the entire project before the system is put into operation, through that, all adverse factors should be eliminated to ensure the safe and reliable operation of the project. This chapter explains the engineering commissioning of CG project as an example.

1 Factory Acceptance Test

Due to the increase in the DC voltage and transmission distance, the operation characteristics of the ± 1100 kV UHV DC transmission system have significantly changed, compared to the ± 800 kV UHV DC project. For example, the fault current rises faster. The coupling influence between the high-voltage and low-voltage valve groups, and between the faulty and healthy pole is stronger. The time delay of communication between the sending and receiving end becomes large, which make the effectiveness of coordinated control weak and DC fault recovery slow. FAT focuses on the characteristics and difficulties of ± 1100 kV UHV DC transmission, reveals the operation and fault rules of ± 1100 kV UHV DC, improves the operation performance of DC system, and lays a technical foundation for the construction of ± 1100 kV UHV DC transmission project.

I. Purpose and Significance of FAT

The FAT of DC control and protection system refers to the power system simulation in a digital or physical way to form a closed-loop test system, with connecting the main equipments of the real DC control and protection system to digital model through interface equipment, such as power amplifier. Since the combination test comprehensively and realistically reflects the actual operation conditions of DC control and protection system, it can check the interface characteristics of all components and overall function and performance of DC control and protection system. Before the DC control and protection system arrives at the site, the internal function modules and interfaces can be tested. The advantages, such as fault reappearance, easy-to-meet test conditions and friendly working environment, can be well to quickly solve the problems in the combination test, which can greatly reduce the on-site commissioning time and shorten the project construction cycle.

On one hand, the characteristics of CG project, such as ultra-high voltage, ultra-long line, ultra-large capacity, layered connection to different voltage AC power grid, and coupling effects on DC side, bring new challenges and high requirements to the DC control and protection system design. Before the project is put into operation, it is necessary to fully verify the functions of the control and protection system. On the other hand, there are many suppliers of secondary equipment for CG project, and the interface design is complex. Detailed interface tests need to be carried out to ensure the correct interface functions between the control and protection system and other secondary equipment. The FAT aims to comprehensively improve the reliability and further optimize the function and performance of the control and protection system of CG project, which strives to prevent problems caused by the control and protection devices during the system commissioning.

II. FAT Items and Scheme

1. General Introduction

The equipments for the FAT include the DC control and protection system, valve base electronics (VBE), midpoint DC voltage divider (Guquan converter station), valve cooling control and protection system, station master clock, emergency and stability control, inter-station communication switching device, and optical current transformer (TA) interface equipment, etc.

The FAT is divided into 24 major items, involving system self-monitoring and switching, sequential control and interlocking, control system dynamic response, reactive power control, fault and protection (bipolar/pole/converter protection), VBE equipment interface, dual control host crashing measures, etc. The CG project is an UHV DC transmission project that is layered connected to 500 kV/1000 kV AC power grids. Considering the characteristics, the test items about reverse operation and layered connection are conducted.

The control and protection equipment for this FAT totals 232 cabinets, including AC station control, converter control, converter protection, pole control, pole protection, VBE, SCADA and inter-station communication, etc. The equipment connection diagram is shown in Fig. 1.

2. FAT Items

1. DC control test

The DC control test includes sequential control and interlocking test, initial state check test, open line test, tap control test, active power control test, reactive power control test, self-monitoring and switching test, external characteristic test, voltage

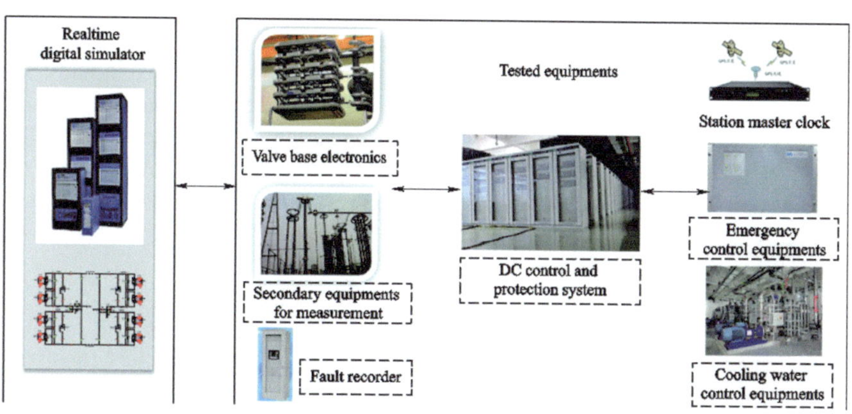

Fig. 1 The connection diagram of equipments in the FAT of CG project

balance control test, converter entry/exit test, additional control test, midpoint voltage divider fault test, etc.

2. DC protection test

The DC protection test items includes: ① Protection blocking sequential control; ② Converter fault test; ③ Pole fault test; ④ Bipolar fault test; ⑤ AC system fault test; ⑥ DC filter fault test; ⑦ DC line protection test.

The fault location diagram of DC protection test is shown in Fig. 2.

(1) The converter fault types are shown in Table 1.
(2) The pole fault types are shown in Table 2.
(3) The bipolar fault types are shown in Table 3.
(4) The AC system fault types are shown in Table 4.
(5) The DC filter fault types are shown in Table 5.

3. Interface Test

   (1) VBE interface test.
   (2) Stability control interface test.
   (3) Valve cooling control and protection interface test.
   (4) Optical TA interface test.

4. Special Test

   (1) Specific test.
   (2) Supplementary test.
   (3) Operation test.
   (4) System commissioning pre-verification test.

III. FAT Results

(I) General Introduction

To achieve the goal of "fully verifying the control protection performance and the on-site system commissioning items, ensuring that no on-site system commissioning problems and long-term security and stability problems are caused by insufficient control and protection system test", 2368 test items were completed, including 1690 function, performance and interface tests, 58 research tests, 523 system commissioning pre-verification tests, and 155 operation and maintenance function tests. The FAT covered all 45 operating modes of the DC system, and key strategy changes were thoroughly and repeatedly verified under various operating modes. The test fully verified the function and performance of DC system, and properly handled all kinds of defects and problems.

(II) Problems Found and Solved in the FAT

More than 30 problems were found and solved in the FAT of the CG project, involving system platform, application software, equipment hardware, and equipment interface. Some typical problems are listed as follows.

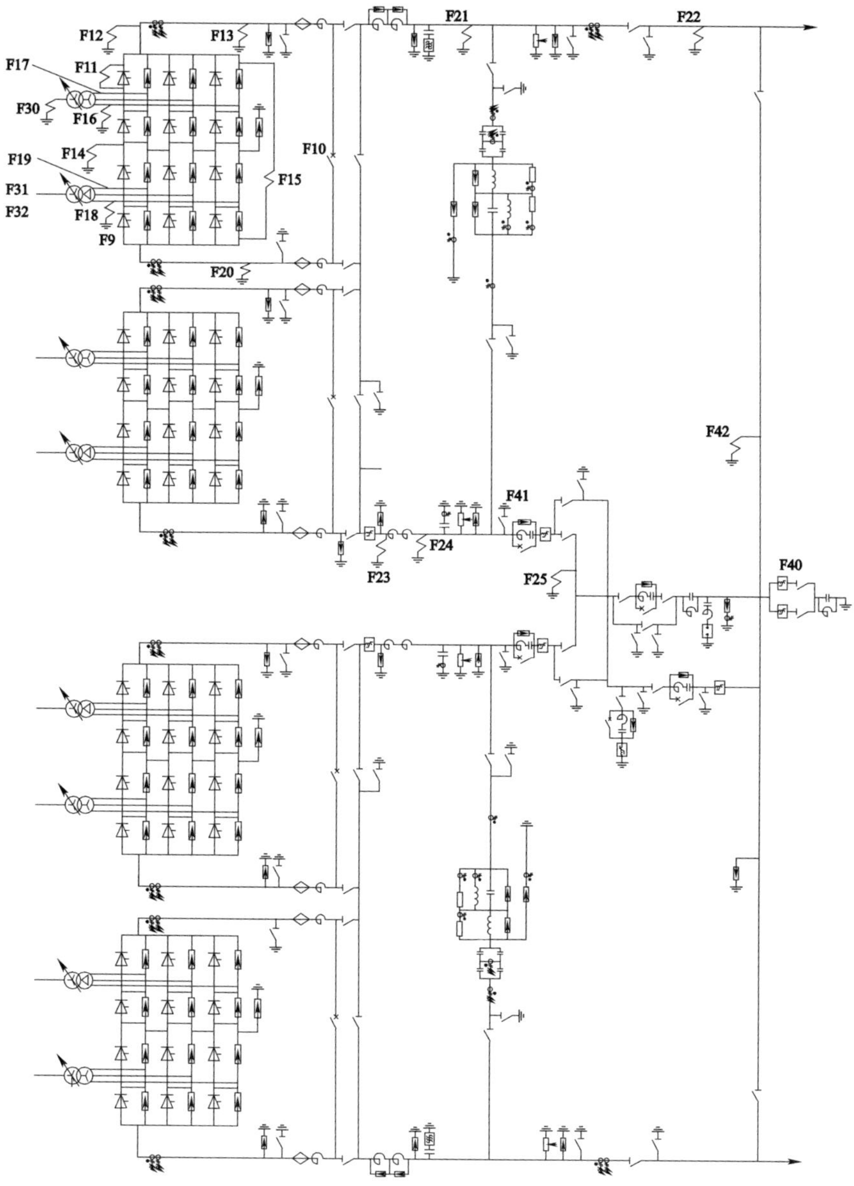

Fig. 2 Schematic diagram of fault point in DC protection test

Table 1 Converter fault types

No.	Fault types	Fault location
1	Plus loss fault	F9
2	Bypass breaker fault during the closing process	F10
3	Bypass breaker fault during the opening process	F10
4	Valve short circuit	F11
5	Vavle short circuit backup protection	F11
6	Ground fault on the high voltage side of converter	F12
7	Ground fault at the neutral-point of converter	F14
8	Converter short circuit	F15
9	YY converter transformer single-phase fault	F16
10	YY converter transformer phase-phase fault	F17
11	YD converter transformer single-phase fault	F18
12	YD converter transformer phase-phase fault	F19

Table 2 Pole fault types

No.	Fault types	Fault location	Note
1	Plus loss fault	F9	
2	Pole bus ground fault 1	F13	
3	Ground fault backup protection at the neutral-point of converter	F14	
4	Ground fault at the neutral-point of pole	F20	
5	Ground fault backup protection at the neutral-point of pole	F20	
6	Pole bus groud fault 2	F21	
7	DC line fault	F22	Faults can be categorized into short-duration, long-duration, and permanent faults
8	Neutral bus groud fault 1	F23	
9	Neutral bus groud fault 2	F24	
10	Grounding electrode line open circuit	F40	
11	NBS fault	F41	
12	Converter exiting without communication		
13	No-load energization fault		

Table 3 Bipolar fault types

No.	Fault types	Fault location	Note
1	Ground fault on the neutral bus of bipolar system	F25	Faults can be categorized into short-duration, long-duration, and permanent faults
2	Unbalanced current on the grounding electrode line	F40	
3	Station ground overcurrent protection		
4	Ground fault backup protection on the neutral bus of bipolar system	F25	
5	Metallic return line ground fault	F42	
6	Transfer switch protection	NBGS fault	
7		GRTS fault	
8		MRTB fault	

Table 4 AC system fault types

No.	Fault types	Fault location	Note
1	Phase-to-earth fault	F30	Fault duration is categorized as short-term faults and long-term faults, with communication/without communication, and reclosure successful/unsuccessful
2	Phase-phase fault	F31	Fault duration is categorized as short-term faults and long-term faults, with communication/without communication
3	Three-phase fault	F32	Fault duration is categorized as short-term faults and long-term faults, with communication/without communication

Table 5 DC filter fault types

No.	Fault types	No.	Fault types
1	Ground fault on the high voltage side	5	High voltage capacitor breakdown fault
2	Short circuit on the bridge arm of high voltage capacitor	6	DC filter resistor thermal overload fault
3	High voltage capacitor groud fault	7	DC filter reactor thermal overload fault
4	Groud fault on the low voltage side		

1. Sending-end single valve group protection block, DC line low voltage protection misoperation

(1) Fault phenomenon: The bipolar operates on full voltage, the short circuit fault occurs on the valve in high-voltage group at the rectifier station of pole 1. Then the valve short circuit protection of high-voltage group acts, valve group conducts X blocking, the DC line low voltage protection of pole 1 acts, the phase shift restarts. The waveforms of high and low voltage valve groups are shown in Figs. 3 and 4.

(2) Cause analysis. The X blocking sequence of valve group at the rectifier station is as follows: the faulty valve group immediately conducts phase shift and is blocked, the bypass switch is closed, and the AC switch is tripped. The phase shift happens on another valve group of the same pole (the firing angle rapidly increases from 15° to 164°). After the bypass switch of faulty valve group is closed, the phase shift is released, and the firing angle quickly drops from 164° to 110°, then the current controller restores the angle at a certain rate. After the faulty valve group on inverter side receives the blocking signal from the

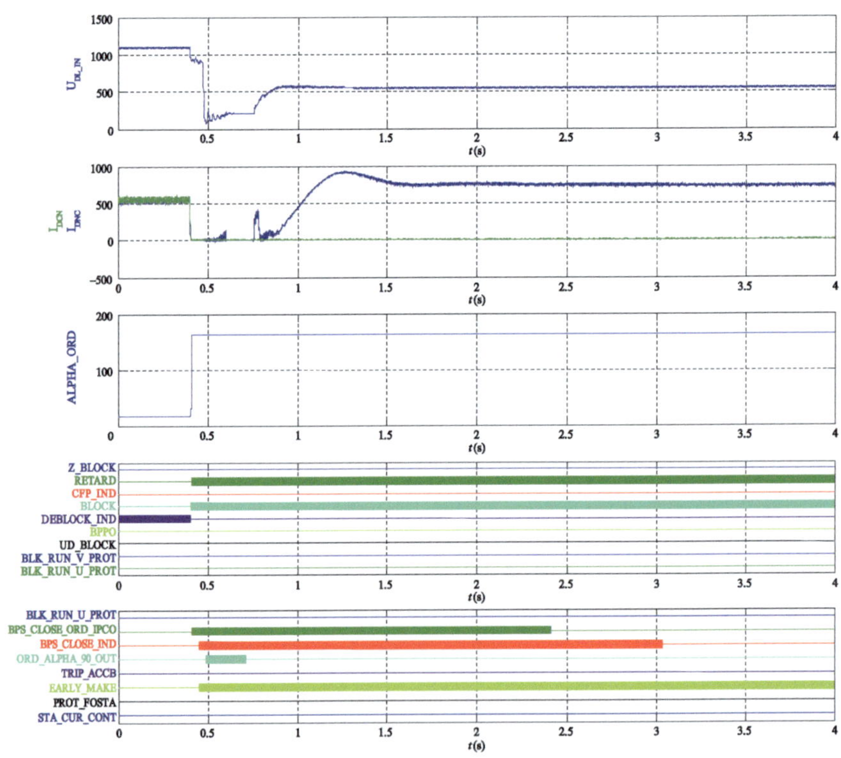

Fig. 3 High voltage valve waveforms of CG project

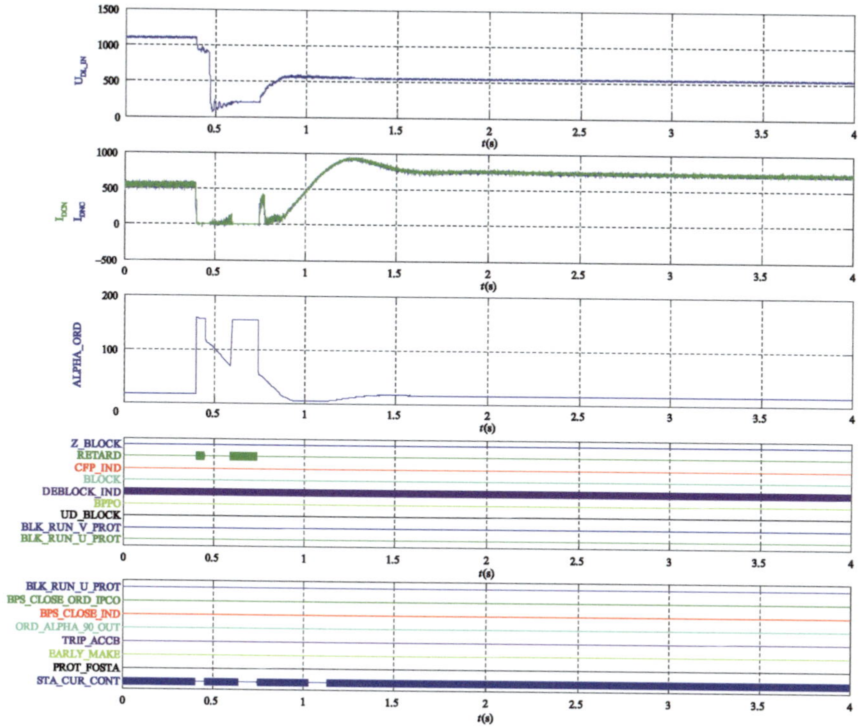

Fig. 4 Low voltage valve waveforms of CG project

rectifier station, it executes Alpha90, closes the bypass switch, then the bypass is connected and the valve group is blocked. Another valve group of the same pole automatically adjust itself.

During the test, after the high voltage valve group of the rectifier station conducts X blocking, the low voltage valve group releases the phase shift and restores the angle. During this process, the DC line low voltage protection unexpectedly acts, and the phase shift restarts. The criterion for DC line low voltage protection is that the DC voltage is lower than 0.35 (per unit value), and the phase shift restarts after a delay of 80 ms. After checking the wave recording, the criterion for DC line low voltage protection is met, and the protection action is correct.

After reviewing the waveforms of the FAT of the ± 800 kV UHV DC SL project, it was found that during the process of phase shift elimination and angle recovery of the non-faulty valve group, the voltage establishment speed of the SL project is higher than that of the CG project. Therefore, under the same test conditions, the DC voltage of the CG project rises slowly, reaching the set value of DC line low voltage protection, and the protection action is triggered. The waveform is shown in Figs. 5 and 6.

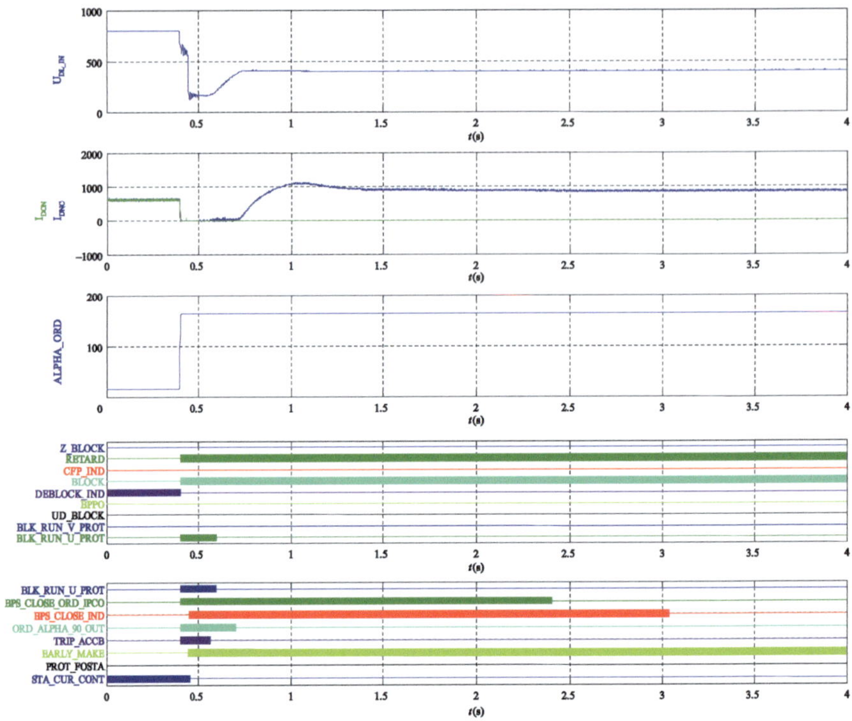

Fig. 5 High voltage valve waveforms of SL project

The line length of the SL project is 1220 km, and that of the CG project is 3324 km. After analysis, it is confirmed that the different line lengths of two projects affect the DC voltage establishment speed. Further explanation is given as follows.

Bipolar operates on full voltage, the power is 0.1 (per unit value), the DC line metallic grounding fault happens on the pole 1 for 100 ms. The DC line fault component protection and traveling wave protection act, phase shift successfully restarts, and the DC line returns to normal operation. The waveform is shown in Figs. 7 and 8.

From Figs. 7 and 8, it can be seen that the converter is still in the phase shift state, after the fault is cleared in 100 ms. At this time, the voltage on the line of the SL project oscillates around 40 kV, and that of the CG project oscillates around 200 kV. From the end of the phase shift to the time when the voltage rises to 90%, SL project takes 115 ms, and CG project takes 168 ms. It can be seen that the line of CG project is long and the line-to-ground capacitance is large, thus the voltage establishment speed is slower under the same fault conditions.

The line length of the CG project is further shortened in the RTDS model, and the same short circuit blocking test of HV valve group is simulated. After the HV valve group is blocked, the LV valve group releases the phase shift, restores the angle, and establishes voltage and current. Due to the short DC line, the voltage is

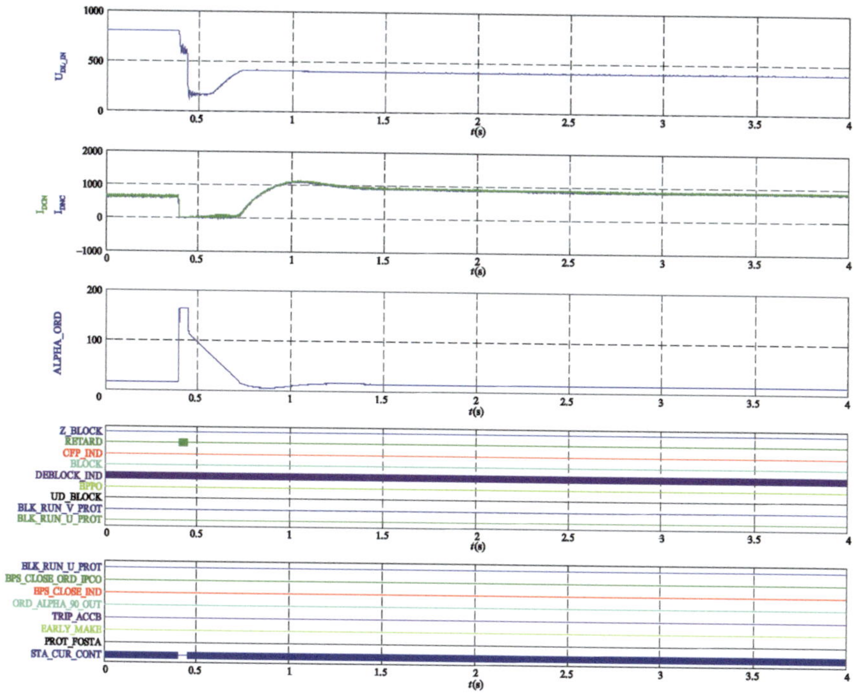

Fig. 6 Low voltage valve waveforms of SL project

established quickly, and the misoperation of DC line low voltage protection does not occur during this process. The waveforms can be seen in Figs. 9 and 10.

(3) Solution. When the single valve conducts X blocking, after the phase shift of the non-fault valve group of this pole ends, the angle recovery speed is accelerated to quickly establish the DC voltage. The measure is to accelerate the angle recovery rate from 164° to 90° after phase shift (the adjustment from 90° to 15° is controlled by the current controller and the rate remains unchanged), which can accelerate the establishment of voltage and avoid DC line low voltage protection misoperation. After modifying the strategy, the test was conducted again, and no DC line low voltage protection misoperation occurred. The waveforms can be seen in Figs. 11 and 12.

2. Inverter side single valve Y Block, healthy valve group commutation failure

(1) Fault phenomenon. Bipolar operates on full voltage, Y block of the pole 1 high voltage valve group on the inverter side is simulated through presetting technique, the low voltage valve group of this pole has commutation failure. The waveforms can be seen in Figs. 13 and 14.

(2) Cause analysis. The Y blocking action sequence of single valve on the inverter side is as follows: The Alpha-90 is immediately executed, the switch is tripped, the bypass switch is closed, and the bypass pair put into operation. After the

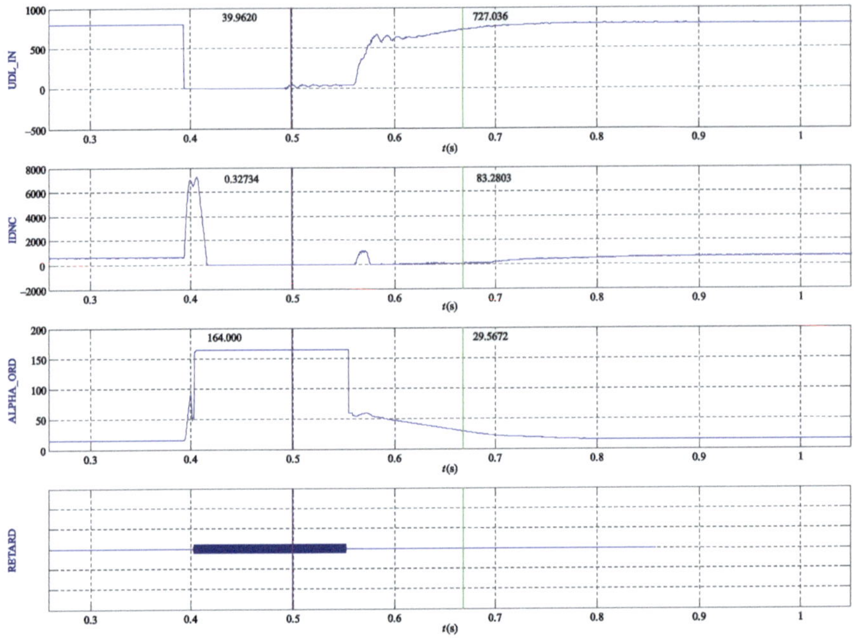

Fig. 7 DC line fault waveform of SL project

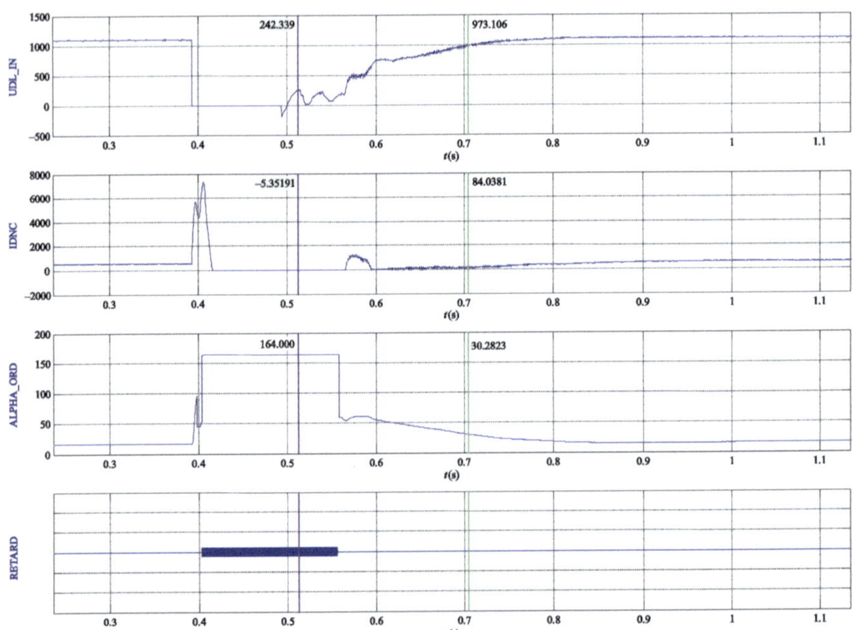

Fig. 8 DC line fault waveform of CG project

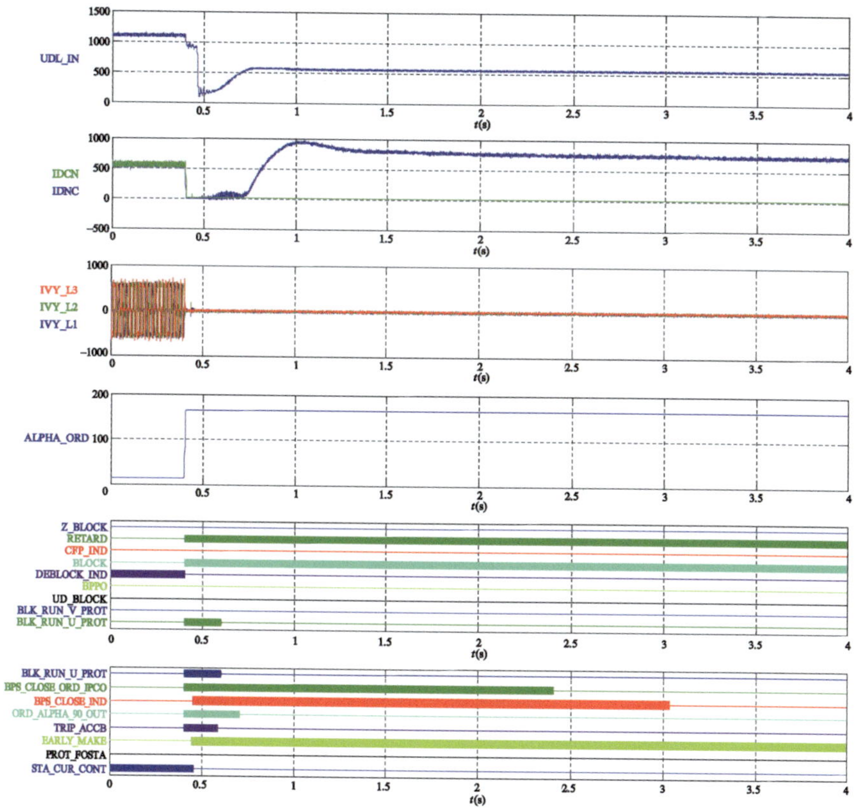

Fig. 9 Valve group fault test for CG project: HV valve group waveform (I)

bypass switch is closed, the converter valve is blocked. Another valve group of this pole automatically adjusts itself. After the rectifier side receives the action signal from the inverter side, it executes Alpha-90, puts into the bypass pair, closes the bypass switch, and blocks the converter valve after the bypass switch is closed. Another valve group of this pole automatically adjusts itself.

During the test, because the DC line is long, the communication delay between the two stations is large (about 35 ms). The high voltage valve group on the inverter side executes Y blocking, gradually puts into the bypass pair, and closes the bypass switch. During this process, the rectifier station has not yet received the blocking signal from the inverter station and cannot execute the blocking sequence, which causes the DC current to rise rapidly, leads to commutation failure of the low voltage valve group on the inverter side.

(3) Solution. When the faulty valve group on the inverter side executes Y block, it delays 10 ms to put into the bypass pair, to shorten the time difference between the actions on the rectifier and inverter sides, which is conducive to the rectifier

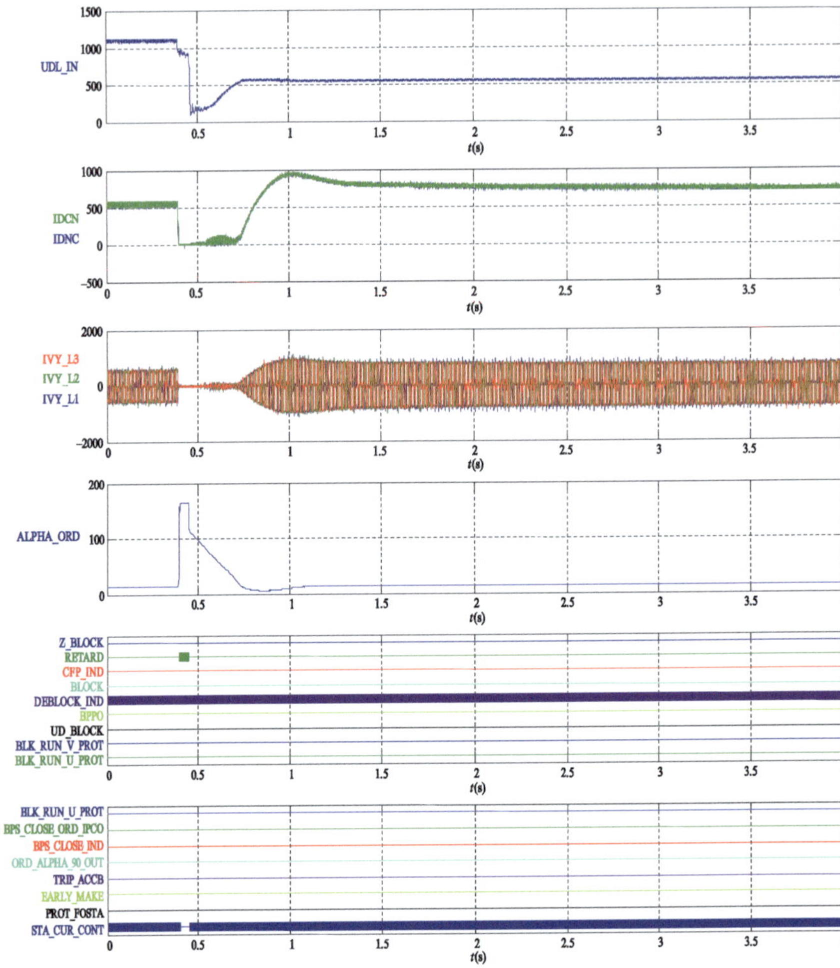

Fig. 10 Valve group fault test for CG project: LV valve group waveform (I)

side helping inverter side suppress the DC current rise and avoid commutation failure. After modifying the strategy, the test was conducted again, and the low voltage valve group did not have commutation failure. The waveforms can be seen in Figs. 15 and 16.

The Y blocking action sequence of valve groups on the inverter side with the modified strategy is as follows: The Alpha-90 is immediately executed, the switch is tripped, the bypass switch is closed. It delays 10 ms to put into the bypass pair, and blocks the converter valve after the bypass switch is closed.

The DC protection that causes the Y block of valve group on the inverter side is as follows:

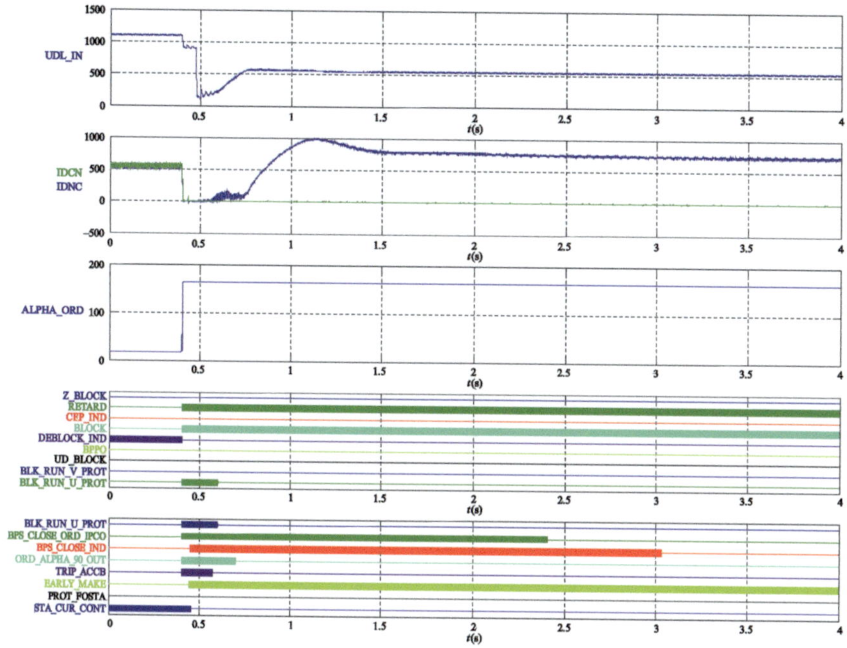

Fig. 11 Valve group fault test for CG project: HV valve group waveform (II)

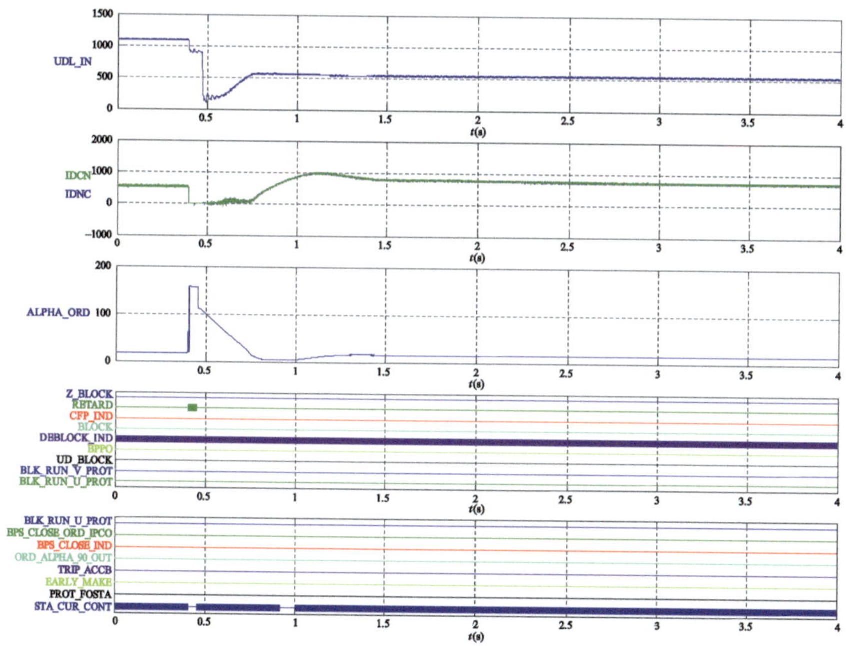

Fig. 12 Valve group fault test for CG project: LV valve group waveform (II)

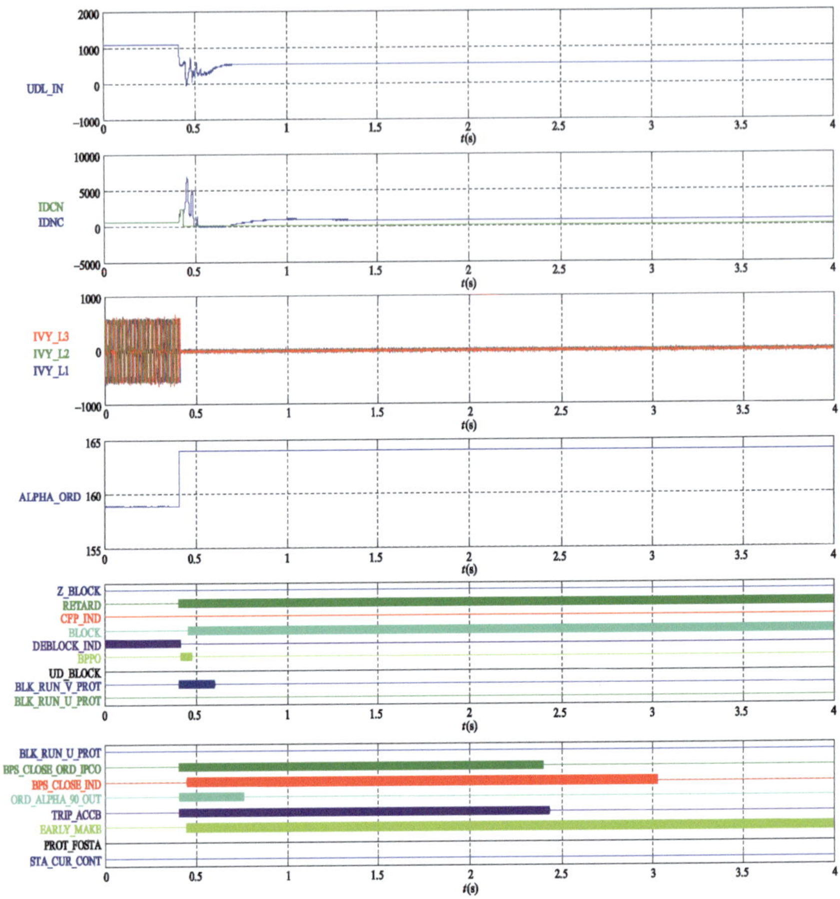

Fig. 13 Inverter station valve group fault test for CG project: high voltage valve group waveform (1)

1) Converter overcurrent protection II, III, IV;
2) Dual bridge commutation failure protection;
3) Bypass switch protection;
4) Bypass pair overload protection;
5) Converter DC overvoltage protection I;
6) Valve group harmonic protection;
7) Voltage overstress protection.

The delay of the above protection to lock the valve is more than several hundred milliseconds, and the Alpha-90 has been already executed before the bypass is put

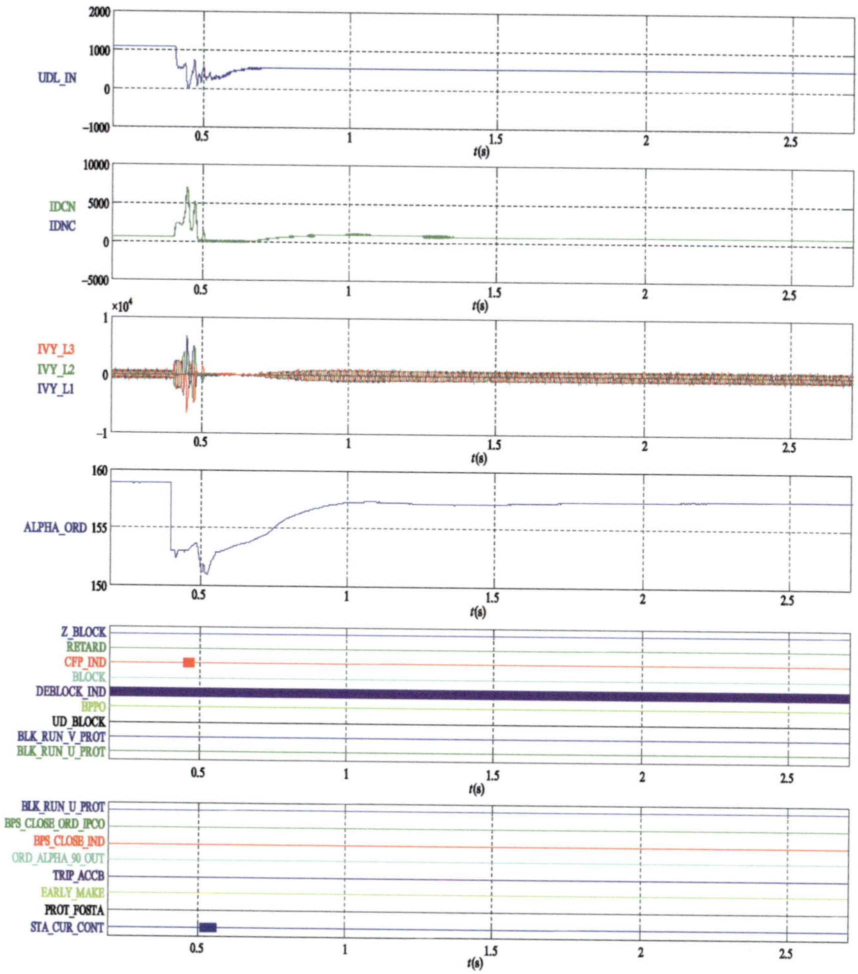

Fig. 14 Inverter station valve group fault test for CG project: low voltage valve group waveform (1)

in, thus the delay of 10 ms to put in the bypass has no significant impact on the above protection and equipments.

In addition, compared with the XS and JS projects, when the valve group on the inverter side has commutation failure and valve short circuit fault to execute S block, there is also a delay to put in the bypass pair (that needs to wait for the early-make signal from AC incoming line switch to put in the bypass pair, the delay is estimated with 5 ~ 10 ms), therefore considering that the serious faults such as valve short

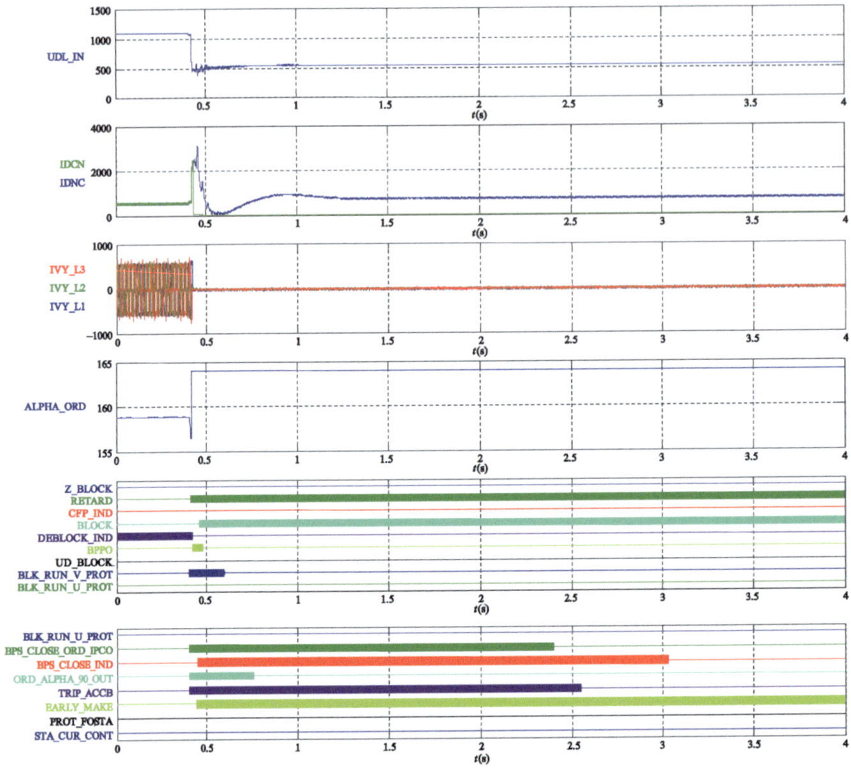

Fig. 15 Inverter station valve group fault test for CG project: high voltage valve group waveform (2)

circuit have adopted delay in puting in the bypass pair, it can be inferred that the risk to add 10 ms delay to other faults is low, and the overall strategy is feasible.

3. A pole fault causes another pole commutation failure

(1) Fault phenomenon. The bipolar operates, the ground fault happens on the inverter side or DC line of one pole, which causes the commutation failure happens on another pole. The waveform is shown in Fig. 17.

(2) Cause analysis. When a grounding fault happens on the inverter side or DC line of one pole, the current increases, and the voltage changes abruptly. Due to the mutual inductance, that causes another pole's current and voltage to change abruptly. If the rise rate of current is large enough, commutation failure will occur. Through experimental research and comparison with the SL project, it

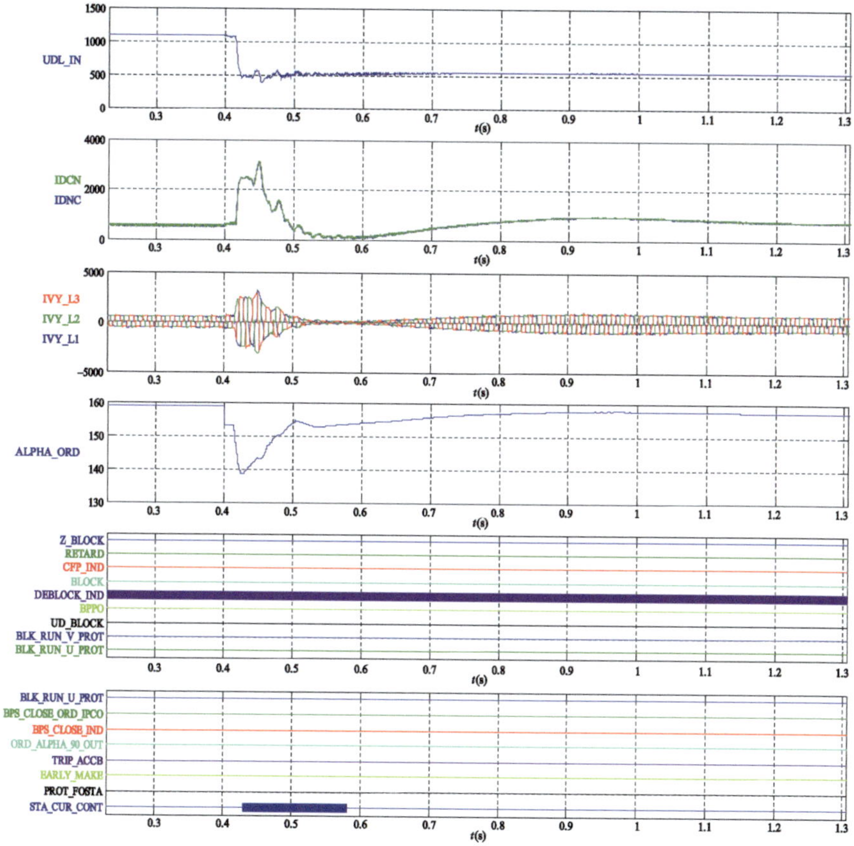

Fig. 16 Inverter station valve group fault test for CG project: low voltage valve group waveform (2)

is judged that the high voltage level is the reason of the increase in the current rise rate of another pole caused by mutual inductance.

Regularity analysis

1) Compared with high DC power operation, the same current increment during low DC power operation results in a larger variation of commutation angle, which leads to insufficient extinction angle and is more likely to cause commutation failure. The relationship between the variation of commutation angle and the DC current is shown in Fig. 18.
2) The closer the fault point is to the inverter station, the more likely it is for the opposite pole to have a commutation failure. The relationship between the variation of DC current and the position of the fault point is shown in Fig. 19.
3) The higher the DC voltage is, the larger the absolute value and variation of the short-circuit current generated by the fault is. The higher the voltage level of the fault pole is, the larger the current generated on the non-fault pole is. The

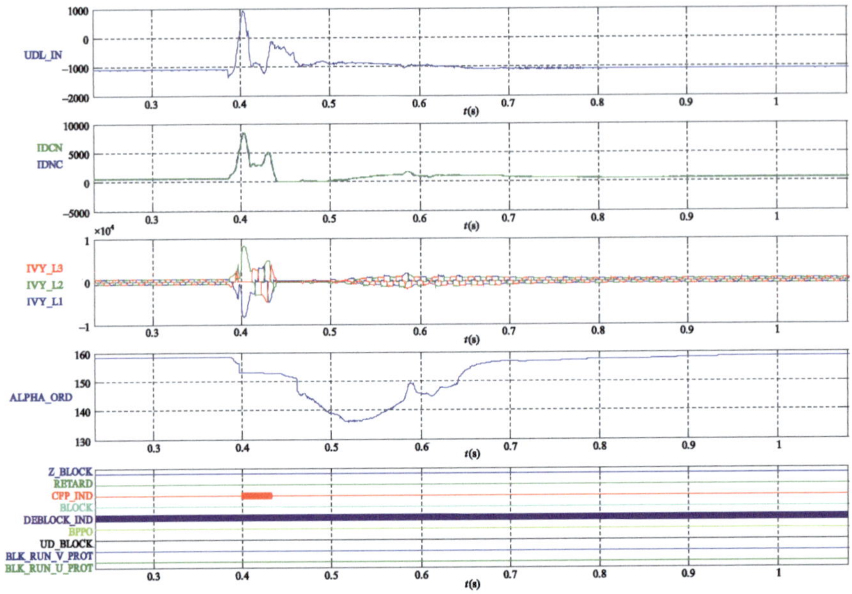

Fig. 17 Waveform for commutation failure caused by mutual inductance

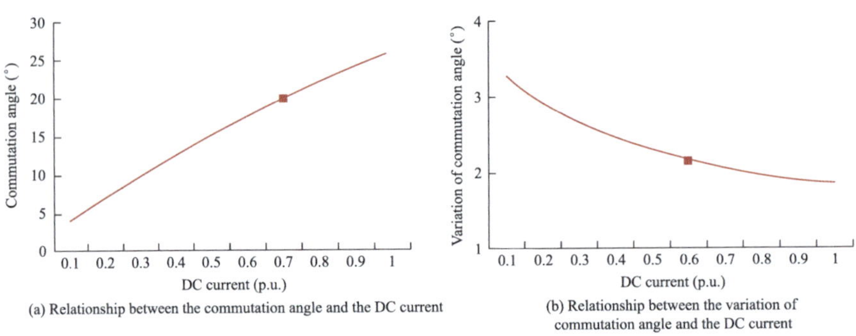

Fig. 18 Relationship between commutation angle, its variation, and DC current

relationship between the fault pole/non-fault pole current and voltage is shown in Figs. 20 and 21.

(3) Solution. By analyzing the regular pattern of voltage and current sudden changes caused by mutual inductance, the action criteria is introduced based on the sudden changes in voltage and current, and the solution is to increase the extinction angle in advance to avoid the commutation failure caused by mutual inductance.

The DC control system detects the DC line voltage U_{DL} measurement value in real-time, and calculates the variation in U_{DL} within 1 ms. When the variation is greater

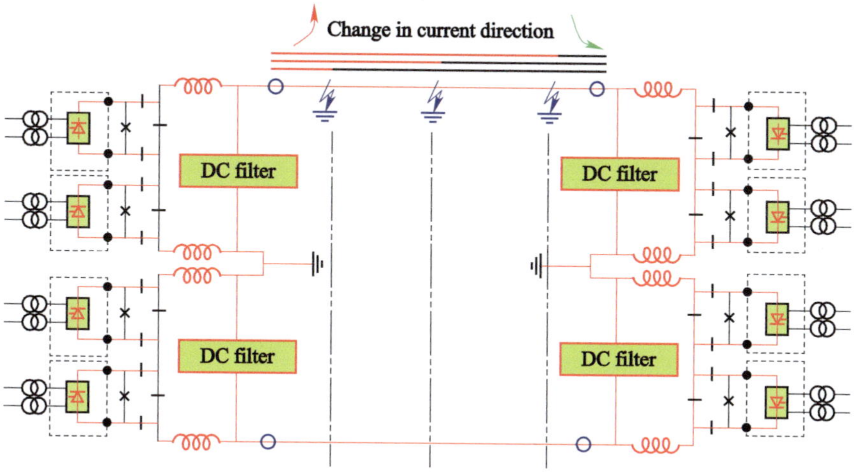

Fig. 19 Relationship between the variation of DC current and the position of the fault

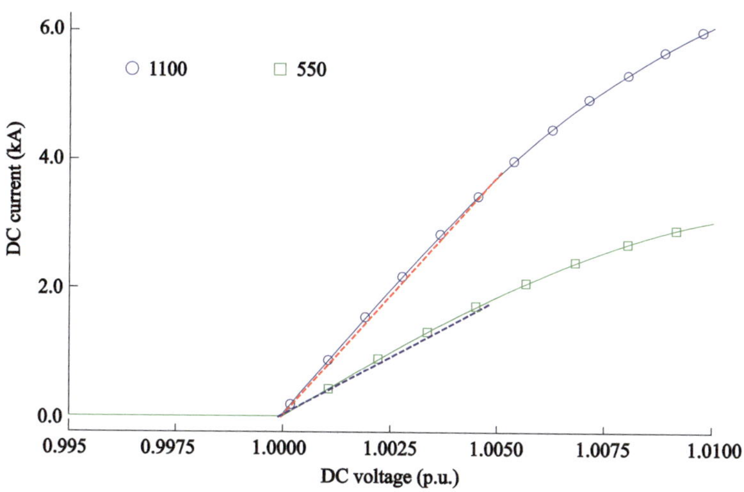

Fig. 20 Relationship between the fault pole/non-fault pole current and voltage (1)

than the threshold, the commutation failure prediction function immediately starts. The waveform without commutation failure with the increment of du/dt is shown in Fig. 22.

Because the different faults cause the different performances of mutual inductance, the specific parameter settings for the prediction of commutation failure caused by voltage sudden change commutation failure prediction have been determined with a large number of experiments.

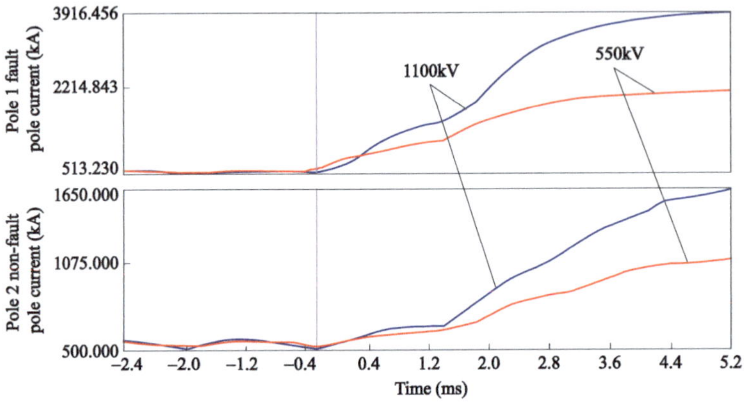

Fig. 21 Relationship between the fault pole/non-fault pole current and voltage (2)

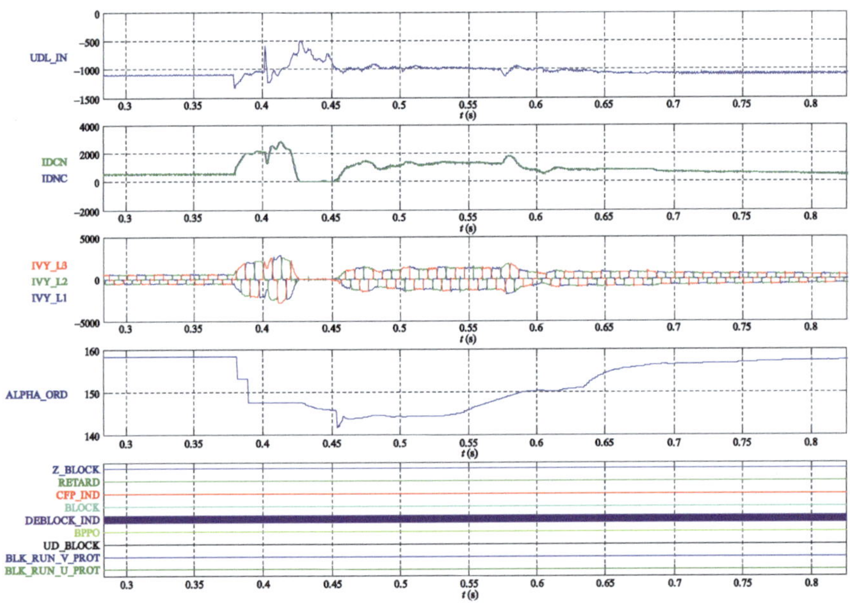

Fig. 22 Waveform without commutation failure with the increment of d*u*/d*t*

4. For bipolar power control, one pole operates in reducing voltage and another pole operates in full voltage. During the process of power increasing, there is a large deviation in the current on the double poles.

Cause analysis: When one pole starts in reducing voltage and another pole starts in full voltage, the initial IO commands of the both poles are the minimum current with 545A, which meets the conditions for entering the BC_CONFIRMED state. The reducing voltage pole enters this state firstly, annd quickly updates the IO command of this pole to IO_AMB, which causes the IO difference between the double poles to be greater than 50A. This case cannot meet the conditions for entering the BC_CONFIRMED state, at this time the full voltage pole cannot enter the BC_CONFIRMED state, thus the full voltage pole can only update IO at a slower speed. During the process of power increasing, there is a large deviation in the current on the double poles.

Solution: After one pole enters the BC_CONFIRMED state, the IO is slowly updated in a short time to ensure that the other pole can meet the conditions for entering the BC_CONFIRMED state, so that it can be controlled according to the bipolar power control mode.

5. The absolute minimum filter does not meet the requirement of power reduction, the power reduction is too large.

Cause analysis: When the absolute minimum filter does not meet the requirement of power reduction, the control protection program will compare the current power and power command, and take the smaller one as the new power command. During the power reduction process, the overshoot causes that the actual power is less than the power command, and the new power command is also refreshed to the actual smaller power value.

Solution: the strategy is modified to only update the IO command within 20 ms after the reactive power reduction signal is triggered, thereby it can avoid mistakenly updating IO command after the power reduction overshoots, which can ensure that the IO command is only updated once in each reactive power reduction process.

6. When the VBE of Pole 2 at the Guquan converter station restores the active signal fiber of the backup system, a trigger pulse is lost.

Cause analysis: When the active signal fiber is plugged in, the hand trembling causes the optical power to oscillate unstably, which leads the backup system to misjudge the active signal as valid, and the pulses are lost during system switch.

Solution: The measure is to optimize the demodulation standard for the active signal.

7. When the VBE of Pole 2 at the Guquan converter station restores the power of the backup system, the control and protection misreports a pulse lossing event.

Cause analysis: The initialization is completed in about 1.5 s after the VBE produced by XD group CO., LD. is powered on., and the VBE-OK signal becomes 1. During this period, the FP signal sent by VBE to CCP is 1 MHz without a 16 μs high level, which causes CCP to misreport a a pulse lossing event.

Solution: After the VBE-OK signal becomes 1, VBE sends the normal FP signal to CCP, and CCP starts the pulse losing detection after it detects that the VBE-OK signal has become 1.

2 Equipment Commissioning

Equipment commissioning refers to the on-site handover test after the equipments are installed at the converter station. The purpose is to confirm the equipment integrity during transportation, to check the status and installation quality of the equipment, to test whether the equipment can safely be charged, loaded and start, and whether the performance of the equipment can meet the requirements of the contract and technical specifications. Equipment commissioning is divided into conventional equipment commissioning and special equipment tests. The conventional equipment commissioning in the converter station is completed based on relevant equipment commissioning standards. The special equipment tests are based on relevant standards, combined with the characteristics of the project and equipments, to compile an equipment commissioning outline, solve various technical problems found during the test, and complete special equipment tests.

I. Routine Test Items

The equipment routine tests of DC project are the responsibility of the installation and construction entities, who test the performance of the electrical equipments in the converter station, handle the problems, and ensure that the equipment performance meets the technical specification requirements and the equipments are ready for subsystem commissioning.

The scope of equipment routine tests includes converter transformers, smoothing reactors, thyristor valves, DC current transformers, DC voltage dividers, wall bushings, AC/DC circuit breakers, AC/DC disconnectors and grounding switches, AC/DC filters, carrier wave devices and noise filters, parallel capacitor banks, air core reactors, optic-electric current transformers, zinc oxide surge arresters, grounding electrode devices, station power transformers, AC high voltage parallel reactors, AC transformers, power cables, GIS, etc.

During the equipment commissioning, the standards will be strictly followed, such as GB 50150—2016 "Electric equipment installation engineering-standard for hand-over test of electric equipment", GB/T 50832—2013 "Standard for acceptance test of electric equipment of 1000 kV system electric equipment installation engineering". And the standard Q/GDW 11743—2017 "Hand-over tests for ± 1100 kV UHVDC electric equipment" is refered. The electrical equipment test entities will compile the test plan and complete the equipment commissioning test.

II. Special Test Items

The scope of equipment special tests includes converter transformers, gas-insulated metal-enclosed switchgear (GIS), AC/DC circuit breakers, zinc oxide surge arresters, grounding electrode devices, AC transformers, power cables, etc.

During the equipment commissioning, the standards will be strictly followed, such as GB 50150—2016 "Electric equipment installation engineering-standard for hand-over test of electric equipment", GB/T 50832—2013 "Standard for acceptance test of electric equipment of 1000 kV system electric equipment installation engineering".

And the standard Q/GDW 11743—2017 "Hand-over tests for ± 1100 kV UHVDC electric equipment" is refered. The electrical equipment test entities will compile the test plan and complete the equipment commissioning test.

Compared to the special field tests of ± 800 kV UHV DC transmission project, the ± 1100 kV UHV DC transmission project, due to its higher voltage level, larger power transmission capacity, and longer transmission distance, has adopted new test plans in the following equipment special tests:

(1) Long-duration induced voltage test of the converter transformer with partial discharge measurement.
(2) AC voltage withstand test and partial discharge test on the valve side of the converter transformer.
(3) On-site measurement of the oscillation characteristics of the DC transfer switch.
(4) Measurement of the grounding network of the converter station.
(5) DC line parameter test.
(6) DC arrester special test.

III. ± 1100 kV UHV DC Equipment Commissioning Handover Test

Through the study of the ± 1100 kV UHV DC equipment commissioning field handover test items, referring to the standard Q/GDW 11743—2017, the handover test of main equipments, such as converter transformer, converter valve, primary equipment in the DC yard, AC filter, DC line and grounding electrode, was compiled. The following items are special tests that have adopted new test plans.

(I) Long-duration Induced Voltage Test of the Converter Transformer with Partial Discharge Measurement Test

1. Test purpose

The ± 1100 kV converter transformer is the key equipment of the CG project and also the most difficult to develop. According to the requirements of standard Q/GDW 11743—2017, a long-duration induced voltage test with partial discharge measurement (partial discharge test) must be conducted on site. This test is a quality assurance test to inspect the installation quality of the converter transformer and is an important guarantee for the successful operation of the converter transformer for one time.

2. Content and requirements

The ± 1100 kV Guquan converter station adopts a layered connection mode, with the grid-side voltage of the LV converter transformer being 1000 kV, and the grid-side voltage of the HV converter transformer being 500 kV. The grid-side voltage levels of the HV and LV converter transformers of the ± 1100 kV Changji converter station are both 750 kV. The following test methods are mainly introduced for the transformer parameters of Guquan converter station. The main content of this test is to conduct a partial discharge test on the grid-side and valve-side windings of converter transformer.

The stricter acceptance criteria for the partial discharge test of the ± 1100 kV converter transformer is proposed based on that of ± 800 kV converter transformer, with raising the acceptance criteria for the partial discharge test from 300 to 100 pC at 1.3 $U_m/\sqrt{3}$ test voltage. The test requirements is shown as follows:

(1) The test voltage does not cause a sudden drop.
(2) During the long-duration test at 1.3 $U_m/\sqrt{3}$ voltage, the continuous level of partial discharge of the grid-side winding is not more than 100 pC, and the continuous level of partial discharge of the valve-side winding is not more than 300 pC.
(3) At 1.3 $U_m/\sqrt{3}$ voltage, the partial discharge does not show a continuous growth trend, and occasional higher amplitude pulses can be ignored.
(4) At 1.1 $U_m/\sqrt{3}$ voltage, the continuous level of apparent charges is not more than 100 pC.
(5) After the partial discharge test, the oil chromatography analysis results meet the requirements, and there is no significant difference from that before the test.

3. Test method

There are two types of test wiring during the on-site partial discharge test of the ± 800 kV converter transformer, namely applied voltage on single terminal and applied voltage on both terminals. The wiring diagram of applied voltage on single terminal is shown in Fig. 23, and the wiring diagram of applied voltage on both terminals is shown in Fig. 24.

The common test wiring for ± 800 kV converter transformer is the applied voltage on single terminal for Y-connection and the applied voltage on both terminals for D-connection. The Y-connection converter transformer of ± 1100 kV Guquan converter station also uses the applied voltage on single terminal, but if the low-voltage D-connection converter transformer uses the applied voltage on both terminals, it will exceed the withstand voltage value of plant test.

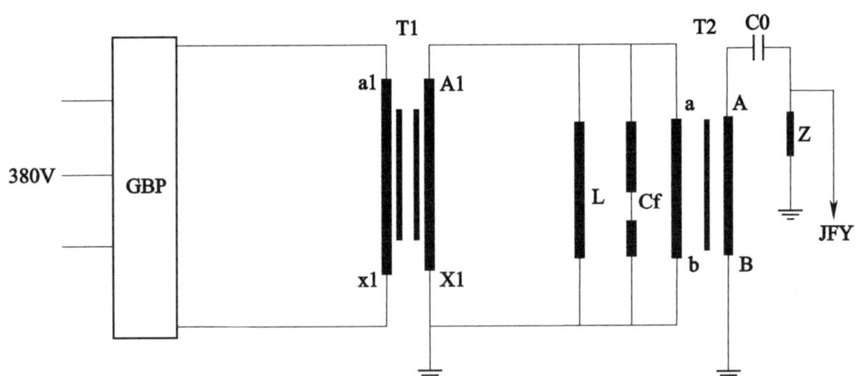

Fig. 23 Wiring diagram of applied voltage on single terminal. GBP—Frequency converter; T1—Excitation transformer; T2—Converter transformer; L—Shunt reactor; Cf—Capacitive voltage divider; C0—Bushing capacitor; JFY—Partial discharge tester; Z—Detection impedance for partial discharge test

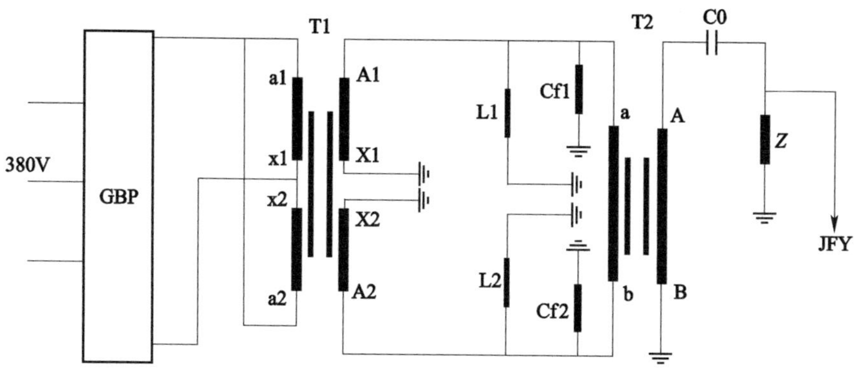

Fig. 24 Wiring diagram of applied voltage on both terminals

According to Figs. 23 and 24 and the main parameters of the low-voltage D-connection converter transformer at the Guquan converter station, the voltages from the windings on each side to the ground and among each other winding can be respectively calculated under the applied voltage on single terminal and on both terminals. If the applied voltage on both terminals is used, the voltage difference between A terminal on the grid side and b terminal on the valve side is 1131.8 kV under the $1.5U_m/\sqrt{3}$ test voltage, which exceeds the withstand voltage value of plant test (1100 kV). The voltage difference between A terminal on the grid side and b terminal on the valve side is 980.8 kV and under the $1.3U_m/\sqrt{3}$ test voltage, which exceeds 80% of the withstand voltage value of plant test (880 kV), and the time is long (1 h). That does not meet the transformer requirement, which is stipulated in GB/T 1094.3—2017 "Power transformers—Part 3: Insulation levels, dielectric tests and external clearances in air". for example, the voltage in the repeated dielectric test should not exceed 80% of the plant test value.

When the applied voltage on single terminal is adopted, the voltage difference between A terminal on the grid side and b terminal on the valve side is 952.7 kV under the $1.5U_m/\sqrt{3}$ test voltage, which exceeds 80% of the withstand voltage value of plant test (880 kV), but the overvoltage time is short ($\leq$ 1 min), and the manufacturer approves this test writing.

Therefore, the on-site partial discharge test of the low-voltage D-connection converter transformer at the ± 1100 kV Guquan converter station should not use the applied voltage on both terminals, but should use the applied voltage on single terminal used in the plant test of the equipment.

According to the above calculation method, the high-voltage D-connection converter transformer at the ± 1100 kV Guquan converter station can use either the applied voltage on single terminal or on both terminals. But in order to save equipment investment, the high-voltage D-connection converter transformer also uses the applied voltage on single terminal.

4. Engineering application

   (1) The partial discharge test results of the converter transformers at the ± 1100 kV Guquan converter station and the ± 1100 kV Changji converter station meet the relevant requirements.
   (2) The feasibility to use the applied voltage on single terminal for the on-site partial discharge test of UHV converter transformer with D-connection has been verified, and it has met the voltage requirements of the relevant regulations.

(II) AC Withstand Voltage and Partial Discharge Test on the Valve Side of the Converter Transformer

1. Test purpose

Due to the high difficulty in the design and manufacture of the ± 1100 kV HV converter transformer, after completing the plant test, a reinstallation of bushing was carried out on the valve side of the converter transformer. Therefore, it is necessary to conduct the AC withstand voltage and partial discharge test on the valve side to assess the insulation performance of the valve side of the converter transformer. This test is the first time in China to conduct the AC withstand voltage and partial discharge test on the valve side of an UHV converter transformer on-site in the valve hall. The test voltage is conducted at 100% of the plant voltage. The test voltage for the Y-connection and D-connection converter transformer respectively reaches up to 1297 and 987 kV. According to the plant test standards, the partial discharge of the converter transformer was monitored on-site for the first time.

2. Content and requirements

According to the standard GB/T 1094.3—2017, GB/T 18494.2—2007 "Converter transformers—Part 2: Transformers for HVDC applications" and Q/GDW 11743—2017 standards, this test was conducted on two high-voltage D-connection converter transformers and two high-voltage Y-connection converter transformers in the CG Project.

The test applies power frequency voltage to the valve side winding from the a or b bushing terminal of the valve side. The D-connection converter transformer is subjected to a voltage of 100% of the plant voltage value of 987 kV, the withstand voltage time is 60 min, and the partial discharge is monitored. The Y-connection converter transformer is subjected to a voltage withstand test at 100% of the plant voltage value of 1297 kV for 1 min, then the voltage drops to 80% of the plant voltage value of 1037 kV for 60 min, and the partial discharge is monited. According to the partial discharge assessment requirements of the on-site handover test, the partial discharge on the valve side should not exceed 300 pC, and there should be no obvious increasing trend.

3. Test method

   (1) This test is the first time to be conducted in the valve hall. Taking the D-connection converter transformer as an example, under the test voltage of

987 kV, whether the safety insulation distance of equipments in the valve hall, such as insulation pillars, conductors, arresters, and valve towers, and the safety performance of the converter valve can meet the requirements of the test environment and background partial discharge level is an important factor in assessing whether this test can be successfully completed. After the electric field calculation in the valve hall, it was determined that only the insulation pillar and conductor fittings in front of the converter transformer were removed, and the valve tower and valve arrester were not moved. The minimum insulation clearance is 7.5 m between the corona ring of the equipment and the valve arrester. Before the installation of the converter transformer, several open line tests were conducted in the valve hall using the test equipment and with simulating the relative spatial position of converter transformer. The background partial discharge of valve hall environment was detected in advance, and the value was less than 100 pC, proving that the set of test equipments and the valve hall environment can meet the test requirements.

(2) This test uses the adjustable frequency series resonance method, which is different from the adjustable inductance series resonance used in the plant test. The test frequency of field test should meet the requirements of the converter transformer manufacturer. The final test frequency is 81 Hz, which meets the relevant test requirements.

(3) This test uses the UHV GIS device-type insulation test platform. Compared with the traditional multi-section tower structure, this platform does not require a crane to enter the valve hall for hoisting, and can fully meet the control requirements of the exhaust environment in the valve hall.

4. Engineering application

(1) The AC withstand voltage and partial discharge tests on the valve side of the converter transformer were respectively conducted at the Changji and Guquan converter stations, and both passed the assessment The partial discharge was no more than 150 pC with meeting the relevant requirements.

(2) The test environment in the valve hall was subjected to electric field calculations. The calculation and test results show that the air clearance should not be less than 7.5 m, 6 m and 5 m for the withstand voltage and partial discharge tests of 1000 kV, 750 kV and 500 kV voltage level in the valve hall respectively. The field strength should be less than 15 kV/cm. For the withstand voltage test under 1297 kV, the minimum air clearance should not be less than 8.5 m. The above reference values can reduce equipment removing work and avoid repetitive work.

(III) On-site Handover Test of Converter Station and Grounding Electrode Grounding System

1. Handover Test Items

The commissioning items of the converter station and grounding electrode grounding system should include: ① Grounding resistance measurement; ② Current distribution measurement of the current guiding cable and grounding electrode components; ③ Step and contact potential measurement.

2. Test standards and content

   (1) Test standards. The grounding network and DC grounding electrode of converter station are tested according to the following standards:
      1) DL/T 437—2012 "Technical guide for HVDC earth electrode system".
      2) DL/T 475—2017 "Guide for measurement of grounding connection parameters".
   (2) Test content
      1) Grounding resistance (impedance) test. The overall condition of the grounding network of converter station is understood through the grounding resistance measurement of the grounding network and grounding electrode.
      2) Step voltage. Step voltage (potential difference) refers to the potential difference between two points on the ground at a horizontal distance of 0.8 m when the grounding short-circuit (fault) current flows through the grounding device.
      3) Contact potential. Contact potential (potential difference) refers to the potential difference between two points at a horizontal distance of 0.8 m from the equipment on the ground and at a vertical distance of 1.8 m from the ground along the equipment shell, frame or wall when the grounding short-circuit (fault) current flows through the grounding device.
3. Engineering application

For the first time in China, the power frequency inverse phasor increment method recommended in the standard DL/T 475—2017 was used for measurement, and the non power frequency method was used for measurement at the same time. The two methods verify each other and can more accurately obtain the true value of the power frequency characteristic parameters. For shunt measurement, the vector measurement method is used. The measurement is carried on the each truss branch on the ground in Changji converter station, and the vector value is summed. The Guquan converter station uses climbing to directly measure the external lightning conductor, and the vector value is summed. During the test, the mutual inductance effect of test lines was considered. When wiring, the line was laid in the way with the least mutual inductance effect. Especially the grounding impedance of the Changji converter station was only 25 mΩ, and the total line length outside and inside the station reached 6 km. The method overcomes the mutual inductance effect, accurately measures the shunt vector, and realizes the accurate measurement of the small grounding resistance.

(IV) On-site Handover Test of DC and Grounding Electrode Lines

1. Test purpose

The transmission line length of the CG project reaches 3328 km. The parallel line length along the already-operated Hami-Zhengzhou and Jiuquan-Hunan project exceeds 1000 km, and the interference voltage and current signals will reach a high level. It is necessary to carry out tests on related DC transmission line parameters to check the insulation condition of transmission line, verify the phase and test frequency characteristic, etc.. That is considered as an important basis for protection setting and calculation.

2. Handover test items

According to the standard DL/T 1566—2016 "Guide for measurement of DC transmission line and earth electrode line parameters" and Q/GDW 11090—2013 "Transmission line parameter frequency characteristic measurement guidelines", the test items for DC and grounding electrode lines should include: ① Induced voltage test; ② Polarity check; ③ Insulation resistance measurement; ④ DC resistance measurement of the line; ⑤ Frequency characteristic measurement of line parameters.

3. Induced voltage test

Tests should be conducted for induced voltage under open-circuit conditions, induced voltage under short-circuit conditions, and induced current under short-circuit conditions at the end.

4. Polarity check

With grounding the line end of pole 1, the line beginning and end of the pole 2, and adding DC power source at the beginning of pole 1, the current I_1 is measured. Then, with hanging the end of pole 1 and adding DC power source at the beginning of pole 1, the current I_2 is measured.

When $I_1 \neq 0$ and $I_2 \approx 0$, the polarity of the pole 1 line is correct. When $I_1 \approx I_2 \neq 0$, there is a grounding point on the pole 1 line, and the polarity of the pole 1 line may be incorrectly marked or there may be end nodes between the pole 1 and pole 2 lines. When $I_1 \approx I_2 \approx 0$, there is a breakpoint on the pole 1 line.

5. Insulation resistance measurement

The 5000 V DC voltage is added to the line to test the insulation resistance.

6. DC resistance measurement of the line

The DC power source is added between the beginning of two poles, the DC voltage and current at the beginning is measured. The series DC resistance value $R_{1\text{-}2}$ of pole 1 and 2 is calculated, and the DC resistances R_1 and R_2 of pole 1 and 2 are calculated according to the following formula.

$$R_1 = R_2 = \frac{1}{2} R_{1-2}$$

The DC resistance of the test downlead should be deducted from $R_{1\text{–}2}$.

7. Frequency characteristic measurement of line parameters

The open-circuit and short-circuit impedances are measured under the positive sequence and zero sequence respectively. After completing the line parameter tests with all selected frequency, the frequency characteristic curve of the electrical parameters of the DC transmission line is plotted with the frequency f as the horizontal axis and the line parameters of unit length calculated in the test as the vertical axis.

8. Engineering application

The DC line parameters and grounding electrode line parameters measured according to the above test plan have been applied in the CG Project.

(V) On-site Handover Test of DC Arrester

1. Test items

According to the standard GB 11032—2012 "Metal-oxide surge arresters without gaps for a. c. systems", GB/T 22389—2008 "Guidelines for metal oxide surge arresters without gaps for HVDC converter stations", DL/T 377—2010 "The acceptance tests of high voltage direct current equipment", the arrester tests are carried out. The DC reference voltage and leakage current under 0.75 times of the DC reference voltage are measured. The AC continuous current test is carried out and the resistive leakage current is measured.

2. DC reference voltage measurement

The whole or single section of the arrester is measured according to the DC reference current value specified by the manufacturer. The measurement method should comply with the standard GB 11032—2020, and its reference voltage value should not be lower than the contract specified value.

3. Leakage current test under 0.75 times DC reference voltage

The measurement is carried out according to that specified in GB 11032—2020. Under 0.75 times the DC reference voltage, its leakage current should not exceed 50 μA for a single-column arrester, and the leakage current value should not exceed the value specified by the manufacturer's standard for multi-column parallel arresters or the arresters with a rated voltage above 216 kV.

4. AC continuous current test

According to the standard GB/T 22389—2008, for arresters with significant continuous operation voltage, a test voltage with effective value equal to the peak value of continuous operating voltage divided by $\sqrt{2}$ is added to arresters. For arresters without significant continuous operation voltage, a test voltage with effective value equal to the 0.8 times peak value of continuous operating voltage divided by $\sqrt{2}$ is added to arresters. In this test, if the manufacturer's specified value is higher than that of GB/T 22389—2008, the manufacturer's specified value is used, which is more stringent. The test is conducted on the arrester components, and the applied power frequency test voltage is calculated based on the ratio of the rated voltage of the whole arrester to the rated voltage of the components.

5. Engineering application

The proposed on-site handover test plan for the DC arrester was implemented, and the on-site handover tests for the DC arresters at Changji and Guquan converter station were completed. The test results meet the requirements of the engineering technical specifications.

(VI) On-site Handover Test of DC Transfer Switch

1. Handover test items

The DC switch consists of bypass switch, metallic return transfer switch, ground return transfer switch, neutral bus switch, and neutral bus grounding switch. The test items should include: ① Circuit breaker test; ② Auxiliary circuit test; ③ Oscillation circuit oscillation characteristic measurement.

2. Circuit breaker test

The circuit breaker test is conducted in accordance with the relevant sections of GB 50150—2016.

3. Auxiliary circuit test

The auxiliary circuit test includes: ① Capacitor test; ② Reactor test; ③ Surge arrester (non-linear resistor) test.

4. Oscillation circuit oscillation characteristic measurement

(1) Test purpose. The measurement of the oscillation characteristics of DC circuit breaker is an overall characteristics assessment of DC circuit breaker. The measurement of the oscillation frequency and damping resistance of DC circuit breaker transfer circuit is to assess the oscillation characteristics of DC circuit breaker. The principle is to charge a capacitor with a few hundred volts, close the SW1 auxiliary switch. when the circuit oscillates, the circuit current is measured, and then the circuit frequency and damping resistance is calculated. The purpose is to provide the final check before the equipment is put into operation without harming the charging device.

(2) Test standard. The oscillation characteristics of DC circuit breaker are measured in accordance with the following relevant standards:

   1) DL/T 274—2012 “Hand-over test of ± 800 kV HVDC electric equipments”;

   2) DL/T 273—2012 “Preventive test code for ± 800 kV HVDC electric equipments”.

(3) Test method. The measurement circuit is shown in Fig. 25, where TA is a current transformer. The measurement cable is a 75 Ω coaxial cable. The oscilloscope response frequency is 100 MHz. SW1 is an auxiliary switch. E is a DC charging power source.

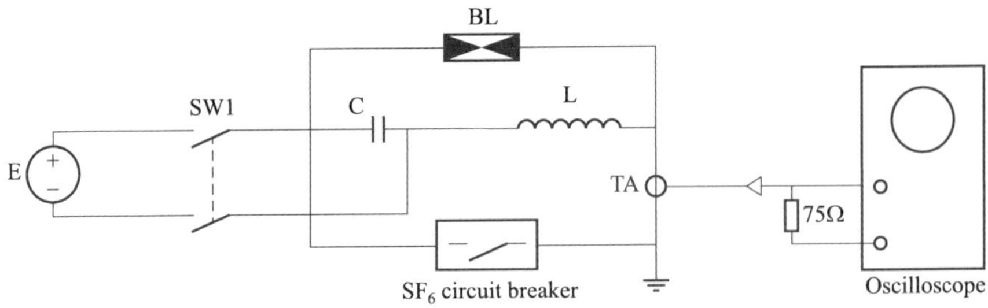

Fig. 25 Wiring diagram of oscillation characteristic measurement for oscillation circuit

The test steps are as follows:

(1) Disconnect the SF_6 circuit breaker, close the auxiliary switch SW1, and charge the capacitor to 300 V.
(2) Disconnect the auxiliary switch SW1.
(3) Electrically close the SF_6 circuit breaker and measure the oscillation waveform.

The oscillation frequency and damping resistance of the DC circuit breaker transfer circuit is calculated based on the current waveform.

5. Engineering application

The measurement test of the oscillation frequency and damping resistance of DC circuit breaker transfer circuit verifies that the oscillation characteristics of DC circuit breaker meet the technical specification requirements.

IV. ± 1100 kV UHV Equipment Field Test Platform and Test Technology Research

After completing the main parameter research, test requirement analysis, and handover test scheme design for ± 1100 kV UHV equipment, the test items and equipment requirements for handover test of ± 1100 kV UHV main equipments are analyzed. Through investigating the test capabilities of relevant companies, the development of key test equipments and platforms and the research of key test technologies are carried out.

(I) Research on the Integrated Insulation Test Platform for UHV GIS

1. Research background

Insulation breakdown faults of UHV GIS occur in handover tests, commissioning, and operation. Insulation test is the most effective method to check these faults, but there are two problems in field implementation:

(1) The test risk is high. Due to the high test voltage, both the reactor and the voltage divider are independent tower structures. The equipment occupies a large space and is close to neighboring equipment, which is prone to air gap discharge. In addition, the voltage divider is top-heavy and has a risk of tipping over, which is more prominent in areas prone to short-term high winds such as the northwest of China.

(2) The test period is long. The test equipment is numerous, large, and heavy. On-site assembly not only involves a large amount of work, but also has a high difficulty in hoisting. A single test cycle can last 2 to 3 days, affecting the power supply recovery time after fault repair.

In response to the above requirements, an integrated test platform based on reactor and voltage divider integration technology, and self-lifting technology is developed. This platform has realized the integration and automation, reduced the floor space, the workload of test preparation and the difficulty of hoisting, lowered the test risk, improved the efficiency of on-site testing work, and met the needs of large-scale construction and rapid response to operation and maintenance in UHV projects.

2. Principle, structure, and performance

This device is an UHV GIS integrated insulation test platform based on oil-immersed reactors. Its transportation and test states are shown in Figs. 26 and 27. It includes insulation test equipment and a movable platform with self-lifting function. The movable platform includes a platform chassis, wheels installed at the bottom of the platform chassis, hydraulic support legs installed on the side of the platform chassis that can prop up the platform chassis vertically, a main hydraulic cylinder, a bracket, and a base on the platform chassis. The insulation test equipment is placed on the movable platform through the bracket and base. The bracket and base can be pushed by the main hydraulic cylinder to complete a 90° flip and stand up. The bracket and base are detachably connected, and after the bracket and base are separated, they can independently complete the lying down and folding. This device does not require on-site equipment assembly and equipment hoisting, completely subverting the traditional preparation method for UHV GIS on-site insulation tests, greatly reducing the equipment footprint, allowing for quick transitions within the substation, and safely and conveniently conducting UHV GIS on-site AC voltage withstand tests.

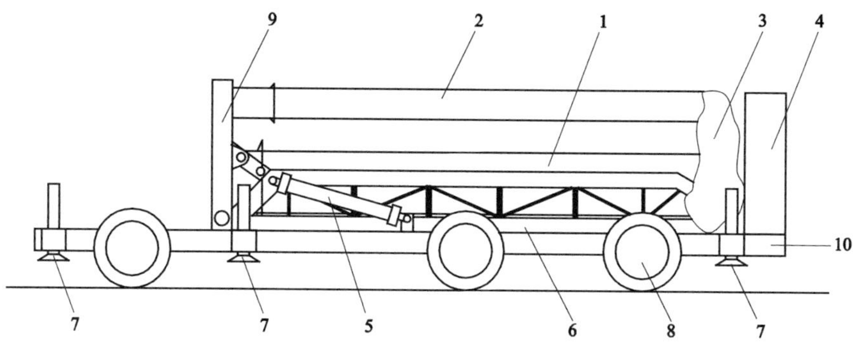

Fig. 26 Schematic diagram of the UHV GIS integrated insulation test platform in transportation state. 1—Resonant reactor; 2—Capacitive voltage divider; 3—Expandable corona ring; 4—Hydraulic pump station; 5—Main hydraulic cylinder; 6—Bracket; 7—Hydraulic support legs; 8—Wheel; 9—Base; 10—Platform chassis

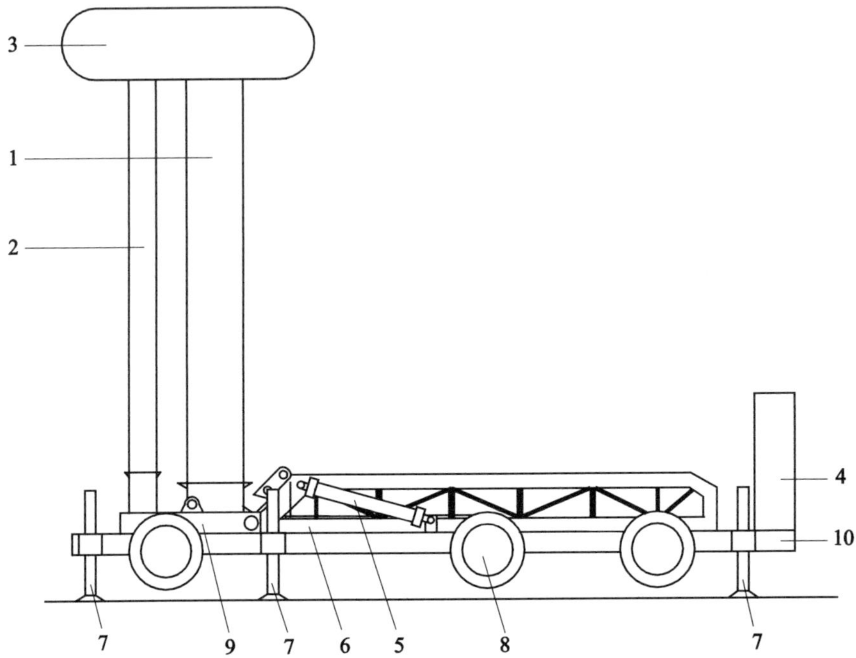

Fig. 27 Schematic diagram of the UHV GIS integrated insulation test platform in operational state. 1—Resonant reactor; 2—Capacitive voltage divider; 3—Expandable corona ring; 4—Hydraulic pump station; 5—Main hydraulic cylinder; 6—Bracket; 7—Hydraulic support legs; 8—Wheel; 9—Base; 10—Platform chassis

The research proposes a self-lifting UHV GIS integrated insulation test platform based on oil-immersed reactors, with rated parameters reaching 1200 kV/12 A, and voltage measurement accuracy not less than 1%. Due to the integrated fixed structure of the resonant reactor, voltage divider, and common expandable corona ring, and the optimized resonant reactor that can be transported lying down, it has a hydraulic self-lifting device that can lift the test equipment from lying down to standing up. This platform does not need to rely on external hoisting equipment to prepare for the 1000 kV GIS on-site insulation test. The platform can also quickly move within the substation, complete valve hall indoor tests without fuel environment requirements, significantly improve the test efficiency, reduce the risks of equipment hoisting and adverse weather, effectively guarantee the progress of substation construction, shorten the maintenance time, and has great promotion prospects and application value.

In the 1000 kV GIS main circuit insulation test, an expandable corona ring is also used. This corona ring uses an inflatable structure, and the outer surface is attached with wear-resistant, burn-resistant conductive cloth. Under the 1100 kV test voltage, this corona ring can still ensure that the quality factor Q of the entire series resonance test system is greater than 80, which can meet the test requirements.

3. Electric field and temperature rise optimization based on multi-physics field coupling simulation model

The finite element method was used to perform electric field simulation and optimization design on the designed test platform, as shown in Fig. 28. In combination with the actual conditions of the test site, the design scheme of the inflatable single corona ring was established, and the maximum surface field strength of the corona ring and bottom flange was controlled within 1.0 kV/mm, in order to minimize local discharge during the test while meeting the basic insulation requirements. The optimization method for the equipment can provide reference for the design of similar equipment.

The method of using fluid-thermal coupling inside the resonant reactor and adopting the convection heat transfer coefficient outside the resonant reactor can efficiently simulate the complete heat exchange process of the reactor, as shown in Fig. 29. At t = 3600 s, the maximum temperature of the reactor shell is 42 °C, the maximum and average temperature of the internal insulation oil is respectively 72.29 °C and 41.24 °C. The maximum and average temperature of the coil is respectively 74.8 °C and 62.72 °C. The hot spots of the reactor coil are mainly distributed in the 52 ~ 59 coils, with temperatures above 70 °C. The insulation oil of the reactor

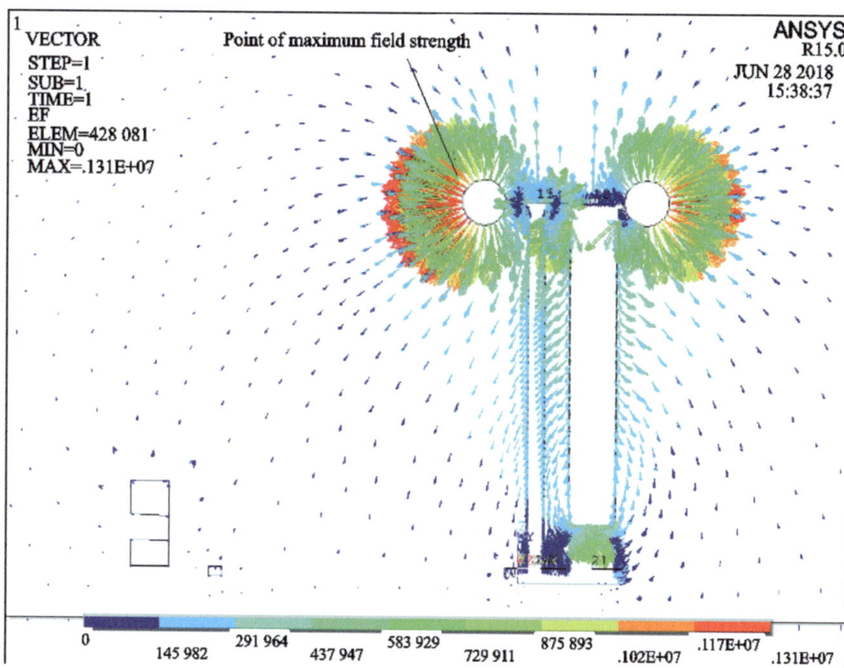

Fig. 28 Simulation results of electric field distribution on the external insulation of the integrated test platform

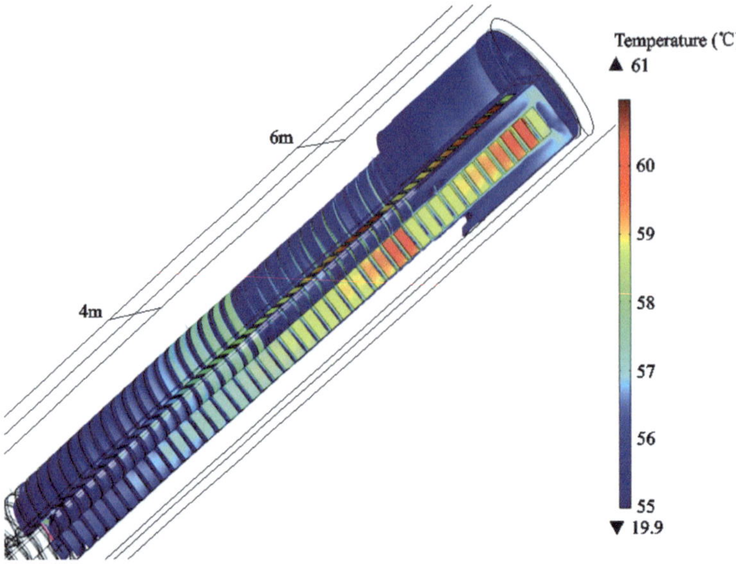

Fig. 29 Simulation results of resonant reactor temperature rise

is mainly distributed in the horizontal oil channels of the upper layer of the 52 ~ 59 coils, with local temperatures above 70 °C.

Based on the analysis of the flow velocity distribution of insulation oil, the flow of insulation oil in the horizontal oil channel of the reactor is mainly axial, resulting in temperature accumulation in the horizontal oil channel and the formation of hot spots. The insulation oil inside and outside the coil mainly flows axially, resulting in uneven radial temperature distribution of the reactor coil and insulation oil.

When the insulation oil with smaller viscosity is chosen, the temperature distribution of the reactor insulation oil is more uniform, the maximum temperature of the reactor decreases by 10.03 °C. The maximum and average temperature of the coil respectively decreases by 9.32 °C and 6.46 °C. While reducing the viscosity of the insulation oil, the top oil channel of the 50 ~ 59 coils is widened to 1.5 times the original, the temperature of the reactor is reduced to below 60 °C, and the internal temperature distribution is more uniform.

4. Field application examples

   (1) The application of 1100 kV GIS field insulation test is shown in Fig. 30. This set of equipments was used to complete the field insulation test of 1100 kV GIS, 1100 kV circuit breaker and 550 kV GIS at the ± 1100 kV Guquan converter station. The test preparation time was reduced to 8 h, and the efficiency increases of 80% compared to traditional test equipments.
   (2) The application of the field AC voltage withstand test on the valve side of HV converter transformers and the partial discharge tests is shown in Fig. 31. This device can also be used for on-site valve side AC voltage

Fig. 30 1100 kV GIS on-site insulation test

withstand tests of converter transformers in the valve hall. Such tests need to be carried out in the valve hall, where cranes and gas exhaust vehicles cannot enter, making it impossible to use traditional multi-section reactors for hoisting to complete the test equipment setup. This device completed the valve side AC withstand voltage and partial discharge tests of the HD4 HV converter transformer in the high-voltage valve hall of the ± 1100 kV Changji converter station, as shown in Fig. 31. The test was strictly in accordance with the plant test standards, the test voltage reached 100% of the plant voltage with 987 kV, the partial discharge was less than 150 pC, and the test results met the requirements.

5. Research results

Using this test platform, no crane is needed during the preparation and transition process of the test. Only after the test equipment is ready, one aerial work platform is used to connect the high voltage expanded diameter conductor. Each test preparation time only takes half a day. The platform greatly improves the test preparation efficiency in the construction of the converter station and reduces the risk of multi-workface crossover and crane hoisting. At the same time, this test platform can also enter the valve hall to carry out valve side AC voltage withstand and partial discharge tests of the converter transformer. Under the premise of meeting the environmental requirements of the valve hall, this test can be carried out efficiently and safely.

(II) Research on the Long-duration Induced Voltage and Partial Discharge Test Platform with Applied Voltage on Single Terminal for Converter Transformer

Fig. 31 Valve side AC withstand voltage test of HV converter transformer

1. Research background

The partial discharge test of the ± 1100 kV HV converter transformer with applied voltage on single terminal is more suitable for the field handover test because it is in line with the design expectations of the converter transformer manufacturer. The test transformer under the applied voltage on single terminal has a higher voltage level and larger capacity, and the equipment weight exceeds the rated load of conventional transport equipment. Due to the strict environmental requirements in the valve hall, it is restricted by various factors such as equipment transportation, transportation height, and layered connection of the converter transformer, thus a suitable test platform needs to be developed.

2. Principle and structure

The core equipment of this test platform is the test transformer, which is a movable self-lifting transformer, as shown in Figs. 32 and 33. The rated parameters are 600 kVA, 380 kV/0.35 kV, which can meet the induction partial discharge test of the ± 1100 kVHV converter transformer under the applied voltage on single terminal. The movable platform includes a platform chassis, wheels installed at the bottom of the platform chassis, hydraulic support legs installed on the side of the platform chassis that can prop up the platform chassis vertically, a main hydraulic cylinder, a bracket, and a base on the platform chassis. The transformer is placed on the movable

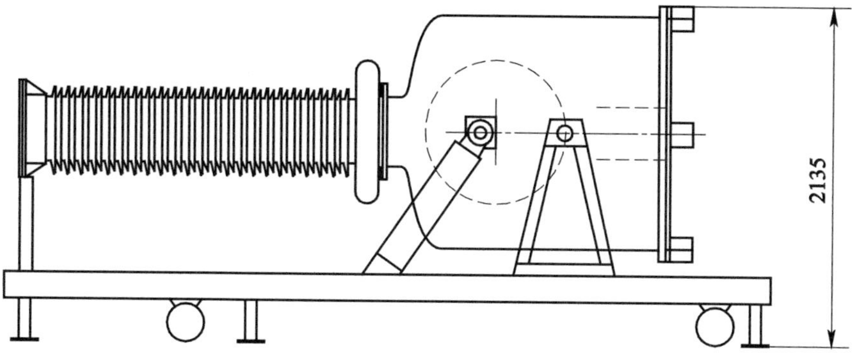

Fig. 32 Schematic diagram of the test transformer in transportation state

platform through the bracket and base, and the bracket and base can be pushed by the main hydraulic cylinder to complete a 90° flip and stand up.

3. Field application example

Using this set of devices, the field partial discharge tests of 28 converter transformer were carried out at the ± 1100 kV Guquan converter station, and the test site is shown in Fig. 34.

4. Research results

The long-duration induction partial discharge with applied voltage on single terminal for the converter transformer is carried out with the application of this test device, which has greatly improved the test preparation efficiency in the construction of converter stations, achieving the research and development goals of this device.

(III) Research on the Standardization Measurement Method of DC Transfer Switch Oscillation Characteristics based on Multiple Parameters

1. Research background

The DC transfer switch needs to interrupt a large DC current during the transfer process. The oscillation circuit is used to form an oscillating current during the current interruption process, the oscillating current is superimposed on the original DC current to have a zero-crossing point, so that the breaking device can reliably open. The oscillation circuit is composed of a capacitor and inductor series circuit connected in parallel between the break points. The resistance, capacitance, inductance, and oscillation frequency parameters of this oscillation circuit have an important impact on the oscillation characteristics, directly determining the reliability of the DC transfer switch breaking. Therefore, it is necessary to measure the above parameters to ensure that they meet the relevant requirements.

At present, the test method is generally to use a DC test power source to charge the capacitors in the oscillation circuit until the test voltage. Then, the test auxiliary switch QD is disconnected, and then the breaking device QB in the DC transfer

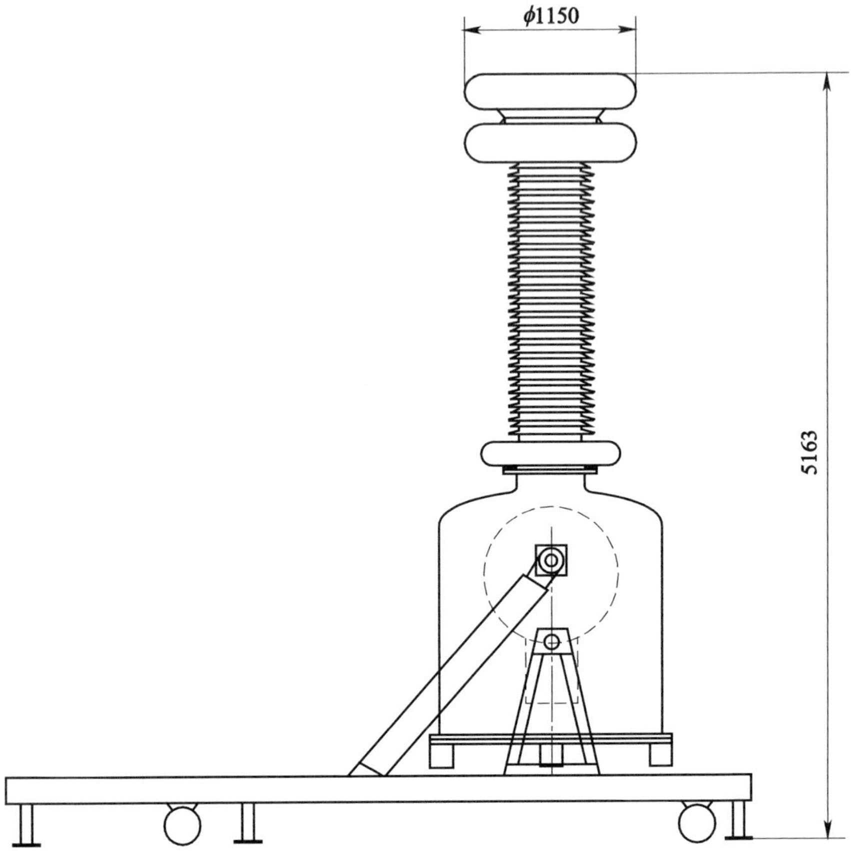

Fig. 33 Schematic diagram of the test transformer in operational state

switch is closed. The current sensor and data acquisition system are used to record the oscillation current waveform of the circuit. Finally, a capacitance measurement device is used to measure the capacitance value of the capacitor bank.

The above measurement method has the following shortcomings:

(1) The oscillation parameters, such as circuit equivalent capacitance, circuit equivalent inductance, circuit damping resistance, and decay time constant, are all calculated based on the collected oscillation current waveform, and the data source is single and cannot be mutually confirmed. Due to the zero-drift phenomenon in the output waveform of the current sensor, which shows that the current sensor will also output a certain signal value when no current passes through, the calculation relying solely on the oscillation current waveform collected by the current sensor has a relatively large error.

(2) The premise for calculating the circuit equivalent inductance and circuit damping resistance value is that the circuit equivalent capacitance must be

Fig. 34 On site partial discharge test for converter transformer with the applied voltage on single terminal

known. But during field test, a capacitance tester is generally used to measure the capacitance value of the capacitor bank. The measured result is the sum of the capacitance of the capacitor group and the parallel capacitance at the breaking point of the breaking device. Compared with the real circuit equivalent capacitance, there is a certain error, which will cause a large error in the subsequent calculation results.

2. Principle, structure, and performance

The test team proposed a DC transfer switch oscillation parameter measurement method based on multiple parameters, which solved the technical problems, such as single source and low accuracy of existing measurement data. The method effectively guided the field DC transfer switch oscillation characteristics measurement work.

The schematic diagram of the DC transfer switch oscillation characteristic measurement is shown in Fig. 35. During the test, while measuring the oscillation current waveform, the oscillation waveform of the voltage on the capacitor is measured and recorded synchronously.

The typical voltage waveform of the DC transfer switch oscillation characteristics is shown in Fig. 36.

According to the recorded voltage waveform, the n and the (n + m) moments on the same peak direction are selected with the time values T_{vp1} and T_{vp2} and amplitude values v_{p1} and v_{p2}. The above measurement values is substituted into formulas (1) and

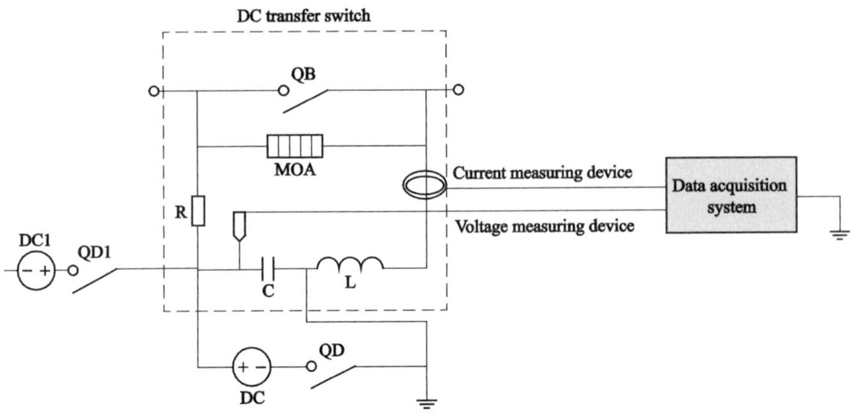

Fig. 35 Schematic diagram of the DC transfer switch oscillation characteristic measurement. QB—Breaking device; MOA—Arrester; C—Circuit capacitance; L—Circuit equivalent inductance; R—Circuit damping resistance; DC—DC test power source; QD—Test auxiliary switch

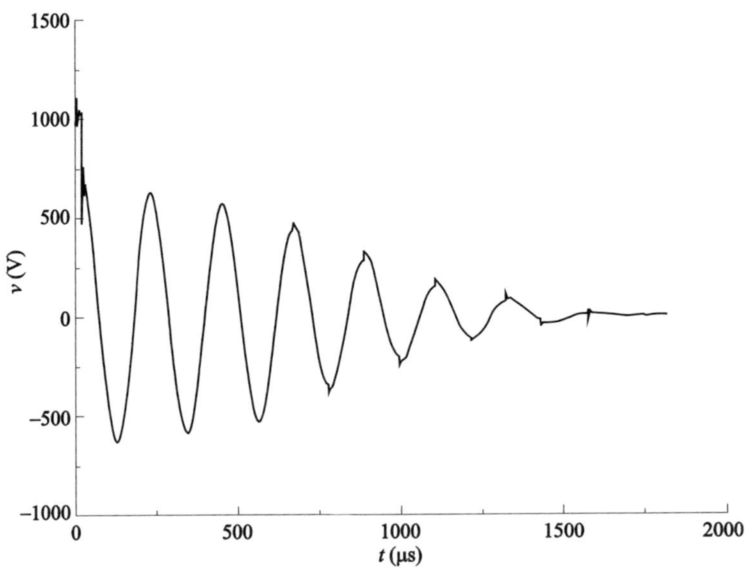

Fig. 36 Typical voltage waveform of DC transfer switch oscillation characteristics

(2), the oscillation period T, oscillation frequency f, the oscillation circuit inductance value L and of the oscillation circuit resistance value R of the DC transfer switch can be calculated respectively.

$$\tau = \frac{T_{\text{vp2}} - T_{\text{vp1}}}{\ln \frac{v_{\text{p1}}}{v_{\text{p2}}}} \tag{1}$$

$$T = \frac{T_{\mathrm{vp2}} - T_{\mathrm{vp1}}}{m} \tag{2}$$

where

- T_{vp1} the time of the *n*th peak value of the oscillation voltage.
- T_{vp2} the time of the $(n + m)$th peak value of the oscillation voltage on the same direction.
- v_{p1} the amplitude of the *n*th peak value of the oscillation voltage;
- v_{p2} the amplitude of the $(n + m)$th peak value of the oscillation voltage on the same direction.

In field measurements, because the capacitance of the capacitor bank is measured using a capacitance measurement device, the measured result is the sum of the capacitance of the capacitor group and the parallel capacitance of the breaking device, which has a certain error relative to the real circuit capacitance, causing a certain impact on the oscillation characteristics calculation. The test team proposes an optimized oscillation characteristic calculation method: The oscillation parameters of the circuit are calculated by measuring the charging voltage U_0.

Take any peak value i_{p1} and the corresponding moment T_1 into formula (3).

$$i_{\mathrm{p1}} = \frac{U_0}{2\pi f L} \mathrm{e}^{-\alpha T_1} \tag{3}$$

By substituting f and τ, L can be solved.

$$L = \frac{(T_2 - T_1) U_0}{2\pi i_{\mathrm{p1}}} \mathrm{e}^{-\frac{\ln(i_{\mathrm{p1}} - i_{\mathrm{p2}})}{T_2 - T_1} T_1} \tag{4}$$

Based on the above calculation method, we can further solve for the circuit damping resistance R and the circuit equivalent capacitance C.

This optimized oscillation characteristic calculation method is entirely based on the actual oscillation current and voltage waveform, which eliminates the error introduced by calculation with using the capacitance value of the capacitor bank instead of the circuit capacitance.

3. Field application examples

    (1) Field application at ± 1100 kV Changji and Guquan converter station. This test method was used to carry out the DC transfer switch oscillation characteristic field test at the ± 1100 kV Changji and Guquan converter station, as shown in Fig. 37. The special tests for 8 DC transfer switches are completed. Based on the multi-parameter DC transfer switch oscillation characteristic measurement method, we can accurately grasp the state information of the DC transfer switch circuit, including oscillation frequency and period, equivalent capacitance, equivalent inductance and damping resistance of the circuit, decay time constant, etc. The results are mutually confirmed and the

Fig. 37 Special field test for DC transfer switch

measurement errors are reduced. This allows for timely and comprehensive analysis and diagnosis of early defects in DC transfer switches, accurate evaluation and judgment of the equipment status and transfer capability of DC transfer switches, ensuring the safe and stable operation of the UHV DC project.

(2) The standard Q/GDW 11871—2018 "Guide for measurement of oscillation characteristics of DC switch" was formed. In the past, the standard in the field of DC transfer switch oscillation parameter measurement is blank.

(IV) Research on the Method to Locate the Temporary Grounding Points of Transmission Lines based on Loop DC Resistance

1. Research background

In order to accurately locate the temporary grounding point of the overhead transmission line, the DC resistance of the transmission line is measured, and the circuit structure of the transmission line is changed to obtain multiple loop DC resistance measurement equations, and then the location of the temporary grounding point of the line and other unknown parameters are solved. Since the design value of the line parameters and the measured value of the loop DC resistance both have errors, this will significantly reduce the positioning accuracy of the grounding point. The error theory is used to analyze the error of the loop DC resistance positioning method, and then the method is improved to propose an optimized method for high-precision temporary grounding point positioning of the line.

This method avoids factors such as power frequency parameters and mutual inductance coupling of line, and is not affected by the value of grounding resistance of temporary grounding point. The improved loop DC resistance method can also solve the line DC resistance as an unknown parameter, eliminating the error caused by line parameters. This method is simple and feasible with the characteristics of anti interference caused by power frequency and high positioning accuracy.

2. The basic principle of the positioning method based on the loop DC resistance

To locate the temporary grounding point, the DC resistance of the open circuit and short circuit are measured respectively. This allows for the acquisition of two DC resistance location equations, which can be used to solve for the two unknowns: the line DC resistance and the grounding resistance of the grounding point. The position of the temporary grounding point can be deduced from the DC resistance per unit length of the line. The basic principle of the DC resistance method is analyzed below, using the example of a transmission line with only one temporary grounding point on a single pole. Based on error analysis theory, the error level of this method and its main influencing factors are assessed, and the improved methods are proposed.

When the end of the transmission line is not grounded, a DC power source is connected at the beginning of the line, and grounded through the grounding network at the beginning of the line. The circuit DC resistance between the line and the ground is measured, as shown in Fig. 38. This resistance includes the grounding resistance of the grounding network at the beginning, the DC resistance of the line at the beginning, and the grounding resistance at the grounding point. The location equation is as follows:

$$r_{o} = r_{hg} + r_{x} + r_{twr} \tag{5}$$

where

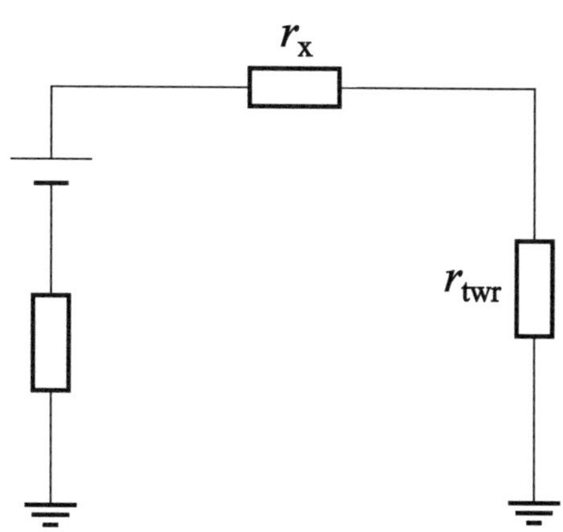

Fig. 38 Open circuit and ungrounded measurement loop at the end of the line

r_o is the measurement value of the circuit DC resistance when the end of line is open.
r_{hg} is the grounding resistance of the grounding network at the beginning of line.
r_x is the DC resistance of the line at the beginning;
r_{twr} is the grounding resistance at the grounding point (mainly includes the resistance of the tower material at the grounding point and the grounding resistance of the grounding device).

When the end of the transmission line is grounded, the circuit DC resistance between the line and the ground is measured at the beginning of the line, as shown in Fig. 39. This DC resistance includes the resistance of the grounding network at the beginning, the DC resistance of the line at the beginning, the grounding resistance at the grounding point, and the DC resistance of the line at the end (including the grounding resistance of the grounding network at the end). The location equation is as follows:

$$r_s = r_{hg} + r_x + \frac{r_{twr}(r_y + r_{eg})}{r_{twr} + r_y + r_{eg}} \tag{6}$$

其中

$$r_y = r_l - r_x$$

where

r_s is the measurement value of the circuit DC resistance when the end of line is grounded;
r_y is the DC resistance of the line at the end;
r_l is the DC resistance of the entire line.

By combining the open circuit location Eq. (5) and the short circuit location Eq. (6), the DC resistance location equations can be obtained.

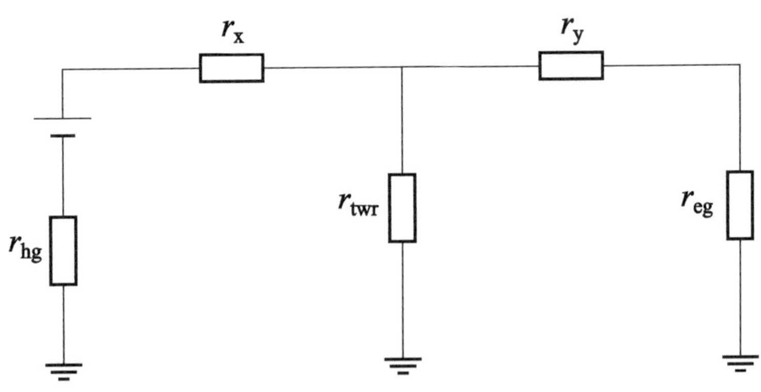

Fig. 39 Short circuit and grounded measurement loop at the end of the line

$$\begin{cases} r_o = r_{hg} + r_x + r_{twr} \\ r_s = r_{hg} + r_x + \frac{r_{twr}(r_y + r_{eg})}{r_{twr} + r_y + r_{eg}} \end{cases} \tag{7}$$

The grounding resistances of the grounding network at the beginning r_{hg} and end r_{eg}, and the DC resistance of the entire line r_l are all designed values and known parameters. The DC resistance at the beginning of the line r_x, and the grounding resistance at the grounding point r_{twr} are unknown parameters, which can be obtained by calculation.

$$\begin{cases} r_x = r_s' - \sqrt{(r_s' - r_{lg})(r_s' - r_o')} \\ r_{twr} = r_o - r_{hg} - r_x \end{cases} \tag{8}$$

where $r_{lg} = r_l + r_{eg}$, $r_o' = r_o - r_{hg}$, $r_s' = r_s - r_{hg}$.

According to the design value of the DC resistance per unit length of the line rrpl, the distance from the grounding point to the beginning of line x can be obtained, thus the position of the temporary grounding point on the line can be determined.

3. Optimization and accuracy improvement of the loop dc resistance positioning method

To eliminate the design error of line parameters, the number of equations can be increased, and the DC resistance of the entire line can be solved as an unknown parameter, thereby the positioning errors caused by line parameter deviations can be avoided. In order to reduce the impact of measurement errors on positioning accuracy, the loop DC resistance parameters of the line are measured at both the beginning and end of line. The main ideas are as follows.

(1) Measurement at the beginning of line: At the beginning of line, measure the DC resistance with open-circuit at the end, the DC resistance of the short-circuit and grounding at the end, and the DC resistance of the constant resistance in series at the end.
(2) Measurement at both the beginning and end of line: At the beginning of line, measure the DC resistance with open-circuit at the end, the DC resistance with short-circuit at the end. At the end of line, measure the DC resistance with open-circuit at the beginning.

4. Improved method 1—the constant resistance is connected at the end of line in series

When the end of the line is open, it is equivalent to an infinite resistance connected at the end. When the end is short-circuited to the ground, it is equivalent to a zero resistance connected at the end. Similarly, an appropriate constant resistance can be connected in series at the end, which can increase the number of positioning equations to solve the DC resistance parameters of the line as unknowns. This method can avoid

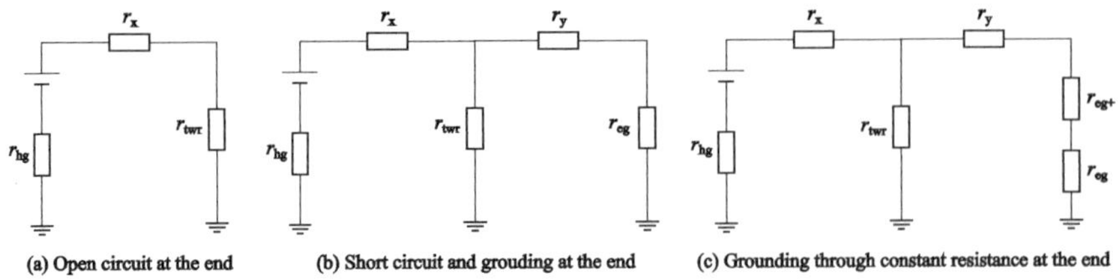

Fig. 40 Schematic diagram of improved loop DC resistance positioning method (1)

errors introduced by the DC resistance parameters of the line. Its circuit structure is shown in Fig. 40.

Its positioning equation is

$$\begin{cases} r_o = r_{hg} + r_x + r_{twr} \\ r_{s1} = r_{hg} + r_x + \dfrac{r_{twr}(r_y + r_{eg})}{r_{twr} + r_y + r_{eg}} \\ r_{s2} = r_{hg} + r_x + \dfrac{r_{twr}(r_y + r_{eg+} + r_{eg})}{r_{twr} + r_y + r_{eg+} + r_{eg}} \end{cases} \tag{9}$$

where

r_{eg+} The constant resistance connected in series at the end of line.

The solution is

$$\begin{cases} r_x = r_o - r_{hg} - r_{twr} \\ r_{twr} = \sqrt{\dfrac{r_{eg+} r'_{s2} r'_{s1}}{r'_{s1} - r'_{s2}}} \\ r_y = \dfrac{r_{twr}^2}{r'_{s1}} - r_{twr} - r_{eg} \end{cases} \tag{10}$$

where $r'_{s1} = r_o - r_{s1}$, $r'_{s2} = r_o - r_{s2}$, $r'_{eg} = r_{eg+} + r_{eg}$.

5. Improved method 2—measurement is conducted at double end of line

Due to the factors such as different environmental temperatures where the line is located, the measurement value error has a certain difference with measuring the DC resistance of the line at the beginning of line, which will cause a significant reduction in positioning accuracy. In order to reduce the impact of measurement errors, the DC resistance of the line is firstly measured at the beginning of line according to the circuit shown in Fig. 40a. The DC resistance of the line is then measured at the end of line according to the circuit shown in Fig. 41a. Finally, the end of line is short-circuited according to Fig. 41b, the DC resistance of the line is measured at the beginning of line. By increasing the redundancy of the equations, the impact of measurement errors is reduced.

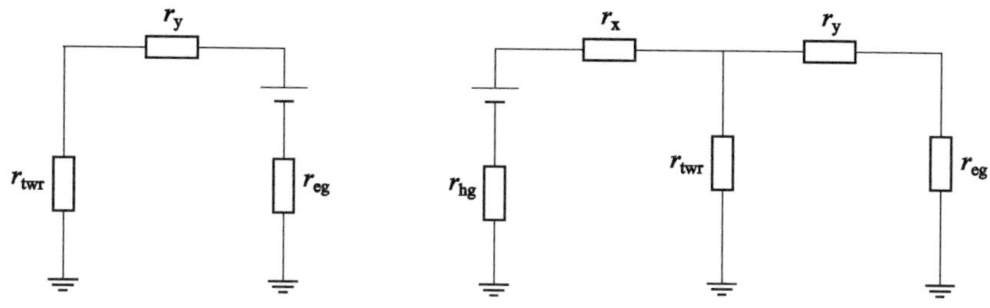

(a) Open circuit at the beginning and measurement at the end

(b) Short circuit at the end and measurement at the beginning

Fig. 41 Schematic diagram of improved loop DC resistance positioning method (2)

Its positioning equation is

$$\begin{cases} r_{o1} = r_{hg} + r_x + r_{twr} \\ r_{o2} = r_g + r_y + r_{twr} \\ r_{s1} = r_{hg} + r_x + \dfrac{r_{twr}(r_y + r_g)}{r_{twr} + r_y + r_g} \end{cases} \tag{11}$$

where

r_{o1} DC resistance measured at the beginning of line with the open-circuit at the end of line;

r_{o2} DC resistance measured at the end of line with the open-circuit at the beginning of line;

R_{s1} DC resistance measured at the beginning of line with the short-circuit at the end of line;

The solution is

$$\begin{cases} r_{twr} = \sqrt{r_{o2}(r_{o1} - r_{s1})} \\ r_x = r_{o1} - r_{twr} - r_{hg} \\ r_y = r_{o2} - r_{twr} - r_g \end{cases} \tag{12}$$

6. Research results

(1) The positioning accuracy of the improved method 2 is the highest, and it has a good effect to reduce the random errors of DC resistance measurement values and measurement errors caused by environmental temperature.
(2) When the relative errors of the DC resistance measurement values are equal, the improved method 1 has higher positioning accuracy.
(3) According to the factors and characteristics of positioning errors, the application scope of the above methods is shown as follows:

1) Loop DC resistance positioning method—improved method 1 is suitable for shorter lines with less interference, and can be used to supplement and verification of improved method 1.
2) Loop DC resistance positioning method—improved method 2 is suitable for lines that have measurement conditions at both ends of line. This method has the highest positioning accuracy and can be used preferentially.

V. Main Research Results

Based on the review of the main parameters of ± 1100 kV UHV equipments and handover test plans, this chapter further analyzes the test equipment requirements for each equipment handover test and the test capabilities of each companies. Some key equipments that cannot yet meet test requirements and some key tests that lack experiences were found. Thus the following research results have been formed in the standardization of test methods and the development of test equipments:

(1) In the long-duration induced voltage and partial discharge test of HV converter transformers, the single side applied voltage on single terminal was adopted. A self-lifting test transformer suitable for the single side applied voltage method of 1100 kV converter transformers was developed for this test method and successfully applied in all tests.
(2) A UHV GIS integrated insulation test platform was developed, adopting an integrated structure of resonant reactors and voltage dividers, combined with a style of self-standing and lifting mechanical assembly, achieving rapid test preparation and free hoisting of the platform. The external insulation electric field distribution of the platform and the temperature rise of the resonant reactor were optimized based on a multi-physics field coupling simulation model, and the special test of 1100 kV GIS and the AC applied voltage and partial discharge test on the valve side of HV converter transformer were successfully completed.
(3) In the oscillation parameter measurement test of DC transfer switch, the measurement method based on multiple parameters was proposed, to solve the technical problem of single source of existing measurement data and low accuracy. The standard Q/GDW 11871—2018 "Guide for measurement of oscillation characteristics of DC switch" was formed, and the standard can effectively guide the oscillation characteristics measurement work of DC transfer switch.
(4) In the transmission line parameter measurement method, an improved positioning method of line temporary grounding point based on loop DC resistance was proposed. For different conditions of transmission line, two improved methods were proposed respectively, effectively improving the positioning accuracy, and providing troubleshooting methods for line parameter measurement tests.

3 Subsystem Commissioning

Subsystem commissioning is the energizing or startup test of all independent subsystems in the converter station. Its purpose is to check whether a single device can operate safely after being connected to the system with other components as a subsystem, and whether its performance meets the requirements of the contract and engineering technical specifications.

I. Test Scope and Items

The subsystem commissioning of the converter station is divided into eight parts: converter valve subsystem commissioning, converter transformer subsystem commissioning, AC yard subsystem commissioning, AC filter field subsystem commissioning, DC yard subsystem commissioning, station power subsystem commissioning, auxiliary system and other subsystem commissioning, control and protection equipment subsystem commissioning. Subsystem commissioning is carried out separately according to these eight parts, covering all subsystems in the converter station, mainly including AC yard equipments, AC filter field equipments, converter transformers, converter valves, DC voltage dividers, current transformers, DC control and protection systems, remote communication systems, metering systems, fault recording systems, protection information management substations, AC station power systems, station DC power supplies, valve cooling systems, uninterruptible power supplies (UPS), air conditioning systems, ventilation systems, fire detection and fire fighting systems, closed-circuit television and infrared surveillance systems, etc.

II. AC Yard Subsystem Commissioning

(I) AC Yard Subsystem Test

(1) Test purpose. The content of AC yard subsystem commissioning in the UHV converter station is explained. The scope of AC yard subsystem test includes the circuit breaker, disconnector, and grounding switch of the AC yard. Through the subsystem test, it can be detailedly checked whether the interface between the above equipments and thc secondary system and the interlocking function between the switch equipments are correct.

(2) Test contents:

1) Switch quantity output signal combination test.
2) Local/remote trip and close operation.
3) Synchronization function verification.
4) Trip and drive test.

(3) Test result record.

(II) AC Disconnector Subsystem Test

(1) Test purpose. The AC disconnector subsystem test is a combination test of AC disconnector and the control protection system in the converter

station. Its purpose is to verify that the interface function of the disconnector and the secondary system is normal, and to check whether its performance meets the requirements of the contract, relevant standards and specifications.

(2) Test content:
   1) Switch quantity output signal combination test.
   2) Local/remote trip and close operation.

(3) Test record.

(III) AC Grounding Switch Subsystem Test

(1) Test purpose. The AC grounding switch subsystem test is a combination test of the AC grounding switch and the control protection system in the converter station. Its purpose is to verify that the interface function of the grounding switch and the secondary system is normal, and to check whether its performance meets the requirements of the contract, relevant standards and specifications.

(2) Test content
   1) Switch quantity output signal combination test.
   2) Local/remote trip and close operation.
   (3) Test record.

(IV) Interlocking Function Inspection

(1) Test purpose. The AC switch field interlocking function inspection is to check whether the logic interlocking function is correct and complete, to prevent accidents caused by operator errors.

(2) Test content. For each circuit breaker and disconnector, according to the relevant design documents, the closing operations are conducted when the interlocking conditions are met or not met. And the potential on the two terminals allowed for operation in the control box is measured according to the signal table to verify whether the interlocking function is correct.

(3) Test record.

(V) Engineering Application

According to the proposed AC grounding switch subsystem test plan, the opening and closing performance and interlocking function of AC yard circuit breakers, disconnectors, grounding switches and other electrical equipment have been verified to be correct.

III. Station Transformer Commissioning and 66 kV/35 kV Subsystem Commissioning

(I) Station Transformer Subsystem Test

(1) Test purpose. The station transformer subsystem test is a combination test of the station transformer and the control protection system. Its purpose is to verify that the interface function of the station transformer and the

secondary system is normal, and to check whether its performance meets the requirements of the contract, relevant standards and specifications.

(2) Test content:
   1) Switch quantity output signal combination test.
   2) Analog output signal combination test.
   3) Station transformer tap position indication.
   4) Station transformer tap control.
   5) Station transformer cooler switching operation.
   (3) Test record.

(II) 66 kV/35 kV AC Circuit Breaker Subsystem Test

(1) Test purpose. The 66 kV/35 kV AC circuit breaker subsystem test is a combination test of the AC circuit breaker and the control protection system in the converter station. Its purpose is to verify that the interface function of the circuit breaker and the secondary system is normal, and to check whether its performance meets the requirements of the contract, relevant standards and specifications.
(2) Test content
   1) Switch quantity output signal combination test.
   2) Local/remote trip and close operation.
   3) Synchronization function verification.
   4) Trip and drive test.
(3) Test record.

(III) 66 kV/35 kV AC Disconnector Subsystem Test

(1) Test purpose. The 66 kV/35 kV AC disconnector subsystem test is a combination test of the AC disconnector and the control protection system in the converter station. Its purpose is to verify that the interface function of the disconnector and the secondary system is normal, and to check whether its performance meets the requirements of the contract, relevant standards and specifications.
(2) Test content
   1) Switch quantity output signal combination test.
   2) Local/remote trip and close operation.
(3) Test record.

(IV) 66 kV/35 kV AC Grounding Switch Subsystem Test

(1) Test purpose. The 66 kV/35 kV AC grounding switch subsystem test is a combination test of the AC grounding switch and the control protection system in the converter station. Its purpose is to verify that the interface function of the grounding switch and the secondary system is normal, and

to check whether its performance meets the requirements of the contract, relevant standards and specifications.

(2) Test content

1) Switch quantity output signal combination test.

2) Local/remote trip and close operation.

(3) Test record.

(V) Interlocking Function Verification

(1) Test purpose. For each circuit breaker and disconnector, according to the relevant design documents, the closing operations are conducted when the interlocking conditions are met or not met. And the potential on the two terminals allowed for operation in the control box is measured according to the signal table, to verify whether the interlocking function is correct.

(2) Test content. For each circuit breaker and disconnector, according to the relevant design documents, the closing operations are conducted when the interlocking conditions are met or not met. And the potential on the two terminals allowed for operation in the control box is measured according to the signal table, to verify whether the interlocking function is correct.

(3) Test record.

(VI) Engineering Application

According to the commissioning plan of the station transformer and the 66 kV/35 kV subsystem, the performance of the station transformer and the 66 kV/35 kV AC yard equipments were checked, and the switching and interlocking function of equipment were verified.

IV. AC Filter Field Subsystem Commissioning

(I) AC Filter Field Disconnector Subsystem Commissioning

(1) Test purpose. The AC filter field disconnector subsystem test is a combination test of the AC filter field disconnector and the control protection system in the converter station. Its purpose is to verify that the interface function of the disconnector and the secondary system is normal, and to check whether its performance meets the requirements of the contract, relevant standards and specifications.

(2) Test content

1) Switch quantity output signal combination test.

2) Local/remote trip and close operation.

(3) Test record.

(II) AC Filter Field Grounding Switch Subsystem Test

(1) Test purpose. The AC filter field grounding switch subsystem test is a combination test of the AC filter field grounding switch and the control

protection system in the converter station. Its purpose is to verify that the interface function of the disconnector and the secondary system is normal, and to check whether its performance meets the requirements of the contract, relevant standards and specifications.

(2) Test content
  1) Switch quantity output signal combination test.
  2) Local/remote trip and close operation.

(3) Test record.

(III) Interlocking Function Test

(1) Test purpose. The AC filter field interlocking function test is to inspect whether the logic interlocking function is correct and complete, to prevent accidents caused by operator errors.

(2) Test content. For each circuit breaker and disconnector, according to the relevant design documents, the closing operations are conducted when the interlocking conditions are met or not met. And the potential on the two terminals allowed for operation in the control box is measured according to the signal table, to verify whether the interlocking function is correct.

(3) Test record.

(IV) AC Filter Bank Tuning Test

(1) Test purpose:
  1) Adjust the capacitance values of the 4 arms of the capacitor so that their unbalanced current is within the allowable range;
  2) Adjust the AC filter impedance (the device is adjustable) to make the filtering points of each phase meet the design requirements;
  3) Measure and plot the amplitude-frequency curve and phase-frequency characteristics of filter on each phase and reactive power compensation capacitor.

(2) Test content
  1) Capacitor imbalance adjustment.
  2) Filter tuning.

(3) Test record.

(V) Phase Selection Closing Device Subsystem Test of AC Filter Inlet Switch

(1) Test purpose. The phase selection closing device subsystem test of the AC filter inlet switch is a combination test of the phase selection closing device of AC filter inlet switch and the control protection system. Its purpose is to verify whether the phase selection closing function of AC

filter inlet switch meets the contract, relevant standards and specifications, and whether its interface function with the secondary system is normal.

(2) Test content
   1) Switch quantity output signal combination test.
   2) Local/remote trip and close operation.
(3) Test record.

(VI) Middle Switch Control Function Subsystem Commissioning

(1) Test purpose. The middle switch control function subsystem test is a test of the DC system bipolar operation without electricity, mainly to verify whether the middle switch interlocking control is normal.
(2) Test content. Tests such as the converter transformer and AC line in series, middle switch interlocking control; converter transformer and AC filter in series, middle switch interlocking control; AC filter and AC line in series, middle switch interlocking control, etc.
(3) Test record.

(VII) Engineering Application

The AC filter field subsystem commissioning scheme was proposed, which verified the AC filter field switch equipment, AC filter, and middle switch interlocking performance, ensuring that these devices operate normally before electrification.

V. Converter Transformer Subsystem Commissioning
(I) Test Purpose

The converter transformer subsystem commissioning is a combination test of the converter transformer and the control protection system. Its purpose is to verify that the interface function of the converter transformer and the secondary system is normal, and to check whether its performance meets the contract, relevant standards and specifications.

(II) Test Content

(1) Switch quantity output signal combination test.
(2) Analog output signal combination test.
(3) Tap position indication.
(4) Tap control.
(5) Cooler switching operation.
(6) Remote transmission of converter transformer electronic control (TEC).

(III) Engineering Application

According to the proposed commissioning scheme of the converter transformer subsystem, the interface function of the converter transformer and the secondary system is verified to be normal, ensuring that its performance meets the requirements of the contract, relevant standards and specifications.

VI. DC Yard and Valve Hall Subsystem Commissioning

(I) Converter Valve Subsystem Test

(1) Test purpose. The converter valve subsystem test is a combination test of the converter valve and the control protection system. Its purpose is to verify that the interface function of the converter valve and the secondary system is normal, and to check whether its performance meets the requirements of the contract, relevant standards and specifications.
(2) Test content
  1) Switch quantity output signal combination test.
  2) Input signal combination test.
  3) Converter valve leakage test.
  4) Signal check of arrester action time.
(3) Test record.

(II) DC Circuit Breaker Subsystem Test

(1) Test purpose. The DC circuit breaker subsystem test is a combination test of the DC circuit breaker and the control protection system at the converter station. Its purpose is to verify that the interface function of the circuit breaker and the secondary system is normal, and to check whether its performance meets the requirements of the contract, relevant standards and specifications.
(2) Test content
  1) Switch quantity output signal combination test.
  2) Local/remote trip and close operation.

(III) DC Disconnector Subsystem Test

(1) Test purpose. The DC disconnector subsystem test is a combination test of the DC disconnector and the control protection system at the converter station. Its purpose is to verify that the interface function of the disconnector and the secondary system is normal, and to check whether its performance meets the requirements of the contract, relevant standards and specifications.
(2) Test content
  1) Switch quantity output signal combination test.
  2) Local/remote trip and close operation.
  3) Interlocking function.

(IV) DC Grounding Switch Subsystem Test

(1) Test purpose. The DC grounding switch subsystem test is a combination test of the DC grounding switch and the control protection system at the converter station. Its purpose is to verify that the interface function of the DC grounding switch and the secondary system is normal, and to check

whether its performance meets the requirements of the contract, relevant standards and specifications.

(2) Test content
   1) Switch quantity output signal combination test.
   2) Local/remote trip and close operation.
   3) Interlocking function.

(3) Test record.

(V) Low Voltage Energization Test of Converter Valve

(1) Test purpose. Before the converter valve is charged at high voltage, the low voltage energization test of the converter transformer with valve must be completed first. The purpose is to check the correctness of the primary connection of the converter transformer, the correctness of the voltage indication of the end shield divider of the converter transformer, the correctness of the trigger synchronous voltage of the converter valve, the correctness of the converter firing control voltage, and to check the phase sequence of the primary voltage and the trigger sequence of the valve group.

(2) Test content. Generally, this test directly connects an AC power source of about 400 V to the grid side of the converter transformer to meet the voltage requirements for the thyristor valve conduction. In some cases, an autotransformer is also needed to make the voltage supplied to the converter firing control (CFC) meet its requirements.

TextBreak="Yes"It should be noted that the phase sequence to the CFC must be correct. The secondary of the autotransformer and the test power source must correctly correspond to the primary phase sequence of the converter transformer, which must be carefully checked when the temporary connections are established.

The on-load tap changer of the converter transformer is adjusted to output the maximum valve-side voltage U_{dio}.

(VI) DC Filter Bank Tuning Test

(1) Test purpose
   1) Adjust the capacitance values of the 4 arms of the capacitor so that their unbalanced current is within the allowable range;
   2) Adjust the filter impedance to make the filtering points of each phase meet the design requirements;
   3) Measure and plot the amplitude-frequency and phase-frequency characteristics curve of filter on each phase.

(2) Test Content
   1) Capacitor imbalance adjustment.
   2) Filter tuning.

(VII) Engineering Application

The DC yard and valve hall subsystem commissioning plan was proposed, verifying that the interface function between the primary and secondary equipments are normal, such as converter valve, DC yard switch equipments, DC filters, etc. The low voltage energization test of the converter valve was completed, checking the correctness of the trigger synchronous voltage of the converter valve, the correctness of the CFC voltage, and checking the phase sequence of the primary voltage and the trigger sequence of the valve.

VII. Signal detection of DC line fault location system

(I) Test Purpose

The signal detection of DC line fault location system is a test to check the working status of the fault location device. Its purpose is to ensure that the DC line fault location system device functions normally and to check whether its performance meets the requirements of the contract, relevant standards and specifications.

(II) Test Content

(1) Apply voltage and current to the device with simulating line faults.
(2) Calculate the impedance value of the fault and compare it with the distance measurement of device.
(3) Close the switch of device power supply, OWS message is normal, set value does not change.
(4) Check that the remote communication of device is normal.

(III) Test Results and Test Records

The DC line fault location system device functions are verified and the test data is recorded.

VIII. Inspection of Grounding Electrode Impedance Monitoring Device

(I) Test Purpose

The grounding electrode impedance monitoring device inspection is a test conducted on the converter station's grounding electrode line condition monitoring device. Its purpose is to verify whether the device functions normally and complies with the contract requirements, as well as relevant standards and specifications.

(II) Test Content

(1) Check the correct connection of the secondary circuit for the grounding electrode line impedance.
(2) Check the monitoring function of grounding electrode line impedance.

(III) On-site Testing and Recording

The grounding electrode impedance monitoring device was verified to be functioning normally, and the monitoring data was recorded.

IX. AC Protection Subsystem Commissioning

(I) AC Bus Protection Subsystem Test

(1) Test purpose. The AC bus protection subsystem commissioning is a combination test of the converter station circuit breaker and AC bus protection. Its purpose is to verify that the interface function of the circuit breaker and protection is normal, and to check whether its performance meets the requirements of the contract, relevant standards and specifications.
(2) Test content:
    1) Switch quantity input signal verification.
    2) Switch quantity output signal verification.
    3) Trip and drive test.
(3) Test record.

(II) Optical Fiber Differential Protection Subsystem Test

(1) Test purpose. The optical fiber differential protection subsystem test is a combination test of the converter station circuit breaker and line protection. Its purpose is to verify that the interface function of the circuit breaker and protection is normal, and to check whether its performance meets the requirements of the contract, relevant standards and specifications.
(2) Test content
    1) Switch quantity input signal verification.
    2) Switch quantity output signal verification.
    3) Trip and drive test.

(III) Circuit Breaker Protection Subsystem Test

(1) Test purpose. The circuit breaker protection subsystem test is a combination test of the converter station circuit breakers and other protections. Its purpose is to verify that the interface function of the circuit breaker and protection is normal, and to check whether its performance meets the requirements of the contract, relevant standards and specifications.
(2) Test content
    1) Switch quantity input signal verification.
    2) Switch quantity output signal verification.
    3) Trip and drive test.

(IV) Station Transformer Protection Subsystem Test

(1) Test purpose. The station transformer protection subsystem test is a combination test of the converter station circuit breaker and station transformer protection. Its purpose is to verify that the interface function of the circuit breaker and protection is normal, and to check whether its performance meets the requirements of the contract, relevant standards and specifications.
(2) Test content
    1) Switch quantity input signal verification.
    2) Switch quantity output signal verification.
    3) Trip and drive test.

(3) Test record.

(V) AC Filter Protection Subsystem Test

(1) Test purpose. The AC filter protection subsystem test is a combination test of the converter station circuit breaker and AC filter protection. Its purpose is to verify that the interface function of the circuit breaker and protection is normal, and to check whether its performance meets the requirements of the contract, relevant standards and specifications.
(2) Test content
1) Switch quantity input signal verification.
2) Switch quantity output signal verification.
3) Trip and drive test.

(VI) Converter Transformer Protection Subsystem Test

(1) Test purpose. The rectifier transformer protection subsystem test is a combination test of the converter station circuit breaker and converter transformer protection. Its purpose is to verify that the interface function of the circuit breaker and protection is normal, and to check whether its performance meets the requirements of the contract, relevant standards and specifications.
(2) Test content
1) Switch quantity input signal verification.
2) Switch quantity output signal verification.
3) Trip and drive test.

(VII) Engineering Application

According to the AC protection subsystem commissioning scheme, the interface function of the AC yard circuit breaker and protection was verified to be normal.

X. DC Control and Protection Subsystem Commissioning

(I) DC Pole Control and Protection Subsystem Test

(1) Test purpose. The DC pole control and protection subsystem test is a combination test of the converter station circuit breaker and DC pole control and protection. Its purpose is to verify that the interface function of the circuit breaker and protection is normal, and to check whether its performance meets the requirements of the contract, relevant standards and specifications.
(2) Test content
1) Switch quantity input signal verification.
2) Switch quantity output signal verification.
3) Whole group drive test.
(3) Test record.

(II) DC Filter Protection Subsystem Test

(1) Test purpose. The DC filter protection subsystem test is a combination test of the fast disconnector and DC filter protection at the converter station.

Its purpose is to verify that the interface function of the fast disconnector and protection is normal, and to check whether its performance meets the requirements of the contract, relevant standards and specifications.

(2) Test content
   1) Switch quantity input signal verification.
   2) Switch quantity output signal verification.
   3) Transmission test.

(3) Test record.

(III) Sequential Control Test

(1) Test purpose. The sequential control test is to check the automatic sequential process of the pole control system. Its purpose is to verify that the sequential control function is normal, and to check whether its performance meets the requirements of the contract, relevant standards and specifications.

(2) Test content
   1) Pole connection/isolation.
   2) Pole isolation and grounding/pole without grounding.
   3) Connection/isolation of DC filter.
   4) Charging/power off.
   5) Coordination of metallic /ground return line.

(IV) Engineering Application

According to the DC control and protection subsystem commissioning scheme, the interface function of the DC yard circuit breaker and protection was verified to be normal.

XI. Remote Control Subsystem Commissioning

(1) Test purpose.

Remote control subsystem commissioning is a communication combination test between the converter station DC control protection system and the national dispatch, network dispatch, and provincial dispatch centers. Its purpose is to verify that the national dispatch center can correctly display and control the status of the converter station equipment. Additionally, it ensures that the relevant network dispatch and provincial dispatch centers can also accurately display the status of the equipment.

(2) Test content
   1) Switch quantity verification.
   2) Analog quantity verification.

(3) Test results. According to the remote control subsystem commissioning plan, it was verified that the national dispatch center can correctly display and control the status of the converter station equipment. The relevant network

and provincial dispatch centers can also correctly display the status of the equipment.

XII. Protection Information Management Subsystem Commissioning

(1) Test Purpose. The protection information management subsystem commissioning is a communication combination test of the converter station DC control protection system, AC protection and protection information substation. Its purpose is to verify that the converter station's AC and DC protection information is correctly accessed by the protection substation, and that the substation accurately forwards the AC protection information to the DC control protection system.

(2) Test content

1) T For each AC and DC protection signal, perform the signal combination test individually, and check and confirm that the protection information management substation and the operator workstation display the status correctly.

2) T For each AC and DC protection's value setting and recovery, check and confirm that the protection information management substation operates correctly.

(3) Engineering application. According to the proposed protection information management subsystem commissioning plan, it was verified that the converter station's AC and DC protection information is correctly accessed by the protection substation, and that the substation accurately forwards the AC protection information to the DC control protection system.

XIII. Other Subsystem Commissioning

(I) Fault Recording System Test

(1) Test purpose. The fault recording system test is a combination test of the converter station DC control protection system and the fault recording system. Its purpose is to verify that the fault recording system can correctly display the status of switch quantity and analog quantity, and that it can start recording correctly.

(2) Test content

1) Switch quantity verification.

2) Analog quantity verification.

(3) Test record.

(II) Station Master Clock Subsystem Test

(1) Test purpose. The station master clock subsystem test is a combination test of the station master clock and the secondary systems of the converter station. Its purpose is to verify that the secondary systems of the converter station can correctly receive the timing signal from the station master clock.

(2) Test content. For each screen cabinet, ensure it receives the station master clock timing signal. Check and confirm that the time display on each cabinet is correct.

(III) Engineering Application

According to the commissioning plan, it was verified that the fault recording system can correctly display the status of switch quantities and analog values, and that it can start recording accurately. Additionally, the converter station's secondary systems were confirmed to correctly receive the timing signal from the station master clock.

XIV. Auxiliary System Commissioning

(I) Station AC Power Supply System Test

(1) Test purpose. The station AC power supply system test is a combination test of the station AC power supply system and the control protection system. Its purpose is to verify that the interface function of the station AC power supply system and the secondary system is normal, and to check whether its performance meets the requirements of the contract, relevant standards and specifications.
(2) Test content
 1) Switch quantity input signal verification.
 2) Switch quantity output signal combination test.
 3) Analog quantity output signal combination test.
 4) Automatic bus transfer function check.

(II) DC Auxiliary Power Supply Subsystem Test

(1) Test purpose. The DC auxiliary power supply system test is a combination test of the DC auxiliary power supply system and the control protection system, the purpose of which is to verify that the interface function of the DC auxiliary power supply system and the secondary system is normal, and to check whether its performance meets the requirements of the contract, relevant standards and specifications.
(2) Test content
 1) Switch output signal combination test.
 2) Analog output signal combination test.
(3) Test records.

(III) Uninterruptible Power Supply (UPS) Subsystem Test

(1) Test purpose. The UPS subsystem test is a combination test of the UPS system and the control protection system. Its purpose is to verify that the interface functions of the UPS system and the secondary system are normal, and to check whether their performance meets the requirements of the contract, relevant standards and specifications.

(2) Test content
    1) Switch output signal combination test.
    2) Analog output signal combination test.
(3) Test records.

(IV) Valve Cooling Control Subsystem Test

(1) Test purpose. The valve cooling control subsystem test is a combination test of the converter valve cooling control system and the control protection system. Its purpose is to verify that the interface functions of the valve cooling control system and the secondary system are normal, and to check whether their performance meets the requirements of the contract, relevant standards and specifications.
(2) Test content
    1) Internal cooling system commissioning. Check the cooling system before power transmission. Manually start the electromechanical equipment. The valve cooling system will automatically start to simulate the test.
    2) External cooling system commissioning. Manually start and stop each equipment unit, and automatically start and stop each equipment unit. Additionally, initiate and cease operation of the electric heater.
(3) Test records.

(V) Fire-Fighting Subsystem Test

(1) Test purpose. The fire-fighting subsystem test is a combination test of the converter station fire-fighting system and the control protection system. Its purpose is to verify that the interface functions of the fire-fighting system and the secondary system are normal, and to check whether their performance meets the requirements of the contract, relevant standards and specifications.
(2) Test content
    1) Switch output signal combination test.
    2) Local/remote start and stop pump operation.
    3) Emergency start-up test.
(3) Test records.

(VI) Air Conditioning Subsystem Test

(1) Test purpose. The air conditioning subsystem test is a combination test of the converter station air conditioning system and the control protection system. Its purpose is to verify that the interface functions of the air conditioning system and the secondary system are normal, and to check whether their performance meets the requirements of the contract, relevant standards and specifications.

(2) Test content: Switch output signal combination test.

(VII) Fire Detection System Test

(1) Test purpose. The fire detection system test is a combination test of the converter station fire detection system and the control protection system. Its purpose is to verify that the interface functions of the fire detection system and the secondary system are normal, and to check whether their performance meets the requirements of the contract, relevant standards and specifications.
(2) Test content: Switch output signal combination test.
(3) Test records.

(VIII) Integrated Auxiliary Monitoring Subsystem Test

(1) Test purpose. The integrated auxiliary monitoring subsystem test is a combination test of the safety monitoring system and the control protection system. Its purpose is to verify that the interface functions of the integrated auxiliary monitoring system and the secondary system are normal, and to check whether their performance meets the requirements of the contract, relevant standards and specifications.
(2) Test content
    1) When test conditions can be met, the safety monitoring system actually sends a signal. When test conditions cannot be met, a simulated signal is generated, which mimics the action of shorting the two ends of the junction point.
    2) Observe the signal event list on the operator's workstation for any such signal event. If a signal event appears, the test passes, and you should proceed to the next test. If no signal event is present, check the connection, identify and correct the reason, then repeat the aforementioned steps.
(3) Test record.

(IX) Emergency Control System Test

(1) Test purpose:
    1) Check if the connections for the emergency control system and the associated protection and measurement devices meet the specified requirements.
    2) Check whether the action of the emergency control system meets the requirements.
    3) Check all relevant analog quantities and digital quantities.
(2) Test content
    1) Check the switch quantity provided by the DC pole control system to the emergency control system.
    2) Check the analog quantity from the emergency control system to the DC pole control system.

3) Check the switch quantity from the emergency control system to the DC pole control system.

(3) Test record.

(X) Integrated Online Monitoring Subsystem Test

(1) Test purpose. The integrated online monitoring subsystem test is a combination test of the converter station security monitoring system and the control protection system. Its purpose is to verify that the interface function of the integrated online monitoring system and the secondary system is normal, and to check whether its performance meets the requirements of the contract, relevant standards and specifications.

(2) Test content

1) When test conditions can be met, the monitoring system actually sends a signal. When test conditions cannot be met, a simulated signal is generated, which mimics the action of shorting the two ends of the junction point.

2) Check whether the monitoring status on the integrated online monitoring list matches the actual value on site. If it is consistent, the test passes, and proceed to the next test. If not, check the connection to find the reason, correct it, and then repeat the aforementioned step.

(3) Test record.

(XI) Engineering Application

According to the commissioning plan for the auxiliary subsystems, the interface functions between each auxiliary system and the secondary system have been verified to be normal.

XV. Injection and Energization Test

(I) 1000 kV/750 kV/500 kV AC TA Primary Injection Test

(1) Test purpose

1) Check the ratio of the TA without performing accuracy calibration, and confirm that the installation position of the TA is correct.

2) Check the connectivity of the secondary circuit of the TA to ensure they are not open.

3) Verify that the grounding of the secondary winding of the TA and associated protection and measurement devices complies with the specified requirements.

4) Verify that the installation polarity of the TA adheres to the system's requirements.

5) Check all relevant analog quantities and digital quantities.

(2) Test content

1) Measure each group of TA circuits. Avoid opening the circuit. Verify that each group of TA has a single-point grounding and that there is good insulation between the TA and the ground.

2) Adopt the test wiring using 2 or 3 in serials.

3) Start the equipment and power on TA.
4) Verify that the ratio and polarity of the TA are correct.
5) Halt any ongoing outputs and perform the necessary switching operations to prepare for the upcoming series of TA tests.

(3) Test records.

(II) 1000 kV/750 kV/500 kV AC TV Secondary Voltage Test

(1) Test purpose
1) Verify that the grounding of the secondary winding of the TV and associated protection and measurement devices complies with the specified requirements.
2) Verify that the installation polarity of the TV adheres to the system's requirements.
3) Check all relevant analog quantities and digital quantities.

(2) Test content
1) Measure each group of TV circuits. Avoid opening and shorting the circuit. Verify that each group of TV has a single-point grounding and that there is good insulation between the TV and the ground.
2) Utilize a relay protection tester to apply a range of voltage values to the secondary circuit at the TV terminal box.
3) The secondary voltage range encompasses all voltage circuits, including all secondary devices within the system.
4) Check all protection devices, measurement devices, and system monitors that share a coil with the specified protection device, and confirm that the connections of the secondary circuits are correct.

(3) Test records.

(III) AC filter field TA Primary Current Injection Test

(1) Test purpose
1) Check the ratio of the TA without performing accuracy calibration, and confirm that the installation position of the TA is correct.
2) Check the connectivity of the secondary circuit of the TA to ensure they are not open.
3) Verify that the grounding of the secondary winding of the TA and associated protection and measurement devices complies with the specified requirements.
4) Verify that the installation polarity of the TA adheres to the system's requirements.
5) Check all relevant analog quantities and digital quantities.

(2) Test content
1) Check whether the primary circuit of the TA is reliably and firmly grounded. Verify that each group of TA has a single-point grounding and that there is good insulation between the TA and the ground.
2) Adjust the primary output current while continuously monitoring it to maintain the current at the predetermined value.

3) Verify the correctness of the TA's transformation ratio and check the connectivity of the secondary current circuit.

(3) Test records.

(IV) Station Transformer AC TA Primary Side Current Injection Test

(1) Test purpose
1) Check the ratio of the TA without performing accuracy calibration, and confirm that the installation position of the TA is correct.
2) Check the connectivity of the secondary circuit of the TA to ensure they are not open.
3) Verify that the grounding of the secondary winding of the TA and associated protection and measurement devices complies with the specified requirements.
4) Verify that the installation polarity of the TA adheres to the system's requirements.
5) Check the transformer primary connection for reliability and correctness.
6) Check all relevant analog quantities and digital quantities.

(2) Test content
1) Measure each group of TA circuits. Avoid opening the circuit. Verify that each group of TA has a single-point grounding and that there is good insulation between the TA and the ground.
2) Implement a three-phase power supply method on the medium voltage side of the circuit breaker.
3) Start the equipment and power on TA.
4) Halt any ongoing outputs and perform the necessary switching operations to prepare for the upcoming series of TA tests.

(3) Test records.

(V) Converter Transformer TA Primary Side Injection Test

(1) Test Purpose
1) Check the ratio of the TA without performing accuracy calibration, and confirm that the installation position of the TA is correct.
2) Check the connectivity of the secondary circuit of the TA to ensure they are not open.
3) Verify that the grounding of the secondary winding of the TA and associated protection and measurement devices complies with the specified requirements.
4) Verify that the installation polarity of the TA adheres to the system's requirements.
5) Check all relevant analog quantities and digital quantities.

(2) Test Content
1) Disconnect the leads on both the network and valve sides of the converter transformer.

2) Measure each group of TA circuits. Avoid opening the circuit. Verify that each group of TA has a single-point grounding and that there is good insulation between the TA and the ground.
3) For each injection, consider separately energizing the three star-connected and delta-connected converter transformers.
4) Start the equipment and power on TA.
5) Verify that the ratio and polarity of the TA are correct.
6) Halt any ongoing outputs, and then remove the test wiring to restore the original connection of the converter transformer.

(3) Test Record.

(VI) DC Yard and Valve Hall DC Current Measurement Device Primary Side Injection Test

(1) Test Purpose
1) Check the ratio of the TA without performing accuracy calibration, and confirm that the installation position of the TA is correct.
2) Check the connectivity of the secondary circuit of the TA to ensure they are not open.
3) Verify that the installation polarity of the TA adheres to the system's requirements.
4) Check all relevant analog quantities and digital quantities.

(2) Test Content
1) Confirm that the secondary circuit of the DC TA is correctly connected.
2) Check whether the primary circuit of the TA is reliably and firmly grounded.
3) Connect the output end of the large current generator to the positive and negative terminals of the transformer using the cable.
4) Monitor and adjust the primary output current to maintain it at the predetermined value during the test, as the current is large.
5) If no current output or only a very small current is detected during the injection process, stop the injection immediately. Then, shut down the instrument to investigate the cause.
6) Verify the correctness of the TA's transformation ratio.

(3) Test Record.

(VII) Primary Side Energization Test of DC Voltage Divider in DC Yard

(1) Test Purpose
1) Check the transformation ratio of the DC voltage divider, and confirm that the installation position, the wiring of the secondary circuit, and the overall setup meet the design requirements.
2) Verify that the secondary protection measurement device of the voltage divider can correctly collect voltage, and then ensure that the grounding of the measurement device meets the specified requirements.

(2) Test Content
  1) Verify that the secondary circuit of the DC voltage divider is correctly connected.
  2) Ensure that the primary lead of the energized DC voltage divider is safely disconnected from all other equipment.
  3) Prepare the plant test report, incorporating all pertinent data for reference purposes.
  4) Apply DC voltage to the high voltage end of the DC voltage divider.
  5) Compare the applied voltage with the voltage values on the protection cabinet and measuring device, and the applied voltage should be consistent with these values.

(3) Test records.

(VIII) Engineering application

As per the commissioning plan, the interface functions of both the current and voltage transformer secondary circuits with the control and protection system have been confirmed to be normal.

XVI. Research Achievements

By analyzing the technical content and technical specifications of the ± 1100 kV converter station subsystem commissioning, we compiled a subsystem commissioning plan. This process yielded the following research achievements:

(1) The commissioning content of the AC yard subsystem in the UHV converter station has been elaborated. It includes tests for the AC yard circuit breaker, disconnector, and grounding switch. Through subsystem testing, detailed checks can be performed to ensure that the interfaces between the equipment and the secondary system, as well as the interlocking functions of the switchgear, are correct.
(2) The commissioning content for the station transformer and the 66 kV/35 kV area subsystem in the UHV converter station has been elaborated. It includes tests for the station transformer, 66 kV/35 kV AC yard circuit breaker, disconnector, and grounding switch. Through subsystem testing, detailed checks can be performed to ensure that the interfaces between the equipment and the secondary system, as well as the interlocking functions of the switchgear, are correct.
(3) The commissioning content of the AC filter field subsystem in the UHV converter station has been elaborated. It includes tests for the AC filter field circuit breaker, disconnector, grounding switch, phase selection closing device, and AC filter banks. Through subsystem testing, detailed checks can be performed to ensure that the interfaces between the equipment and the secondary system are correct, the interlocking functions between the switchgear are functioning properly, the middle switch interlocking function is normal, and the tuning performance of each filter meets the design requirements.

(4) The commissioning content of the converter transformer subsystem in the UHV converter station has been elaborated. Through subsystem testing, detailed checks can be performed to ensure that the interface between the converter transformer and the secondary system is correct, assess the functionality of the tap changer and cooler controls, and validate the accuracy of the information sent by the TEC device to the integrated online monitoring backend.

(5) The commissioning content of the DC yard and valve hall subsystem in the UHV converter station has been elaborated. It includes tests for the converter valve, DC circuit breaker, DC disconnector, DC grounding switch, converter valve low voltage energization, DC filter bank tuning, DC line fault location system, grounding electrode impedance monitoring device. Through subsystem testing, detailed checks can be performed to ensure that the interface between the equipment and the secondary system, as well as the interlocking function of the switchgear, are correct. Through the converter valve low voltage energization test, we can effectively verify the correctness of the measurement signal primary wiring and the triggering sequence related to the converter transformer and valve. The DC filter tuning test allows us to check and correct the tuning performance of the DC filter. The DC line fault location system test and the grounding electrode impedance detection device test aim to verify the correct functioning of these equipment components.

(6) The commissioning content of the AC protection subsystem in the UHV converter station has been elaborated. It includes tests for the AC bus protection, fiber optic differential protection, circuit breaker protection, station transformer protection, AC filter protection, converter transformer protection. Through subsystem testing, we can verify the correctness of the interfaces between each device and the protection device.

(7) The commissioning content of the DC control protection subsystem in the UHV converter station has been elaborated. It includes tests for the DC pole control protection, DC filter protection, and sequential control test. Through subsystem testing, we will verify the correctness of the control functions of the switchgear related to protection and confirm that the sequential control functions are operating normally.

(8) The commissioning content of the protection information management substation in the UHV converter station has been elaborated. Through subsystem testing, we can verify whether the AC/DC protection information from the converter station can be correctly accessed by the protection substation, and whether the protection substation can accurately forward the AC protection information to the DC control protection system.

(9) The commissioning content of the fault recording system and the station master clock subsystem in the UHV converter station has been elaborated. Through subsystem testing, we can verify whether the fault recording system functions normally to display the status of switch quantities and analog quantities, whether the recording can start correctly, and whether the secondary system of the converter station can receive the station master clock's timing signals correctly.

(10) The commissioning content of the auxiliary system subsystem in the UHV converter station has been elaborated. It includes tests for the station AC power supply system, DC auxiliary power supply system, UPS system, valve cooling control system, fire system, air conditioning system, fire detection system, integrated auxiliary monitoring system, emergency control system, integrated online monitoring system. These devices and subsystems, while not affecting the operation of the DC system, come in many types and involve complex specialties. Through subsystem testing, we can verify in detail whether the interfaces of these devices with the secondary system are correct.

(11) The commissioning content of the current injection and energization test in the UHV converter station has been elaborated. This special subsystem test requires the use of an AC large current generator with an output of at least 300 A and a DC voltage of at least 10 kV. Through this test, the wiring, ratio, polarity, and grounding of the measuring device can be effectively inspected.

(12) According to the compiled subsystem commissioning plan, the subsystem commissioning at Changji and Guquan converter stations was successfully completed, establishing a solid foundation for the subsequent station and system commissioning processes. Throughout the entire testing phase, no critical issues were encountered; specifically, there were no instances of reversed polarity in TA, uncontrollable circuit breakers or disconnectors, incorrect telemetry displays, or misoperation of protection triggering. The subsystem commissioning items are complete, the test results have been verified as accurate.

4 Station Auxiliary System Commissioning

The purpose of the station auxiliary system commissioning for the UHV DC transmission project is to verify the functionality of each converter station according to the contract and engineering technical specifications. This commissioning is conducted separately at both converter stations in preparation for the end-to-end system commissioning.

I. Station System Commissioning Items and Contents

Before commencing the station system commissioning, the energization test of the 750 kV/500 kV/1000 kV AC yard (including the AC bus) at the converter station must be completed. The station system commissioning includes the following items: ① AC bus and AC yard energization test; ② AC filter energization test; ③ Sequential operation test; ④ Trip test; ⑤ Converter transformer charging test; ⑥ Anti-interference test; ⑦ DC open line test.

Station system commissioning is carried out separately at both converter stations, and each test item's content includes test purpose, test conditions, test steps, test content, and acceptance technical specifications.

The AC bus charging test is conducted at both stations, with each bus being charged for 2 h. During each bus charging, the converter station's AC protection is calibrated. Following the completion of the charging, the AC bus undergoes a 24-h trial operation.

The AC filter charging test is conducted for each filter sub-bank, with each sub-bank being charged for 2 h until all filter sub-banks charging tests are completed. During each charging period, the filter protection check is completed.

For the CG Project, there are four types of single converter metallic/ground return connection modes, two types of monopole metallic /ground return connection modes, four types of bipolar single converter connection modes, four types of bipolar unbalanced converter connection methods, and one type of bipolar double converter connection method. In addition, there is an open line test and a de-icing connection mode.

The sequential operation test must verify the various connection modes and check the interlocking relationships between the circuit breakers and disconnectors. Compared to a $\pm$ 800 kV DC project's station system commissioning tests, the content of the sequential operation test is fundamentally the same.

According to the hierarchical structure of DC control and protection and the principle of the three-out-of-two design for protection, trip tests must be conducted separately on the converter control, pole control, and bipolar control protection devices. It is necessary to simulate faults on two protection devices at the same level to verify the trip function and the subsequent operation sequence of the system after a trip occurs.

The UHV DC monopole is configured with two converters in series, and the bipolar is configured with four converters in series. According to the design and technical specifications, each converter transformer must undergo a charging test. This process continues until the charging tests for all converter transformers are completed.

There are 16 types of DC line open line test (OLT) connection modes. Through optimization research, it has been determined that the OLT can be conducted at one converter station with the DC line connected, and at the other converter station with the DC line disconnected. This optimization reduces the total number of distinct OLT modes from 16 to 8. Consequently, the DC voltage of the single converter OLT increases to 550 kV, and the DC voltage of the monopole OLT increases to 1100 kV.

The anti-interference test is conducted on each cabinet within the DC secondary system to ensure the electromagnetic interference resistance performance of the cabinets during operation.

II. Changji Converter Station System Commissioning Scheme

(I) DC Sequential Operation Test

(1) Test purpose. This test confirms the correct execution of operation sequences and electrical interlocks, ensures the DC equipment remains safe in the event of an incomplete operation sequence due to a fault, and

validates the operation and execution of each step under both manual and automatic control modes.

(2) Test content.

1) Monopole double converter connection mode:
   a. Pole 1 double converter ground return connection mode;
   b. Pole 2 double converter ground return connection mode;
   c. Pole 1 double converter metallic return connection mode;
   d. Pole 2 double converter metallic return connection mode.
2) Monopole single converter connection mode:
   a. Pole 1 LV converter ground return connection mode;
   b. Pole 2 LV converter ground return connection mode;
   c. Pole 1 HV converter ground return connection mode;
   d. Pole 2 HV converter ground return connection mode;
   e. Pole 1 LV converter metallic return connection mode;
   f. Pole 2 LV converter metallic return connection mode;
   g. Pole 1 HV converter metallic return connection mode;
   h. Pole 2 HV converter metallic return connection mode.
3) Monopole OLT connection mode:
   a. Pole 1 HV converter with DC line disconnected;
   b. Pole 2 HV converter with DC line disconnected;
   c. Pole 1 LV converter with DC line disconnected;
   d. Pole 2 LV converter with DC line disconnected;
   e. Pole 1 HV converter with DC line connected;
   f. Pole 2 HV converter with DC line connected;
   g. Pole 1 LV converter with DC line connectede;
   h. Pole 2 LV converter with DC line connected;
   i. Pole 1 with DC line disconnected;
   j. Pole 2 with DC line disconnected;
   k. Pole 1 with DC line connected;
   l. Pole 2 with DC line connected.
4) For the bipolar single converter connection mode, each pole features a 12-pulse converter connection structure. Each converter station offers four distinct connection and operation modes:
   a. Pole 1 LV converter in operation, Pole 2 LV converter in operation;
   b. Pole 1 LV converter in operation, Pole 2 HV converter in operation;
   c. Pole 1 HV converter in operation, Pole 2 LV converter in operation;
   d. Pole 1 HV converter in operation, Pole 2 HV converter in operation.
5) In a bipolar system with an unbalanced 3/4 converter configuration, one pole features a double converter connection structure, while the other pole features a single converter connection structure. This setup allows for four types of connection and operation modes:

a. Pole 1 double converter in operation, Pole 2 LV converter in operation;
b. Pole 1 double converter in operation, Pole 2 HV converter in operation;
c. Pole 2 double converter in operation, Pole 1 LV converter in operation;
d. Pole 2 double converter in operation, Pole 1 HV converter in operation.

6) In the bipolar double converter (4 converter) connection mode, there is only one operation mode for the bipolar system with double converters.
7) In the bipolar connection mode, monopole is operated as OLT connection mode:
a. Pole 1 in operation, Pole 2 with DC line disconnected;
b. Pole 1 in operation, Pole 2 with DC line connected;
c. Pole 2 in operation, Pole 1 with DC lined is connected;
d. Pole 2 in operation, Pole 1 with DC line connected.

(3) Record the test results.

(II) Last Trip Test

(1) Test purpose. The last trip test should be conducted either before the converter station is energized or after any modifications have been made to the control protection circuit, including software. All test items in the last trip test must be executed at least once. Select one or more protections to initiate the protective tripping to ensure that all tripping actions, including the actions of the protection tripping matrix circuit, are correct.
(2) Test content. Choose 1–2 of the following for testing.
1) The operator manually presses the emergency stop button in the control room.
2) Mechanical relay protection;
3) Control and protection system, including software protection;
4) Converter transformer (gas relay or pressure release relay);
5) Smoothing reactor (gas relay or pressure release relay);
6) Fire-fighting system.
(3) Record the test results.

(III) Changji Converter Station 750 kV AC Yard Equipment Charging Test

(1) Test purpose
1) Check the insulation status of the 750 kV switch and related primary equipment at Changji Converter Station, and also verify the operation of the arrester.

2) Check the phase sequence and the secondary circuit of the voltage transformer for the related bus at the 750 kV side of the Changji converter station.

(2) Test content

1) The 750 kV AC yard switch at Changji converter station is energized in sequence.
2) Check the effective value and phase angle of the secondary voltage for the 750 kV bus voltage transformer at Changji converter station.
3) Monitor the operation of the arrester.
4) Check the insulation status of related primary equipment.

(3) Record the test results.

(IV) Changji Converter Station 750 kV AC Filter Group and Capacitor Bank Charging Test

(1) Test purpose

1) Examine the switching inrush current and overvoltage of the AC filter and parallel capacitors banks to ensure they are within the allowable range.
2) Examine the circuit breaker's ability to cut off the AC filter and parallel capacitor banks. Additionally, check whether it can smoothly interrupt the capacitive current and prevent reignition.
3) Examine the function of the on-time switching device that controls the circuit breaker. Verify that it can effectively manage the activation of reactive power equipment at the correct time. Specifically, check if the circuit breaker can select the closing moment when the voltage between its contacts is zero to avoid closing all three phases simultaneously.
4) Inspect the secondary circuit of the protection related to the AC filter and high-voltage parallel capacitor bank.

(2) Test content

1) For each of the 4 banks of AC filters and parallel capacitors at Changji converter station, switch them on and off, and repeat this process 3 times.
2) First bank of AC filters and capacitor group charging test.
3) Second bank of AC filters and capacitor group charging test.
4) Third bank of AC filters and capacitor group charging test.
5) Fourth bank of AC filters and capacitor group charging test.
6) Verify the protection for the AC filter, the AC filter bus, and the 750 kV switch.

(3) Record the test results.

(V) Changji Converter Station Substation Transformer Impulse Closing Test

(1) Test purpose
   1) Verify whether the connection of voltage transformer on the high and low voltage sides of the No.1 and No.2 station transformer is correct.
   2) Test the ability of the No.1 and No.2 station transformers to withstand the closing voltage.
   3) Assess the ability of the 750 kV switch to engage and disengage the No.1 and No.2 station transformers under no-load conditions.
   4) Assess the ability of the protection devices for the No.1 and No.2 station transformers to withstand the inrush current during closure under no-load conditions.

(2) Test content
   1) The 7515 switch at Changji converter station is to engage and disengage each of the No.1 and No.2 station transformers under no-load conditions, repeating the process 5 times for each transformer.
   2) When the 7515 switch at Changji converter station is charging, 750 kV busbar I is operating normally, while busbar II serves as the test busbar.
   3) The charging protection for switches 7512 and 7513 at Changji converter station have been set and put into operation.

(3) Record the test results.

(VI) Changji Converter Station Transformer 10 kV Side Reactor Switching Test

(1) Test purpose
   1) Verify whether the connection of voltage transformers on the low side of the No.1 and No.2 station transformer is correct.
   2) Assess the ability of the 10 kV switch to engage and disengage inductive loads.
   3) Measure if the overvoltage generated by the switch when engaging and disengaging inductive loads meets the required standards.
   4) Measure the harmonics before and after the reactor is activated, and assess their impact on the voltage levels on each side.
   5) Check the correctness of the secondary circuits on both the 750 kV side and the 10 kV side of the differential protection for the No. 1 and No. 2 station transformers. Additionally, verify the correctness of the secondary circuit for the 10 kV side reactor protection and confirm the polarity of the busbar protection.

(2) Test content
   1) Switches 1021 through 1026 and 1011 through 1016 at the Changji converter station each engage and disengage the reactor three times.
   2) Verify the polarity of the differential protection for the No.1 and No.2 station transformers.
   3) Check the polarity of the 750 kV busbar differential protection (7515, 7514 switches).

4) Check the polarity of the 10 kV side No.12 busbar differential protection.

(3) Record the test results.

(VII) Changjji Converter Station Harmonic Measurement

(1) Test purpose. Measure the changes and distribution of system harmonics both before and after the energization of the station transformer, as well as before and after the operation of a single AC filter.
(2) Test content. Measure the system background harmonics, and then assess the changes in system harmonics under the following conditions: before and after the station transformer is energized, and before and after a single AC filter is activated.
(3) Record the test results.

(VIII) Changji Converter Station Switch Interlocking Test

(1) Test purpose. Verify if the switches in the strings with 3/2 connection at AC yard have the capability to automatically trip the middle breaker once both side circuit breakers have been opened.
(2) Test content. Conduct switch interlocking tests for the following strings: the 3rd, 4th, and 5th strings that are shared by the converter transformer and the AC line; the 1st string that is shared by the converter transformer and the AC filter; and the 2nd, 6th, and 8th strings that are shared by the AC filter and the AC line.
(3) Record the test results.

(IX) Changji Converter Station Converter Transformer and Blocking Valve Group Charging Test

(1) Test Purpose. Verify that the inrush current during the switching of the converter transformer is within the specified limit, consistent with previous design research results, and that the resonance is adequately suppressed by the damping characteristics. Additionally, confirm that the triggering phase of the thyristor component is correct.
(2) Test Content
   1) Perform a charging test on the converter transformer and blocking valve group at the Changji converter station.
   2) Both the HV and LV converter transformers for Poles 1 and 2 are energized, with each transformer undergoing the energization process five times.
(3) Record the test results.

(X) DC Line Open Line Test at Changji Converter Station

(1) Test Purpose. Check the DC voltage control function, the relationship between the converter valve firing angle and DC voltage, the voltage tolerance of the valve, and the voltage tolerance of both the DC filter

and the DC transmission line. Additionally, verify the correctness of the open line test sequence control and the protection system.

(2) Test Content. The UHV DC transmission project includes 16 distinct connection modes for conducting open circuit tests:
   1) 4 types for single converter with DC line disconnected, 4 types for single converter with DC line connected.
   2) 2 types for single pole with DC line disconnected, 2 types for single pole with DC line connected.
   3) 2 types for one pole operation and the other pole with DC line disconnected, 2 types for one pole operation and the other pole with DC line connected.

(3) Record the test results.

XI Anti-interference Test at Changji Converter Station

(1) Test Purpose. Verify whether the AC/DC protection and control equipment will misoperate when switching the bus and using the walkie-talkie and mobile phone.
(2) Test Content. Conduct an anti-interference test on secondary equipment, including VBE and the protection cabinet, pole control and its protection cabinet, station control systems, relay houses, and the monitoring system, etc.
(3) Record the test results.

(XII) Research Results

The commissioning plan for the Changji converter station system has been compiled, including tests such as the AC bus and AC yard energization, AC filter energization, sequential operation, non-energized trip, converter transformer charging, anti-interference, and the DC open line test. These tests aim to verify the functionality of the converter station and provide technical support for on-site system commissioning of the project.

III. Guquan Converter Station System Commissioning Plan

(I) DC Sequential Operation Test

Same as Sect. 2, (I).

(II) LastTrip Test

Same as Sect. 2, (II).

(III) Charging Test of Wancheng I Line and 1000 kV 1st and 3rd Strings of AC Yard at Guquan Converter Station

(1) Test Purpose
   1) Complete the energized operation test of the 1000 kV Wancheng I line circuit breaker at Wannan converter station, and subsequently check the status of the line equipment.

2) Complete the charging test for Buses I and II of the 1000 kV AC yard at the Guquan converter station.
3) At Guquan converter station, complete the charging test for the 1st and 3rd string equipment in the 1000 kV AC yard, as well as the TV phase sequence checking test.
4) Conduct a polarity check on the 1000 kV Wancheng I line protection.
5) At Guquan converter station, complete the voltage circuit checks for the equipment protection, measurement, and metering equipment of both the 1st and 3rd strings in the 1000 kV AC yard.
6) The switch protection polarity check for the 1st and 3rd strings at the 1000 kV AC yard of the Guquan converter station has been completed. Additionally, the branch current polarity check has been conducted when the side switch is connected to the I and II busbar protections of the 1000 kV AC yard.
7) Perform a test to measure the background harmonics.

(2) Test content

1) Conduct a charging test on the 1061 switch at the Wannan converter station for the 1000 kV Wancheng I line.
2) Conduct a charging test for the 3rd string, as well as for the I bus, II bus, and 62 M bus at the 1000 kV Guquan converter station.
3) Conduct a charging test for the 1st string and 61 M bus at the 1000 kV Guquan converter station.
4) Conduct the 1000 kV Wancheng I line load test (I).
5) Conduct the 1000 kV Wancheng I line load test (II).
6) Convert the operation mode and recover the protection setting values for the 1000 kV Wancheng I line.

(3) Record test results.

(IV) 1000 kV Wancheng II line and 2nd string charging test at the AC Yard of Guquan Converter Station

(1) Test purpose

1) Complete the energized operation test of the 1000 kV Wancheng II line circuit breaker at Wannan converter station, and subsequently check the status of the line equipment.
2) At Guquan converter station, complete the charging test for the 2st string equipment in the 1000 kV AC yard, as well as the TV phase sequence checking test.
3) Conduct a polarity check on the 1000 kV Wancheng II line protection.
4) At Guquan converter station, complete the voltage circuit checks for the equipment protection, measurement, and metering equipment of the 2st string in the 1000 kV AC yard.
5) The switch protection polarity check for the 2st string at the 1000 kV AC yard of the Guquan converter station has been completed.

Additionally, the branch current polarity check has been conducted when the side switch is connected to the I and II busbar protections of the 1000 kV AC yard.

6) Perform a test to measure the background harmonics.

(2) Test content

1) Conduct a charging test on the 1053 switch at the Wannan converter station for the 1000 kV Wancheng II line.
2) Conduct a charging test for the 2nd string and 1000 kV Wancheng II line at the Guquan converter station.
3) Conduct the 1000 kV Wancheng II line load test (I).
4) Conduct the 1000 kV Wancheng II line load test (II).
5) Convert the operation mode and recover the protection setting values for the 1000 kV Wancheng II line.

(3) Record test results.

(V) 1000 kV AC Filter and Capacitor Bank Switching Test at the Guquan Converter Station

(1) Test purpose

1) Examine the switching inrush current and overvoltage of the AC filter and parallel capacitor bank to ensure they are within the allowable range.
2) Test the switch's ability to cut off capacitive current.
3) Measure the operating overvoltage during the activation of the AC filter and parallel capacitors bank.
4) Complete the load test for the AC filter and the reactive power compensation capacitor bank, then check the operational status of the high-voltage equipment, and finally, inspect for any abnormal heating in the connections.
5) Verify the correctness of current and voltage for the protection, control, and measurement devices associated with the AC filter and reactive power compensation capacitor bank. Additionally, ensure that the unbalanced current of the capacitors, including filter capacitors, remains within permissible limits.
6) Verify the correctness of the current and voltage of the filter bus differential protection.
7) Perform a test to measure the background harmonics.

(2) Test content

1) Conduct switching and energization tests on the AC filter in the 1st bank on bus 61.
2) Conduct switching and energization tests on the AC filter in the 2st bank on bus 62.
3) Record the test results.

(VI) 500 kV Chengsan I Line and Guquan Converter Station AC Yard 1st and 3rd Strings Charging Test

(1) Test Purpose
   1) Complete the energized operation test of the 500 kV Chengsan I line circuit breaker at Wusan converter station, and subsequently check the status of the line equipment.
   2) Complete the charging test for Buses I and II of the 500 kV AC yard at the Guquan converter station.
   3) Conduct an energized operation test on the 500 kV Chengsan II Line (Guquan side) circuit breaker at the Wusan converter station, and verify the status of the line equipment.
   4) At Guquan converter station, complete the charging test for the 8st and 9rd string equipment in the 500 kV AC yard, as well as the TV phase sequence checking test.
   5) Conduct a polarity check on the 500 kV Chengsan I line protection.
   6) At Guquan converter station, complete the voltage circuit checks for the equipment protection, measurement, and metering equipment of both the 8st and 9rd strings in the 500 kV AC yard.
   7) The switch protection polarity check for the 8st and 9rd strings at the 500 kV AC yard of the Guquan converter station has been completed. Additionally, the branch current polarity check has been conducted when the side switch is connected to the I and II busbar protections of the 500 kV AC yard.
   8) Perform a test to measure the background harmonics.

(2) Test content
   1) Conduct a charging test on the 5041 switch at the Wusan converter station for the 500 kV Chengsan I line.
   2) Conduct a charging test for the 500 kV 8rd string, as well as for the I bus and II bus at the Guquan converter station.
   3) Conduct a charging test for the 500 kV 9st string, as well as for Chengsan II and 65 M bus at the Guquan converter station.
   4) Conduct the 500 kV Chengsan I line load test (I).
   5) Conduct the 500 kV Chengsan I line load test (II).
   6) Convert the operation mode and recover the protection setting values for the 500 kV Chengsan I line.

(3) Record test results.

(VII) Guquan Converter Station AC Yard Charging and Load Test

(1) Test purpose
   1) At Guquan converter station, complete the charging of the 1st, 2nd, 5th, 6th, and 7th string equipment in the 500 kV AC yard, perform the TV phase sequence checking test, and conduct the voltage circuit inspection for the protection, measurement, and metering equipment in the 500 kV AC yard.
   2) At Guquan converter Station, complete the protection polarity check for the 1st, 2nd, 5th, 6th, and 7th string switches of the 500 kV AC yard, and perform the branch current polarity check when the side

switch is connected to the I and II busbar protections of the 500 kV AC yard.
3) Perform a test to measure the background harmonics.

(2) Test content
1) Conduct the 500 kV 1st string charging test at Guquan converter station.
2) Conduct the 500 kV 2th string charging test at Guquan converter station.
3) Conduct the 500 kV 5th string charging test at Guquan converter station.
4) Conduct the 500 kV 6th string charging test at Guquan converter station.
5) Conduct the 500 kV 7th string charging test at Guquan converter station.
6) Conduct the 500 kV 7th string load test at Guquan converter station.
7) Conduct the 500 kV 6th string load test at Guquan converter station.
8) Check the protection polarity for 500 kV 5th string at the Guquan converter station (load test).
9) Conduct the 500 kV 2th string load test at Guquan converter station.
10) Conduct the 500 kV 1st string load test at Guquan converter station.
11) Convert the operation mode.

(3) Record test results.

(VIII) The 500 kV Chengsan II Line Charging and Load Test

(1) Test purpose
1) Complete the energized operation test of the 500 kV Chengsan II line circuit breaker at Wusan converter station, and subsequently check the status of the line equipment.
2) Complete the phase sequence checking test for the TV of the Chengsan I and II transmission lines.
3) Complete the protection polarity check of the 500 kV Chengsan II line.
4) Perform a test to measure the background harmonics.

(2) Test content
1) Conduct a charging test on the 5053 switch at the Wusan converter station for the 500 kV Chengsan II line.
2) Conduct the 500 kV Chengsan II line load test (I).
3) Conduct the 500 kV Chengsan II line load test (II).
4) Convert the operation mode and recover the protection setting values for the 500 kV Chengsan II line.

(3) Record test results.

(IX) Guquan Converter Station 511B and 512B Station Transformers Switching Test

(1) Test purpose
   1) Check the withstand impulse switching performance of the 500 kV station transformer.
   2) Check the operating condition of the 500 kV station transformer.
   3) Check the impact of the inrush current of the 500 kV station transformer on the protection.
   4) Check the correctness of the voltage and current of the protection, measurement, and metering system of the 500 kV station transformer.
   5) Perform a secondary phase check on both the high and low voltage sides of the 500 kV station transformer, and verify the accuracy of the secondary voltage readings.
   6) When conditions allow, check the correctness of the 500 kV bus protection current connected into the 511B and 512B station transformer branch current; check the correctness of the current and voltage of the 500 kV station transformer protection.
   7) Perform a test to measure the background harmonics.

(2) Test content
   1) Conduct the 511B station transformer switching test;
   2) Conduct the 511B station transformer load test;
   3) Conduct the 512B station transformer switching test;
   4) Conduct the 512B station transformer load test;

(3) Record test results.

(X) Switching test of the 500 kV AC filter and capacitor bank at the Guquan Converter Station

(1) Test purpose
   1) Examine the switching inrush current of the AC filter and parallel capacitor bank to ensure it is within the allowable range.
   2) Test the switch's ability to cut off capacitive current.
   3) Measure the operating overvoltage during the activation of the AC filter and parallel capacitors bank.
   4) Complete the load test for the AC filter and the reactive power compensation capacitor, then check the operational status of the high-voltage equipment, and finally, inspect for any abnormal heating in the connections.
   5) Verify the correctness of current and voltage for the protection, control, and measurement devices associated with the AC filter and reactive power compensation capacitor. Additionally, ensure that the unbalanced current of the capacitors, including filter capacitors, remains within permissible limits.

6) Verify the correctness of the current and voltage of the filter bus differential protection.
7) Perform a test to measure the background harmonics.

(2) Test content
1) Conduct switching and energization tests on the AC filter in the 3st bank on bus 63.
2) Conduct switching and energization tests on the AC filter in the 4st bank on bus 64.
3) Conduct switching and energization tests on the AC filter in the 5st bank on bus 65.

(3) Record the test results.

(XI) Guquan Converter Station Auxiliary Power Automatic Switching Test

(1) Test purpose
1) Ensure that the automatic switching function of the 10 kV station's auxiliary power system matches the original design specifications.
2) Ensure that the automatic switching function of the 400 V station's auxiliary power system matches the original design specifications.
3) Verify the timing coordination and logic of the automatic switching systems for both 10 kV and 400 kV.

(2) Test content
1) Perform a verification of the 10 kV automatic switching function.
2) Perform a verification of the 400 V automatic switching function.
3) Conduct a thorough verification of the timing coordination and logic for the automatic switching systems at both 10 kV and 400 V.

(3) Record the test results.

(XII) Guquan Converter Station Switch Interlocking Test

(1) Test purpose. Verify if the switches in the strings with 3/2 connection at AC yard have the capability to automatically trip the middle breaker once both side circuit breakers have been opened.
(2) Test content. Conduct switch interlocking tests for the following strings: the 2th string that is shared by the converter transformer and the AC line; the 6th string that is shared by the converter transformer and the AC filter; and the 1st, 3rd, 4th, 5th, 7th, 8th and 9th strings that are shared by the AC filter and the AC line.
(3) Record the test results.

(XIII) Guquan Converter Station Background Harmonic Measurement

(1) Test purpose. Measure the changes and distribution of system harmonics both before and after the loop-in of Wancheng I and II lines, as well as before and after the operation of a single AC filter and reactor.
(2) Test content.
1) The 1000 kV bus voltage and harmonic of Guquan converter station.

2) The voltage, current, and harmonic of the 1000 kV Wancheng I and II lines.
3) The 500 kV bus voltage and harmonic content of Guquan converter station.
4) The 500 kV AC line current and harmonic.

(3) Record the test results.

(XIV) Guquan Converter Station Converter Transformer and Blocking Valve Group Charging Test

(1) Test Purpose. Verify that the inrush current during the switching of the converter transformer is within the specified limit, consistent with previous design research results, and that the resonance is adequately suppressed by the damping characteristics. Additionally, confirm that the triggering phase of the thyristor component is correct.

(2) Test Content
1) Perform a charging test on the 500 kV converter transformer and blocking valve group at the Guquan converter station.
2) Perform a charging test on the 1000 kV converter transformer and blocking valve group at the Guquan converter station.
3) The LV converter transformers for Poles 1 and 2 are energized, with each transformer undergoing the energization process five times.
4) The HV converter transformers for Poles 1 and 2 are energized, with each transformer undergoing the energization process five times.

(3) Record the test results.

(XV) DC Line Open Line Test at Guquan Converter Station

(1) Test Purpose. Check the DC voltage control function, the relationship between the converter valve firing angle and DC voltage, the voltage tolerance of the valve, and the voltage tolerance of both the DC filter and the DC transmission line. Additionally, verify the correctness of the open line test sequence control and the protection system.

(2) Test Content
1) Open line test of a single converter with DC line disconnected at Guquan converter station.
2) Open line test of a single converter with DC line connected in manual mode at Guquan converter station.
3) Open line test of a single converter with DC line connected in automatic mode at Guquan converter station.
4) Open line test of a single pole with DC line disconnected.
5) Open line test of a single pole with DC line connected in automatic mode.
6) When Pole 1 is operating and power is being transmitted in the forward direction, the open line test of the single converter of Pole 2 is conducted at Guquan converter station.

7) When Pole 1 is operating and power is being transmitted in the forward direction, the open line test is conducted for Pole 2 with DC line connected.

(3) Record the test results.

(XVI) Anti-Interference Test at Guquan Converter Station

Same as Sect. 2, (XI).

(XVII) Research Results

The commissioning plan for the Guquani converter station system has been compiled, including tests such as the 500 kV AC bus and AC yard energization, 500 kV AC filter energization, 1000 kV AC bus and AC yard energization, 1000 kV AC filter energization, sequential operation, non-energized trip, converter transformer charging, anti-interference, and the DC open line test. These tests aim to verify the functionality of the converter station and provide technical support for on-site system commissioning of the project.

IV. Engineering Application

Following the proposed Changji and Guquan converter station system commissioning plan, we completed the system commissioning at both stations, addressed various technical issues encountered during the process, and laid the groundwork for seamless end-to-end system commissioning.

V. Main Research Achievements

By conducting a technical content and specification analysis of the ± 1100 kV converter station system commissioning, a system commissioning plan was developed. The following research outcomes were achieved.

(1) The commissioning items for the station system were clarified and include the following: AC bus and AC yard energization tests, AC filter energization tests, sequential operation tests, trip tests, converter transformer charging tests, anti-interference tests, and DC open line tests.
(2) The results of the sequential control tests for single converter, single pole, and bipolar connection modes in the converter station have been analyzed. This analysis ensures that if a fault occurs and interrupts the execution of an operation sequence, the DC equipment remains in a safe state. Sequential control can be tested in both manual and automatic control modes to verify that each step of the operation and its execution conform to the design requirements.
(3) The last trip test of the converter station was analyzed, detailing the operation steps, test locations, simulation principles of the trip test, and acceptance standards. It is essential to check whether the DC protection action is correct, including the protection trip matrix circuit.
(4) The energization tests for the 750 kV AC yard at Changji converter station, as well as for the 1000 and 500 kV AC yard at Guquan converter station, were analyzed. The test content, conditions, operation steps, and acceptance

standards were provided. The test encompasses several aspects: checking the bus charging phase, verifying the effective value and phase of the secondary voltage of the bus voltage transformer; inspecting the insulation condition of the associated equipment; and examining the wiring of the secondary circuit for the corresponding switch synchronizing voltage. Additionally, a scheme for verifying the polarity of switch protection and bus differential protection was provided.

(5) An energization scheme for the filter was proposed, encompassing verification schemes for AC filter protection, AC filter bus protection, and related switch protection.

(6) The station transformer was charged five times. During the charging process, the 750 kV bus I at the Changji converter station was operating normally, while bus II was utilized as the test bus. The test schemes, including the station transformer load test, 500 kV station transformer protection checks, and polarity checks for the metering system's voltage and current, were compiled.

(7) The automatic switching performance and test scheme were studied. As a result, the following schemes were proposed: the 10 kV automatic switching function verification test, the 400 V automatic switching function verification test, and the time coordination and logic verification test for the interaction between the 10 kV and 400 V automatic switching systems.

(8) The circuit breaker interlocking tests at Changji and Guquan converter stations were analyzed. A middle circuit breaker interlocking test scheme was compiled to verify that the interlocking functions of the middle breaker on the strings shared by the filter and the line, the filter and the converter transformer, and the converter and the line comply with the technical specification requirements.

(9) The harmonic measurements at the converter station were analyzed, and a test scheme was compiled to include the system background harmonic measurements, the system harmonic changes before and after the station transformer is energized, and the harmonic changes before and after a set of AC filters is energized.

(10) The energization test of the converter transformer and blocking valve group at the converter station was analyzed, detailing the test content, conditions, operation steps, and acceptance standards for the test results. The test content included 5 energization tests for each of the 24 converter transformers on the bipolar system.

(11) The DC open line test in the converter station was analyzed, detailing the test content, conditions, operation steps, and acceptance standards for the test results. The test content includes open line tests for a single converter with the line both disconnected and connected, open line tests for a single pole and double converter with the line both disconnected and connected, and tests with one pole operating while the other pole is in an open line state. Additionally, the open line test can be conducted in either manual or automatic mode. During the open line test, it is essential to verify several key aspects: the DC voltage control function, the relationship between the converter valve firing angle and the DC voltage, the voltage tolerance of the valve, and the voltage tolerance of both

the DC filter and the DC transmission line. Additionally, the correctness of the open line test sequence control and the protection system must be confirmed.

(12) The anti-interference test was analyzed, with the test content, conditions, operation steps, and acceptance standards for the test results. The test encompasses anti-interference checks for secondary equipment such as VBE and protection cabinets, pole control and protection cabinets, station control systems, relay houses, and monitoring systems, ensuring that all equipment complies with the specified design requirements.

5 System Commissioning

The purpose of system commissioning is to comprehensively assess all equipment and their functions within the DC transmission project. It verifies whether the performance indicators of the DC transmission system meet the contractual and technical specifications. Commissioning ensures the security and reliability of the equipment and system once the project is operational. It also helps to understand and master the operational performance of the interconnected power grid.

I. System Commissioning Content

The system commissioning at each stage includes the following content:

(1) System calculation analysis. Establish a digital calculation model for the DC transmission system to conduct a detailed analysis and research on the commissioning system of the DC transmission project. This includes an analysis of the operation mode for system commissioning, an analysis of the commissioning items, a security and stability analysis under the commissioning mode along with research into accident prevention measures, an analysis of the dynamic characteristics and control protection features of the system, and an analysis of the system's electromagnetic transient overvoltage performance.

(2) The tasks include writing system commissioning plans, dispatching plans, on-site implementation plans, and testing plans:

    1) Based on the results of calculation analysis and simulation tests, and in conjunction with the main circuit connection operation mode of the UHV DC transmission project, formulate a system commissioning plan.
    2) According to the system commissioning plan and the specific test items, the power dispatching center is assisted in drafting the dispatching plan for system commissioning.
    3) According to the test items determined by the system commissioning plan, formulate the system commissioning implementation plan.
    4) According to the content of the system commissioning plan and test items, write the system commissioning test plan, which includes: DC system testing, overvoltage testing, AC system testing, AC harmonic testing, DC harmonic testing, electromagnetic environment testing, noise testing, infrared temperature measurement, etc.

(3) On-site system commissioning. According to the system commissioning plan and the on-site test implementation plan, organize the completion of system commissioning test items on site. Execute the system commissioning and testing plan, analyze the test results, supervise the defect elimination, and carry out the tracking calculation analysis of on-site commissioning.
(4) System commissioning summary. The content of the system commissioning summary includes:

    1) Sorting and archiving of on-site commissioning test results, records, waveform diagrams, and related materials.
    2) Analyzing the system commissioning results and writing and publishing the system commissioning reports.
    3) Writing a summary report on the system commissioning work and reporting it to the project owner.

(5) Test work schedule at each stage. Formulate the project test work plan according to the progress of the DC project construction. The test work plan and test content should be adjusted based on the project's progress and the start-up committee's decisions.

II. System Characteristics Analysis of the ± 1100 kV UHV DC Transmission Project

(I) Short Circuit Ratio

In the early stage of the ± 1100 kV UHV DC transmission project's operation, the electrical connection between the converter station and the main network is weak, especially when there is no effective power support at the sending end. However, the effective short-circuit ratio of the sending end system is greater than 3.5, due to the close electrical distance between the planned Changji converter station and the nearby 750 kV substation. Whether the Guquan converter station at the receiving end is directly connected to the 1000 kV UHV substation or the high and LV converters are layered connected to the 500 kV/1000 kV power grid, the effective short-circuit ratio remains greater than 3.5. This meets the operation requirements of the ± 1100 kV DC system bipolar rated power of 12,000 MW.

(II) Voltage Control with Single Converter Deblocking/locking

Due to insufficient power support at the Changji converter station, the transmission power with a single converter deblocking and blocking is limited to only 300 MW. To meet the minimum requirements for the filter bank, when two AC filter banks are put into or taken out of operation during DC converter deblocking or banking, the voltage change on the 750 kV bus is relatively large. Appropriate measures must be taken to control the bus voltage fluctuations.

(III) Impact of Grounding Pole Current on the System

Based on the experience from the Hami-Zhengzhou project system commissioning, in the ground return operation mode, the DC bias of the main transformer at the substation near the grounding pole of the Changji converter station may be relatively large, potentially affecting system stability. Therefore, during the early stages of system commissioning, a DC bias test plan is developed. This plan includes testing the main transformer at the substation near the grounding pole for DC bias. The characteristics of each test point are analyzed to establish a model that estimates the DC bias value when the rated current flows through the ground return line. This provides measurement data and technical materials for controlling DC bias. Subsequently, a Neutral DC Current Blocking Device (NCBD) is installed at the neutral point of the relevant transformer to mitigate the impact of the DC component on the transformer and ensure system stability.

(IV) Converter Station Secondary System

The control system and the DC protection system of the CG project are independent of each other. The DC control system is designed according to the principles of a layered structure and redundancy. In principle, the control layer comprises several hierarchical levels: the station control layer, the bipolar control layer, the pole control layer, the valve group control layer, and the valve base electronic equipment control layer. Current commands and control signals are primarily exchanged between the bipolar/pole layer control host and the valve group layer control host. Closed-loop control of DC current, DC voltage, and extinction angle, as well as the deblocking and locking functions of the converter, are all managed at the valve group layer control host.

The DC protection and converter transformer protection systems employ a three-out-of-two logic scheme. This scheme indicates that if two out of three DC protection devices detect a fault, the protection will be triggered. AC filter protection, along with AC bus, switch, line, and other protections, are designed to be fully redundant. The DC protection system encompasses a range of protections, including converter protection, pole protection, bipolar protection, DC filter protection, AC bus overvoltage protection, and AC filter protection.

The converter station's operation control and monitoring adhere to the principle of layered control. The SCADA system is utilized to execute start/stop operations, sequence control, and control over all circuit breakers, disconnectors, and other equipment within the AC/DC switchyard. Additionally, it performs monitoring, alarming, recording, and transmission of all analog signals, switch position signals, equipment faults, device abnormal signals, and protection action signals. The system is also capable of setting and modifying various DC control parameters and set values.

(V) Converter Valve

Drawing from the experience of commissioning the $\pm$ 800 kV UHV DC system, the relatively complex issue at the interface between the converter valve and the DC control protection has been analyzed and studied. This complexity arises due to the use of thyristor converter valves with varying technical approaches. Different test methods are employed to verify the interface performance of converter valves from various manufacturers, as well as the control protection systems. It is essential that their performance conforms to the technical specifications.

(VI) Analysis of HV/LV Converter Layered Connected to 500 kV/1000 kV Power Grid

The HV/LV converters at the Guquan converter station, which is the receiving end of the ± 1100 kV UHV DC transmission project, are layered connected to the 500 kV and 1000 kV AC power grids. Consequently, the adjustment step sizes of the tap changers for the HV/LV converter transformers differ. Specifically, one step of the tap changer for the 500 kV converter transformer is 1.25%, while for the 1000 kV converter transformer, it is 0.65%. The voltages and phase angles of the two AC systems vary, which can lead to DC voltage imbalances between the HV and LV converters. To accommodate these variations, the DC control strategy deviates significantly from those of previous UHV projects, where the sending and receiving ends were connected to the same AC grid.

III. ± 1100 kV UHV DC System Commissioning

(I) Calculation and Analysis of DC Project Connecting to the AC Power Grid

According to the grid planning for the ± 1100 kV DC transmission system, the converter station on the sending end is connected to the 750 kV AC power grid. On the receiving end, the converter station's high and low voltage sides are respectively connected to the 500 kV and 1000 kV AC power grids. A simulation model of the ± 1100 kV DC project, which connects to both the Xinjiang and East China power grids, has been established to perform stability calculations and analysis.

1. Analysis on sending end connecting to the grid

In the early stages of operation for the ± 1100 kV UHV DC transmission project, the electrical connection between the converter station and the main grid is weak due to the lack of effective power support at the sending end. However, this connection satisfies the requirement of maintaining an effective operating short-circuit ratio greater than 3.5.

Under various conditions where the converter station lacks a supporting power supply, the voltage disturbance resistance of the converter station bus varies. This variation occurs because a group of AC filters is being switched on or off on the 750 kV bus of the converter station, resulting in a maximum bus voltage change of approximately 12 kV. Consequently, during system commissioning, it is essential to implement appropriate measures to prevent bus voltage fluctuations that may arise from the simultaneous switching on or off of several sub-banks of filters when the DC system undergoes deblocking and blocking. For instance, measures could include the switching on or off of low-voltage reactors during DC deblocking and blocking.

2. Analysis on receiving end connecting to the grid

The receiving end of the ± 1100 kV UHV DC transmission project is connected in a layered manner to the 500 kV and 1000 kV AC power grids. In the scenario where the converter directly connects to the UHV substation, the short-circuit ratio is 3.86.

In contrast, for the layered connection scheme, the short-circuit ratios are 5.37 on the 1000 kV side and 5.16 on the 500 kV side. Both schemes represent robust multi-feed AC-DC systems. However, the layered connection scheme exhibits a higher short-circuit ratio, indicating that the AC system provides stronger support to the DC system.

In the layered connection scheme where the HV/LV converters of the converter station are connected to the 500 and 1000 kV power grids, the maximum change in bus voltage does not exceed 3 kV when a group of AC filters is switched on or off on the 500 kV bus. Similarly, when a group of AC filters is switched on or off on the 1000 kV bus, the maximum change in bus voltage does not exceed 3.6 kV.

(II) Overvoltage Simulation Calculation

1. Overvoltage simulation calculation of converter transformer activation

(1) Under the condition that the circuit breaker is not equipped with a closing resistor, no significant resonant overvoltage occurs when the converter transformers on both the rectification and inverter sides of the converter station are switched on, and the 750 kV/500 kV AC bus arresters at the converter station do not activate. However, when the YD converter transformer on the inverter side is switched on, it can easily lead to commutation failure of the normal operation pole.

(2) When the circuit breaker of the converter transformer at both the rectification and inverter sides of the converter station is equipped with a closing resistor ranging from 1200 to 1500 Ω, and this resistor is engaged for a duration of 11 to 13 ms, it effectively suppresses resonant overvoltage. Under these conditions, resonant overvoltage does not occur when the converter transformer is energized at both sides of the converter station. Furthermore, the maximum inrush currents at the rectification and inverter sides do not exceed 1500 A and 1000 A, respectively, which helps to prevent commutation failure in the opposite pole.

(3) Through calculation and analysis, it has been determined that in the light load flow mode, the short-circuit current is 20 kA. The closing resistance of the incoming line switch for the converter transformer is 1500 Ω. When the converter transformer at the rectifier station is switched on, the overvoltage is 1.14 p.u., and when the converter transformer at the inverter station is switched on, the overvoltage is 1.129 p.u.. These values comply with the technical specifications of the ± 1100 kV DC transmission system.

2. DC overvoltage simulation and calculation

(1) Typical fault overvoltage calculation. Under various operating conditions and modes of the DC system, simulation calculations have been conducted to assess the overvoltage levels at the converter station during the disturbance processes associated with typical fault types. These include faults in the DC pole lines at both the sending and receiving ends, neutral busbars, converter valves, smoothing reactors, and so on. By applying the insulation coordination principles established for the ± 800 kV UHV DC

transmission project, it has been confirmed that the overvoltage levels at the converter station and the energy consumption of the arresters are both within permissible limits.

(2) Protection coordination and blocking sequence overvoltage calculation. The protection blocking sequence and coordination principles established for the ± 800 kV UHV DC transmission project are deemed applicable to the ± 1100 kV UHV DC transmission project as well. In line with the setting principles of the ± 800 kV UHV DC control protection, it is assumed that if the protection operates correctly under various fault conditions, the overvoltage level at the converter station for the ± 1100 kV UHV DC transmission project will be within the equipment's tolerance range. Furthermore, the overvoltage levels at all other points on the DC side of the converter station, with the exception of the converter valve and neutral busbar area, also comply with the necessary requirements.

(3) Calculation of DC system control strategy's impact on overvoltage.The Voltage Dependent Current Order Limiter (VDCOL) significantly affects the overvoltage on the secondary side of the converter transformer at the converter station. For the ± 1100 kV DC transmission project, the VDCOL parameters can be optimized as needed to reduce overvoltage levels on the secondary side.

As the DC current or the leakage reactance of the converter transformer increases, the likelihood of commutation failure during the DC line's fault and recovery period also rises. Consequently, the overvoltage level at the converter station increases. Adjusting the controllers of both the fault and non-fault poles can stabilize the restart of the affected pole and prevent commutation failure in the non-affected pole, thereby helping to reduce overvoltage.

(III) System Stability Calculation

By calculating and analyzing the stability characteristics and operational control technologies of the ± 1100 kV UHV DC transmission project when connected to the power grid, we can propose the necessary requirements for the sending and receiving end systems, as well as security defense measures for the DC connection.

1. AC fault analysis

N-1 Fault: For the N-1 fault scenario of the Xinjiang 750 kV AC line, East China UHV AC line, and 500 kV AC line, no measures need to be taken as the system remains stable when the power receiving end of the project is situated in the southern part of Anhui province.

N-2 Fault: In the event of a power transmission channel loss scenario caused by an N-2 fault in the 1000 kV UHV AC line or the 500 kV AC line, no measures are required as the system maintains stability when the power receiving end of the project is situated in the southern part of Anhui province.

2. DC fault analysis

When the ± 1100 kV UHV DC system experiences a single pole blocking, each 500 kV tidal power flow does not result in an overload, and the receiving end does not require any measures, as the system remains stable. This stability is observed when the power receiving end of the project is located in the southern part of Anhui province.

Following a ± 1100 kV UHV DC bipolar blocking fault, the generator power angle at the sending end exhibits rapid swings, leading to system instability. Stability is restored after deactivating a 5600 MW generation unit at the sending end, as illustrated in Fig. 42. Additionally, the 1000 kV UHV and 500 kV lines near the receiving end have not experienced voltage instability, as demonstrated in Fig. 43.

(IV) DC Control Protection Calculation

1. AC line fault on the inverter side

A fault on the 500 kV AC line at the inverter side of the ± 1100 kV DC project results in a recovery time for DC power that is comparable to that of the XS project. However, during commutation failure, the DC current is significantly higher. When a single-phase fault occurs at the inverter station, the reactive power absorbed from the system during the commutation failure increases. Furthermore, the reactive power injected into the AC system during the recovery from commutation failure also increases, leading to a greater impact on the AC system during both the fault and the recovery phases.

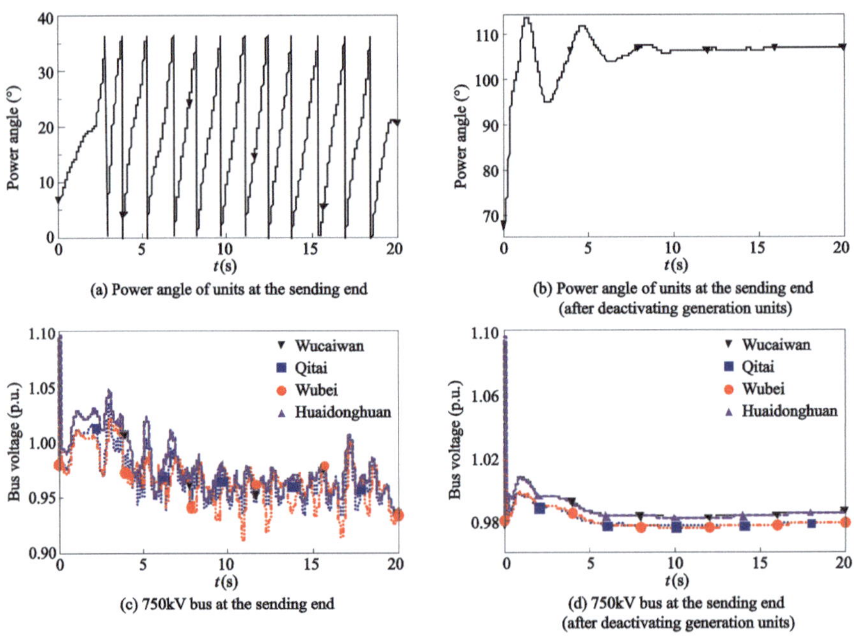

Fig. 42 System stability after deactivating 5600 MW generation units at the sending end for bipolar blocking fault

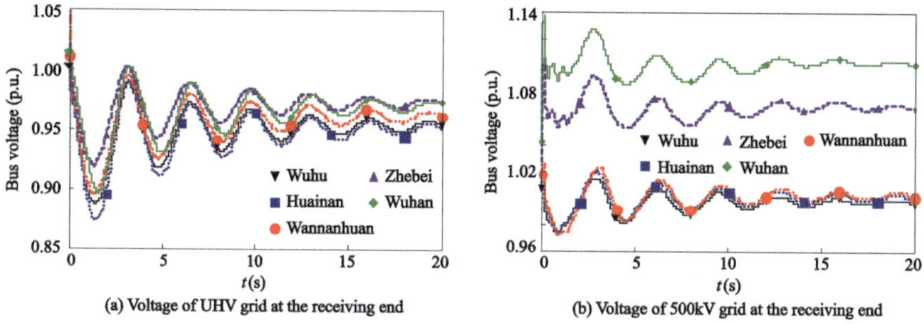

Fig. 43 Voltage stability of the 1000 and 500 kV grid near the receiving end for bipolar blocking fault

(1) During commutation failure, the reactive power absorbed is related to both the amplitude and the firing angle of the DC current. Similarly, the reactive power injected into the AC system during the recovery phase of the commutation failure is associated with the amplitude of the DC current that was present during the fault. This relationship is illustrated in Figs. 44 and 45.

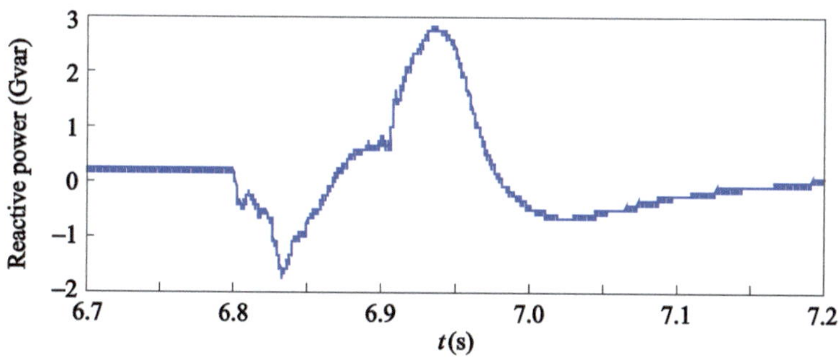

Fig. 44 Waveform of reactive power exchanged between ± 800 kV inverter station and AC system

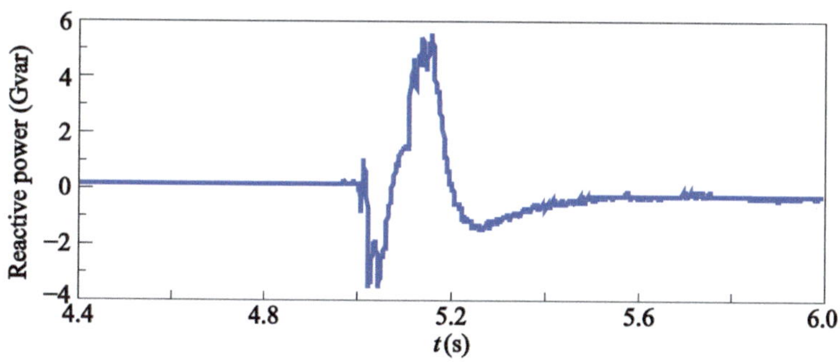

Fig. 45 Waveform of reactive power exchanged between ± 1100 kV inverter station and AC system

(2) When a single-phase fault occurs at the inverter station and the effective short-circuit ratio falls to 2.1, the reactive power absorbed during the DC commutation failure leads to a significant decrease in the non-fault phase voltage of the converter bus. Additionally, during the recovery process from the commutation failure, there is a significant increase in the converter bus voltage.

2. AC line fault on the rectifier side

However, during the three-phase fault and its recovery period, the rectifier station injects increased reactive power into the AC system, as illustrated in Figs. 46 and 47. Consequently, the impact on the AC system is more pronounced during both the fault and recovery phases.

3. DC line fault

During the rated power operation of the DC bipolar system, an instantaneous ground fault occurs in Pole 1's DC line. Consequently, the power is transferred to Pole 2, which experiences an overload of 1.2 times its capacity for 3 s. During the recovery

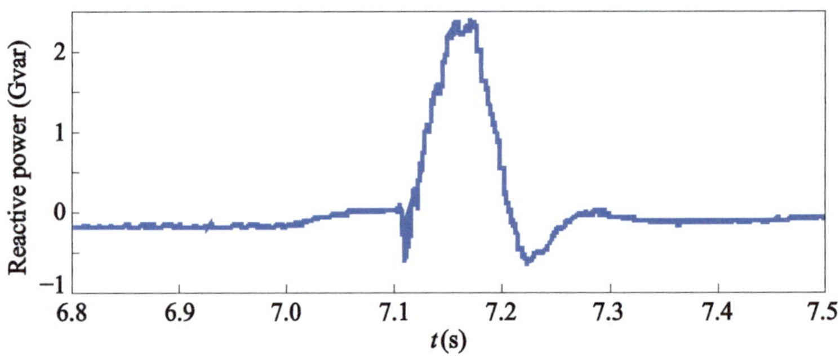

Fig. 46 Waveform of reactive power exchanged between ± 800 kV converter station and AC system for three phase fault on the rectifier side

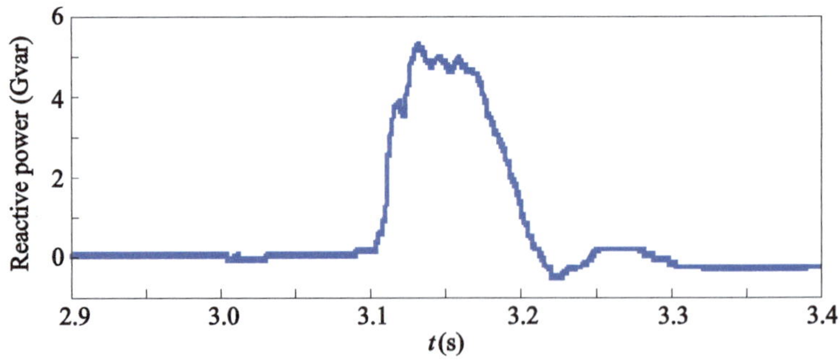

Fig. 47 Waveform of reactive power exchanged between ± 1100 kV converter station and AC system for three phase fault on the rectifier side

process of Pole 1's fault, a commutation failure occurs on the inverter side of Pole 2. At this time, the voltage rise time parameter of the VDCOL is recorded as 0.04 s. The cause is attributed to the rapid recovery of Pole 1, which leads to a drop in the AC voltage and, in turn, excessive current in Pole 2. Simulation studies indicate that the DC voltage rise filtering time parameter of the VDCOL could be increased appropriately to control the current recovery speed, thereby improving the post-fault recovery characteristics to some extent. However, the potential impact of a slower power recovery on the system frequency must be taken into account. Therefore, it has been ultimately determined that the VDCOL voltage rise time parameter should be set at 0.15 s to prevent commutation failure in Pole 2.

4. Reactive power control strategy optimization during DC power modulation process

Due to the large transmission capacity of the ± 1100 kV UHV DC project, the slow switching of reactive power devices during DC emergency power or frequency modulation processes leads to significant absorption or injection of reactive power by the converter station from or to the AC system. This can result in overvoltage or low voltage on the converter bus. Simulation results demonstrate that decreasing the activation delay of AC filter under reactive power control from 3 s to 0.5 s can shorten the duration to absorb a large amount of reactive power from the AC system. This adjustment also helps to reduce the disturbance duration of both the AC and DC systems during the emergency process.

(V) Analysis of System Commissioning Scheme

1. Classification of main circuit connection modes and system commissioning schemes

   (1) The ± 1100 kV UHV DC transmission project features 45 distinct connection line modes, setting it apart from the conventional ± 800 kV UHV DC transmission projects. This includes the unique configuration where the ± 550 kV LV converters on the inverter side are connected to the 1000 kV AC grid, and the ± 1100 kV converters are connected to the 500 kV AC grid, thereby forming a layered connection to both the 500 kV and 1000 kV AC grids with HV and LV converters. The voltage level of the DC converter in this project exceeds that of the ± 800 kV, necessitating higher performance standards for the primary equipment.
   (2) Classification of system commissioning schemes. The commissioning plan for the ± 1100 kV UHV DC transmission project can be divided into three main phases. The first phase involves the commissioning of a single converter system. The second phase encompasses the commissioning of a monopole dual converter system, including the commissioning of the DC bias measurement system. The third phase comprises the commissioning of a bipolar system, including the commissioning for the layered connection mode.

2. Preparation of system commissioning scheme

(1) Basic test items. Since the main circuit connection mode is identical for both the ± 1100 kV and ± 800 kV UHV DC transmission projects, the fundamental test items are also consistent. The primary components of the basic commissioning schemes for single converter, monopolar dual converter, and bipolar systems encompass: DC start/stop operations, initial operation tests, basic control mode testing, protection tripping tests, DC blocking tests, steady-state operation tests, verification of DC control functions, steady-state DC operation control, dynamic characteristic assessments, disturbance response tests, DC line fault simulations, transitions between metallic and ground returns, tests for rated load and overload conditions, reactive power control, and grounding electrode tests, among others.

The system commissioning for the ± 1100 kV UHV DC transmission project at the receiving end, where the HV and LV converters are layered connected to the 500 and 1000 kV grids, primarily includes the following verification tests: ① Power forwarding and basic control protection function verification. ② DC voltage control function verification. ③ Converter transformer tap switch control function verification. ④ Reactive power control function verification. ⑤ Additional power control, including verification of the coordination function of the emergency control device. ⑥ Power transfer and distribution function verification. ⑦ Online converter switching function verification. ⑧ Power reversing and basic control protection function verification.

3. New test items

(1) Based on the configuration of DC control protection, we have developed a plan for verifying the basic control functions and device performance. Additionally, a separate plan has been compiled to verify the performance of protection trips. This approach takes into account the DC blocking performance tests at various levels, including the single converter level, the monopole dual converter level, and the bipolar dual converter level.
(2) Utilizing the hierarchical control principle for operation control and monitoring at the converter station, a test plan has been developed to verify the information exchange bus between pole control and VBE. In compiling the control system bus fault and system switching test plan, considerations are integrated at multiple levels: the station control level, the pole control level, and the converter control level.
(3) Based on the design and functionality of the auxiliary power source and other auxiliary systems at the converter station, the verification tests for auxiliary system functions are conducted at both the pole control level and the converter control level. This approach is taken because the station's auxiliary system design consists of a dedicated set of auxiliary systems for each converter.
(4) Ground/metallic transfer and neutral bus switch protection tests are essential components of the verification process. During ground/metallic transfer tests

conducted with the rated current in both the HZ project and the XZ project, incidents of arrester damage were reported. The rated current for the ± 1100 kV UHV DC project is 5455A, which exceeds that of the HZ and XZ projects by 545A. Consequently, when performing the ground/metallic transfer test with the rated current for the ± 1100 kV DC project, the switch protection settings must be established based on the secondary equipment FAT results and the performance of the MRTB and GRTS switches. This is to ensure that both the equipment performance and control protection functions are thoroughly verified, and to guarantee that no equipment damage occurs during the test process.

Drawing from the neutral bus switch failure (NBSF) principle and the test outcomes in both the Hami-Zhengzhou and Xiluodu-Jinhua projects, we have utilized the DC equipment parameters and the NBSF protection strategy specified in the design to formulate the NBSF test plan, and executed the NBSF system test for the ± 1100 kV UHV DC transmission project.

(5) Single converter online energization and de-energization. Building on the experience from system tests in the HZ and XZ projects, the procedures for online energization and de-energization of the single converter within the ± 1100 kV UHV DC system have been optimized. When the inter-station communication mode is not available, online de-energization of the converter on the rectification side is prohibited, as is the online energization of the converter on the inversion side. Furthermore, due to issues with converters layered connected to the receiving grid, the coordination between the energization and de-energization of the converter and the strategy for emergency control devices also requires careful consideration.

(6) Converter control and valve base electronic interface (VBE) test. If the two converter stations within the ± 1100 kV UHV DC system utilize converter valves from different manufacturers, this can complicate the DC control protection and the interface between the converter valve and the VBE. Consequently, an additional test item for verifying the performance of the pole control and VBE has been incorporated.

(7) Layered connection test. As previously mentioned, the HV/LV converters at the receiving end are layered connected to the AC grid. This presents a novel technical challenge not addressed in the HZ and XZ projects. It is essential to verify the performance of the layered connection mode and to consider the inclusion of a system test item for layered connection when developing the system test plan.

(8) Converter transformer gas-producing verification test. Following the discovery of gas production due to thermal effects under high current in a converter transformer at the Yibin converter station during the DC system test of the XZ project, it is recommended to incorporate a gas-producing verification test for the converter transformer. This test should be conducted after the completion of the rated load operation test for a single converter or the pilot run of the DC system. The purpose is to ascertain whether the

converter transformer is capable of operating continuously at its rated power without producing excessive gas.

(9) Other tests.Other tests encompass the DC bias test, ice melting system test, converter station auxiliary system test, rated power and overload operation test, and the emergency control device FAT, etc.

Based on the aforementioned analysis, the commissioning items for the system have been established, thereby laying the groundwork for the development of the system commissioning plan for the ± 1100 kV UHV DC transmission project.

IV. System Commissioning of Monopolar and Single Converter Mode

(I) Single Converter Main Loop Connection

According to the design requirements for the UHV DC main loop connection, there are 16 possible connection modes for a system configured with a monopolar and a single converter:

(1) The commissioning items for low-voltage converter of Pole 1 with ground return.
(2) The commissioning items for low-voltage converter of Pole 1 with metallic return.
(3) The commissioning items for HV converter of Pole 1 with ground return.
(4) The commissioning items for HV converter of Pole 1 with metallic return.
(5) The commissioning items for low-voltage converter of Pole 2 with ground return.
(6) The commissioning items for low-voltage converter of Pole 2 with metallic return.
(7) The commissioning items for HV converter of Pole 2 with ground return.
(8) The commissioning items for HV converter of Pole 2 with metallic return.
(9) The commissioning items for the LV converter on the rectifier side and the HV converter on the inverter side of Pole 1, operating with ground return.
(10) The commissioning items for the LV converter on the rectifier side and the HV converter on the inverter side of Pole 1, operating with metallic return.
(11) The commissioning items for the HV converter on the rectifier side and the LV converter on the inverter side of Pole 1, operating with ground return.
(12) The commissioning items for the HV converter on the rectifier side and the LV converter on the inverter side of Pole 1, operating with metallic return.
(13) The commissioning items for the LV converter on the rectifier side and the HV converter on the inverter side of Pole 2, operating with ground return.
(14) The commissioning items for the LV converter on the rectifier side and the HV converter on the inverter side of Pole 2, operating with metallic return.
(15) The commissioning items for the HV converter on the rectifier side and the LV converter on the inverter side of Pole 2, operating with ground return.
(16) The commissioning items for the HV converter on the rectifier side and the LV converter on the inverter side of Pole 2, operating with metallic return.

(II) Classification of Single Converter System Commissioning Items

If the test items for ground return and metallic return are combined, there are 8 connection modes available for the test items.

The basic connection mode for the monopolar and single converter setup includes four types of metallic or ground return connection modes for both Pole 1 and Pole 2's low-voltage and HV converters. This configuration is used for system commissioning to test the performance of single converter units at each station. With its low DC voltage and minimal power transmission, this mode is suitable for conducting high-power tests such as thermal operation, overload, and metallic-to-earth transfer at rated current, minimizing the impact on the AC system in case of a DC system failure. The commissioning plan for these modes, which serves as the fundamental framework, considers the control and protection performance under the monopolar single converter connection. It includes tests for DC start/stop operations, control modes, protection trips, steady-state and dynamic characteristics, DC line faults, metallic-to-earth transfers, rated and overload, reactive power control, and grounding electrode test.

The remaining four connection modes (metallic/earth) are considered cross-connection modes for the monopolar and single-converter system. These modes are typically not used under normal system operation but may be employed if a converter fails. According to technical specifications, an operator can establish a cross-connection mode for the single converter. During the system commissioning process for the monopolar single-converter, the performance of both primary and secondary devices of each converter has been thoroughly tested. Whether testing the cross-connection mode or the standard single-converter mode, the key difference lies in the altered connection between converters, necessitating adjustments to DC control parameters and the protection input/output parameters, which also require testing. When drafting the system commissioning plan for these modes, focus on the basic control mode tests. The main content should include DC start/stop operations, basic control mode tests, basic protection trips, and steady-state operation tests, forming a derivative plan for the monopolar single-converter system commissioning.

(III) Monopolar single-converter System Commissioning Plan

Based on the above research results, the monopolar single-converter system commissioning items are as follows.

(1) Power forward, initial operation test:

   1) Pole start and stop, manual blocking.
   2) Manual switching of the control system.
   3) Emergency stop test.
   4) Analog input signal check.
   5) Initial operation test with metallic return.

(2) Power forward, protection trip, X, Y, Z blocking:

   1) With communication, rectifier station simulates grounding electrode open circuit zone III protection trip X blocking.

2) With communication, rectifier station simulates bipolar neutral bus differential protection trip Y blocking.
3) With communication, rectifier station simulates pole bus differential zone II protection trip Z blocking.
4) Without communication, rectifier station simulates DC overvoltage zone II protection trip X locking.
5) Without communication, rectifier station simulates bypass pair overload protection trip Y blocking.
6) Without communication, rectifier station simulates pole bus differential zone II protection trip Z blocking.
7) With communication, inverter station simulates valve group differential protection trip X blocking.
8) With communication, inverter station simulates bipolar neutral bus differential protection trip Y blocking.
9) With communication, inverter station simulates grounding electrode open circuit zone I protection trip Z locking.
10) Without communication, inverter station simulates bipolar neutral bus differential protection trip Y blocking.
11) Without communication, inverter station simulates pole bus differential zoene II protection trip Z blocking.
12) With communication, rectifier station simulates valve short circuit (VSC) protection trip X blocking.
13) Without communication, rectifier station simulates valve overcurrent protection zone II trip Y blocking.
14) Without communication, inverter station simulates valve DC differential protection zone II trip Y blocking.
15) VBE fault and trip on rectifier side.
16) VCE fault and trip on inverter side.
17) Valve cooling system fault and trip on rectifier side.
18) Valve cooling system fault and trip on inverter side.

(3) Power forward, steady-state operation, basic monitoring function check:

1) Power forward, steady-state operation, power failure simulation of on-duty system.
2) Power forward, steady-state operation, simulation of host board card crash.
3) Power forward, steady-state operation, check the utilization rate of host (GR or MR).
4) Power forward, steady-state operation, simulation of DC line fault (applicable only to Changji converter station).
5) Control-LAN control bus fault.
6) Station-LAN bus fault.
7) 60044-8 bus fault.

(4) Power forward, steady-state operation, joint current control

1) Control of current rise and fall, as well as stopping the current rise and fall.
2) Control system switch during current rise and fall phases.
3) Main control station transfer.
4) Converter transformer tap switch control, manually change tap switch position.
5) Current command step.
6) Control mode switch, current margin compensation and current command step.

(5) Power forward, steady-state operation, joint power control:

1) Converter start and stop.
2) Power rise and fall.
3) System switch during power rise and fall phases.
4) Power command step.
5) Communication failure during power rise and fall test.
6) Mode transition, current margin compensation.
7) Transition between Power control and joint current control.

(6) Power forward, communication failure, independent current control test:

1) Converter start and stop.
2) Emergency stop.
3) Current rise and fall.
4) System switch during current rise and fall phases.

(7) Power forward, reactive power control:

1) Manual switching of filters.
2) Filter requirements (GR or MR).
3) Filter replacement.
4) Reactive power control—Q mode.
5) Reactive power control—U mode.
6) $U_{\max}$ control.

(8) Power reverse, initial operation test

1) Converter start and stop, manual blocking.
2) Emergency stop test.
3) Without communication, emergency stop test.
4) Analog input signal check.
5) Initial operation test with metallic return.

(9) Power reverse, protection trip, X, Y, Z blocking:

1) With communication, rectifier station protection trip X blocking.
2) With communication, rectifier station protection trip Y blocking.
3) With communication, rectifier station protection trip Z blocking.

4) Without communication, rectifier station protection trip Y blocking.

(10) Power reverse, steady-state operation, joint current control:

1) Control of current rise and fall, as well as stopping the current rise and fall.
2) Control system switch and communication failure during current rise and fall phases.
3) Main control station transfer.

(11) Power reverse, steady-state operation, joint power control:

1) Power rise and fall.
2) During the power rise and fall test, simulate a communication failure.
3) Power command step.

(12) Power forward, pulse lose fault:

1) Ground return, single pulse lose fault on inverter side.
2) Ground return, multiple pulse loss fault on inverter side (more than 5 times).
3) metallic return, multiple pulse loss fault on inverter side (more than 5 times).
4) metallic return, no communication, multiple pulse loss fault on inverter side (more than 5 times).

(13) Power reverse, pulse loss fault. metallic return, multiple pulse loss fault on inverter side.

(14) Disturbance test

1) Ground return, simulate a DC line fault near the rectifier station.
2) Ground return line, simulate a DC line fault near the inverter station.
3) metallic return, simulate a DC line fault near the inverter station.
4) metallic return, simulate a DC line fault near the rectifier station.
5) Loss of DC 110 V power system C.
6) Loss of DC 110 V power system A.
7) Loss of DC 110 V power system B.
8) AC 400 V auxiliary power switch.
9) Rectifier station simulates I_{DNC} loss fault.
10) Inverter station simulates I_{DNC} loss fault.

(15) Power forward, power control, high power test:

1) Converter startup.
2) Converter power control.

(16) Power forward, current control, high power test

1) Current rise and drop.

2) Tap switch control, manual adjustment of tap switch.
3) Earth/metallic return transfer (300 MW).
4) metallic/ground return transfer (300 MW).
5) Earth/metallic return transfer (3000 MW).
6) metallic/ground return transfer (3000 MW).

(17) Power forward, ground return, thermal operation and overload test:

1) Preliminary detection of equivalent disturbing current.
2) Preliminary detection of AC harmonics.
3) Audibility noise measurement.
4) Power loss measurement of station auxiliary system.
5) Ground return, standby cooler not in operation, thermal operation test at rated power.
6) Ground return, standby cooler in operation, thermal operation test at 1.05 times rated power.

(18) Power forward, reactive power control: ground return operation, reactive control.
(19) Grounding electrode test.
(20) DC bias measurement:

1) DC current at 0.1 p.u..
2) DC current at 0.2 p.u..
3) DC current at 0.3 p.u..
4) DC blocking, predict the bias current at each test point when the rated current is expected.

(IV) Monopolar single converter system commissioning derivative scheme

Based on the aforementioned research findings, the derivative connection mode for the monopolar and single converter system is the cross-connection mode. The system commissioning items focus solely on the basic control mode and the protection trip test items. The primary content encompasses DC start/stop, basic control mode testing, fundamental protection tripping, steady-state operation testing, and other related test items.

Generally, the single converter derivative connection mode is a cross-connection mode that forms when either a high-voltage or LV converter fails and ceases operation at one end of the converter station, and a group of converters also ceases operation at the other end. This constitutes an abnormal operation mode. Consequently, the system commissioning test items are limited to the basic control mode and the protection trip tests.

V. Monopolar Dual Converter System Commissioning Scheme

According to the design requirements of the UHV DC transmission project, the monopolar dual converter has 4 connection modes:

(1) Pole 1 dual converter operation mode, ground return.

(2) Pole 2 dual converter operation mode, ground return.
(3) Pole 1 dual converter operation mode, metallic return.
(4) Pole 2 dual converter operation mode, metallic return.

If the test items of the ground return and the metallic return are combined, there are 2 test items for the monopolar dual converter (metallic/earth) connection mode, namely the pole 1 (metallic/earth) connection mode and the pole 2 (metallic/earth) connection mode.

The monopolar dual converter connection mode is a prevalent operational mode in UHV DC transmission projects. The aim of system commissioning under this mode is to evaluate the performance of control protection and auxiliary equipment, as well as to assess the performance of primary equipment in monopolar operation. Consequently, when developing the system commissioning plan for this mode, it is essential to thoroughly consider all control protection functions and equipment performance test items applicable to the monopolar operation mode of the UHV DC system. The primary components of the system commissioning plan encompass DC start/stop, basic control mode testing, protection trip testing, metallic/earth transfer testing, steady-state operation evaluation, dynamic characteristic assessment, DC line fault simulation, reactive power control performance testing, and rated load testing, among others.

The monopolar double converter commissioning items are as follows.

(1) Power forward, initial operation test:
   1) Pole start/stop, manual blocking.
   2) Manual switching of the control system.
   3) Emergency stop test.
   4) Analog input signal check.
   5) Initial operation test, metallic return.

(2) Power forward, protection trip, X, Y, Z blocking:
   1) With communication, rectifier station simulates grounding electrode open circuit zone III protection trip X blocking.
   2) With communication, rectifier station simulates bipolar neutral bus differential protection trip Y blocking.
   3) With communication, rectifier station simulates pole bus differential zone II protection trip Z blocking.
   4) Without communication, rectifier station simulates the converter differential protection trip X blocking.
   5) Without communication, rectifier station simulates the bipolar neutral bus differential protection trip Y blocking.
   6) Without communication, rectifier station simulates the pole differential protection trip Z blocking.
   7) With communication, inverter station simulates the grounding electrode open circuit zone III protection trip X blocking.

8) With communication, inverter station simulates the bipolar neutral bus differential protection trip Y blocking.
9) With communication, inverter station simulates the pole bus differential zone II protection trip Z blocking.
10) Without communication, inverter station simulates the DC overvoltage protection trip X blocking.
11) Without communication, inverter station simulates the 50 Hz protection trip Y blocking.
12) Without communication, inverter station simulates the pole bus differential zone II protection trip Z blocking.
13) With communication, rectifier station simulates the valve short circuit protection trip X blocking.
14) With communication, inverter station simulates the valve overcurrent protection trip Y blocking.

(3) Power forward, steady-state operation, basic monitoring function check

1) Power forward, steady-state operation, power failure simulation of on-duty system.
2) Simulate DC line fault (only rectifier station).
3) Simulate DC TA fault.
4) Control LAN bus fault.
5) Control host-PMI-60044- 8) Bus fault.
6) Control host-PMI-TDM bus fault.

(4) Power forward, steady-state operation, joint current control

1) Control of current rise and fall, as well as stopping the current rise and fall.
2) Control system switch during current rise and fall phases.
3) Main control station transfer.
4) Converter transformer tap switch control, manually change tap switch position.
5) Current command step.
6) Voltage command step.
7) Extinction angle step (γ).
8) Control mode switch, current margin compensation and current command step.

(5) Power forward, steady-state operation, joint power control:

1) Pole start and stop.
2) Power rise and fall.
3) System switch during power rise and fall phases.
4) Power command step.
5) Communication failure during power rise and fall test.
6) Mode transition, current margin compensation.
7) Transition between Power control and joint current control.

(6) Power forward, communication failure, independent current control test:

1) Pole start and stop.
2) Emergency stop.
3) Current rise and fall.
4) System switch during current rise and fall phases.
5) Independent current control/combined current control/power control transfer.

(7) Power forward, normal voltage/reduced voltage operation

1) Voltage reduction start/stop.
2) Manually initiate voltage reduction.
3) Initiate voltage reduction through protection system.
4) Current control, voltage reduction start/stop.
5) metallic return, power control, voltage reduction start/stop.
6) metallic return, current control, voltage reduction start/stop.
7) Transformer tap switch control, manually change tap switch position.
8) Power rise/fall.
9) Power command step.
10) Communication failure.
11) Power control/combined current control transfer.

(8) Power forward, reactive power control

1) Manual switching of filters.
2) Filter requirements.
3) Filter replacement.
4) Reactive power control.
5) Voltage control.
6) U_{max} control.
7) Q_{pc} control.
8) Gamma voltage regulation (Gamma Kick) function.

(9) Power forward, low power, earth/metallic return transfer

1) Earth/metallic return transfer (600 MW).
2) Metallic/ground return transfer (600 MW).
3) Metallic return, grounding operation test utilizing the station grounding network at the inverter station.

(10) Power reverse, initial operation test

1) Pole start/stop, manual blocking.
2) Emergency stop test.
3) Without communication, emergency stop test.
4) Analog input signal check.
5) Initial operation test, metallic return.

(11) Power reverse, protection trip, X, Y, Z blocking

1) With communication, rectifier station simulates grounding electrode open circuit zone III protection trip X blocking.
2) With communication, rectifier station simulates bipolar neutral bus differential protection trip Y blocking.
3) With communication, rectifier station simulates pole bus differential zone II protection trip Z blocking.
4) Without communication, rectifier station simulates bipolar neutral bus differential protection trip Y blocking.
5) With communication, inverter station simulates grounding electrode open circuit zone III protection trip X blocking.
6) With communication, inverter station simulates bipolar neutral bus differential protection trip Y blocking.
7) With communication, inverter station simulates pole bus differential zone II protection trip Z blocking.
8) Without communication, inverter station simulates pole bus differential zone II protection trip Z blocking.

(12) Power reverse, steady-state operation, joint current control:

1) Control of current rise and fall, as well as stopping the current rise and fall.
2) Control system switch and communication failure during current rise and fall phases.
3) Main control station transfer.

(13) Power reverse, steady-state operation, joint power control:

1) Power rise/fall.
2) Communication failure during power rise and fall test.

(14) Power forward, pulse lost fault

1) Ground return, single pulse lose fault on inverter side.
2) Ground return, multiple pulse loss fault on inverter side (more than 5 times).
3) Metallic return, multiple pulse loss fault on inverter side (more than 5 times).
4) Metallic return, single pulse lose fault on rectifier side.
5) Metallic return, multiple pulse loss fault on rectifier side (more than 5 times).
6) Metallic return, no communication, multiple pulse loss fault on inverter side (more than 5 times).
7) Metallic return, no communication, multiple pulse loss fault on rectifier side (more than 5 times).

(15) Power reverse, pulse loss fault: Metallic return, multiple pulse loss fault on inverter side.

(16) Disturbance test

1) Simulate a DC line fault near the rectifier station.

2) Simulate a DC line fault near the inverter station.
3) Under the reduced voltage operation mode, simulate a DC line fault near the inverter station.
4) Under the reduced voltage operation mode, simulate a DC line fault near the rectifier station.
5) Under the metallic return mode, simulate a DC line fault near the inverter station.
6) Under the metallic return mode, simulate a DC line fault near the rectifier station.
7) Power reverse, simulate a DC line fault near the rectifier side.
8) Power reverse, simulate a DC line fault near the inverter side.
9) Under the ground return mode, simulate a high-resistance fault within the DC line protection functionality.
10) Automatically restart the single converter test in the event of an unsuccessful DC line fault restart attempt.
11) In the event of an AC auxiliary power failure, switch off the AC 400 V power for Pole 1's HV converter.
12) In the event of an AC auxiliary power failure, switch off the AC 400 V power for Pole 2's HV converter..
13) Loss of 110 V DC power C for pole control.
14) Loss of 110 V DC power A for pole control.
15) Loss of 110 V DC power B for pole control.
16) Loss of 110 V DC power C for station control.
17) Loss of 110 V DC power A for station control.
18) Loss of 110 V DC power B for station control.
19) Lose of UPS for station control.
20) Simulate transmitter power failure of DC TA.
21) Converter transformer TV open circuit fault test.
22) DC yard TV open circuit fault test.
23) Valve base electronic system (VBE, VCU and VCE) switch test.
24) Internal cooling pump switch test of valve cooling system.
25) Normal operation, high-voltage VBE loses Under Voltage signal.
26) HV converter VBE trip circuit self-check function verification.
27) HV converter control protection and VBE interface function verification.
28) HV converter thyristor level fault trip.
29) Normal operation, manual switching of DC filter at rectifier station.
30) Normal operation, manual switching of DC filter at inverter station.

(17) Control location change test

1) Local/remote control transfer test, remote control start/stop test.
2) Local/remote control transfer test, remote control power rise and fall test.
3) Operate on the backup panel, start/stop test.
4) Operate on the backup panel, power rise and fall test.

(18) Power forward, combined power control, high power test:

1) Pole start.
2) Pole power control.
3) Ground return/metallic return transfer.
4) Metallic return/ground return transfer.

(19) Power forward, pole current control:

1) Current rise/fall.
2) Tap switch control, manual adjustment of the tap switch.
3) Mode transfer, current control on the inverter side.

(20) Power forward, thermal operation test:

1) Special measurement, preliminary detection of equivalent disturbing current.
2) Special measurement, preliminary detection of AC harmonics.
3) Special measurement, radio interference measurement.
4) Special measurement, audibility noise measurement.
5) Special measurement, power loss measurement of station auxiliary system.
6) Ground return, standby cooler not in operation, rated load operation test.
7) Ground return, standby cooler in operation, overload operation test at 1.05 times rated power.
8) Converter transformer tap switch control, manual change of tap switch position.
9) Metallic return, standby cooler not in operation, rated load operation test.
10) Metallic return, standby cooler in operation, overload operation test at 1.05 times rated power.

(21) Power forward, reactive power control

1) Ground return operation, reactive power control.
2) Metallic return operation, reactive power control.
3) Metallic return reduced voltage operation, reactive power control.
4) Ground return operation, voltage control.
5) Metallic return operation, voltage control.
6) Metallic return reduced voltage operation, voltage control.

VI. Bipolar System Commissioning Scheme

(I) Classification of Bipolar System Commissioning Schemes

The UHV DC transmission project's bipolar connection mode encompasses three distinct connection modes. The first type is the bipolar double converter connection mode, which becomes the most commonly utilized operation mode once the project is operational. This mode is unique. The second type is the bipolar single converter per pole connection mode, offering 16 unique connection operation modes. The third type is the bipolar unbalanced converter connection mode—characterized by one pole with a double converter and the other pole with a single converter—which presents 8 unique connection operation modes. Consequently, the bipolar system commissioning scheme is categorized into three separate commissioning schemes, each aligned with a different connection mode.

(1) The bipolar double converter connection mode only has one operation mode, becoming the most commonly utilized operation mode once the project is operational. Consequently, when developing the system commissioning scheme for this connection mode, it is imperative to thoroughly consider the performance of control protection and auxiliary equipment, as well as the performance of primary equipment under the bipolar connection mode. The system commissioning scheme primarily encompasses DC bipolar start/stop, basic control mode test, protection trip, steady-state operation, dynamic characteristics, various types of faults, reactive power control, rated load test, overload test, and other relevant components, thereby constituting the foundational commissioning strategy for the bipolar system.
(2) The bipolar single converter per pole connection mode comprises 16 distinct operation modes. As stipulated by the technical specifications, the performance of the bipolar single converter connection mode must be meticulously verified during the system commissioning process. It is important to note that in the commissioning schemes for both the monopolar single converter and the monopolar double converter, as well as the bipolar double converter, the performance of both primary and secondary equipment within the UHV DC system has already been rigorously tested. Therefore, when formulating the system commissioning scheme for this mode, the focus is solely on the basic control mode test items of the system. The principal components of this scheme include DC start/stop, basic control mode test, basic protection trip, steady-state operation, reactive power control test, among others. These elements collectively constitute the derivative scheme 1 of the bipolar system commissioning—namely, the bipolar single converter system commissioning scheme.
(3) The bipolar unbalanced converter connection mode is characterized by one pole with a double converter and the other pole with a single converter, with a total of 8 operation modes. As stipulated by the technical specifications, the performance of the bipolar unbalanced converter connection mode must be meticulously verified during the system commissioning process. It is important to note that in the commissioning schemes for both the monopolar single converter and the monopolar double converter, as well as the bipolar double converter, the performance of both primary and secondary equipment within the UHV DC system has already been rigorously tested. Therefore, when formulating the system commissioning scheme for this mode, the focus is solely on the basic control mode test items of the system. The principal components of this scheme include DC start/stop, basic control mode test, basic protection trip, steady-state operation, reactive power control test, among others. These elements collectively constitute the derivative scheme 2 of the bipolar system commissioning—namely, the bipolar unbalanced converter system commissioning scheme.

(II) Bipolar Double Converter System Commissioning Scheme

The commissioning of the bipolar double converter system constitutes the fundamental scheme for the bipolar system commissioning process. The system commissioning items are as follows:

(1) Power forward, bipolar start/stop test:

   1) Bipolar simultaneous start/stop, manual blocking.
   2) One pole operation, another pole start/stop test.

(2) Bipolar operation, power rise and fall, main control right transfer

   1) Power forward, bipolar power rise and fall.
   2) Power forward, bipolar operation, pole 1 power/current rise and fall.
   3) Power forward, one pole operation, another pole start/stop.
   4) Power forward, bipolar operation, single pole communication failure, single pole independent control, current rise/fall.
   5) Power forward, bipolar operation, bipolar power control/power control/current control transfer.
   6) Power forward, bipolar operation, main control right transfer.

(3) Automatic power control

   1) Power forward, automatic power control.
   2) Power forward, manually switch control system during automatic power rise/fall process.
   3) Power forward, manually emergency stop converter during automatic power rise/fall process.

(4) Pole trip, power compensation

   1) Power forward, pole 1 trip, power compensation.
   2) Power forward, pole 2 trip, power compensation.

(5) Current command step: Power forward, one pole current step test.
(6) Grounding electrode balance test

   1) Power forward, grounding electrode balance test.
   2) start/stop test utilizing the station grounding network for grouding purposes at both rectifier and inverter station.

(7) Reduced voltage operation

   1) Power forward, bipolar reduced voltage operation test.
   2) Power forward, bipolar reduced voltage start/stop and stable operation.
   3) Power reverse, reduced voltage operation test.

(8) Disturbance test

   1) Grounding electrode line open circuit.
   2) Rectifier side grounding, HV converter exit operation test.

3) Rectifier side grounding, HV converter put into operation test.
4) Rectifier side grounding, monopolar fault test.
5) Grounding mode transfer between station and grounding electrode.
6) Power forward, AC line fault.
7) Power reverse, AC line fault.
8) 10 kV AC power source switch.
9) Pole 1 operation, Pole 2 turn to maintenance test.
10) Pole 2 maintenance, Pole 1 start/stop test.
11) Full voltage operation, 0300 and 05000 (WN-Q17) replacement test.
12) Reduced voltage operation, 0300 and 05000 (WN-Q17) replacement test.
13) Switch position signal loss test for the neutral area DC yard.
14) Host power off test for valve group control protection.
15) Communication interruption test for high and low voltage valve group.
16) Pole 2 high-voltage VBE duty (ACTIVE) signal loss test.
17) AC yard No.1 and No.2 bus CVT line broken fault test.
18) DC yard NBS protection overlap fault test at rectifier station.
19) DC yard NBS protection overlap fault test at inverter station.
20) Test for non-fault converter restart in the Event of Pole Blocking by the converter differential protection at the rectifier station.
21) Test for non-fault converter restart in the Event of Pole Blocking by the converter differential protection at the inverter station.
22) Pole 1 operation, power forward, Pole 2 performs OLT test at the rectifier station.
23) Pole 2 operation, power forward, Pole 1 performs OLT test at the inverter station.
24) test for valve base electronics (VBE) system's reception of signals from the converter control and protection (CCP) host.
25) Filter sub-bank low current disconnect test.
26) Valve cooling system operation CPU online replacement.
27) Bipolar asymmetric operation, healthy pole DC line fault test.
28) Simulate network storm test.

(9) Control location change test

1) Remote control start/stop test (National dispatch center).
2) Remote control power rise and fall test (National dispatch center).
3) Operate on the backup panel, start/stop test.
4) Operate on the backup panel, power rise and fall test.

(10) DC modulation function test

1) Power reduction.
2) Power increase.
3) Test for simulating the AC system frequency control function.
4) Test for the signal additional control function.
5) Test for the system stability control device interface.

(11) Bipolar power rise/fall, high load test

1) Manual control of power rise/fall.
2) Automatic control of power rise/fall.

(12) Reactive power control test

1) Power forward, reactive power control—Q mode.
2) Reactive power control—U mode.
3) Power forward, manual removal of a bank of AC filters at the inverter station.

(13) Reduced voltage operation test

1) Bipolar operation, pole 1 operation in reduced voltage.
2) Bipolar operation, pole 2 operation in reduced voltage.
3) During the operation in reduced voltage, check the reactive power performance.

(14) Rated power operation test and overload test

1) Bipolar rated load operation test.
2) Bipolar rated load operation, pole 1 overload test.
3) Bipolar rated load operation, pole 2 overload test.
4) Special measurement, preliminary check of equivalent disturbing current (I_{eq}).
5) Special measurement, preliminary check of AC harmonic measurement (THFF).
6) Special measurement, interference measurement.
7) Special measurement, audibility noise check.
8) Special measurement, power loss of station auxiliary equipment.
9) Reducted voltage operation, rated power operation test.

(III) Bipolar Single Converter System Commissioning Scheme

Based on the aforementioned research findings, the bipolar single converter encompasses 16 operational modes. Given that the connection mode of the bipolar single converter is not the standard operational mode, the commissioning scheme for the bipolar single converter system will focus solely on the basic control mode test items. The primary components of this scheme include bipolar start/stop, basic control mode test, basic protection trip, steady-state operation, reactive control, and additional relevant tests.

(IV) Bipolar Unbalanced Converter System Commissioning Scheme

Based on the aforementioned research results, the bipolar unbalanced converter system possesses 8 operational modes. As the connection mode of the bipolar unbalanced converter is not the standard operational mode, the commissioning scheme for this system will concentrate exclusively on the basic control mode test items. The principal elements of the commissioning scheme include bipolar start/stop, basic control mode test, basic protection trip, steady-state operation, reactive control, and additional relevant tests.

VII. Switch on–off Test Scheme for Single Converter

Based on the main circuit connection mode of the CG project, the single converter is capable of independent access and exit operations. Following analysis and research, a test scheme for the independent access and exit of the single converter has been developed.

Prior to the commissioning of the bipolar system, in accordance with the system commissioning plan, all tests for single converter, monopolar, bipolar, and cross-connection modes at both stations have been completed.

(1) Bipolar full voltage operation, with communication, single converter online switching test

1) Bipolar operation, online switch off test for the converter at the high voltage side of pole 1 on the rectification side.
2) Bipolar operation, online switch on access test for the converter at the high voltage side of pole 1 on the rectification side.
3) Bipolar operation, online switch off test for the converter at the high voltage side of pole 2 on the rectification side.
4) Bipolar operation, online switch on access test of the converter at the high voltage side of pole 2 on the rectification side.
5) Bipolar operation, online switch off test for the converter at the low voltage side of pole 1 on the rectification side.
6) Bipolar operation, online switch on access test of the converter at the low voltage side of pole 1 on the rectification side.
7) Bipolar operation, online switch off test of the converter at the low voltage side of pole 2 on the rectification side.
8) Bipolar operation, online switch on access test of the converter at the low voltage side of pole 2 on the rectification side.
9) Bipolar operation, online switch off test of the converter at the high voltage side of pole 1 on the inverter side.
10) Bipolar operation, online switch on access test of the converter at the high voltage side of pole 1 on the inverter side.
11) Bipolar operation, online switch off test of the converter at the high voltage side of pole 2 on the inverter side.
12) Bipolar operation, online switch on access test of the converter at the high voltage side of pole 2 on the inverter side.
13) Bipolar operation, online switch off test of the converter at the low voltage side of pole 1 on the inverter side.
14) Bipolar operation, online switch on access test of the converter at the low voltage side of pole 1 on the inverter side.
15) Bipolar operation, online switch off test of the converter at the low voltage side of pole 2 on the inverter side.

16) Bipolar operation, online switch on access test of the converter at the low voltage side of pole 2 on the inverter side.
17) Bipolar operation, pole 1 voltage reduction by 80%, online switch off test of the HV converter on the rectification side.
18) Bipolar operation, pole 1 voltage reduction by 80%, online switch on access test of the HV converter on the rectification side.
19) Bipolar operation, pole 1 voltage reduction by 80%, online switch off test of the HV converter on the inverter side.
20) Bipolar operation, pole 1 voltage reduction by 80%, online switch on access test of the HV converter on the inverter side.
21) Bipolar operation, pole 2 voltage reduction by 80%, online switch off test of the LV converter on the rectification side.
22) Bipolar operation, pole 2 voltage reduction by 80%, online switch on access test of the LV converter on the rectification side.
23) Bipolar operation, pole 2 voltage reduction by 80%, online switch off test of the LV converter on the inverter side.
24) Bipolar operation, pole 2 voltage reduction by 80%, online switch on access test of the LV converter on the inverter side.
25) Bipolar operation, current control, online switch test of the HV converter on the rectification side.
26) Bipolar operation, current control, voltage reduction by 80%, online switch test of the HV converter on the rectification side.
27) High power bipolar operation, online switch on access test of the LV converter at pole 1 on the rectification side.
28) High power bipolar operation, online switch test of the HV converter at pole 1 on the rectification side.
29) High power bipolar operation, online switch test of the LV converter at pole 2 on the rectification side.

(2) Bipolar asymmetric converter operation, with communication, single converter online switching test

1) Pole 1 dual converter operation, online switch off test for the LV converter on the rectifier side.
2) Pole 1 dual converter operation, online switch on access test for the LV converter on the rectifier side.
3) Pole 1 dual converter operation, online switch off test for the HV converter on the rectifier side.
4) Pole 1 dual converter operation, online switch on access test for the HV converter on the rectifier side.
5) Pole 2 dual converter operation, online switch off test for the HV converter on the rectifier side.
6) Pole 2 dual converter operation, online switch on access test for the HV converter on the rectifier side.
7) Pole 2 dual converter operation, online switch off test for the LV converter on the rectifier side.

8) Pole 2 dual converter operation, online switch on access test for the LV converter on the rectifier side.
9) Pole 1 dual converter operation, online switch off test for the LV converter on the inverter side.
10) Pole 1 dual converter operation, online switch on access test for the LV converter on the inverter side.
11) Pole 1 dual converter operation, online switch off test for the HV converter on the inverter side.
12) Pole 1 dual converter operation, online switch on access test for the HV converter on the inverter side.
13) Pole 2 dual converter operation, online switch off test for the HV converter on the inverter side.
14) Pole 2 dual converter operation, online switch on access test for the HV converter on the inverter side.
15) Pole 2 dual converter operation, online switch off test for the LV converter on the inverter side.
16) Pole 2 dual converter operation, online switch on access test for the LV converter on the inverter side.
17) Pole 2 dual converter operation, current control, online switch off test for the HV converter on the inverter side.
18) Pole 2 dual converter operation, online switch on–off test for LV converter.
19) Pole 1 dual converter operation, Pole 1 voltage reduction by 80%, online switch on–off test for HV converter.
20) Pole 1 dual converter operation, Pole 1 voltage reduction by 80%, online switch on–off test for LV converter.

(3) Monopolar dual converter operation, with communication, single converter online switching test

1) Pole 1 dual converter operation, online switch off test for the HV converter on the rectifier side.
2) Pole 1 dual converter operation, online switch on access test for the HV converter on the rectifier side.
3) Pole 2 dual converter operation, online switch off test for the HV converter on the rectifier side.
4) Pole 2 dual converter operation, online switch on access test for the HV converter on the rectifier side.
5) Pole 1 dual converter operation, online switch off test for the HV converter on the inverter side.
6) Pole 1 dual converter operation, online switch on access test for the HV converter on the inverter side.
7) Pole 2 dual converter operation, online switch off test for the LV converter on the inverter side.
8) Pole 2 dual converter operation, online switch on access test for the LV converter on the inverter side.

9) Pole 1 dual converter operation at 80% voltage reduction, online switch on access test for the LV converter on the rectifier side.
10) Pole 2 dual converter operation at 80% voltage reduction, online switch on access test for the LV converter on the inverter side.

(4) No communication between stations, single converter online switching test

1) Bipolar operation, no communication, online switch off test for the HV converter of pole 1.
2) Bipolar operation, no communication, online switch off test for the LV converter of pole 2.

VIII. Commissioning Scheme of Converters Layered Connected to 500 kV/1000 kV Power Grid

In accordance with the design requirements of the CG project, this book has examined the system commissioning scheme for the layered connection of HV/LV converters to the 500 and 1000 kV grids. It has also established the test items for the system commissioning scheme in this hierarchical connection mode.

(I) System Commissioning of Layered Connection Mode

At the receiving end of the ± 1100 kV UHV DC transmission project, the HV and LV converters are layered connected to the 500 and 1000 kV grids. This setup necessitates specialized research into the testing and verification of DC control protection, reactive power control, power transfer, and the coordination with security and stability control devices in the layered connection mode.

From the above analysis, it is evident that the DC yard main circuit connection of the ± 1100 kV UHV DC transmission project mirrors that of previously operational UHV DC transmission projects. However, a distinct feature of the receiving end in this project is the adoption of a layered connection for both HV and LV converters to the 500 and 1000 kV grids, which differs from the connections in the UHV DC transmission projects already in operation. Research has revealed that this layered connection mode introduces certain differences compared to the conventional UHV connection modes.

(II) System Commissioning Items of Layered Connection Mode

(1) HV/LV converters are connected layered to the 500 and 1000 kV grids. The initiation or cessation of a single converter's operation can impact the functioning of the other converter on the same pole.
(2) The control of the HV/LV converter transformer tap switch is different.
(3) The control modes of the HV and LV converters are different.
(4) The DC neutral line voltage U_{dm} has been added.
(5) The reactive power control of the HV/LV converters is different.
(6) The additional power control function is different.

Accordingly, the system commissioning scheme for the layered connection mode is developed with a focus on these six key aspects. The commissioning items for the layered connection mode are as follows.

(1) Layered connection mode, initial operation test

1) One pole operation, the other pole start/stop test.
2) Power forward, emergency stops for different converters on the bipolar.

(2) Layered connection mode, steady-state operation test:

1) Bipolar single converter, current control, one pole power rise and fal test, and system switching test.
2) Monopolar power control on one pole, single converter current rise and fall control test, and system switching test on the other pole.
3) Bipolar power control on one pole, power/current rise and fall control test, and system switching test on the other pole.

(3) Layered connection mode, protection trip test

1) Bipolar operation, simulate HV converter transformer trip test.
2) With communication, simulate commutation failure of Y bridge, protection trip, and X blocking at the inverter station.
3) Without communication, simulate the pole bus differential zone II protection trip Z blocking test at the inverter station.

(4) Layered connected mode, reduced voltage operation test:

1) Bipolar reduced voltage operation, the other pole power/current rise and fall test, and system switching test.
2) Dual converter voltage reduction operation on one pole, single converter power/current rise and fall test, and system switching test on the other pole.
3) Single converter operation on one pole, dual converter protection voltage reduction test on the other pole.
4) Pole 1 metallic return operation, protection voltage reduction test.

(5) Layered connection mode, disturbance test

1) With communication, metallic return, single trigger pulse loss test for the HV converter of pole 1 on the inverter side.
2) Without communication, metallic return, multiple trigger pulse loss test for the LV converter of pole 1 on the inverter side.
3) Bipolar asymmetric operation, DC line fault of full voltage pole.
4) Bipolar operation, station grounding, simulate single converter trip.
5) Bipolar operation, station grounding, simulate one pole trip.

(6) Layered connection mode, DC voltage control test:

1) Bipolar operation, simulate U_{dm} fault of pole 1 on the inverter side.
2) Bipolar operation, simulate U_{dm} fault of pole 2 on the inverter side.
3) DC voltage control function verification.

4) Converter transformer tap switch control function verification.
5) AC/DC harmonic test.

(7) Layered connection mode, reactive power control test

1) Reactive power control verification test.
2) A sub-bank of filters on the high voltage side of the inverter side exits.
3) A bank of filters on the high voltage side of the inverter side exits.

(8) Layered connection mode, DC modulation test

1) Power reduction.
2) Power transfer and distribution function verification.
3) Security and stability control device combination test.

(9) Layered connection mode, power reverse delivery test

1) Power reverse, basic control protection function verification.
2) Power reverse, DC power rise and fall.
3) Power reverse, bipolar operation, pole 1 power/current rise and fall.
4) Power reverse, current control, bipolar voltage reduction, current rise and fall.
5) Power reverse, metallic return, with communication, loss of a single pulse for the HV converter on the inverter side.
6) Power reverse, metallic return, no communication, loss of a single pulse for the HV converter on the inverter side.

IX. Engineering Application
(I) System Commissioning

In accordance with the proposed end-to-end system commissioning scheme, the system commissioning for the CG Project has been completed. The test items that were successfully completed include the establishment of the initial operation mode, DC system protection trip functions, system monitoring capabilities, current control, power control, reactive and voltage control, earth/metallic return transfer tests, loss of trigger pulse scenarios, AC/DC auxiliary power switching, disturbance tests, and single converter high-load and overload operation tests. The outcomes of these tests are satisfactory.

During the system commissioning process, the performance of primary equipment, including converter transformers, converter valves, smoothing reactors, AC filters, capacitors, switches, and the DC control protection system, was subjected to rigorous inspection. For any issues identified, the commissioning command post conducts daily technical analyses and communicates promptly with both the equipment manufacturers and the installation companies. This ensures that problems are addressed in a timely manner, thereby enhancing and refining the functionality of the DC system.

(II) Special Test Results

(1) DC bias test. During the system commissioning period, a DC bias test was conducted on the main transformers of the substations surrounding the grounding electrodes at both ends. The test points exhibiting larger DC biases were addressed. The outcomes of these treatments were subsequently verified during the single converter high-load test and were determined to be within normal parameters.
(2) DC parameters and harmonic test results. Throughout the system commissioning process, each task necessitated the testing of key DC parameters, including voltage, current, and firing angle, all of which were confirmed to be within the acceptable range. During the operation of a single converter at its rated load, AC and DC harmonics were assessed, with the results complying with the project's technical specifications.
(3) Electromagnetic environment test results. During the system commissioning period, audible noise tests were conducted at various key locations within the converter station, including the station's perimeter, the converter transformer area, the valve hall inspection channel, the AC filter banks on the AC side, the indoor DC area, and the outdoor DC area. The test results indicated that the noise levels at the station's perimeter complied with the specified requirements.

X. Main Research Results

In association with the CG Project, a system commissioning plan has been developed, taking into account the connection mode and the performance characteristics of the UHV DC transmission project's equipment.

(1) The content of the system commissioning work for the UHV DC transmission project has been thoroughly analyzed. This analysis included pre-commissioning system calculations and assessments, as well as the development of system commissioning plans, dispatching protocols, and testing regimens. Based on the confirmed system commissioning scheme, an on-site system commissioning test plan was formulated. Following the system commissioning scheme and test plan, the on-site commissioning activities were meticulously organized. Upon completion of the system commissioning, the test outcomes were compiled, and a comprehensive technical report on the system commissioning was drafted.
(2) The system conditions for the CG Project's system commissioning were meticulously analyzed, and recommendations were formulated for system security and operational modes during the commissioning phase in 2020. These recommendations were designed to meet the stringent requirements of the $\pm$ 1100 kV UHV DC transmission project's system commissioning process.
(3) The UHV DC transmission project features 45 distinct connection modes, categorized into three primary types: the single converter connection mode, the monopolar connection mode, and the bipolar connection mode. Accordingly, the system commissioning plan for the UHV DC transmission project is structured

into three main components. The first part details the single converter system commissioning scheme, encompassing both the basic and derived connection mode commissioning schemes. The second part outlines the monopolar system commissioning scheme, which includes the monopolar system commissioning scheme and the DC bias measurement scheme. The third part addresses the bipolar system commissioning scheme, a comprehensive plan that includes the bipolar double converter system commissioning scheme, the bipolar single converter system commissioning scheme, the bipolar asymmetric converter system commissioning scheme, the online access and removal system commissioning scheme for the single converter, and the system commissioning scheme for the layered connection mode.

(4) The single converter connection mode encompasses either Pole 1 or Pole 2 operating in LV converter mode and HV converter mode, which is referred to as the basic operation mode of the single converter. All other modes, involving cross-connection of the LV converter or the HV converter, are termed derived operation modes of the single converter. Accordingly, the single converter system commissioning scheme is bifurcated into basic and derived schemes. The basic scheme includes all test items for this mode, whereas the derived scheme comprises only the fundamental test items applicable under the single converter connection mode.

(5) The monopolar connection mode is categorized into two distinct modes: the Pole 1 dual converter connection mode and the Pole 2 dual converter connection mode. Given that the commissioning test items for both Pole 1 and Pole 2 are identical, a single monopolar system commissioning scheme is devised, precluding the need for any derived schemes.

(6) The bipolar connection mode comprises three distinct modes: the bipolar single converter connection mode, the bipolar unbalanced converter connection mode, and the bipolar dual converter connection mode. The bipolar dual converter connection mode is considered the basic connection mode, while the bipolar single converter connection mode is categorized as the first derived connection mode, and the bipolar unbalanced converter connection mode is categorized as the second derived connection mode. Accordingly, the bipolar system commissioning scheme is structured into three components: a basic scheme, a derived scheme 1, and a derived scheme 2. The basic scheme encompasses all test items for this connection mode, whereas the derived schemes include only the fundamental test items applicable under the bipolar system connection mode.

(7) Considering the connection mode characteristics, with HV/LV converters at the Guquan station of the CG project layered connected to the 500 and 1000 kV power grids, a system commissioning test scheme for the layered connection mode has been developed.

References

Tu Jingzhe, Yi Jun, Wang Chao, et al. Analysis on instability mechanism of power angle and voltage interaction considering dynamic characteristics of photovoltaic generator. Automation of Electric Power System, 2020, 44(13): 157–165.

Xu Shiyun, He Jingbo, Fan Mingjian, et al. Stability and Control of Receiving-end Power System After Hierarchical Connection of ±1100kV UHVDC. Power System Technology, 2019, 43(5): 1683–1689.

Wang Yating, Zhang Yichi, Guo Xiaojiang, et al. A study on system connection schemes for ±1100kV UHVDC sending and receiving ends. Power System Technology, 2016, 40(7): 1927–1933.

Xiong Yongxin, Shen Yu, Yao Wei, et al. Improved additional power coordinated control of±1100kV UHVDC system With hierarchical connection mode. Power System Technology, 2017, 41(11): 3448–3456.

Shen Yu, Xiong Yongxin, Yao Wei, et al. Comprehensive evaluation for receiving end connection scheme of ±1100kV UHVDC power transmission, 2018, 38(8): 195–202.

Wu Guifang. Electromagnetic environment problem brought about by ±500kV DC transmission project in China. Power System Technology, 2005, 29(11): 5–8.

Wan Baoquan, Zhang Xiaowu, Zhang Guangzhou, et al. Study on busbar corona characteristic test of ±800kV Yunguang HVDC converter substation. High Voltage Engineering, 2008, 34(9): 1788–1791.

Yue Yunfeng, Peng Guanyan, Wang Yan, et al. Research on electric field characteristics of busbar in HVDC converter substation. Electric Power, 2013, 46(11): 26–29 + 51.

Lu Jiayu, Ju Yong. Study on the limits of electromagnetic environment of±800kV DC transmission lines. Electric Power, 2006, 39(10): 37–42.

Maruvada P S, Janischewky W. Analysis of corona loss on DC transmission lines: II—bipolar lines. IEEE Transactions on Power Apparatus and System, 1969, 88(10): 1476–1491.

Zhao Wanjun. High voltage DC transmission engineering technology. Beijing: China Electric Power Press, 2004.

Yang Wankai, Li Xinnian, Yin Yonghua, et al. Analysis of ±1100kV UHVDC system commissioning test technology Analysis. Proceedings of the CSEE, 2015, 35, 8–14.

Yang Wankai, Yin Yonghua, Ban Liangeng, et al. Study on ±1100kV UHVDC system commissioning test program. Power System Technology, 2015, 39(10), 2815–2821.

Tang Hao, Jia Pengfei, Li Jinzhong, et al. Testing technology of multi-harmonic characteristic parameters and its application for UHVDC dry-type smoothing reactors. High Voltage Engineering, 2017, 43(03), 859–865.

Z. Liu, *±1100kV UHV DC Power Transmission Technology*,
https://doi.org/10.1007/978-981-96-3712-6

MIX
Papier aus verantwortungsvollen Quellen
Paper from responsible sources
FSC® C105338

If you have any concerns about our products,
you can contact us on
ProductSafety@springernature.com

In case Publisher is established outside the EU,
the EU authorized representative is:
Springer Nature Customer Service Center GmbH
Europaplatz 3, 69115 Heidelberg, Germany

Printed by Libri Plureos GmbH
in Hamburg, Germany